Coronavirus (COVID-19) Outbreaks, Environment and Human Behaviour

Rais Akhtar

Editor

Coronavirus (COVID-19) Outbreaks, Environment and Human Behaviour

International Case Studies

 Springer

Editor
Rais Akhtar
International Institute of Health Management
and Research
New Delhi, India

ISBN 978-3-030-68122-7   ISBN 978-3-030-68120-3   (eBook)
https://doi.org/10.1007/978-3-030-68120-3

Cover image credit: Michael Scherer/shutterstock

This Springer imprint is published by the registered company Springer Nature Switzerland AG
The registered company address is: Gewerbestrasse 11, 6330 Cham, Switzerland

Foreword

The world is in the second year of the COVID-19 pandemic. The pandemic and the virus continue to evolve in significant ways. There is a plethora of perspectives from which to view the pandemic, gauge its impacts, and consider potential interventions. My perspective is that of a trained academic geographer. Thus, I view the pandemic, and the chapters in this volume, through a geographical lens. Does geography matter in terms of the pandemic?

It has been argued by writers such as Alvin Toffler and Thomas Friedman that the instantaneous communication, rapid transportation, intercontinental movements of goods and people that are the hallmarks of globalization have, in many cases, rendered geographical differences mote. Friedman's book on this is titled "The World is Flat." In an ironic twist, the COVID-19 pandemic both supports and refutes this thesis. On the one hand, the remarkably rapid spread of COVID-19 to virtually every corner of the globe seems to have overrun geography. The symptoms of the virus were reported from Wuhan, China, in early December 2019. By mid-January, cases were reported in other Asian countries, and by late January, cases were appearing in Australia, Europe, Canada, and the USA. By early spring 2020, over 180 nations reported the presence of the virus. Much of this spread occurred despite countries implementing travel restrictions, traveler quarantines, etc. In recent weeks have seen similar rapid and global diffusion of new COVID-19 variants. In terms of the global spread of the COVID-19 virus, the world has indeed proven to be flat.

In contrast to the apparently limited role of geography in influencing the initial spread of the virus, we see much evidence of significant geographical differences in the impacts of COVID-19 and the attempts to mitigate the disease. Per capita infection rates and death rates show remarkable variation from nation to nation. New Zealand, by virtue of geographic isolation, and a strong proactive policy of border restrictions and internal measures, has brought the infection rate to near zero. However, geographic isolation has not a sole, or even consistent, explanatory variable of such differences. When the chapters of this book were first written in 2020, the island nation of the UK had a COVID-19 infection rate per-million inhabitants that were approximately twenty times greater than the per-million infection rate in the island nation of Japan. Today, in February 2021, that difference persists with the infection rate in the UK being eighteen times greater than Japan. Mortality rates per million in the UK are thirty-three times that of Japan. Clearly, geographical differences in demography, culture, governance, and economy are playing a significant role in COVID-19 infection and mortality rates. However, at this point, such geographic differences still resist easy explanations by variables such as GDP. For example, the USA has the world's highest nominal GDP but also suffers from the world's highest number of COVID-19 deaths, with over 463,000 fatalities and counting. If the impacts of COVID-19 have varied geographically, so too have mitigation policies. Public health efforts to stem the spread of the virus have varied by nation and by subnational jurisdictions. Now, we are seeing wide variations between countries and within countries in terms of vaccine accessibility. We are also seeing differences in the effect of vaccines on the various newly evolving geographic strains of the virus.

Understanding the influence of geography on the ultimate impacts and mitigation strategies for COVID-19 is confounded at this stage by the evolving nature of the pandemic and the virus. Scientific understanding of many aspects of the disease and its long-term health effects are still developing. On a national scale, we saw important differences in the early part of the time series of infections and deaths. Some countries, such as South Korea, had relatively high infection rates early in the pandemic and then marked subsidence until the Fall of 2020. In contrast, the USA has had three distinct peaks in infection rates. The latest surge in the Fall of 2020 and early Winter of 2021 have been far higher than the previous two, with over 250,000 new cases per day at its peak. This recent zenith in infections is essentially a global phenomenon. How national and subnational policies on the pandemic are influencing the evolving trajectory of infection rates is a question of vital importance. This is a desperate race to both stem the spread and to vaccinate populations. The greater the number of people infected, the higher the probability of new strains evolving and resisting current vaccines.

The differential geographies of national COVID-19 impacts and policies are why this volume of collected studies is so important. Here, we are presented with perspectives from different countries provided by scientists working within each of those countries. Through his wide-ranging knowledge of the field of health geography, previous experience as Professor and Head, Department of Geography and Regional Development, at the University of Kashmir, affiliation with the International Institute of Health Management Research in New Delhi, and Editor of Health Care Patterns and Planning in Developing Countries, and Extreme Weather Events and Human Health, Rais Akhtar, has been able to draw together perspectives from countries as diverse as Argentina, France, Italy, Malaysia, Nigeria, South Africa, Thailand, Taiwan, India, Malaysia, Australia, Mexico, Brazil, Peru USA, UK, Canada, Russia, Pacific Islands, and elsewhere. After my initial writing of this Foreword, seeing the USA remained unrepresented in the volume, Professor Akhtar, kindly invited me to add a geographical perspective from the USA. Not only do the chapters provide the views of scientists working directly in these geographically diverse settings, but they also reflect a variety of expertise. The chapter authors not only include those in geography departments working on various aspects of health and environment but also include those in health sciences departments working in areas such as epidemiology, pediatrics and child health, and social and preventive medicine.

Given the wide range of countries covered in the volume, and the differing research perspectives brought to bear, it is not surprising that the chapters present a varied suite of approaches and insights. These approaches include geospatial analytic treatments that seek to identify statistically the relations between disease spread and spatial factors such as transportation networks or distance from airports, or place-based factors such as population density, socioeconomic status, or political affiliation. These approaches are illustrated in the chapters on Russia, Hong Kong, South Korea, and the USA. Other chapters provide broad overviews of the introduction and spread of the disease, and highlight the public health strategies that have been enacted. The chapter on Malaysia, for example, examines the early success of that country's movement control order in tempering the initial spread of COVID-19. In the case of Argentina, the chapter provides local-scale details of the virus through analysis of its spread and the demography of its impacts in the City of Buenos Aires. The chapter on Italy seeks to take the broadest approach and considers COVID-19 not only in terms of its global spread, but in the context of earlier pandemics such as the plague and the Spanish flu. Many of the chapters provide policy prescriptions based on the empirical evidence available at the time of writing. The chapter on South Africa takes a hard look at responses there and provides a concise set of thoughts on the way forward in what is termed the "new normal." Not only do the chapters here illustrate profound international geographical differences in COVID-19 impacts and mitigation strategies but also the very often reveal important regional differences within the countries being studied. Geography certainly does matter.

In the end though, we must acknowledge that the COVID-19 pandemic is still evolving. The story of COVID-19 continues to have a geographical component of international differences in impacts and mitigation strategies that remain to be resolved. No single volume can hope to elucidate all of this. In addition, anything written today is a snapshot in time of a dynamically fluid situation. The chapters herein still provide an important service of informing us of the situation here and now in a diversity of countries—and how the pandemic is being considered by contemporary scholars. The chapters are like the dispatches home from a raging battle—critical information for today, and of historical importance to future generations contemplating the successes and failures of the responses to the COVID-19 pandemic.

Glen MacDonald, FRSC
UCLA Endowed Chair in Geography of
California and the American West UCLA
Distinguished Professor Director, UC White
Mountain Research Center, Chair, UCLA
Canadian Studies Program

Preface

This book is the outcome of geographical understanding of spatial pattern and diffusion of coronavirus from foci region in China to neighboring countries and how it diffused to various regions of the world and become a pandemic that resulted **as of 1st April 2021, in the death of 2.8 million worldwide and more than 5,45,887 in the USA alone**.

Undoubtedly, COVID-19 pandemic has proved to be the most disastrous globally that has devastated the socioeconomic scenario of the world. Faulty response in several countries including the USA, Brazil, and the United Kingdom caused greater human sufferings.

In terms of spatial patterns of COVID-19 cases as described by WHO, Americas ranks number 1 followed by Europe, South-east Asia, Eastern Mediterranean, Africa, and Western Pacific. By country, USA ranks number one followed by Brazil, India, Russia, France, UK, Italy, Turkey, Spain, Germany, Colombia, and Argentina. It is astonishing to note that China, the foci of the COVID-19 infection, has only 92,490 cases with only 4634 deaths and stands at 60 in terms of total number of COVID-19 cases.

Such regional variation of COVID-19 infections and deaths encourages me to prepare a volume covering selected countries from different continents with wide socioeconomic inequalities in order to understand the environment and human behaviour which are determinants of intensity of COVID-19.

The book comprises 27 chapters and divided into four parts–Introductory, Australasia, Europe, Russia & Africa and Americas. Chapter "Introduction: Coronavirus, Environment and Human Behaviour" discusses the historical pandemic scenario in general and spatial pattern of COVID-19 with focus on early diffusion of COVID-19 and present situation of the disease. The chapter also highlights inequalities and poverty in developed and developing countries due to COVID-19 and focuses on the politics of vaccine in the context of its accessibility to vulnerable people. Chapter "Introduction: Coronavirus, Environment and Human Behaviour" also appraises developed countries to strengthen the required health infrastructure for newer virus and vector-borne diseases owing to climate change. With stress and anxiety on the rise as a result of lockdown and job insecurity, mental health services are particularly critical in ensuring the overall well-being and adjustment of our nation. Chapter "Coping During Covid-19 Public Health Crises" describes the impact of stress and trauma on daily functioning as well as strategies deemed helpful for coping during the pandemic. The second part on Australasia includes 9 chapters on Pacific Islands, Taiwan, Thailand, Malaysia, Indonesia, South Korea and Hong Kong, India and Australia. Chapter "COVID-19 Pandemics in the Pacific Island Countries and Territories" on Pacific Islands gives a brief reflection on historical experience with epidemics in the Pacific Island region. The remainder of the chapter looks at the epidemiology of the COVID-19 pandemics, the 'who, when, and where' of COVID-19 in the Pacific Islands, its associated risk factors, and possible ways to control the pandemics. Taiwan was predicted to be one of the most affected countries because of its economic activities and transportation exchanges with China (Chapter "Proactive Strategies, Social Innovation, and Community Engagement in Relation to COVID-19 in Taiwan"). Even as the COVID-19 continues to spread around the globe, everyday life and daily activities have mostly continued as normal, with only a few cases of local transmission occurring in Taiwan. Taiwan has adopted proactive strategies for epidemic surveillance, border controls, community-based prevention and control, stockpiling of medical supplies, and health education and outreach to prevent the spread of the disease and maintain the health and safety of the Taiwanese population. Chapter "Healthcare Resources in the Context of COVID-19 Outbreak—Thailand, 2020" on Thailand talks about measures which led to swiftly control and curb the spread of the COVID-19 scourge. This investigation assessed the health effects of COVID-19 and the adequacy of the available health care resources in responding to the threats posed by the COVID-19 pandemic in Thailand. Chapter "How Malaysia Counters Coronavirus Disease (COVID-19): Challenges and Recommendations" on COVID-19 in Malaysia highlighted the crucial role of a multilateral partnership led by the Malaysian government in handling an outbreak of highly infectious disease. The enforcement of the Movement Control Order (MCO) is the main reason that the chain of COVID-19 infection was able to be halt and control over a short period. The MCO is also proven to be an effective strategy to identify the clusters in which the virus spread has originated or affected. Indonesia confirmed the first COVID-19 cases on March 2, 2020. When the first cases appeared in Depok, West Java, the President, and Ministry of Health Republic of Indonesia urged the societies not to panic in facing COVID-19 as the country is ready in tackling the virus (Chapter "COVID-19 in Indonesia: Geo-Ecology and People's Behaviour"). However, the authors assert a great portion of the blame also lies in unproductive behaviour from the government and its officials, as they also show a lack of leadership in the pandemic in that they produce many contradictory policies and many government personnel do not themselves adhere to COVID-19 protocol, such as not wearing a mask properly, which in combination communicates a lack of seriousness in the handling of the pandemic and a failure to produce a sense of crisis. In chapter "Does Socioeconomic Status Matter? A Study on the Spatial Patterns of COVID-19 in Hong Kong and South Korea", the authors argue that while people with lower socioeconomic status (SES) had been reported as most vulnerable in Hong Kong and South Korea, we found no significant correlations between the COVID-19 infection and the SES at the local level. Instead, we found a positive association between the COVID-19 infection and a local SES measure in South Korea, contrary to the USA. The findings imply the potentially vulnerable areas of COVID-19 may vary by country. During COVID-19 pandemic, in India where a large chunk of labourers and workers in the informal sector resides, migrants from urban clusters returned to their rural areas in some of the states and thus creating many possible vulnerabilities. In the absence of a fixed place for vending activity, mobile vendors, who are the poorest of the poor in urban India, face challenges to do their economic activity in local trains and also take care of their families with their meager earnings (Chapter "Lives of Migrants During COVID-19 Pandemic: A Study of Mobile Vendors in Mumbai"). To make the city of Mumbai to be more inclusive for the urban poor, therefore there is a need to provide suitable opportunities through which, not only can we enhance their livelihood skills, but also make them to be the contributors to the overall development of the society.

In chapter "Beating Back COVID-19 in Mumbai" on Mumbai, the author contends that the government has been able to reduce COVID-19-related deaths despite the increase in number of cases. The COVID-19 crisis is far from over. The pandemic has laid bare inadequacies across science, emergency response, healthy systems strengthening as well as policy across the globe. Yet, this provides a great opportunity for policy makers to address deep-rooted structural challenges. Lessons from this experience will help in overcoming future pandemics. Finally, chapter "COVID-19: The Australian Experience" in this part presents a case study of Australia. This chapter examines various aspects of the COVID-19 pandemic in Australia from January to late October 2020. It shows how, by comparison with most other developed nations, the country fared well in terms of both cases and deaths suffered in the disease outbreak. The chapter conveys both an epidemiological perspective on the pandemic and related social, economic, political, and behavioural dimensions. Mistakes made and painful lessons learnt from dealing with the outbreak are described. The pandemic still has a considerable way to go in Australia before any return to totally 'normal' life can be anticipated, the chapter concluding with a warning of the dangers of complacency toward the still present viral threat. The scattered outbreaks of COVID-19 in eastern Australia in early March 2021 are a relevant reminder of this.

The third part on Europe, Russia and Africa encompasses eight chapters. In chapter "COVID-19 in France: The Revenge of the Countryside" on France, the author argued that the country was very poorly prepared, had to resort to very strict lookdown measures so that the health system can ensure the care of the many patients. The virus is still circulating and establishing a comprehensive review of this pandemic which has killed more than 90,000 people is still surging. The territory and the population have been very unequally affected by the virus which has reached, as for any epidemic, the most densely populated areas. This is how, for the first time in decades, metropolises are losing their appeal and are calling for new land use planning more favourable to rural areas.

Chapter "Past Major Infectious Diseases and Recent COVID-19 Pandemic: Health and Social Problems in Italy" focuses on Italy; the country was the first in Europe impacted by COVID-19, particularly the northern Italy. In this study, the various significant infectious diseases that have affected Italy are discussed. Most of these diseases have affected not only Italy, but also a large part of the world. Italy, as a destination for population movements for many centuries, has had to face infections from the east and the south. Pandemics today have a greater virulence and expansion, due to the greater speed of the means of communication and the numerous points of gathering of people, such as tourist resorts, large cities, and reckless behaviour regardless of precise rules. In chapter "COVID-19 in Italy in the Context of the Pandemic Induced by SARS-Cov-2. Is There a Relationship Between COVID-19 and Atmospheric Air Pollution?", the authors contend that in Italy, the possibility of performing autopsies or post-mortem diagnostic studies on confirmed COVID-19 cases has been intensively debated; however, while post-mortem pathological analysis of COVID-19 patients in China has shown findings consistent with interstitial pneumonia and in some cases of acute respiratory distress syndrome (ARDS), in Italian autopsies, the involvement of endothelial, prevalently of microvascular system has been identified for the recruitment of multiple cytokine-activated inflammatory cell lineages. The UK (Chapter "United Kingdom: Anatomy of a Public Policy Failure") has one of the highest COVID-19 fatality rates due to legacy problems in government and health care management, slowness in recognizing the severity of the outbreak that allowed the disease to gain hold across the country (despite scientific warnings), and serious public policy failings, including delayed lockdown, insufficient provision of personal protective equipment, and errors in introducing a test and trace system. The disease has disproportionately affected vulnerable populations, especially care home staff and residents, key workers, people from poorer socioeconomic groups and black, Asian and minority ethnic (BAME) citizens. The pandemic has had a serious effect on the UK economy, including unemployment and a record level of public debt.

In chapter "COVID-19 Spread in the Iberian Peninsula during the "First Wave": Spatiotemporal Analysis" on Iberian Peninsula, authors identify areas with significant high COVID-19 incidence from January until June, and retrospective spatial-temporal cluster and a spatial variation in temporal trends analysis were conducted. The research reveals important spatial disparities across (and within) the Iberian Peninsula regions, established during the first months of 2020. In the period analyzed, the COVID-19 crude incidence rate was of 385.2 cases per 100,000 inhabitants in Portugal and 608.5 in Spain. The spread was significantly faster in Spain where by the end of March, 63% of cases have been identified (23% in Portugal); a second increasing period was found in Portugal in June, but not in Spain. The spatial-temporal analysis of COVID-19 incidence allowed the identification of six high risk clusters which account for 50% of the total cases analyzed; the clusters identified in March include regions from Spain and north of Portugal (one cluster groups the north of Portugal and the north-western regions of Spain), in June one cluster was identified in Portugal, in the Lisbon Metropolitan Area. Chapter "COVID-19 in the Russian Federation: Regional Differences and Public Health Response" on Russia presents a medical-geographical analysis of the SARS CoV-19 pandemic development in the Russian Federation as by the end of August 2020. In general, the initial course of pandemic in Russia was characterized by a basic reproductive ratio (R_0) of 2.41 (2.22–2.60), which is relatively low as compared to some most affected countries. A spatial regression analysis demonstrated that the onset of the epidemics by the regions of Russia was determined by their proximity to major international airports and connectivity of transportation network, while morbidity and mortality rates show a pronounced relationship with the population density, urban population proportion and proportion of the population over working age. Chapter "COVID-19 in South Africa" is devoted to South Africa. The author submits COVID-19 compounds a deep underlying South African health crisis. The fragmented and inequitable health system was unable to carry the enormous burden of disease. Health outcomes were already worse than in other upper-middle-income countries. Massive income and wealth inequality, poor governance, and endemic corruption had destroyed social solidarity and led to distrust in government and normal politics. When COVID arrived, the state responded rapidly. However, the lockdown burdened poor people further and impaired access to routine health care. The crisis galvanized civil society in diverse and creative ways. The future depends on whether a broad social movement comes together to work for a common goal of social solidarity and equality.

Chapter "Geographical Dynamics of COVID-19 in Nigeria" focused on Nigeria. The chapter on Nigeria examines the entry and spread of COVID-19 in Nigeria. It also discusses some policy responses of government as well as some of the consequences of the disease, and the perceptions and attitudes of the public. Government moved to establish or expand testing laboratories, and isolation and treatment centers. Available data indicate that the spread is on an upward trajectory, testing capacity is inadequate, and the number of cases is not a true reflection of the pandemic situation. Ignorance, skepticism, mistrust of government, and economic hardship are bedeviling efforts to combat the disease.

The fourth part on Americas incorporates seven chapters on Mexico, Argentina, Brazil, Peru, Canada, and the USA. COVID-19 pandemics in Mexico (Chapter "Political Miscalculation: The Size and Trend of the COVID-19 Pandemic in Mexico") have had a significant effect in the country, in terms of cases, deaths, and economic impact by various standards. Within Latin America, Mexico is the country with highest lethality after Brazil. We analyze the differences in the trends of the pandemics at the national and state levels and relate them to variables that have been posed as driving forces of the COVID-19 contagion velocity, namely population density, water security, and the adoption of public policies intended to curb transmission.

In chapter "Evolution of the Coronavirus Disease (Covid-19) in Argentine Territory. Implications for Mental Health and Social and Economic Impacts", COVID-19 situation of the disease in Argentina, according to the Ministry of Health of the Nation, awakens the urgent need to respond with local inquiries to the knowledge of the dynamics of this pathology that at the moment is far from over. This document aims to make an analytical summary of what has happened so far in the country with COVID-19 and to take into account the results of monitoring the pandemic with the data obtained from official reports from the Ministry of Health. Aspects of its evolution, its spatial distribution, and the temporal diffusion of this pathology have been investigated. In addition, relevant scientific evidence is added in relation to issues of mental health and social isolation that are emerging from this pandemic.

The first case of COVID-19 in Brazil (Chapter "Early Stages of the Coronavirus Disease (COVID-19) Pandemic in Brazil: National and Regional Contexts") occurred in February 2020, affecting mainly the country's large urban centers homogeneous manner throughout the territory, reaching small and medium-sized cities in the country and overloading local health services, which in many places reached their maximum capacity. There is still no clear trend toward stabilization of the epidemic curve, since the data available to estimate the behaviour of infection in the country are lagged by the low testing rate. The chapter on Peru (Chapter "Impact of Coronavirus Disease (COVID-19) on Public Health, Environment and Economy: Analysis and Evidence from Peru") examines the consequences of the pandemic caused by the novel coronavirus in Peru from a socioeconomic, health, and environmental perspective. The pre-pandemic situation is detailed, as well as future gaps and opportunities. Peru, reacted early with the adoption of strict prevention measures, but did not have the expected effect, deficiencies in the health sector became evident; likewise, due to its climate variability and main economic activities focused on agriculture and commercialization, it was affected both environmentally and economically with a forced halt to the great economic progress achieved.

Chapter "The Spread of COVID-19 Throughout Canada and the Possible Effects on Air Pollution" on Canada COVID-19 virus spread throughout nearly all of Canada's provinces and territories in the month of March. Although travelers to Canada from Asia and the Middle East tested positive in January and February, far more positive travelers arrived from the USA and Europe, suggesting a greater impact of COVID-19 in Canada. Following the surge of COVID-19 in March, Canada's most populated areas predominantly saw the highest absolute numbers of cases as well as cases per population. Smaller communities, and Indigenous communities in particular, saw fewer cases. Comparing 2020 air pollution levels to 2019 has suggested a marginal decrease from 2019 to 2020 in O_3, PM_{10}, and $PM_{2.5}$ but not in CO and SO_2 following the pandemic shutdown in March. However, low numbers of monitoring stations and other factors hamper our ability to draw conclusions with great certainty. Health professionals have warned about the threats of flu (influenza)-like pandemics, for decades. (Chapter "Public Health Perspective of Racial and Ethnic Disparities During SARS-CoV-2 Pandemic"). The USA counts for one-fourth of all infection cases worldwide. In the USA, racial and ethnic minority groups, such as African Americans and Hispanics, have been disproportionately impacted compared to their white counterparts due to their work and living conditions, lack of health care access and lower health insurance, long-standing structural forms of discrimination, pre-existing economic and social inequalities, and high rates of underlying chronic conditions. The inability to socially distance due to crowded living and working conditions, inability to take time off from work or work from home, and the lack of adequate testing and tracing protocols have put low-income communities, including service workers, farm workers, and frontline workers, at high risks of infection. Some countries, however, have been less impacted than others due to good public health practice and early implementation of social measures, such as the ban on international travels, rigorous testing, implementation of contact tracing protocols, and social distancing. For many vulnerable communities, these measures must be rethought of and adapt to the specific local community.

COVID-19 arrived in the United States of America (USA) early from multiple sources (Chapter "Geographic Patterns of the Pandemic in the United States: Covid-19 Response Within a Disunified Federal System"). Although federal border restrictions were implemented, the magnitude of international travel into the country rendered it impossible to stop the arrival of the virus at multiple locations. Community spread began quickly, and the high personal mobility of US via air travel and personal vehicles allowed for rapid dispersal across the 50 US states. The subsequent per-capita infections and mortality has varied markedly for individual states. Former President Trump abrogated much of the pandemic public health responsibilities to the states and provided mixed-messaging on the pandemic. This contributed to a disunified and politicized response, with individual states having differing policies. Public opinion and behaviour also became divided. A unified, federally led, science-based response may have reduced COVID-19 cases and deaths. Chapter "Conclusion and Suggestions" is devoted to conclusion and possible suggestions to understand changing socioeconmic, environmental and human behavioural determinants and their impacts on human health scenario in both developed and developing countries.

New Delhi, India   Rais Akhtar

Acknowledgements

In the process of writing, editing, and preparing this book, there have been many people who have encouraged, helped, and supported us with their skill, thoughtful evaluation of chapters and constructive criticism.

First of all, we are indebted to all the contributors of chapters from both developed and developing countries for providing the scholarly and innovative scientific piece of research to make this book a reality. We are also thankful to the reviewers who carefully and timely reviewed the manuscripts.

We are also grateful to Prof. Glen MacDonald, University of California, Los Angeles, for writing Foreword, which adds greatly to the book with his thoughtful insights.

I am thankful to my family—wife Dr.Nilofar Izhar, daughter Dr. Shirin Rais, Assistant Professor, AMU, and my son-in-law Dr. Wasim Ahmad, Associate Professor, IIT Kanpur, who encouraged and sustained me in developing the structure of the book and editing tasks, and I am deeply grateful for their support and indulgence.

Finally and most essentially, we are deeply obliged to the Springer and the entire publishing team, without whose patience, immense competence, and support this book would not have come to fruition. We specially thank Dr. Robert K. Doe whose energizing leadership ensured that this book would indeed translate to reality. I wrote to Dr. Robert sometime in early April 2020 about a book on coronavirus encompassing environmental and human behaviour in selected international case studies. Dr. Robert Doe wrote back expressing his interest in such book.

We are also thankful to Mr. Arumugam Deivasigamani and Ritu Chandwani, for their constant cooperation.

New Delhi/Aligarh, India   Rais Akhtar

Contents

About the Editor

Rais Akhtar is presently Adjunct Faculty, International Institute of Health Management and Research, New Delhi and formerly Professor of Geography, and Dean, University of Kashmir, Srinagar, National Fellow and Emeritus Scientist (CSIR), CSRD, Jawaharlal Nehru University, New Delhi, and Visiting Professor Department of Geology, AMU, Aligarh He has taught at the Jawaharlal Nehru University, New Delhi, University of Zambia, Lusaka, and the University of Kashmir, Srinagar. He is recipient of a number of international fellowships including Leverhulme Fellowship (University of Liverpool), Henry Chapman Fellowship (University of London), and Visiting Fellowship, (University of Sussex), Royal Society Fellowship, University of Oxford, and Visiting Professorship, University of Paris-10. He was elected Fellow of Royal Geographical Society, London, and Royal Academy of Overseas Sciences, Brussels.

He delivered invited lectures in about 55 universities geography departments and medical colleges including London School of Hygiene and Tropical Medicine, Liverpool School of Tropical Medicine, Leeds Institute of Health Sciences, Maggiore Hospital in Bologna, and University of Pisa, Institute of Public Health, Academy of Medical Sciences, Perugia, Italy, Dept. of Geography, University of Edinburgh, and School of Public Health, Johns Hopkins University, Baltimore.

He was Lead Author (1999–2007), on the Intergovernmental Panel on Climate Change (IPCC), which is the joint winner of Nobel Peace Prize for 2007. He is the recipient of Nobel Memento.

He has to his credit 91 research papers and 19 books published from India, UK, USA, Germany, and the Netherlands. His books include *Malaria in South Asia*, Springer, 2010 (with A. K. Dutt and V. Wadhwa): *Climate Change and Human Health Scenario in South and Southeast Asia*, published in 2016 by Springer, and his other books include *Climate Change and Air Pollution* (with C. Palagiano) published in 2018 by Springer, and *Geographical Aspects of Health and Disease in India* (with Andrew Learmonth) published in 2018 by Concept (New Delhi). His latest book entitled: *Extreme Weather Events and Human Health*, was published by Springer in early 2020.

He is Member of Expert Group on Climate Change and Human Health of the Ministry of Health and Family Welfare, Government of India. He has been appointed as Corresponding Member of Italian Geographical Society in October 2018.

Introductory

Introduction: Coronavirus, Environment and Human Behaviour

Rais Akhtar

Abstract COVID-19 pandemic has devastated the world's socio-economically, and nearly half of the world's 3.3 billion global workforce are at the risk of losing their livelihood. The pandemic has also widened socio-economic inequalities in both developed and developing countries. This chapter discusses epidemics and pandemics briefly since the time of Hippocrates. The chapter highlights the present COVID-19 scenario and, throw light on early diffusion pattern, possible association, if any, between temperature and COVID-19, inequalities and poverty, politics behind the availability of vaccine, climate change and virusborne and vectorborne diseases, and air pollution and COVID-19.

Keywords Diffusion pattern · Hippocrates · Air pollution · COVID-19 vaccines · The lancet · Altitudinal rise of diseases · Dr. Anthony Fauci · Vaccine

Historical Perspective

Epidemics and pandemics of diseases with high mortality had always been a major menace for humanity. They posed the greatest challenge to human existence. Such Pandemics raging the world since the time of Hippocrates whose book *Of the Epidemics* written 400 B.C.E, described epidemics of diseases specifically fevers with characteristics of seasons, "the rains were abundant, constant, and soft, with southerly winds; the winter southerly, the northerly winds faint, droughts; on the whole, the winter having the character of spring. The spring was southerly, cool, rains small in quantity. Summer, for the most part, cloudy, no rain, the Etesian winds, rare and small, blew in an irregular manner. The whole constitution of the season being thus inclined to the southerly, and with droughts early in the spring, from the preceding opposite and northerly state, ardent fevers occurred in a few instances, and these very mild, being rarely attended with haemorrhage, and never proving fatal" (Hippocrates, 400 B.C.E.).

R. Akhtar (✉)
International Institute of Health Management and Research, New Delhi, India
e-mail: raisakhtar@gmail.com

In England in 1849, cholera claimed 5308 lives in the major port city of Liverpool, an embarkation point for immigrants to North America, and 1834 in Hull, and for the whole of Great Britain cholera claimed 23,000 lives. According to Melvyn Howe, "Environmental improvements, particularly in housing, nutrition and education together with dramatic advances in medicine, have resulted in a far greater life expectancy for the average person" (Howe 1972). In the USA, it is believed more than 150,000 Americans died during the two pandemics of cholera between 1832 and 1849.

According to CDC, the Spanish influenza pandemic was the most severe pandemic, spreading worldwide during 1918–1919. It is estimated that about 500 million people or one-third of the world's population became infected with this virus. The number of deaths was estimated to be at least 50 million worldwide with about 675,000 occurring in the USA.

Present Scenario

Unquestionably the COVID-19 pandemic has proved to be the most disastrous globally that has devastated the socio-economic scenario of the world. Faulty response in several countries including the United States and the United Kingdom caused greater human sufferings, with 30 million cases and more than 0.54 million, deaths in the USA (as on 1st April 2021). "These deaths reflect a true measure of the human cost of the Great Pandemic of 2020, These deaths far exceed the number of US deaths from some armed conflicts, such as the Korean War and the Vietnam War, and deaths from the 2009 H1N1 (swine flu) pandemic, and approach the number of deaths from World War II" (quoted from Beusekom 2020). COVID-19 hospitalization reached an all-time high in the month of November 2020.

Former President Trump's faulty and unscientific ideas and policies on the virus and later on vaccine have created an unscientific environment that not only damaged the science but also confused citizens about the severity of the virus and the measures to be adopted, for instance, the use of face masks and social distancing. Also, Trump ignored the worsening COVID-19 situation in the USA (Collinson 2020). Jeff Tollefson, therefore, rightly postulates "The US president's actions have exacerbated the pandemic that has killed more than 200,000 people in the United States, rolled back environmental and public-health regulations and undermined science and scientific institutions. Some of the harm could be permanent" (Tollefson 2020). It was this inconsistent and illogical behaviour that on 29 May 2020, Donald Trump announced the USA would sever its relationship with WHO and redirect funds to US global health priorities (Gostin et al. 2020). The USA reported for the first time more than 84,000 cases on 23rd October since the pandemic began.

Dr. Anthony Fauci has now been appointed by US President Joe Biden as Chief Medical Advisor, and has praised Melbourne's response to the coronavirus, saying he "wished" the USA could adopt the same mentality. Fauci said that Melbourne's lockdown and mandatory mask-wearing had struck the right balance between public health and opening up the economy. There is universal mask-wearing in Melbourne, and you are fined a thousand dollars if you are outside not wearing them. Some 99.9% of people wear masks. Fauci advises that America should adopt the same mentality (Zhou 2020). Anthony Fauci has been awarded 2021 Dan David prize for 'defending science'.

Beginning of COVID-19

Although no study has been conducted at a global scale on the environmental and human behaviour aspects of coronavirus (COVID-19) infection, its origin and diffusion from the epicentre in China revealed a strange pattern. From the geographical distribution, it looks that in the beginning major coronavirus hotspots—Wuhan (China), Iran, Northern Italy, France, Spain, UK and the USA—are located in temperate cooler regions and rarely in warm tropical areas. For instance, the temperature in these regions ranged between 12 and 15 °C during late March and early April 2020.

It is evident that COVID-19 is spread by respiratory droplets produced when people breathe, talk, sneeze or cough. The virus survives better in cold, dry conditions typical of temperate winters. Low humidity also promotes evaporation of virus droplets into tiny aerosol particles that linger in the air, increasing the risk of airborne transmission in winter. Anthony Fauci, an expert on virus diseases, and earlier part of COVID-19 team of the Trump administration, warned that COVID-19 infection rates are too high heading into winter (Chevez et al. 2020).

In India, the Head of India's National Expert Group on Vaccine Administration for COVID-19 asserted that India could see a second wave of coronavirus infections in the winters as he cited the rising cases in European countries—UK, France, the Netherlands, Belgium and Germany (H.T. 2020, Singh 2020). However, declining COVID-19 cases earlier in the Indian states of Maharashtra, Tamil Nadu, Gujarat and particularly in Delhi indicated that India might not witness a second wave, something many developed nations have been grappling with as a result of the new variants of COVID-19. Nevertheless, since late February and early March 2021, Maharashtra, Tamil Nadu, Delhi and Kerala states of India are witnessing a second wave of the virus.

Early Diffusion Pattern

In the beginning, the diffusion patterns of COVID-19 were to some extent similar to the diffusion theory by Swedish geographer Torsten Hagerstrand, who propounded his theory of types of disease diffusions, and the earlier trends in coronavirus outbreaks reflect the expansion and relocation diffusions. The virus expanded from foci areas to neighbouring parts of China and Hong Kong, and relocated to far-off areas in Iran and Italy, particularly Northern Italy (Milan, Bologna and Venice), which are closer to the Alps. France, Spain and the UK are other European countries that suffered from the COVID-19. However, the COVID-19 spread in the USA, Brazil, Australia and other Latin American countries defeated this hypothesis that COVID-19 impacted countries above the equator. Undoubtedly, the USA suffered heavily and ranks number one in terms of both morbidity and mortality. Redlener, and colleagues assert that about 1,30,000 to 2,10,000 deaths could have been saved in the US, had the appropriate strategies to combat COVID-19 been adopted with a strong political will (Redlener et al. 2020).

Larry Cooley and Johannes Linn have rightly asserted, "The COVID-19 crisis presents the world with a huge challenge: Everyone and everything is affected, and the response has to be both quick and global. While it is first and foremost a health crisis, it is also an education crisis, an employment and economic crisis, a crisis of hunger and a poverty and, in some countries, a crisis of governance and political stability. According to World Bank estimates, the global impacts will be profound and long-lasting. For developing countries with much larger populations at risk, fewer resources, and less capacity, the pressure to develop innovative approaches, test them quickly, and deliver them at scale is especially great" (Cooley and Linn 2020).

It is evident from the diffusion of COVID-19 that a cooler climate is congenial for its spread. The opposite is true of malaria, Plasmodium, which prefers a warmer climate and enhanced warming causing occurrence of altitudinal rise of malaria in highlands and sub-mountainous areas. Based on the rainfall worldwide, Brazilian researchers confirm COVID-19 cases increase with greater precipitation. For each average inch per day of rain, there was an increase of 56 COVID-19 cases per day. However, no association was obtained between rainfall and COVID-19 deaths (Guite 2020). Links between COVID-19 cases and temperature are less certain. Some studies reported link between temperature and COVID-19, while others have not. However, higher temperatures are associated with a lower number of cases in Turkey, Mexico, Brazil and the USA (Guite 2020). A new study by researchers at the University of Cambridge suggests that climate change may have played a role in coronavirus pandemic. "Increases in temperature, sunlight and carbon dioxide, which affect the growth of plants and trees, have shifted the makeup of vegetation in southern China, turning tropical shrubland into tropical savannah and deciduous woodland. This type of forest, the authors contend, is more suitable to bat species" (Berardelli 2021).

Having stated that, the question arises as to why western developed countries are more vulnerable than developing countries with the exception of South American countries? This may be because of the epidemiological transition that occurred in developed countries. The epidemiological transition is a process by which the pattern of mortality and disease is transformed from one of high mortalities due to communicable vectorborne/virusborne epidemics affecting all age groups.

The developed countries which once suffered from communicable/virusborne diseases such as cholera, plague, typhus and smallpox up until end of nineteenth century and influenza in the twentieth century were able to control and eradicate, and transformed the scenario from communicable into degenerative/lifestyle diseases. Even hospitals were transformed. In England, "Hospital units that treat children and very sick babies are having to shut their doors temporarily to new patients because they are 'dangerously' short of specialist staff—paediatric doctors and nurses" (Campbell 2017).

As a result of this shift, there are not sufficient provisions/facilities to treat virus-borne or vectorborne diseases. The crisis in European countries included Italy, France, Spain, UK and the USA. These countries totally lack sufficient testing laboratories, PPEs and ventilators and have resulted in a large number of deaths in these countries. The New York Times reported on 23 November that "Now the U.S. Has Lots of Ventilators, but Too Few Specialists to Operate Them" (New York Times 2020). Such is the shortage of specialists in the most developed country.

Developing countries including India do have facilities for testing as their focus lies on public health with attention to both communicable and degenerative (lifestyle) diseases. Thus, there is no serious crisis in terms of essential equipments. Neglect of public health specialization in developed countries resulted in poor understanding of the COVID-19 disease. It was reported that medical workers are being asked to do the work which they never did. Masks, ventilators and PPE kits were in short supply leading to huge mortality in the USA, UK, Italy, France and Spain. WHO, CDC and other health agencies advised people to use face masks, but in some developed countries including the USA, some people are against the use of masks. "In the midst of pandemic, a small piece of cloth has incited a nationwide feud about public health, civil liberties and personal freedom. Some Americans refuse to wear a facial covering out of principle. Others in this country are enraged by the way that people flout the mask mandates". In France, face masks are now compulsory in most workplaces where there is more than one employee present, as it struggles with a resurgence in coronavirus cases (BBC 2020). In early April, 2021 both Italy and France begin partial lockdown. In France special medical planes dispatched patients from overrun Paris intensive care units to less saturated regions.

The Lancet Study

A study published in the Lancet asserts that, "The failure of governments to tackle a three-decade rise in preventable diseases such as obesity and type 2 diabetes has fuelled the COVID-19 pandemic and is stalling life expectancy around the world, a comprehensive study has found" (Boseley 2020). In support of this Richard Horton, editor-in-chief of the Lancet, elaborated in the context of United Kingdom, that," the areas where life expectancy was lowest – the north-east, north-west, Yorkshire and Humberside were the areas hardest hit by Covid. ...I don't think it's a coincidence, adding that COVID-19 was not a single pandemic, but a synthesis of a coronavirus and an epidemic of non-communicable diseases on a background of poverty and inequality.... It is the interaction of the virus with people living with other diseases-that is the challenge that we face, especially when you factor in the social gradient issue. So I think governments, if they focus only on trying to reduce the prevalence of a virus, this is a strategy that in the long term will fail," Horton asserts (Boseley 2020).

Inequalities and Poverty in Developed and Developing Countries

As is evident, the COVID-19 pandemic is the unprecedented in its spatial distribution and intensity. It has spread to almost every continent, with the exception of most Pacific Islands and some countries in the central Asia and northern Africa. It has infected some 128 millions and killed 2.8 millions as on 1st April, 2021. Based on the new estimates by the UNDP, "as a combined measure of the world's education, health and living standards—is on course to decline this year for the first time since the concept was developed in 1990. The decline is expected across the majority of countries—rich and poor—in every region" (UNDP 2020). International Monetary Fund rightly asserts that inequitable distribution of coronavirus vaccines risks exacerbating financial vulnerabilities. To focus on inequitable accessibility of COVID-19 vaccine on global health disaster. Olivier Wouters, and his colleagues warned in their paper published in The Lancet on February 12, 2021, that, "COVID-19 pandemic is unlikely to end until there is global rollout of vaccines that protect against severe disease and preferably drive herd immunity" (Wouters et al. 2021).

Politics of Vaccine Accessibility

Inequalities are likely to occur in the access to safe vaccine in poor countries. "Wealthy countries have struck deals to buy more than two billion doses of coronavirus vaccine is a scramble that could leave limited supplies in the coming year" (Callaway 2020). For example, the USA has made deals worth upwards of US$6 billion with several firms. An analysis by the international charity Oxfam finds that, even if all five of the most-advanced vaccine candidates succeed, there will not be enough vaccine for most of the world's people until 2022. Oxfam International warned that nations representing just 13 per cent of the world's population have already cornered more than half (51 per cent) of the promised doses of leading COVID-19 vaccine candidates. In order to address this issue of fair distribution of vaccine, COVAX has been created which is co-led by Gavi, the Coalition for Epidemic Preparedness Innovations (CEPI) and WHO is aimed at working for global equitable access to COVID-19 vaccine. Pharmaceutical companies have invested heavily in vaccine development in the USA and in the UK. Pfizer and Moderna vaccines are ready for emergency-use authorization. In fact, UK has already authorized Pfizer for its emergency use. The other aspect is the difficulties in distribution of vaccine in developing countries, particularly in Africa. In this regard, the need is to strengthen existing distribution systems in these countries to deliver vaccines efficiently. Nkengasong and colleagues assert that "For instance, refrigerated bottles of Coca-Cola are available in even the remotest areas of Africa. Our health systems should learn from, and even partner with, such commercial systems. We also need innovative technology to track distribution" (Nkengasong et al. 2020). The end of

2020 and early 2021 period saw the arrival of several vaccines from both from the developed and developing countries including USA, UK, China, Russia and India. Some few of these countries have been providing vaccines to the selected developed and developing countries. Pfizer and Moderna are being used mostly in developed countries. Oxford vaccine AstraZeneca is being used in both developed and developing countries including India.

The Russian vaccine Sputnik V is being used in several Latin American countries, including Mexico, Paraguay, Venezuela, Bolivia and Argentina. Indian vaccine is being exported to Brazil, Morocco, Bangladesh, Myanmar, Saudi Arabia, Bahrain, South Africa, Canada, Mongolia, Bhutan, Nepal, Maldives, Mauritius, Seychelles and Sri Lanka. Several Asian countries including Pakistan, Singapore, Malaysia and the Philippines have signed deals with the Chinese Sinovac vaccine, and Turkey has also approved the Sinovac vaccine for emergency use. Hungary becomes first European Union nation to use Chinese vaccine against COVID-19.

In January 2021, Johnson & Johnson brought out a new single-dose vaccine, and it has now been approved for use in the USA and other countries. However it is astonishing to note that a vaccine trade war has erupted with emphasis on vaccine nationalism, the EU decided not to permit vaccine export until the requirements of European Union countries are fulfilled. Italy was the first country to block the vaccine shipment to Australia.

Despite the availability of different vaccines and their accessibility, a section of population in both developed and developing countries is sceptical about its efficacy and likely side effects. About 25% of US population is still suspicious about it (Engler 2021, Whatley and Shodia 2020). In India about 59 percent people are skeptical about Covid vaccine (Indiatvnews 2020; Acharjee 2021).

Altitudinal Rise and Emergence of Vectorborne Diseases: A Warning for Developed Countries

This new disease pattern must be a warning to developed countries to take serious note of focusing on both vectorborne/virusborne and chronic diseases, as climate change is also associated with the expansion of important disease vectors in Europe: Aedes albopictus (the Asian tiger mosquito), which transmits diseases such as dengue and chikungunya, and Phlebotomus sand-fly species, which transmits diseases including Leishmaniasis (Akhtar 2020; Semenza and Suk 2018). Initially, it was thought that mountain communities are free from COVID-19. From the Andes to Tibet, the coronavirus seems to be sparing populations at high altitudes, but the current incidence of COVID-19 in Peru, Chile and the Kashmir mountains in India shows that even high mountain dwellers are being impacted by the virus.

Air Pollution and COVID-19

Air pollution is yet another dimension as some studies suggest that increased air pollution may lead to an increase in COVID-19 intensity. Focusing on climate change impacts on air pollution, particularly ozone pollution, IPCC has also clearly stressed that "pollen, smoke and ozone levels are likely to increase in a warming world, affecting health of residents in major cities. Rising temperatures will worsen air quality through a combination of more ozone in cities, bigger wild fires and worse pollen outbreaks, according to a major UN climate report" (IPCC 2018).

Scientists have urged that in the face of future climate change, stronger emission controls are enforced to avoid worsening air pollution and the associated exacerbation of health problems, especially in more populated regions including megalopolises of the world encompassing both developing and developed countries. By 2050, it could be 6.6 million premature deaths every year worldwide (Ansari 2015). The American Lung Association's "state of the air" finds 166 millions Americans are living in unhealthy ozone or particle pollution with health risks (Milman 2016; American Lung Association 2016). Another research highlights that "while the number of unhealthy polluted days has dropped in the past year, more than half of the US population lives in areas with potentially dangerous air pollution, and about six out of 10 of the top cities for air pollution in the USA are located in the state of California" (McHugh 2016).

Brazil, Russia, India, China, South Africa (BRICS) and Mexico have been drawing special attention due to the pollution emissions released into the atmosphere by their increasing number of industries and their exaggerated consumption of products (Akhtar and Palagiano 2018).

The intensity of air pollution in developed and developing countries opens up an opportunity to conduct research at the association of air pollution and the intensity of COVID-19 in these countries. In India, a report suggests that although there is no conclusive study to establish the causal link between air pollution and COVID-19 cases the deterioration of air quality particularly in metro cities including Delhi may result in increased risk of lung infections (The E.T. 2020).

In another study based on Sir Ganga Ram Hospital in New Delhi, India, Arvind Kumar asserts that "Virus particles can piggyback on PM 2.5 pollutants, leading to a deadly cocktail of toxic air in winter", and Kumar opined "the record-high number of new COVID-19 cases being reported in Delhi is a direct effect of air pollution" (Ethiraj 2020).

Conclusion

However, there is a need to undertake intensive regional research in order to understand the global pattern of diffusion of coronavirus focusing on infection migration and indigenous origin that has caused tremendous global economic and social disaster. Understanding of peoples' behaviour is crucial towards safety measures against infection, as COVID-19 impacted socio-economic and cultural life of population in all corners of the world. Most countries including developed and developing one have been in the lockdown status for months. Some European countries, Australia and the state of California introduced lockdown, curfew, and school closure in order to fight COVID-19 resurgence.

Finally, one must recall what was said by Anthony Fauci in 2008, in a Foreword to the book: *Encyclopedia of Pestilence Pandemics and Plagues*, edited by Joseph F. Byrne. Fauci asserts, "There are many lessons to be learned, but among the most important is this: the next pandemic waits in the wings for some convergence of critical determinants not yet imagined by any of us. How we respond may make a difference not only for ourselves, but for the rest of the world as well" (Fauci 2008). The global deaths of 2.8 million and 0.54 million deaths in USA alone have demonstrated that we have not heeded the advice of Dr. Anthony Fauci.

References

Acharjee, S. (2021). Five reasons why health workers are sceptical of covid vaccine. India Today, January 25.

Akhtar, R., & Palagiano, C. (Eds.), (2018). *Climate change and air pollution: The impact on human health in developed and developing countries*. Springer.

Akhtar, R. (Ed), (2020). *Extreme weather events and human health: International case studies*. Springer.

American Lung Association. (2016). *State of the Air*, 2016). www.lung.org.

Ansari, A. (2015, September 16). Study: More than six million could die early from air pollution every year. *CNN News*.

BBC. (2020, August 18). *Coronavirus: France to make face masks mandatory in most workplaces*.

Berardelli, J. (2021, February 5). Climate change "may have played a key role" in coronavirus pandemic, study says. *CBS News*.

Boseley, S. (2020, October 15). Thirty-year failure to tackle preventable disease fuelling global Covid pandemic. *The Guardian*.

Callaway, E. (2020, August). The unequal scramble for coronavirus vaccines—By the numbers. *Nature*, 584(7822): 506–507. https://doi.org/10.1038/d41586-020-02450-x.

Chevez, N., Maxouris, C., & Hanna, J. (2020, October 17). Fauci warns that Covid-19 infection rates are too high heading into winter. *CNN*.

Cooley, L., & Linn, J. (2020, June 4). Developing countries can respond to COVID-19 in ways that are swift, at scale, and successful. *Future Development*, Thursday.

Campbell, D. (2017, April 27). Children's hospital units forced to close to new patients due to staff shortages. *The Guardian*.

Engler, M. (2021). *The politics of vaccines and vaccinations* (https://www.morningsidecenter.org/teachable-moment/lessons/politics-vaccines-and-vaccinations) January 1.

Ethiraj, G. (2020, November 14). Coronavirus: The record high number of new cases in Delhi is a 'direct effect of air pollution'. *IndiaSpend.com*.

Fauci, A. S. (2008). Foreword. In, J. P. Byrne (Ed.), *Encyclopedia of pestilence, pandemics, and plagues*, Greenwood Press,Westport, pp. XIX–XX.

Gostin, L. O., Koh, H. H., Williams, M. et al. (2020). Withdrawal from WHO is unlawful and threatens global and US health security, The Lancet, July 9, https://doi.org/10.1016/S0140-6736(20)31527-0

Guite, H. (2020, August 16). How does weather affect COVID-19? *Medical News Today*.

Hindustan Times. (2020, October 18). Cannot rule out possibility of a second wave of covid-19 during winters: NITI Aayog's V.K. Paul, New Delhi.

Howe, G. M. (1972) Man, Environment and disease in Britain: A Medical Geography of Britain through the Ages. Harmondsworth.

Indiatvnews. (2020). 59% India skeptical about COVID-19 vaccine, say they won't rush to take it: Survey, Indiatvnews.com, December 3.

Lambert, L. (2020). The coronavirus has now killed more Americans than every war since the start of the Korean War—combined Fortune, June. https://fortune.com/2020/06/10/coronavirus-dea ths-us-covid-19-killed-more-americans-korea(n-war-vietnam-iraq-persian-gulf-combined-how-many-died).

McHugh, J. (2016, April 20). US air pollution. Worst cities for clean air are in California: Report says, www.ibtimes.com/us-air-pollution-worst-cities-clean-air-are-california-report-says-23.

Milman, O. (2016, April 20). More than half US population lives amid dangerous air pollution, report warns. *The Guardian*.

New York Times. (2020, November 23). *Now the U.S. has lots of ventilators, but too few specialists to operate them*.

Nkengasong, J. N., Ndembi, N., Tshangela, A. & Raji, T. (2020, October 6). COVID-19 vaccines: How to ensure Africa has access. *Nature*.

Redlener, I., Sachs, J., Hansen, S., &, Hupert, N. (2020, October 21). *130,000–210,000 avoidable Covid-19 deaths and counting in the U.S.* (https://ncdp.columbia.edu/custom-content/uploads/2020/10/Avoidable-COVID-19-Deaths-US-NCDP.pdf).

Semenza, J. C., & Suk, J. E. (2018, February 1). Vector-borne diseases and climate change: A European perspective. *FEMS Microbiology Letters, 365*(2). https://doi.org/10.1093/femsle/fnx244. Review.

Singh, N. (2020, October 22). Just like Swine Flu, Covid-19 may see spike during winter: AIIMS Director. H.T., New Delhi.

The Economic Times. (2020, October 27). Air pollution may hinder India's fight against Covid-19, say scientists. New Delhi.

Tollefson, J. (2020). How Trump damaged science—and why it could take decades to recover. Nature, October 5.

UNDP. (2020). Coronavirus vs. inequality, UNDP. https://feature.undp.org/coronavirus-vs-inequa lity.

Whatley, Z., & Shodia. T. (2020). Why so many Americans are skeptical of a coronavirus vaccine, Scientific American, October 12.

Wouters, O. J., Shadlen, K. C., Salcher-Konrad, M., Pollard, A. J., Larson, H. J., Teerawattananon, Y., Jit, M. l. (2021) *Challenges in ensuring global access to COVID-19 vaccines: Production, affordability, allocation, and deployment*. The Lancet, February 12, https://doi.org/10.1016/S0140-673 6(21)00306.

Zhou, N. (2020). Dr. Fauci praises Australia's coronavirus responseand Melbourne's face mask rules, The Guardian, October, 29.

Coping During Covid-19 Public Health Crises

Vanessa N. Dominguez and Muge Akpinar-Elci

Abstract The COVID-19 global health pandemic underlined the importance of all communities having access to the highest quality public health services. With its impact reaching across individuals, relationships, the economy, and environment, COVID-19, as a public crisis from the USA and global perspective, draws attention to policy development for management, specifically under the public sector. With stress and anxiety on the rise, mental health services are particularly critical in ensuring the overall well-being and adjustment of our nation. This chapter describes the impact of stress and trauma on daily functioning as well as strategies deemed helpful for coping during the pandemic.

Keywords Pandemic · Stress · Trauma · Coping · Isolation · Public health

Introduction

According to Fauci et al. (2020), the coronavirus pandemic has massively changed different generations' businesses included. An epidemic that has resulted in disruptions, continual upheavals, and global shifts in ends linked to technology transformation, digitalization, geopolitics transformations, and business model's evolution has facilitated the introduction of collective callus curb its uncertainties. Information further backed up by Heymann and Shindo (2020) elaborates on public health's role in reducing the COVID-19 crisis. Fauci et al. (2020) have argued through a study conducted in health premises that essential health services to be provided at this sadden period should be linked to mental health issues, gender-based violence, non-communicable diseases, and health financing. That also adds to nutrition and food safety and the care of the older people who are at high risk of contracting COVID-19. Hence, this chapter aims to descript the USA and global health perspective of COVID-19 pandemic and to provide strategies on how to cope with uncertainties during this public health crisis.

V. N. Dominguez (✉) · M. Akpinar-Elci
Counseling & Human Services, 2100 New Education Building, 4301 Hampton Blvd., Norfolk, VA 23529, USA
e-mail: vdomingu@odu.edu

© The Author(s), under exclusive license to Springer Nature Switzerland AG 2021
R. Akhtar (ed.), *Coronavirus (COVID-19) Outbreaks, Environment and Human Behaviour*, https://doi.org/10.1007/978-3-030-68120-3_2

13

COVID-19 as a Public Health Crisis

Lipsitch et al. (2020) define "*COVID-19 is a mild yet severe respiratory condition facilitated by acute respiratory syndrome Beta-coronavirus of genus Betacoronavirus* (p. 1195)." According to USA and the global perspective, the pandemic has necessitated specific outcomes. As presented under the History and Etymology of COVID-19, have direct and indirect adverse side effects. It has not only an impact on children but also the youth and elderly mental and physical health, which indeed is a public health crisis as presented by Wenham et al. (2020). The World Health Organization (WHO) further incorporated the urge that the COVID-19 pandemic will always remain a public health emergency of international concern. A committee invented by WHO in their third meeting under Dr. Tedros Adhanom stated that global nations must come up with emergency procedures to survive during this time of the crisis. The public health sector, for instance, should focus hard on resources and expertise in addressing COVID-19 based on its impact on communities. The Public Health Institute in their ability to discuss how the pandemic has transformed global health and development as ascribed under development cooperation policy denotes the various adjustments made in the health sector as a result of the epidemic.

COVID-19, as a public crisis from the USA and global perspective, calls for the attention of policy options for management, specifically under the public sector. The comprehensive view of researchers under Watkins (2020) in their urge to prevent the epidemic calls for further frameworks linked to public health regulations. Public sectors apart from that are expected to generate critical preparedness, response, and readiness to deal with the crisis (Shi et al. 2020). COVID-19 as a crisis has heavily impacted well-being, health, psychosocial and socioeconomic status of many communities in many geography regions. When individuals are mentally and emotionally well, they are more likely to earn higher incomes which has a positive impact on the economy, relationships, and overall well-being of a nation. However, when our communities are suffering from depression and anxiety, this impact could create a ripple effect that leaves people with job loss, housing insecurity, food insecurity, and lack of health insurance. Therefore, it stresses the importance of all communities having access to the highest quality of mental health services.

Social Isolation and Stress During Uncertainty

Stress and anxiety are common emotions to the human experience. They help us problem solve, motivate us to pursue our goals, and protect us from danger, which ultimately help us live the life we want to lead. With that being said, when humans endure chronic stress and anxiety, the impact of these experiences can become debilitating and demoralizing. The perceptions and emotions associated with trauma and stress, such as anxiety, fear, and worry, actually keep us from leading a balanced and more present life (Myawaki 2015; Matosin et al. 2017). Over the course of human development, human beings have adapted to stress in order to survive throughout the centuries. The reactions we have developed to navigate stressful situations are normal and natural; however, they are not always helpful and productive. Furthermore, chronic stress and trauma have been found to have lasting biological influences including brain structure and function, behaviour, and the immune and endocrine systems (Bennett et al. 2016).

Induced psychosocial uncertainty based on the study implemented by Bekemeier et al. (2015) defines underlying consequences linked to the pandemic. In addition to the high fatality and infectivity rates related to the virus, the crisis has resulted in universal psychosocial outcomes linked to economic burdens, mass hysteria, and financial losses. The mass fear for COVID-19, commonly known as *"coronaphobia"* has constructed incidences of a plethora of psychiatric manifestations (Lipsitch et al. 2020). A study conducted through the use of Google Scholar and PubMed searchers on a topic relative to "pandemic," "mental health," and "quarantine" among others showed that COVID-19 pandemic resulted in paranoia, depression, hoarding, anxiety, acute panic, and post-traumatic stress disorder (PTSD) traits of uncertainty. The research also indicated that the health practitioners are at the forefront of contracting the virus in addition to adverse psychological outcomes via burnout, fear of transmitting infections, and conflicting feelings.

The COVID-19 global health pandemic has demonstrated the many ways in which stress reaches a tipping point in human beings, creating a ripple effect individually, relationally, economically, and environmentally (Duane et al. 2020). We all react in similar and distinct ways, and it is natural to think, feel, and behave in the ways we are experiencing during a crisis. The stress and worries we experience during any unexpected change are like a chain of thoughts and emotions that build upon each other. Often the hardest changes to understand and adjust to are the ones that are unexpected and out of our control—a recession, a global pandemic, or a major disaster, for example. Changes of this magnitude can be difficult to come to terms with, but the experience of them can improve or worsen depending on our reaction and attitude (Schaefer et al. 2013; Valand et al. 2020).

As previously mentioned, during times of stress and/or unexpected change, worrying allows us to anticipate obstacles or problems, and gives us the opportunity to plan solutions. A normal amount of worry helps us to overcome *real worry problems* (Whalley and Kaur 2020). During the pandemic, some of these real worry problems can look like tackling the bills, time management, managing work obligations from home, overseeing your kids' online schooling, and relationship conflict to name a few. Often times, when we encounter these daily stressors we can feel overwhelmed by the growing to-do list as well as feel helpless in our ability to get everything done. The social messages we receive about productivity and mental health lead us to frequently conceal and/or avoid the uncomfortable or unwanted experiences associated with stress because we may feel like we are failing or simply because they are distressing. However, avoiding or neglecting our mental and emotional health leads to lasting biological and emotional consequences (Madanes et al. 2020; Miyawaki 2015; Saltzman et al. 2020; Steffani 2020).

According to Dr. Bessel Van der Kolk (2015), trauma is defined as an event that overwhelms the central nervous system, altering the way we process and recall memories. Essentially, a traumatic event shapes our temperament, or the automatic thoughts, feelings, and reactions we utilize to navigate and make sense of the world. This conditioning is stored in our bodies and results in a survival disposition, also known as fight, flight, or freeze. If we experience prolonged stress or trauma, it will shape the way our brain navigates stress as well as influence how our bodies regulate these intense experiences. Typically, the antidote to this disposition is to encounter experiences that directly contradict the conditioning we developed through the trauma, like exercises that help us regain a sense of power and control, and learning skills to help our brains and bodies feel safe, grounded, and in the present moment (Van der Kolk 2015). Below we will examine the parts of our brains impacted by stress and trauma and their functioning.

When undergoing a stressful and/or traumatic situation, three main areas of the brain become activated: the prefrontal cortex (PFC), the anterior cingulate cortex (ACC), and the amygdala (Arnsten 2009; Bennett et al. 2016; Matosin et al. 2017; Shields et al. 2016). When the PFC is activated, we are able to think clearly about different situations and develop the best solutions to resolve them. More importantly, we are less self-focused and are able to communicate our own needs while also acknowledging the needs of others. The ACC is activated when we are feeling triggered—meaning, the perceptions and meaning we attribute to a particular situation illicit an emotional response. If this area is well developed, we are able to determine if our actions would be helpful or unhelpful in responding to the situation. Many different situations can cause us to feel triggered, from receiving an unexpected email from your boss to sensing your partner is feeling withdrawn, to seeing violent content on the news/social media. If our ACC is underdeveloped and/or we are not consciously aware that we are feeling triggered, we may react in ways that are more harmful. For example, if you receive that distressing email from your boss, you may immediately reply without thinking your response through. Ultimately, the ACC will help us manage our emotions so that we do not do things we later regret.

Finally, the amygdala is the fear response center of our brain. Its function is largely outside of our conscious awareness and operates like a 24/7 surveillance to determine if something is a threat. Through our senses, the amygdala detects if something dangerous is present, producing fear in us. When this area is activated, we feel afraid, reactive, and vigilant. This part of our brain evolved to respond to threats in our environment (Sweeton 2017). Threats previously consisted of potentially dealing with other tribes of people or animals encroaching on our resources, navigating uninhabitable environments, etc., that would require us to react quickly and survive potentially life-threatening situations. In modern times, our threats are typically a lot subtler—like pending deadlines, paying the bills on time, dealing with conflict, or even driving on the road. Even though these situations are sometimes a lot less life-threatening, our amygdala will scan and react to stress inducing situations in the same way.

The very real fears we experience in our automatic reactions are developed through this conditioning of survival functioning, or the fight–flight–freeze response, developed through our amygdala. Over time, our stress and worries stemming from this conditioning can lead us to feeling more alienated, disconnected, and depressed. Reactivity can show up individually, in our relationships, but also be reflected in how our governments, military, and health care determine how our resources are allocated during a global traumatic event. The health of a nation is dependent on the overall well-being of its people, especially during a global health crisis (Henderson et al. 2013). Providing resources that aid in trauma and stress management will help people to successfully and smoothly navigate stressful experiences, such as this pandemic, without creating more damage for ourselves, others, or the environment in the long run.

How to Cope with Uncertainty During the COVID-19 Public Health Crisis

According to Adams and Walls (2020), there are various ways of dealing with uncertainty during the COVID-19 public health crisis. The news, which has brought fear, anxiety, grief, and dozens of lockdowns around the world, came to be dealt with in specific ways. Despite the challenges on learning to deal with uncertainty which has promoted constant stress activation on humans' body, the World Health Organization (WHO) identifies four critical forms of dealing with change during the pandemic. One of them includes *"Trigger Identification."* Individuals are expected to identify their triggers through processes such as reading scaremongering stories from social media, which tackles adverse case sceneries (Gostin and Wiley 2016). In the process, one would spend time understanding how to recognize triggers and take action to eliminate or reduce exposure to life triggers. *Eating healthily and exercising* is also one of the ways of coping with uncertainty during COVID-19. Eating healthily, for instance, is vital in maintaining energy levels in addition to preventing mood swings. One can also avoid processed and sugary foods and opt for hydration. On the other hand, as exercising is a natural way of relieving stress as endorphins are released in the body, individuals tend to leave aside triggers and instead focus on strengthening one's hearth through exercises and improving blood flow. *Seeking help* is another process of coping with uncertainty during the crisis. As different people have their ways of dealing with anxiety and uncertainty, especially now that there is no manual to deal with the pandemic. It is vital to seek help when the process is overwhelming. In the process, one would get a different level of care and aid their journey of managing incidences associated with the pandemic. Considering seeking professional help fosters not only an understanding of self but also necessitating cognitive development (Watkins 2020). *Giving Mindfulnessa Change* is also a technique for coping with COVID-19 uncertainty. Mindfulness and meditation help in the restoration of sense control. One can observe and recognize emotions and thoroughly analyze brain signals that might smooth the system. Mindfulness strategies will be explained further below.

Hence, despite uncertainty connected to COVID-19, coping strategies to cope with the situation is vital. The uncertainty caused by putting plans on hold, fear of treatment, and side effects in addition to dying or losing a loved such as contracting COVID-19 can be dealt with via a systematic decision-making process. An individual needs to reduce uncertainty via assumed-based reasoning, and weighing pros and cons linked to the uncertainty. Groups can also create alternatives that suppress and forestall their uncertainties.

Stress Management and Coping Mechanisms with COVID-19

Dan Siegel and Bryson (2012) developed a unique way of describing *whole brain functioning*, or when the prefrontal cortex, anterior cingulate cortex, and amygdala are working in unison. This is also known as *integration*, where information flows easily throughout our brain allowing us to respond to stressful situations in a more present and thoughtful way. However, prolonged stress and trauma cause us to "flip our lid" (Siegel and Bryson 2012; p. 63), creating a disconnect in the part of our brain that allows for rational thinking and empathy from the parts of our brain that illicit emotions and adapt quickly to stressors. For example, imagine you are driving on the highway and someone cuts you off. When our whole brain is working together, we are able to simultaneously step on the breaks to evade an accident while also reassure ourselves that we are safe—"Whew, that was a close one…" However, when our lid is flipped, or the parts of our brain are not communicating in tandem, it is more difficult to soothe ourselves and come down from the stress of being cut off on the highway. This reaction could be in the form of angry outbursts, panic attack, to feelings of overwhelmed and defeat.

The pandemic is considered a global traumatic event (Saltzman et al. 2020; Valand et al. 2020), which means in addition to the daily stressors we tackle in the foreground of our lives, there is the pervasive uncertainty and ambiguity of COVID-19 ongoing in the background. This persistent, added layer of stress will result in more reactivity and less tolerance to typical stimuli day-to-day. The key to calming an overactive fear response center is to practice self-regulating skills like deep breathing, awareness, and compassion. Mindfulness is a stress-reduction-based strategy that has proven to be effective for coping with stress, anxiety, and trauma (Nagy et al. 2020). You can think of mindfulness as becoming a watchful observer of your experiences without overly identifying with them. Its main advantage is to help you live in the present and accept the moment as it is, whether that moment is stressful, joyful, angering, or surprising. Practicing mindfulness will move you into wise mind—or a balance between your overly emotional and overly intellectual sides of our functioning (McKay et al. 2019).

Mindfulness is helpful in regulating emotions, getting through a crisis without making things worse, and successfully resolving interpersonal conflicts. As you practice mindfulness, it calms your mind and body, and you will feel an increase in peace and gratitude. Other strategies include reframing unhelpful thoughts, such as catastrophizing, should or must thinking, or all or nothing thinking, to a more balanced view or perspective. Reframing shifts our view of a particular problem, event, or person and assumes that when individuals are able to view a situation from another perspective, opportunities for finding alternative, acceptable solutions to their problems increase. A few moments of turning inward, taking a few deep breaths, and using our senses to ground ourselves will aid in our whole brain functioning as well as allow us to feel more connected in life. When plagued with stress, our negative thoughts and emotions often lead to separation between us and other, causing us to feel more isolated and alienated. Practicing mindfulness will help you stay in the present when broaching difficult conversations with loved ones. It is usually our distracted thinking that leads to unhelpful perceptions of a particular situation and detracts from our ability to understand others. Practicing a mindful pause when sensing tension in an interaction will help you to consider how you want it to end.

Gottman (2008) also suggests three different strategies when dealing with conflict in relationships. In order to feel more connected and loved in our relationships, he suggests *turning toward* our loved ones during stressful times. This will communicate love, support, and add to our *emotional love bank* in our relationships. Gottman refers to these as *bids for connection* which can create a buffer within our relationships when we are going through difficult times. Bids for connection looks like facing our loved one when we are talking to each other, unplugging from technology, reaching out to hold hands or offer a hug, or even just a pat on the back. Responding to each other's bids is how we learn to communicate and express needs, feel loved, and supported in our relationships.

Secondly, increasing our fondness and admiration toward one another will help us to create more harmonious interactions when navigating stressful situations. This attitude applies not only to our personal relationships but to our daily interactions during the global pandemic. Dr. Gottman (2008) has found that a lack of fondness and admiration in our relationships, particularly romantic ones, leads to feelings of contempt. Contempt is known as a key relationship killer because it detracts from the love and care we feel toward others. Increasing our fondness toward others will also help us remember that we are all connected during this global pandemic and that each of us is struggling in different ways. Beginning with kindness and compassion toward ourselves leads to more kindness and compassion toward others. Finally, managing conflict by finding ways to broach difficult situations in a gentle manner will help all those involved to soothe the physiological arousal that occurs during conflict. Addressing conflict in soft manner will lead to further de-escalation in arguments that is already exacerbated from the pandemic. Similar to mindfulness, pausing and taking some full deep breaths before broaching a topic will bring more clarity in the conversation and focus on what is being discussed.

Conclusion

The COVID-19 global pandemic has set in motion long-lasting changes to the way our public health sectors address and manage crises. Furthermore, as a society we are currently being shaped by this global trauma in ways that we will not fully understand until after its conclusion. In time, we will be able to make better sense of COVID-19 in terms of how we can learn from this experience, make meaning, and come to terms with our pitfalls in handling global crisis. At this time, we can utilize the strategies provided and take heed in the need to ensure all communities are adequately cared for and have access to the resources that will contribute to the overall well-being of our society.

References

Adams, J. G., & Walls, R. M. (2020). Supporting the health care workforce during the COVID-19 global epidemic. *JAMA, 323*(15), 1439–1440.

Arnsten, A. F. (2009). Stress signalling pathways that impair prefrontal cortex structure and function. *Nature Reviews Neuroscience, 10*(6), 410–422.

Bai, Y., Yao, L., Wei, T., Tian, F., Jin, D. Y., Chen, L., & Wang, M. (2020). Presumed asymptomatic carrier transmission of COVID-19. *JAMA, 323*(14), 1406–1407.

Bekemeier, B., Walker Linderman, T., Kneipp, S., & Zahner, S. J. (2015). Updating the definition and role of public health nursing to advance and guide the specialty. *Public Health Nursing, 32*(1), 50–57.

Bennett, M., Hatton, S., & Lagopoulos, J. (2016). Stress, trauma and PTSD: Translational insights into the core synaptic circuitry and its modulation. *Brain Structure and Function, 221*(5), 2401–2426.

Duane, A., Stokes, K., DeAngelis, C., & Bocknek, E. (2020). Collective trauma and community support: Lessons from Detroit. *Psychological Trauma: Theory, Research, Practice, and Policy, 12*(5), 452–454.

Fauci, A. S., Lane, H. C., & Redfield, R. R. (2020). Covid-19—navigating the uncharted.

Gostin, L. O., & Wiley, L. F. (2016). *Public health law: power, duty, restraint.* Univ of California Press.

Gottman, J. M. (2008). Gottman method couple therapy. *Clinical Handbook of Couple Therapy, 4*(8), 138–164.

Henderson, C., Evans-Lacko, S., & Thornicroft, G. (2013). Mental illness stigma, help seeking, and public health programs. *American Journal of Public Health, 103*(5), 777–780.

Heymann, D. L., & Shindo, N. (2020). COVID-19: What is next for public health? *The Lancet, 395*(10224), 542–545.

Lipsitch, M., Swerdlow, D. L., & Finelli, L. (2020). Defining the epidemiology of Covid-19—Studies needed. *New England Journal of Medicine, 382*(13), 1194–1196.

Madanes, S., Levenson-Palmer, R., Szuhany, K., Malgaroli, M., Jennings, E., Anbarasan, D., & Simon, N. (2020). Acute stress disorder and the COVID-19 pandemic. *Psychiatric Annals, 50*(7), 295–300.

Matosin, N., Cruceanu, C., & Binder, E. (2017). Preclinical and clinical evidence of DNA methylation changes in response to trauma and chronic stress. *Chronic Stress, 1* (June 2017).

McKay, M., Wood, J. C., & Brantley, J. (2019). *The dialectical behaviour therapy skills workbook: Practical DBT exercises for learning mindfulness, interpersonal effectiveness, emotion regulation, and distress tolerance.* New Harbinger Publications.

Miyawaki, C. (2015). *Association of social isolation and health across different racial and ethnic groups of older Americans., 35*(10), 2201–2228.

Nagy, S. M., Pickett, S. M., & Hunsanger, J. A. (2020). *The relationship between mindfulness, PTSD-related sleep disturbance, and sleep quality: Contributions beyond emotion regulation difficulties.* Psychological Trauma: Theory, Research, Practice and Policy. https://doi.org/10.1037/tra 00005721.

Whalley, M. & Kaur, H. (2020, March 19). *Guide to living with worry and anxiety amidst global uncertainty.* Psychology Tools. https://www.psychologytools.com/articles/free-guide-to-living-with-worry-and-anxiety-amidst-global-uncertainty/.

Saltzman, L., Hansel, T., & Bordnick, P. (2020). Loneliness, isolation, and social support factors in post-COVID-19 mental health. *Psychological Trauma: Theory, Research, Practice, and Policy, 12*(S1), S55–S57.

Schaefer, S., Morozink Boylan, J., Van Reekum, C., Lapate, R., Norris, C., Ryff, C., et al. (2013). Purpose in life predicts better emotional recovery from negative stimuli. *PLoS ONE, 8*(11), E80329.

Shi, Y., Wang, Y., Shao, C., Huang, J., Gan, J., Huang, X., et al. (2020). *COVID-19 infection: The perspectives on immune responses.*

Siegel, D. J., & Bryson, T. P. (2012). *The whole-brain child: 12 revolutionary strategies to nurture your child's developing mind.* Bantam.

Steffanni, K. (2020). Coping with stress during ongoing coronavirus crisis. *TCA Regional News.* TCA Regional News, 2020-03-20.

Sweeton, J. (2017, March 13). *How to Heal the Traumatized Brain.* Psychology Today. https://www.psychologytoday.com/us/blog/workings-well-being/201703/how-heal-the-traumatized-brain.

Valand, P., Lloyd, N., Robson, M., & Steele, J. (2020). Trauma transformed: A positive review of change during the COVID-19 pandemic. *Journal of Plastic, Reconstructive & Aesthetic Surgery, 73*(7), 1357–1404.

Van der Kolk, B. A. (2015). *The body keeps the score: Brain, mind, and body in the healing of trauma.* Penguin Books.

Watkins, J. (2020). *Preventing a Covid-19 pandemic.*

Wenham, C., Smith, J., & Morgan, R. (2020). COVID-19: The gendered impacts of the outbreak. *The Lancet, 395*(10227), 846–848.

Wolf, E., & Schnurr, P. (2016). Developing comprehensive models of the effects of stress and trauma on biology, brain, behaviour, and body. *Biological Psychiatry, 80*(1), 6–8.

Australasia

COVID-19 Pandemics in the Pacific Island Countries and Territories

Eberhard Weber, Andreas Kopf, and Milla Vaha

Abstract By the middle of February 2021, the COVID-19 pandemics are still ongoing. Globally, 110 million cases have been recorded. More than 2.4 million people have died from the disease caused by the novel SARS-CoV-2 virus. Figures presented in this chapter are taken from the WHO database (WHO 2021), except those from Hawai'i (New York Times database 2021) and Easter Island (France24 2021). All currencies have been converted into USD using Oanda database (Oanda 2020) as of February 9, 2021. In many Pacific Island Countries and Territories (PICT), however, morbidity and mortality have been low. Most PICs have no cases. Where cases exist, mortality has been low with a few exceptions. That being said, social and economic consequences are severe in all places. The chapter gives a brief reflection on historical experience with epidemics in the Pacific Island region. The remainder of the chapter looks at the epidemiology of the COVID-19 pandemics, the "who, when, and where" of COVID-19 in the Pacific Islands, its associated risk factors, and possible ways to control the pandemics. Medical preparedness of PICT health systems in PICTs, and their material and organizational constitution are covered as well as the support PICT received from outside to deal with the situation. The chapter also provides an account of social and economic impacts of COVID-19, including challenges to care for Pacific Islanders away from home. Some PICs (Fiji, Solomon Islands, Tonga, and Vanuatu) had to deal with severe tropical cyclones just at the time when governments tried to implement measures against COVID-19. To conclude, the chapter provides recommendations of how to strengthen PICT capabilities and capacities to improve preparedness against similar medical treats in the future. In the chapter, Fiji and Papua New Guinea (PNG) stand in the center of analysis. Some other Pacific Islands and Territories were included, where reported cases emerged and/or where important situations arose. In a few instances, the chapter also looks at the situation Pacific Islanders faced outside their countries.

Keywords COVID-19 · Pacific island countries and territories · Epidemiology · Economic impacts · Social impacts · Pandemic

E. Weber (✉) · A. Kopf · M. Vaha
The University of the South Pacific, Suva, Fiji
e-mail: eberhard.weber@usp.ac.fj

© The Author(s), under exclusive license to Springer Nature Switzerland AG 2021
R. Akhtar (ed.), *Coronavirus (COVID-19) Outbreaks, Environment and Human Behaviour*, https://doi.org/10.1007/978-3-030-68120-3_3

Introduction

Pacific Islands, like many other parts of the world, paid a colossal toll when Europeans arrived for the first time on their shores. Europeans brought firearms that triggered/worsened tribal conflicts and warfare. Equally, fights between European conquerors and native people took a heavy toll particularly on the sides of the natives. Last but not least, first "contacts" between Europeans and native populations transferred germs in the form of venereal and other diseases such as syphilis, dysentery, measles, smallpox, and the like. People died like flies and decimated peoples to the brink of extinction in many parts of the (pre) colonial Pacific Islands.

Diseases brought by Europeans have virtually affected all inhabited islands in the Pacific. The impact was so serious that by the end of the nineteenth/early twentieth century fears of a possible extinction of Pacific Island peoples were expressed (Denoon 2017; Rivers 1922). The contact with Europeans indeed led to a severe reduction of indigenous people in the Pacific Islands (and elsewhere) (Diamond 1997). Without going into details, most islands in the Pacific Oceans went through epidemic events that for most islands caused death tolls of at least 10% of their populations. Often the toll was much higher.

During the Spanish flu (Western) Samoa's mortality was one of the world's highest. Some 8500 people died of the disease, more than a fifth of Samoa's population. Separated by the waters of the Pacific Ocean, American Samoa, not even 100 km to the southeast, was one of the few places in the world, which did not suffer a single death from the Spanish flu (McLane 2013). When news of the outbreak in (Western) Samoa reached the Governor of American Samoa, he completely locked down borders by late October 1818 and put the US territory under strict maritime quarantine until 1924. Elsewhere mortality was some 16% in French Polynesia. High rates were also recorded in Nauru, where some 18% of the Nauruan population and 39% of Micronesians from the Gilbert Islands (today Kiribati) died. They were working in phosphate mining. In Tonga, mortality was between 5 and 10% (Shanks et al. 2018). The highest number in deaths (some 9000 deaths) occurred in Fiji, which was about 5.5% of the population (Herda 2000). Based on these innumerous experiences of epidemics, Pacific Island societies have very good reasons to be at highest alert levels when pandemics are emerging.

Epidemiological Characteristics of the COVID-19 Cases in the Pacific Island Region

Since the coronavirus started to create havoc at the end of 2019 in the Chinese city of Wuhan, it found its way to virtually every country on Planet Earth. In the Pacific Islands, which includes 14 Pacific Island Countries and eight territories, health challenges arising from SARS-CoV-2 have been relatively small. Nevertheless, in all PICT COVID-19 has become a major economic and social challenge (Fig. 1).

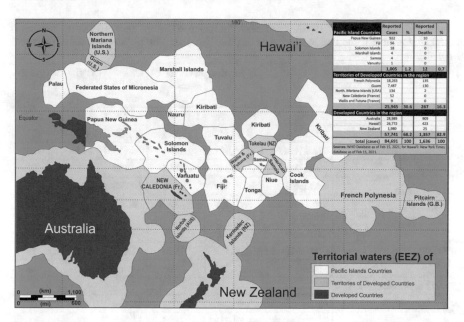

Fig. 1 Pacific Island countries and territories, developed countries at the rim of the Pacific Island region

Suggestions that small isolated places, especially islands, are better protected against infections originating outside are not entirely convincing. The history of epidemics in Pacific Islands suggests otherwise. Today (like then) entry points (air and seaports) are restricted in numbers, they are relatively easy to monitor, especially when numbers of people entering a place are comparatively small. If we look beyond the Pacific Island region, we find many small island states and territories that have big numbers of COVID-19 cases relative to their population (Table 1). This shows that looking solely on size and degree of isolation is too simplistic to explain protection/exposure to the COVID-19 pandemic. Among the PICT (including Hawai'i as US state), 11 out of 25 had not a single COVID-19 case by February 15, 2021. The majority of cases in this part of the world concentrate on three independent countries (Fiji, PNG, and Solomon Islands), four territories of developed countries (French Polynesia [France], Guam [USA], New Caledonia [France], and Northern Mariana Islands [USA]), and one state (Hawaii [USA]) of other countries. In the Pacific Islands, prevalence of COVID-19 is by far highest in politically dependent units (territories and states). Overall in the Pacific Island region, prevalence of COVID-19 has been much lower than in the Caribbean Islands and the island regions of Africa and Asia. A big number of Small Island Developing States (SIDS) without any prevalence of COVID-19 are exclusively in the Pacific Island region. In the Caribbean, four small independent island states (Aruba, Belize, Bahamas, and Dominican Republic) and seven territories of major countries (Sint Maarten (Netherlands), Turks and Caicos Islands (UK), Saint Barthélemy (France), Puerto

Table 1 Reported COVID-19 cases and deaths in island states and territories in the Asian and African, Caribbean, and Pacific island regions (nine countries with highest prevalence per 1 mill population)

SIDS in Africa and Asia		Population	COVID-19 cases	Cases per mill	COVID-19 deaths	Deaths per mill
1	Bahrain	17,01,575	1,12,742	66,257	403	237
2	Maldives	5,40,544	17,828	32,982	58	107
3	Cape Verde	5,55,987	14,741	26,513	139	250
4	Seychelles	98,347	1,892	19,238	8	81
5	Singapore	58,50,342	59,800	10,222	29	5
6	São Tomé and Príncipe	2,19,159	1,482	6,762	19	87
7	Comoros	8,69,602	3,334	3,834	128	147
8	Guinea-Bissau	16,04,528	2,924	1,822	46	29
9	Mauritius	12,71,777	595	468	10	8
SIDS in the Caribbean						
1	Aruba	1,06,766	7,338	68,730	68	637
2	Turks and Caicos Islands (UK)	38,718	1985	51,268	27	697
3	Sint Maarten (Netherlands)	42,882	1969	45,917	12	280
4	Pureto Rico (USA)	28,60,853	97,609	34,119	1,919	671
5	Saint Martin (France)	38,659	1,377	35,619	12	310
6	Belize	3,97,628	12,145	30,544	313	787
7	United States Virgin Islands (USA)	1,04,425	2524	24,170	25	239
8	Guadelope (France)	4,00,124	9,302	23,248	159	397
9	Bahamas	3,93,236	8,311	21,135	178	453
Pacific Island countries and Territories						
1	French Polynesia (France)	2,80,908	18,263	65,014	135	481
2	Guam (USA)	1,68,775	7,487	44,361	130	770
3	Hawai'i (USA)	14,16,202	26,772	18,904	423	299

(continued)

Table 1 (continued)

SIDS in Africa and Asia		Population	COVID-19 cases	Cases per mill	COVID-19 deaths	Deaths per mill
4	Northern Mariana Islands (USA)	57,557	134	2,328	2	35
5	Wallis and Futuna (France)	11,246	9	800	0	0
6	New Caledonia (France)	2,85,498	52	182	0	0
7	Papua New Guinea	89,23,230	922	103	10	1
8	Marshall Islands	59,198	4	68	0	0
9	Fiji	8,96,430	56	62	2	2

Solomon Islands (18), Samoa (4), Vanuatu (1); Cook Islands (0), Fed. States Micronesia (0), Kiribati (0), Nauru (0), Palau (0), Tonga (0), Tuvalu (0), Pitcairn (UK, 0), Tokelau (NZ, 0)
Sources WHO (2021), database as of Feb 15, 2021; for Hawai'i: New York Times (2021), database as of Feb 15, 2021; Only countries/territories with a population of more than 10,000

Rico (USA), Saint Martin (French), United States Virgin Islands, and Guadeloupe (France)) have more than 20000 reported cases per one million of their respective populations. In SIDS of Africa and Asia, there are three countries that reported more than 20000 cases/one million residents (Bahrain, Maldives, and Cape Verde). From a global perspective, Aruba and Bahrain are among the top 15 countries in COVID-19 prevalence and deaths. In the Pacific Island region, only French Polynesia and Guam cross the ratio of more than 20000 reported cases per one million of their population. Four territories (French Polynesia, Guam (USA), Hawaii (USA), Northern Mariana Islands (USA)) top the list. Independent PICs that have cases of COVID-19 are PNG, Fiji, Solomon Islands, Marshall Islands, Samoa, and Vanuatu. The remaining eight Pacific Island Countries did not record a single case as of February 15, 2021.

Prevalence of COVID-19 in Fiji, PNG, and Other PICTs

The first case of COVID-19 in any PICT was confirmed on March 11 in French Polynesia. A member of the French National Assembly had returned from France a few days earlier (RNZ March 13, 2020). By the end of March, most PICTs that have confirmed COVID-19 cases had reported their first case. Except PNG (922 cases), all other political units with more than 100 cases of COVID-19 are Pacific territories of developed countries/or developed countries.

The first case in Fiji relates to a Fiji Airways flight attendant, who had traveled from San Francisco to Nadi International Airport, and then with a flight on the following day to Auckland, New Zealand. He returned to Nadi the same day. Two days later (March 19), he tested positive for SARS-CoV-2. Some cases were detected later (7 cases). Case 9 had arrived from India (via Singapore) on March 22. He had attended a religious function in New Delhi, related to more than 1000 COVID-19 cases (BBC April 16, 2020). Instead of self-isolation for 14 days upon his return, he stayed for a few days in Suva before traveling by ship and bus to Labasa town in the north of Fiji's second largest island Vanua Levu (Fiji Department of Information 2020). There have been eight cases related to case 9. On April 7, Fiji's Police Commissioner announced that criminal investigations have been taken up again the flight attendant (case 1) and the person returning from India (case 9) for allegedly breaching Public Health Regulations and failure to self-isolate when returning to Fiji (Singh 2020). In three other cases, people arriving from overseas introduced the virus to the community.

In the remaining cases, Fiji citizens tested positive during the 14-day quarantine in a government facility. Cases 19–27 had arrived in Fiji on July 1 from New Delhi on an evacuation flight for 112 Fiji citizens. They were all tested negative before boarding the plane in India. On this flight was also a 43-year-old Fiji citizen, who died during the flight of respiratory complications (Fachriansyah and Fachriansyah 2020). Authorities informed the public that the death was not related to COVID-19. Like all others, he had tested negative when boarding the flight in New Delhi. On July 31, Fiji suffered the first casualty from COVID-19, when a 66-year-old man who had arrived to the country by the evacuation flight on July 1 passed away in Lautoka hospital.

Case 28 was that of a 61-year-old person arriving on an evacuation flight from Sacramento, USA (through Auckland), on August 6. He tested positive on August 13 (Pratap 2020). On August 24, he died from complications attributed to COVID-19 in the isolation unit of Nadi hospital. Case 29 was a nurse who contracted the virus during her work in the quarantine facility. Three cases of confirmed SARS-CoV-2 infections (case 29–32) relate to an evacuation flight from India, which reached Fiji on August 27. Two of the passengers tested positive for the virus on September 4, and a third passenger on September 8. The flight proceeded to Christchurch, New Zealand, where five passengers tested positive for SARS-CoV-2 upon arrival (RNZ, September 9, 2020). Another 25 cases happened since then. All were detected during a 14 day quarantine. As of February 9, 2021 Fiji did not report any COVID-19 case in the community since 296 days.

The records of COVID-19 cases in Fiji reveal several interesting facts: It was possible to trace back all transmitted cases in the community to a small number of introduced cases. The health authorities in Fiji were very successful in detecting these transmissions in the community before they could spread widely. Thirty-eight (30) of the cases (including the case of the nurse in a border quarantine facility) were detected before infected people could enter the public. These are close to 90% of the registered cases. Leaving these cases out, Fiji has been free of any cases registered in the public (community transmission and introduced to the public by arriving passengers) since April 20, 2020.

In PNG, the first COVID-19 case was reported on March 20. After the country's eighth case by May 4, the PNG government hoped the worst was over. On June 10, after 44 days without any new reported cases, government announced that "PNG joins countries without COVID-19" (Government of Papua New Guinea 2020). Only 10 days later, it became evident that this move came too early when on June 20 the country's ninth case was confirmed. June saw another case added to the tally, but in July, numbers increased almost daily, with a record of 23 new infections on July 26. August became even worse with record infections on August 12 (73 new infections on a single day). All in August in total 403 new cases were recorded. September saw a relaxation of the situation with 68 new cases. This is still more than double the cases in Fiji. By February 15, 2021, PNG has in total 922 cases (PNG Ministry of Health 2020, 2021).

Most of the COVID-19 cases in PNG are community transmissions (more than 800 reported cases. A very low rate of testing suggests a high number of undetected cases in the communities), followed by infections in testing, health, and quarantine facilities (at least 40 cases). Only a few cases relate to infected persons, who had a relevant travel history to an exposed country. This does not surprise as PNG had locked down air travel at a very early stage (see below). It also does not mean that all infections relate to case 1, but it appears that many (early) cases might have remained undetected, and that tracking from recorded cases to possible sources was not providing better insights into possible sources of infection. This suggests the study of public announcements concerning the spread of COVID-19 in PNG. In Fiji, it was possible to trace every single case of community transmission to a person who had entered the country being infected. One, however, should not forget that PNG has roughly 10 times the population of Fiji (PNG: ≈9 mill.; Fiji: ≈900,000). Sixteen (out of 22) provinces and the National Capital District (NCD) have recorded COVID-19 cases, concentrating in the country's capital Port Moresby (~44%) and the Western Province (~26%). With a 700-km-long, difficult to monitor, border between PNG's Western Province and West Papua (Indonesia), there is possible transborder transmission from West Papua (Blades 2020).

In addition to independent Pacific states, the region also consists of several non-sovereign island territories, which have been more severely affected than PICs. By February 21, 2021, French Polynesia recorded 18263 cases (135 deaths) and Guam 7487 cases (130 deaths). French Polynesia allowed quarantine-free travel to its shores in July, just to witness a peak in cases shortly after (Direction de la Sante 2020). Other territories in the Pacific Island region that have reported COVID-19 cases are the Northern Marianas (134 cases and 2 deaths) and New Caledonia (52 cases, no death). Developed countries at the rim of the Pacific were worst affected with Australia (28989 cases and 909 deaths), Hawai'i (USA) (26772 cases and 423 deaths), and New Zealand (1980 cases and 25 deaths). These are the major air links to places outside the PICs. Most travelers to Pacific Island countries come through either Australia, New Zealand, or the USA.

Prevention and Control Measures Against COVID-19

PNG, the PIC closest to Asia, with frequent flight connections to a number of Asian countries (including China, Hong Kong, Singapore, and the Philippines) and a land border to West Papua (Indonesia), is very vulnerable to become a point of entry for the SARS-CoV-2 virus. Soon after news of a (then) unknown infection in China spread globally PNG Government banned all flights from Asia on January 28, 2020, two days before the World Health Organization (WHO) declared COVID-19 a Public Health Emergency of International Concern (PHEIC). Passengers arriving from other parts of the world had compulsory temperature checks when arriving in the country. At this time, the government also closed the border crossings to West Papua and stopped issuing visas for PNG. This was weeks before the WHO characterized COVID-19 a global pandemic on March 11, 2020. With its lockdown of international travel to PNG, the country was ahead of Italy (January 31) and considerably earlier than Germany (February 28).

Also, the Fiji Government put in place first travel restrictions in the early stages of the pandemic. On February 3, borders were closed to any passengers who had stayed in mainland China within 14 days prior of their arrival to Fiji. At the end of February, the ban was extended to several other countries and the borders were closed to any international travel soon after. Fiji as an important destination for international cruise ships disallowed any cruise liner to report to country's ports with effect of March 15.

At the time of the first reported cases, neither Fiji nor PNG had any testing facility to detect the SARS-CoV-2 virus. Initially, blood samples of suspected COVID-19 cases were sent to Australia for verification. On March 11, Fiji opened a newly refurbished molecular laboratory at the Fiji Centre for Disease Control (CDC) in Suva. Fiji became the fourth place in the Pacific Island region where it was possible to test for the SARS-CoV-2 virus. The other testing facilities were in Hawai'i, New Caledonia, and French Polynesia. This helped to shorten the time to verify suspected SARS-CoV-2 infections considerably. PNG operated long with mobile test kits and sent unclear test samples to Australia.

By early April, Fiji had set up 37 fever clinics in all urban and major rural centers for people with COVID-19 like symptoms. Fever clinics were established near to medical facilities (and in a few cases to schools) but had isolated entry and facilities to avoid that people with COVID-19 symptoms get into contact with other people visiting these facilities. Fiji's health authorities were also very proactive in screening the population for symptoms of COVID-19. According to the Prime Minister Voreque Bainimarama, more than 800,000 persons had been screened by mid-May (RNZ, May 18, 2020). This was about 90% of the country's population.

Government Responses to Contain the Spread of COVID-19

Most South Pacific countries, with or without confirmed cases, have enforced different containment measures as a response to the global pandemic. Immediately after public announcement of the first case in Fiji, containment measures were enforced in Lautoka area where the case was confirmed. The town was isolated from incoming and outgoing traffic and schools and nonessential businesses were closed. Similarly, the greater Suva area was put under a 14-day lockdown on April 3, after the first cases were found in the capital. Schools throughout the country remained closed until early July. In addition, a countrywide curfew during the night was imposed from March 30. By mid-February 2021, the curfew continues. Similarly, in PNG the government declared a nationwide two-week lockdown on March 23 when the first case was confirmed. The lockdown prevented travel between provinces and mandated the closure of all nonessential services. The cross-border transmissions from particularly Indonesia have been a serious concern (Policy Forum 2020).

Most PICs have been quick in declaring the state of emergency (WHO 2020). This allowed governments to enforce measures, including restrictions to individual rights and liberties, which under normal circumstances would not be permitted by law. The PIC governments have, throughout the region, enforced lockdowns, curfews and border closures, closed schools and nonessential businesses, as well as restricted movement of citizens. These measures have been used in countries such as Tonga and Samoa as well, even if there have been no recorded cases of the virus. The state of emergency has been extended in these countries on several occasions, without actual changes in factual circumstances in the number of cases.

Pandemics create a sense of panic and uncertainty. Under these conditions, states have a tendency to turn into a governing mode in which individual rights and liberties are restricted, officially to protect public health. In a democratic country respecting individual civil and political rights, freedoms, and the rule of law, these restrictions must be limited by their scope and justified by nothing else than the health and safety of citizens and enforced only as long and in so far as they are necessary for public safety. The test for any government, the PIC governments included, is if and when these emergency measures are lifted as the acute threat to population eases. Concerned voices in PICs have been calling for governments to avoid the temptation to continue emergency measures without legitimate reasons (see, e.g., McGarry and Newton Cain 2020).

Emergency measures are particularly harmful to certain sections of society. It is widely recognized that pandemics affect especially women, children, minorities, and other vulnerable groups. Pandemics are not gender neutral. Enforced social isolation and quarantine increase the risk of domestic and sexual violence. This is an important concern in the Pacific region, which already has one of the highest rates of physical and sexual violence against women and girls in the world. Women also carry the greatest burden of care work, as well as of housework, to which closures of schools and lockdowns create additional pressures (Pacific Women 2020). Since the beginning of pandemic and the enforcement of emergency measures, the number of calls to domestic violence helplines and centers across the Pacific has increased (UN Women June 11, 2020). Hundreds of people, including minors, have been arrested in Fiji since the nightly restrictions have been imposed, predominantly for breaking the curfew, even months after the government had declared that the country is "COVID-19-free."

The FAO subregional director Eriko Hibi noted that especially in the context of PICs, the pandemic is a *structural* crisis, meaning that despite the low number of actual cases, the impact has already been immense. Disruptions to domestic and global food chains, together with increased food prices, affect the livelihoods of thousands of people. As an example, Hibi mentions how a few days after the Suva lockdown, "the costs of the most consumed vegetables increased between 11 and 36%, in some cases up to 75% (ReliefWeb 2020). The UN estimates that up to 110.000 Fijians could fall below the poverty line due to the pandemic (Mala 2020)—in addition to the 28% of population that already, according to the 2017 census, live below it. In PNG, respectively, 40% of population already live below the poverty line.

As a health concern, all PICs have very limited capacities to treat COVID patients, due to already weak health systems. PNG, for instance, has merely 500 medical doctors, less than 4000 nurses, around 5000 beds in hospitals and health centers, and only 14 ventilators in the country of close to 9 million people (Kabuni 2020). Containment has therefore been a meaningful strategy to protect already vulnerable health systems. Access to care will be a challenge if a number of cases were to increase suddenly. At the same time, economic pressure enforces states reliant on tourism to consider opening their borders to international travelers. National vaccination plans in the Pacific follow different strategies to secure the much needed vaccine stocks vital to reach herd immunity as soon as possible. As in many other parts of the developing world, the COVID-19 Vaccines Global Access (Covax) initiative plays a key role (Wasuka 2021). Governments of PICs have joined the initiative early on in 2020 in order to secure COVID-19 vaccines for their shores. In early February 2021, the Covax facility announced its plan to distribute the first batch of doses to eligible countries in the first quarter of the year (Gavi 2021). This initially included Fiji, Kiribati, Marshall Islands, Federated States of Micronesia, Nauru, PNG, Samoa, Solomon Islands, Tonga, and Tuvalu. In addition, some governments have been engaging in bilateral agreements to receive vaccines. For instance, government officials of the Micronesian states of Guam, Marshall Islands, Northern Marianas, Palau, the Federated States of Micronesia, as well as Polynesia's American Samoa have successfully secured (and already being supplied with) COVID-19 vaccines through their free association with the USA and its Coronavirus Aid, Relief and Economic Security (CARES) Act (US Department of the Interior n.d.). Similarly, the French overseas territories (French Polynesia, New Caledonia, and Wallis and Futuna), as well as New Zealand's freely associated states (Cook Islands, Niue and Tokelau) and its Polynesian neighbouring countries of Samoa, Tonga and Tuvalu receive their vaccines (additionally) from their respective overseas partners (Beaumont 2020; McCulloch 2021). The Fijian government has recently announced to enter bilateral negotiations with Australia and India to secure funding to purchase vaccines in addition to the stocks that will be supplied through Covax (Kumar 2021). While there is uncertainty how the vaccination plans will turn out, Pacific Island governments, particularly those whose economy rely heavily on international travel, push forward to protect their citizens and to reopen their international borders to regain some degree of economic stability and normalcy.

Economic Challenges of COVID-19

Although COVID-19 did fortunately not turn out to become a major health crisis in PICTs, it has led to substantial economic and social disturbances in virtually all PICTs due to a number of unique characteristics. Geographical smallness, remoteness, and fragmentation, together with a significant economic dependency on foreign aid, remittances, and tourism, make PICTs particularly vulnerable to external shocks. Before COVID-19 crippled the global economy, the Pacific Island region experienced an overall positive economic growth of around 3.7% in 2019 (ADB 2020a, b). Since COVID-19 was declared a pandemic, Pacific Island societies have witnessed increasing economic pressures triggered by the halt of international travel and tourist arrival since April 2020, as well as the impacts of the global economic downturn.

At a first glance, it seems that the economic impacts of the pandemic will be less severe in PICTs than in many other countries. Amidst the global crisis, the Asian Development Bank (ADB) released a report in July 2020 forecasting the economic outlook of the Pacific Island region (ADB 2020a, b). The report predicts that Pacific Island economies will experience a 4.3% contraction until the end of the year. Given the globally forecasted 4.9% contraction in real GDP, the decline of the Pacific economies appears to be slightly smaller. However, this should not detract from the fact that the socioeconomic effects for some PICTs are likely to be more severe than for many other nations, particularly the industrialized countries of the Global North.

Popular for secluded white sandy beaches, a number of PICTs have been investing in marketing of these natural resources quite successfully. They have attracted increasing numbers of international travelers and tourists over the past two decades. Accordingly, visitor arrivals have reached a peak of 3.16 million in 2018 (Rex 2019), leading to a significant growth of tourism sectors, which developed into one of the largest sectors to GDP, employment, and foreign exchange in many PICTs. While Australia and New Zealand provide the largest numbers of tourists to Pacific Island destinations, successful marketing strategies have led to an increase in arrivals from the USA, China, UK, and the EU as well. Popular tourist destinations include the Fiji Islands, Cook Islands, Palau, Guam, Vanuatu, Samoa, and French Polynesia. Unsurprisingly, these economies are now worst affected by international travel bans.

Fiji's hitherto growing tourism sector is among the hardest hit in the region. There is a very high economic dependency on foreign exchange earnings and employment connected to the tourism sector (Chambers 2020). Tourism contributes approximately 40% to the country's GDP. In 2019, hopes were high that previous record numbers of 870,309 visitor arrivals in 2018 and 894,389 in 2019 (Fiji Bureau of Statistics

2020) would be topped in 2020. Such hopes led to additional investments in the tourist sector. Fiji's national airline, Fiji Airways, ordered two new aircraft in 2019 to increase services between Fiji and states at the rim of the Pacific Ocean, such as Australia and the USA (Pratibha and Vula 2019). The Fiji National Provident Fund (FNPF) expanded its investment into the local hotel and tourism business (Kumar 2019). The Fiji government passed new tax policies to reduce tourism-related taxes and provide further financial incentives to revitalize Fiji's crippled tourism sector in the 2020/21 national budget. These hopes have vanished for the time being. While it is difficult to predict how and when these investments and tax reforms will eventually play out in an uncertain future under COVID-19, it is obvious that Fiji's economy will continue to contract until international travel bans are lifted and international tourists return. According to the World Bank Fiji's economy contracted by 19% in 2020. The rate of unemployment increased to 27 percent (World Bank 2021). It is the highest economic contraction predicted in the Pacific Island region. Similarly, devastating economic consequences are witnessed in the Cook Islands and French Polynesia, where tourism earns up to 87% of the GDP. In Vanuatu and Niue, tourism contributes up to 40% of the GDP (Rex 2019). Around 25–30% of the GDP results from tourism in Samoa and Tonga (NZ Foreign Affairs & Trade/Pacific Tourism Organization 2020). It is very likely that the economic recession the region experienced in 2020 will well continue into 2021 and 2022. This causes a real possibility of undoing decades of economic development.

As a consequence of the collapse of international travel and trade across the Pacific Islands, the majority of tourism-related businesses and airlines have concluded or minimized their operations for the time being. Employees were either laid off completely or work on reduced hours (and salaries). Affected are not only those directly employed in the sector, but also companies and farmers/fishermen supplying goods and services to hotels and resorts. While official numbers are not available across the region, the Fiji Government, for instance, stated in July 2020 that around a third (115,000) of the country's labour force has lost jobs or work on reduced hours due to the economic downturn (Sayed-Khaiyum 2020). Fiji's Prime Minister has labeled the pandemic as the "job-killer of the century." A similar situation unfolds in the Cook Islands, where the Prime Minister metaphorically referred to the COVID crisis as an "economic tsunami." In Vanuatu, close to 70% employed in tourism have lost their jobs. But not only tourism-dependent economies struggle. In PNG, the region's largest country, tourism plays no major economic role. Restricted international travel and trade has led to a sharp decline in construction, transport, and mining activities resulting in declining revenues and increasing unemployment (ADB 2020a).

Remittances, another major economic pillar for Fiji and many Polynesian and Micronesian PICs, are likely to fall (IMF 2020; World Bank 2020b). Many families rely on remittances sent from family members abroad as a major source of income to meet their daily needs. In countries like Samoa and the Marshall Islands, remittances make up to 15% of the GDP, in Tonga almost 40%. Amidst the corona crisis, remittances to the Pacific are stuttering. The pandemic has globally driven millions of workers into unemployment. Migrant workers are often among the first threatened by job losses. Despite a high demand, seasonal workers from PICs such as Vanuatu or Fiji face difficulties to temporarily migrate to Australia or New Zealand due to global travel restrictions. Travel restrictions in the Pacific Island region have become traps for migrant workers: They cannot work or return to their countries of origin (NZHerold September 7, 2020). The World Bank estimates that remittance flows into the East Asia, and the Pacific Island region is expected to decline by 13% (World Bank 2020b). Cuts in remittances do not only pose serious risks of impoverishment to those affected but have also a negative knock-on effect on local consumption and investment.

Tourism Versus Health

Due to looming loss of billions of dollars caused by the pandemic, a discussion has sparked in some Pacific countries about a possible reopening of borders for tourism. Australia and New Zealand have started discussing the establishment of an exclusive trans-Tasman travel bubble to curb bilateral economic and social exchange in May. Some PICTs, including Fiji, Vanuatu, Cook Islands, Niue, and Tokelau, have been actively lobbying to be included in such a travel bubble to revive their vital tourism sectors (RNZ June 3, 2020). In the meanwhile, such plans have been put on hold due to a sharp increase in COVID-19 cases in Australia's second most populated state of Victoria and are cent COVID-19 reoccurrence in New Zealand's capital Auckland. French Polynesia responded with a bolt move when reopening its borders to much needed tourists on July 15. The result is well known. Within days after borders opened, cruise ship tourists imported the virus to the shores of the archipelago. Fiji is undergoing another way. After New Zealand Prime Minister Jacinda Ardern and her Australian counterpart made it clear that any trans-Tasman or Pacific travel bubble would not happen before the virus is eliminated in their countries, Fiji has been trying to promote alternative ways of getting tourists gradually back. Initiatives like Fiji's Blue Lane aim to promote the country as a safe haven for overseas yachters. The proposed Bula Bubble to establish health protocols in order to allow Australian and NZ tourists visit Fiji's shores is another example of such attempts. The government has been consecutively trying to create confidence among New Zealanders and Australians in the capabilities of the Fijian health system to protect tourists as well as Islanders from any transmission of the virus while visiting the country. Until today, these initiatives have largely been unsuccessful. By February 2021, any travel bubble with Pacific Rim states is unlikely to happen before 2022.

Economic impacts of the pandemic and uncertainties relating to international tourist arrivals have created massive economic challenges in the PICTs. Most PICs do not have economic and administrative capacities to cope with the downturn of their economies like countries of the Global North. Many developing economies in the Pacific Island region can simply not afford to pay larger amounts to citizens in need, to help out businesses with government loans or to finance/implement economic stimulus programs or employment programs to boost local consumption, investment, and employment. More businesses will go bankrupt, and more people are going to lose jobs in the near future. This affects particularly workers in the informal sector where jobs are often cut first as no employment protection measures are in place. In an assessment of the macroeconomic consequences for people of the Pacific, the Development Policy Centre in Australia concluded that the number of people living in extreme poverty could increase up to 40% (Hoy 2020). The slump in revenues is leading to a widening of the current account deficits and rising foreign debt. For many countries, this is likely to lead to budget cuts, as these are heavily dependent on taxes and revenues from tourism and domestic consumption. Cuts in revenue, in turn, may result in budget reductions for infrastructure, education, and other basic resources and services needed for long-term economic growth and social protection.

Social Protection Responses to COVID-19 in Selected Pacific Island Countries

COVID-19 created much unemployment in PICT. Livelihood security has been compromised for a big number of people in the PICTs. State capacities to mitigate the social impacts are far less than in developed countries, where billions or in the case of the USA trillions of US$ have been mobilized to establish at least some kind of social safety nets. Such efforts are essential for social reasons. The danger that a larger section of Pacific Island populations falls below poverty lines has been highlighted above. To support effective demand, however, is also essential for the economy, as people who do not have much purchasing power cannot give stimulus to economic production. In June 2020, the World Bank published a paper on Social Protection and Jobs Responses to COVID-19:A Real-Time Review of Country Measures (World Bank 2020a). Like this chapter, it is actually work in process, adding measures in many countries on a daily basis. The next few paragraphs provide findings as they relate to nine PICs.

In the Cook Islands, parents can get additional funds on child benefit for the time schools are closed (US$ 132 extra per month). People on the welfare list of infirm, destitute, and pensioners can get a one-off cash transfer of US$ 264 to mitigate the economic impact of COVID-19. Paid sick leave is provided to people, who are not sick, but cannot perform work remotely, and have been advised by the Ministry of Health to self-isolate; also, people who need to look after dependents required to self-isolate or sick with COVID-19 can receive paid sick leave. Mainly to support the tourism sector, training measures to upskill employees are supported. For qualified businesses, a training subsidy for up to 35 h/weekly over a period of 3 months is provided for activities not started before from July 1, 2020. The activities must be at least for two months covering not less than 10 h per week. The subsidy is the minimum wage between US$ 176 (full-time) and US$ 88 (part-time) per week.

In the Federated States of Micronesia, there are unemployment benefits for those who became unemployed as direct result of COVID-19. The government has also announced an Economic Stimulus Package (US$ 15 million) to support businesses in the tourism sector. Support comes as subsidies to wages, debt relief, and income and social security tax rebates for affected businesses.

In Fiji, the government announced a one-off payment equivalent to US$ 66 for licensed street traders to mitigate income loss. Fijians in the informal sector who tested positive for the virus can receive a one-off payment equivalent to US$ 443. Banks and hire purchase companies are required to allow a six-month deferral of loan repayments for Fijians who became unemployed or have working hours reduced because of COVID-19. Banks have also been instructed to waive charges on minimum balances in customer's accounts and remove minimum purchasing requirement for electronic transactions. Employees who have tested positive for SARS-CoV-2 can receive 21 days leave paid by the government, provided their annual income is below US$ 14,000. Employees in the tourism sector who have lost their work after February 1, 2020, can initially access US$ 470 from their pension fund. Employees whose work has been affected by physical distancing requirements (incl. those in lockdown areas) and has been put on leave without pay or working hours cut, can access initial US$ 235 from their pension fund. Between April 1, 2020, and December 31, 2020, employees' contribution to the pension fund was reduced to 8% (before 10% of gross income). Employer contribution declined from 10 to 5%.

In PNG, unemployment benefits are provided to all who lost their work because of COVID-19. They can get a one-off payment from their pension fund of up to 20% of their contribution up to US$ 2890.

In Samoa, pensioners received a (one-off) monthly special pension of US$ 120. Companies in the tourism sector could get a six-month moratorium on pension fund payment of their employee. The government in the Solomon Islands provided some relief to its citizens by reducing electricity tariffs, and in Tonga payouts under the Elderly Benefit and Disability Benefit were increased by US$ 45 in April 2020. Some reduction was also provided in water and electricity bills. In Tuvalu, all citizens can expect a payout of US$ 17 per month for the duration of the crisis. The cash transfer has been funded by the government and international donors. In Vanuatu, all school fees for 2020 have been waived. Under the Employment Stabilization Program, employers can get US$ 249 for each employee for up to four months (March–June 2020). Employers can get an additional 15% on the amount of wages paid as additional incentive to maintain employment.

Non-Sovereign Territories and Impact of COVID-19 on Pacific Islanders Away from Home

Pasifika communities across the world have been seriously impacted by the virus. In the USA, for example, Pacific islanders in various states have disproportionate numbers of cases compared to their overall share of population. By early August, some 70 Marshall Islanders have died from COVID-19 in mainland USA. Around

2500 Marshallese have tested positive, some 10%of the Marshall Island community in mainland USA (Johnson 2020). In Spokane county, Washington State, they are 1% of the population, yet they represent 30% of confirmed COVID cases. In Springdale, Northwest Arkansas, many Marshall Islanders work in meat processing plants. There they amount to half of the COVID-19 deaths (Radio New Zealand July 9, 2020). As of September 14, there have been 46 deaths and more than 1,800 infections in a Marshall Islander community of little more than 15,000 people (RNZ August 27, 2020). Also in Hawai'i, Pacific Islanders top the list of COVID-19 cases. By the end of August, they accounted for 31% of all cases, but are just four percent of the population (Fiji Times August 31, 2020). The reasons behind increased vulnerability of Pacific islanders vary from large family groups and closed community ties to higher rates preexisting medical conditions, such as diabetes, as well as low-income jobs allowing no remote work alternative (Cherelle Jackson 2020).

Melanesian countries have much smaller numbers of international emigrants, but also here citizens are affected by COVID-19 outside their countries. Eighteen students from the Solomon Islands have tested positive for COVID-19 in the Philippines (RNZ September 29, 2020). Tensions ran high when the Solomon Island govern-ment allowed more than 80 Chinese businesspeople to enter the country through an evacuation flight reaching Honiara on September 3. Just 21 Solomon Islanders were on the flight. In connection to this flight, the premier of the country's most populous province, Malaita, was calling for an independence referendum as "the nation's leaders were putting their new relationship with Beijing before their own people" (Hollingsworth 2020). Just a year earlier, the country had switched polit-ical alliance and recognition from the Republic of China (Taiwan) to the People's Republic of China in September 2019. Many criticized the government heavily when it did nothing to evacuate the Solomon Island students from the Philippines at the very beginning of the pandemic and exposed them to the risk of becoming infected. They had sent urgent calls for evacuation to their government in early March, when air travel between Manila and Honiara (through Port Moresby) was still open, but nothing had happened (Solomon Star, March 20, 2020). In the meanwhile, some Solomon Island students from the Philippines arrived back home. Among them were two students, who later tested positive while under quarantine in the Solomon Islands. Before boarding their flight back home, they had tested four times negative against the virus. A flight to Honiara to bring home seasonal workers from New Zealand and Australia was canceled due to insufficient quarantine facilities (RNZ, August 21, 2020).

Conclusion

When writing on the chapter, the COVID-19 pandemic was far from over. Hopes are high that also in the months ahead the spread of COVID-19 in PICTs can be contained. Four major aspects already now can be highlighted. As a result of the COVID-19 pandemic, virtually all PICTs experienced an upgrade of their medical facilities to deal with epidemic diseases. Lockdowns of particular areas, where community trans-mission was detected, were rather successful in containing the spread of the disease. At times, it might have been connected to a huge portion of luck that SARS-CoV-2 could not spread to outer islands, e.g., in Fiji. The WHO has been a driving force to support PICTs in their efforts to avoid or contain COVID-19. In close cooperation with many development partners, WHO has made available testing facilities/sets, personal protective equipment, ventilators, and other equipment to provide neces-sary medical care for people infected. In addition, officials of ministries of health, nurses, and doctors received training to upgrade skills in the treatment of COVID-19 patients, especially those where the disease took an aggravated course. The WHO provided such support regardless to political affiliations of PICs. Countries closely connected to the People's Republic of China and countries linked to the Republic of China (Taiwan) received the very same support. Also, in the future it is not realistic to assume that PICs can deal with such huge challenge entirely on their own.

Much more needs to be done to complete the training of medical personnel dealing with highly contagious diseases. Where no or low case numbers exist, experience in treating patients with complications is limited. In many parts of the world, it became evident that the treatment of complicated medical cases improved with time and experience. Fortunately, doctors and nurses in many PICs did not need to make these experiences. Therefore, efforts need to increase in providing training in the operation of ventilators, pharmaceutical aspects in the treatment of patients at high risk, by also the management of epidemics. It certainly is crucial to select a number of (already) well-qualified doctors and nurses to enhance their skills through training overseas, best in circumstances where they can make practical experience.

Economically, the pandemic and the measures to contain the crisis pose major challenges to PICTs. While negative consequences are already felt and will continue to impact the economies, the major goal amidst the crisis is to limit the negative effects. Obviously, the extent of the economic damage depends on the duration of restrictions on international travel and trade, and the level of containment of domestic economic fallout resulting from lockdowns. Pacific governments will need to care-fully consider and evaluate the health risks associated with any lifting of restrictions against any economic benefits. A further spread of the virus within the region might even have more severe impacts on the economy than living with the restrictions of public life.

Fourth, but certainly not least is the restoration of all civic rights once the pandemic has ended. Related to the efforts to avoid/contain COVID-19, many civic rights have been restricted. Fears are that there will not be a return to normality once it would be possible. Pacific Island governments would wear off much of the trust and appreciation earned from a robust handling of the challenges of COVID-19 if they would not return to normality, but misuse and continue with the restrictions of civic rights (Table 2).

Table 2 Beginning of measures to contain COVID-19 in PICT

Pacific Island countries	Start of international travel restrictions	Start of major internal restrictions
Cook Islands	28 March, flights canceled except to/from NZ 15 August, all flights canceled	26 March, public gatherings restricted
Micronesia	3 February, travel to China/other affected countries banned, 5 March, travelers from China/other affected countries banned	18 March, all schools closed
Fiji	3 February, ban of non-Fijians arriving from China 27 February, extended to Italy, Iran, South Korea 20 March, extended to USA, all of Europe, 14-day quarantine 28 February, cruise ships only to Suva/Lautoka 16 March, all cruise ships banned 21 March, most Fiji airways flights grounded 26 March, Nadi airport closed	19 March, social distancing rules introduced, school holidays brought forward by a month 20 March–7 April, Lautoka lockdown 27 March, nationwide curfew from 10 pm to 5 am 29 March, all inter-island shipping stopped 3 April–17 April, Suva lockdown 4 April–2 May, Soasoa (Labasa) lockdown 17 April–2 May, Nabua (Suva) lookdown
Kiribati	1 February visas from China canceled, quarantine for travelers from other affected countries 10 September, borders closed to at least end of 2020	26 March, declaration of state of emergency, social distancing rules apply
Marshall Islands	24 January, ban of travelers been in affected country 14 days prior to arrival in country 18 March, all international travel suspended	7 February, state of emergency, social distancing rules apply, school closure
Nauru	16 March, 14-day quarantine for all arrivals 2 April, travel bans except flights to Brisbane	17 March, state of emergency, closure of border, social distancing rules, 10 people max
Palau	1 February, ban of all flights from China/Hong Kong	20 March, social distancing, 50 people max 23 March, school closure
Papua New Guinea	30 January, flights banned from Asia, lockdown land borders	16 April–3 May, lockdown of capital, curfew 8 pm–6am, social distancing rules, 4 people max

(continued)

Table 2 (continued)

Pacific Island countries	Start of international travel restrictions	Start of major internal restrictions
Samoa	26 January, compulsory screening of all visitor 22 February, ban of all cruise ships 22 March, suspension of air travel with Australia	20 March, state of emergency, border to American Samoa closed, 26 March, social distancing, 5 people mx, inter-island shipping discontinued
Solomon Islands	27 March, all flights into the country suspended	27 March, state of emergency for capital Honiara, social distancing (church remains open) 31 March, closure of all schools
Tonga	20 March, foreign nationals banned from entering, 14-day quarantine for residents, 23 March, all air travel stopped, 28 March, cruise ships/yachts barred from Tonga	20 March, state of emergency, social distancing, 10 persons max, night curfew, 8 pm-6am 29 March, week-long national lockdown
Tuvalu	26 March, all borders closed to any aircraft/ship	26 March, state of emergency, social distancing, school closure, ban of any public meeting
Vanuatu	6 February, no travelers from China, unless 14-day quarantine outside the country + med. certificate	26 March, state of emergency, lockdown of Aneityum Isl., tourist from cruise ship tested positive
Pacific Island territories		
French Polynesia	2 March, travelers need medical certificate upon arrival, cruise ships banned	20 March, a lockdown announced 28 March, curfew 8 pm–5am, in-county travel restricted 12 August, social distancing
Guam	13 March, visitors are check for fever upon arrival 16 March, mandatory quarantine for travelers from affected countries	16 March, social distancing, public schools closed 22 March, state of emergency 28 March, President Trump declares major disaster; makes Guam eligible for federal assistance
New Caledonia	18 March, suspension of all flights	23 March, restrictions on movements, social distancing, no meetings, public events

(continued)

Table 2　(continued)

Pacific Island countries	Start of international travel restrictions	Start of major internal restrictions
Northern Marianas	31 January, USA declares public health emergency, mandatory 14-day quarantine for US citizens visited Hubei Province within last two weeks 1 February, major airlines cancel flights to/from China 6 April, official suspension of all flights	16 March, closure of schools and government offices, 14-day self-quarantine, if recently out of the territory 24 March, limitation of public gatherings 30 March, curfew from 7:00 PM to 6:00 AM 1 April, President Trump approved Major Disaster Declaration

Sources Compiled from various newspaper reports from respective countries, Australia and New Zealand, JHU Web site

References

Asian Development Bank (ADB). (2020a). Asian Development Outlook, June 2020 (Supplement). Manila.

Asian Development Bank (ADB). (2020b). *Pacific economic monitor*, July 2020. https://www.adb.org/sites/default/files/publication/622976/pem-july-2020.pdf.

Beaumont, P. (2020). Covid-19 vaccine: Who are countries prioritising for first doses?. *The Guardian*, November 18, 2020. https://www.theguardian.com/world/2020/nov/18/covid-19-vaccine-who-are-countries-prioritising-for-first-doses.

Blades, J. (2020). Pandemic exposes weakness of PNG's border security. RNZ, April 24, 2020. https://www.rnz.co.nz/international/pacific-news/415003/pandemic-exposes-weakness-of-png-s-border-security.

Chambers, Ch. (2020). Tourism faces competition when borders reopen. *Fiji Sun*, August 8, 2020. https://fijisun.com.fj/2020/08/08/tourism-faces-competition-when-borders-reopen-a-g/.

Cherelle Jackson, L. (2020, July 27). Pacific Islanders in US hospitalised with Covid-19 at up to 10 times the rate of other groups. *The Guardian*.

Denoon, D. (2017). *Pacific island depopulation: Natural or un-natural history? In Peoples of the Pacific* (pp. 395–410). Routledge.

Diamond, J. (1997). *Guns, germs, and steel*. London, UK: Vintage.

Direction de la Sante. (2020). https://www.service-public.pf/dsp/2020/08/10/situation-coronavirus-100820/.

Fachriansyah, F., & Fachriansyah, R. (2020). Fijian citizen dies on chartered Garuda Indonesia flight, stoking COVID-19 fears. *Jakarta Post*, July 2, 2020. https://www.thejakartapost.com/news/2020/07/02/fijian-citizen-dies-on-chartered-garuda-indonesia-flight-stoking-covid-19-fears.html.

Fiji Bureau of Statistics. (2020). *FBoS Release No., 61, 2020.*

Fiji Department of Information. (2020). COVID-19: Fiji has 12 confirmed COVID-19 cases. Fiji has now 12 confirmed cases of COVID-19 as of this afternoon, single largest jump in cases in a day. *Fiji Sun*, April 4, 2020. https://fijisun.com.fj/2020/04/04/covid-19-fiji-has-12-confirmed-covid-19-cases/.

France24 (2021). Easter Island begins vaccinating residents against Covid-19. February 9, 2021. https://www.france24.com/en/live-news/20210208-easter-island-begins-vaccinating-residents-against-covid-19.

Gavi (2021, February 3). The covax facility: Interimdistribution forecast–latest as of 3 February 2021. https://www.gavi.org/sites/default/files/covid/covax/COVAX-Interim-Distribut ion-Forecast.pdf.

Government of Papua New Guinea. (2020). PNG joins countries without COVID-19. PNG Official COVID-19 information website. June 10, 2020. https://covid19.info.gov.pg/index.php/2020/06/ 11/png-joins-countries-without-covid-19/.

Herda, P. S. (2000). Disease and the colonial narrative-The 1918 influenza pandemic in Western Polynesia. *New Zealand Journal of History, 34*(1), 133–144.

Hollingsworth, J. (2020). This Pacific Island province is so frustrated with China's presence that it's pushing for independence. *CNN*, September 18, 2020. https://edition.cnn.com/2020/09/17/ asia/solomon-islands-malaita-intl-hnk-dst/index.html.

Hoy, C. (2020). *Poverty and the pandemic in the Pacific*. Devpolicy@ANU. https://devpolicy.org/ poverty-and-teh-pandemic-in-the-pacific-20200615-2/

International Monetary Fund (IMF). (2020, May 27). Pacific islands threatened by COVID-19. *IMF News*. Washington, D.C.

Johnson, G. (2020). Marshallese struggle with Covid in US. *RNZ*, August 3, 2020. https://www. rnz.co.nz/international/pacific-news/422617/marshallese-struggle-with-covid-in-us.

Kabuni, M. (2020). COVID-19 (Coronavirus) in Papua New Guinea: The state of emergency cannot fix years of negligence. https://dpa.bellschool.anu.edu.au/sites/default/files/publications/attach ments/2020-05/dpa_in_brief_202015_kabuni.pdf.

Kumar, K. (2019). FNPF contributes massively towards hotel and tourism sector, April 15, 2019. https://www.fbcnews.com.fj/news/fnpf-contributes-massively-towards-hotel-and-tou rism-sector/.

Kumar, R. (2021). Expect arrival of COVID-19 vaccine in the first quarter of 2021 – AG. *Fijivil-lage*, February 8, 2021. https://www.fijivillage.com/news/Expect-arrival-of-COVID-19-vaccine-in-the-first-quarter-of-2021--AG-f5xr84/.

Mala, P. (2020). Between 82,000 to 110,000 Fijians expected to fall below the poverty line – UN Women, August 21, 2020. https://www.fijivillage.com/news/Between-82000-to-110000-Fij ians-expected-to-fall-below-poverty-line-according-to-recent-UN-socio-economic-impact-ass essment-5fxr48/.

McCulloch, C. (2021). Covid-19 Pfizer/BioNTech vaccine: New Zealand government gives formal approval. *RNZ*, February10, 2021. https://www.rnz.co.nz/news/national/436132/covid-19-pfizer-biontech-vaccine-new-zealand-government-gives-formal-approval.

McGarry, D., & Newton Cain, T. (2020, August 2). We can't allow Pacific leaders to use coronavirus as a cover for authoritarianism. *The Guardian*. https://www.theguardian.com/global/2020/aug/ 03/we-cant-allow-pacific-leaders-to-use-coronavirus-as-a-cover-for-authoritarianism?fbclid= IwAR3hAIJjtTxEJWZNcGkPy_JkEuaPTkNg55-tti3TYD7JqIQk4WAFsMvISz8.

McLane, J. R. (2013). Paradise locked: The 1918 influenza pandemic in American Samoa. *Sites: A Journal of Social Anthropology and Cultural Studies, 10* (2), 30–51.

Nasiko, R. (2019). Fiji Airways reaches $1bn mark. *Fiji Times*, May 24, 2019. https://www.fijiti mes.com/fiji-airways-reaches-1bn-mark/.

New York Times. (2021). Hawaii Coronavirus map and case count. Last access version February 9, 2021. https://www.nytimes.com/interactive/2020/us/hawaii-coronavirus-cases.html.

New Zealand Foreign Affairs & Trade/Pacific Tourism Organization. (2020). Pacific tourism: COVID 19 IMPACT & RECOVERY sector status report: Phase 1B. https://corporate.southpaci ficislands.travel/wp-content/uploads/2020/05/Pacific-Tourism-Sector-Status-report-Final.pdf.

Oanda. (2020). Currency Converter. https://www1.oanda.com/currency/converter/.

Pacific Women. (2020). |Gender and COVID-19 in the Pacific: Emerging gendered impacts and recommendations for response. https://pacificwomen.org/wp-content/uploads/2020/05/The matic-Brief_Gender-and-COVID19_Pacific-Women-May-2020.pdf.

PNG Ministry of Health. (2020). COVID-19 Pandemics. *Health Situation Report*. https://www.hea lth.gov.pg/subindex.php?news=1.

PNG Ministry of Health. (2021). COVID-19 Pandemics. Health Situation Report. https://www.hea lth.gov.pg/subindex.php?news=1.

Policy Forum. (2020). Covid-19 in Papua New Guinea. https://www.policyforum.net/covid-19-in-papua-new-guinea/.

Pratap, R. (2020). Fiji confirms new COVID-19 case in quarantine. *FBC News*, August 13, 2020. https://www.fbcnews.com.fj/news/covid-19/fiji-confirms-new-covid-19-case-in-quarantine/.

Pratibha, J., & Vula, M. (2019). ADB 2019: Two new State of the art airbus A350 XWB to be added to Fiji airways growing fleet, May 3, 2019. https://fijisun.com.fj/2019/05/03/adb-2019-two-new-state-of-the-art-airbus-a350-xwb-to-be-added-to-fiji-airways-growing-fleet/.

Radio New Zealand. (2020a). Fiji Covid-19 restrictions to remain in full effect. *RNZ*, May 18, 2020. https://www.rnz.co.nz/international/pacific-news/416802/fiji-covid-19-restrictions-to-rem ain-in-full-effect.

Radio New Zealand. (2020b). Four more Solomon Island students test positive for Covid in the Philippines. *RNZ*, September 29, 2020. https://www.rnz.co.nz/international/pacific-news/427 171/four-more-solomon-island-students-test-positive-for-covid-in-the-philippines.

Radio New Zealand. (2020c). New ventilators boost Solomons Covid-19 preparedness. *RNZ*, August 21, 2020. https://www.rnz.co.nz/international/pacific-news/424131/new-ventil ators-boost-solomons-covid-19-preparedness.

Radio New Zealand. (2020d). Pressure mounts on NZ and Aust to include Pacific in 'bubble'. *RNZ*, June 3, 2020. https://www.rnz.co.nz/international/pacific-news/418156/pressure-mounts-on-nz-and-aust-to-include-pacific-in-bubble.

Radio New Zealand. (RNZ, 2020e). Coronavirus: 2 more covid-19 cases in French Polynesia. March 13, 2020. https://www.rnz.co.nz/international/pacific-news/411680/coronavirus-2-more-covid-19-cases-in-french-polynesia.

Radio New Zealand (RNZ, 2020f). Fiji has new Covid-19 case from repatriation flight. September 9, 2020. https://www.rnz.co.nz/international/pacific-news/425604/fiji-has-new-covid-19-case-from-repatriation-flight.

Reliefweb. (2020). Why has COVID-19 turned into a structural crisis in the Pacific? https://relief web.int/report/world/why-has-covid-19-turned-structural-crisis-pacific.

Rex, R. (2019). Regional tourism sector achieves 3.16 million visitor arrivals in 2018. https://cor porate.southpacificislands.travel/regional-tourism-sector-achieves-3-16-million-visitor-arrivals-2018/.

Rivers, W. H. R. (Ed.). (1922). *Essays on the depopulation of Melanesia*. The University Press.

Sayed-Khaiyum, A. (2020, July 17). Budget address. https://www.fiji.gov.fj/Media-Centre/Spe eches/HON-AIYAZ-SAYED-KHAIYUM-S-2020-2021-NATIONAL-BUDGE.

Shanks, G. D., Wilson, N., Kippen, R., & Brundage, J. F. (2018). The unusually diverse mortality patterns in the Pacific region during the 1918–21 influenza pandemic: Reflections at the pandemic's centenary. *The Lancet Infectious Diseases, 18*(10), e323–e332.

Singh, S. (2020). We have traced 834 people who came into contact with the 54y/o man from Labasa – Waqainabete. *FijiVillage*, April 7, 2020. https://www.fijivillage.com/news/We-have-traced-834-people-who-came-into-contact-with-the-54yo-man-from-Labasa---Waqainabete-8r4x5f/.

Solomon Star. (2020). Air Niugini, Virgin announce suspension. *Solomon Star*, March 20, 2020. https://www.solomonstarnews.com/index.php/news/business/item/23093-air-niugini-virgin-announce-suspension.

US Department of the Interior. (n.d.). Federal assistance to the U.S. Territories and freely associated States during the Coronavirus disease 2019 (COVID-19) Pandemic. https://www.doi.gov/oia/cov id19.

UN Women. (2020, June 11). Across the Pacific region, crisis centres respond to increased cases of violence against women amid COVID-19. https://asiapacific.unwomen.org/en/news-and-eve nts/stories/2020/06/across-the-pacific-region-crisis-centres-respond-to-increased-cases-of-vio lence.

Wasuka, E. (2021). Pacific Islands may not finish rolling out COVID vaccines until 2025. *ABC Pacific Beat*, January 21, 2021. https://www.abc.net.au/news/2021-01-21/pacific-islands-covid-rolling-out-vaccines-until-2025/13073054.

WHO. (2020, March 26). COVID-19 Joint external situation report for Pacific Islands, No. 9. https://www.who.int/westernpacific/internal-publications-detail/covid-19-situation-report-for-pacific-islands.

World Bank. (2020a, June 12). Social protection and jobs responses to COVID-19: A real-time review of country measures. "Living paper" version 11. Washington D.C.

World Bank. (2020b, April 22). *World Bank predicts sharpest decline of remittances in recent history*. Press Release.

World Bank. (2021). $102.7million boost to address employment impacts of COVID-19. *Fiji Sun*, February 11, 2021. https://fijisun.com.fj/2021/02/11/102-7million-boost-to-address-employment-impacts-of-covid-19/.

World Health Organization. (2021). WHO coronavirus disease (COVID-19) dashboard. Last accessed version February 9, 2021. https://covid19.who.int/table.

Proactive Strategies, Social Innovation, and Community Engagement in Relation to COVID-19 in Taiwan

Mei-Hui Li

Abstract When the first case of coronavirus disease 2019 (COVID-19) infection occurred in China in late 2019, Taiwan was predicted to be one of the most affected countries because of its economic activities and transportation exchanges with China. On January 20, 2020, the activation of Taiwan's Central Epidemic Command Center (CECC) was established to integrate resource of the administration, the academic, medical and private sectors and formulate guidelines and regulations in the fight against COVID-19. As of February 10, 2021, only less than 1,000 cases of COVID-19 have been confirmed in Taiwan since the first confirmed case on January 21, 2020, and most cases are classified as imported. Tight border controls, rapid mobilization by the public and private sectors, prompt decision-making and implementation, well-coordinated distribution of medical supplies, and transparent information and advanced digital technology are helping Taiwanese society to fight COVID-19. Especially, digital technology provides a new type of social innovation to help making public to easy access transparent information or prevent the transmission of misinformation and disinformation to combat COVID-19 effectively in Taiwanese society.

Keywords Preparedness · Precautionary actions · Social innovation · Digital technology · Leadership

Introduction

Taiwan is 130 km off the coast of mainland China and extensively engages with China in its economic activities and transportation exchanges. A study led by Gardner (2020) from John Hopkins University provided a model (Zlojutro et al. 2019) to analyze the spread of COVID-19 from China and predicted that among 24 countries studied, Taiwan was the country with the second highest risk in terms of the estimated number of imported COVID-19 cases due to its proximity to China and the number

M.-H. Li (✉)
Department of Geography, National Taiwan University, 1, Section 4, Roosevelt Road, Taipei 10617, Taiwan
e-mail: meihuili@ntu.edu.tw

© The Author(s), under exclusive license to Springer Nature Switzerland AG 2021   49
R. Akhtar (ed.), *Coronavirus (COVID-19) Outbreaks, Environment and Human Behaviour*, https://doi.org/10.1007/978-3-030-68120-3_4

of flights linking it to China. In another study, Lai et al. (2020) performed a travel network-based analysis and estimated Taipei City to be in the top three cities at the highest risk based on historical air travel data, connectivity, and the risk of spread between high-risk cities in mainland China and cities in other countries from January to April, 2020. In fact, everyday life and daily activities have mostly continued as normal, with only a few cases of local transmission occurring in Taiwan. Most Taiwanese people trust and are satisfied with Taiwan's Centers for Disease Control (CDC) strategies used during the COVID-19 outbreak (Chen 2020; Huang 2020). This shows that the Taiwanese government's efforts are effective for the prevention of the spread of COVID-19 in local communities. This chapter reviews COVID-19 prevention and control strategies from the government as well as responses from communities in Taiwan.

Epidemiological Characteristics of COVID-19 Cases in Taiwan

As of February 10, 2021, only 935 cases of COVID-19 have been confirmed since the first confirmed case on January 21, 2020. Among these confirmed cases, 819 have been classified as imported, whereas 77 are believed to have been local transmissions, two were from a cargo pilot cluster and one case has unknown sources of infection. The other 36 cases were aboard a Navy vessel on a goodwill Pacific mission to Palau in March, 2020 (Fig. 1) and the Taiwan's CECC announced that this cluster infection originated in Taiwan on May 26, 2020 according to their investigation (Taiwan CDC 2020a). Among the confirmed cases, nine deaths have occurred, all of whom were adults older than 40 years.

Even with only limited cases of local transmission occurring in Taiwan, the possibility of undetected COVID-19 cases in local communities cannot be ruled out in Taiwan. A Japanese student and a Thai migrant worker were tested positive for COVID-19 after returning to their home countries from Taiwan on June 20 and July 25, 2020, respectively. Based on the CECC investigation, none of the over 300 people who came in contact with the Japanese student and Thai worker have tested positive for COVID-19 (Yeh 2020). Furthermore, the 18 Philippine nationals left Taiwan tested positive upon arrival in the Philippines between August 19 and September 11, 2020. Of the 372 people who came in contact with infected individuals in Taiwan, both the results of polymerase chain reaction and antibody tests have come back negative for 202 contacts who have been identified as individuals subjected to COVID-19 testing (Everington 2020). However, the positive tests of the Japanese student, Thai and Philippine workers raised concerns over whether there were still undetected COVID-19 cases in local communities. In addition, the risk of infection posed by travelers coming to Taiwan from abroad remains, and the CECC urges the public to maintain personal preventive practices.

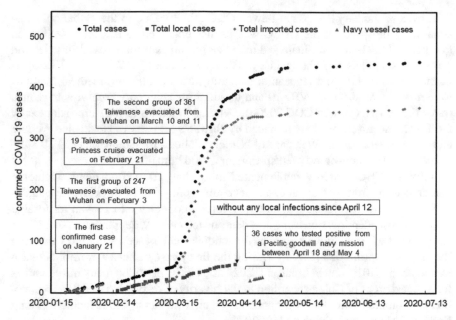

Fig. 1 Cumulative numbers of confirmed COVID-19 cases up to July 15, 2020, and timings of key events in Taiwan (*Source* Taiwan CDC Open Data Portal)

After 253 days without local transmission, Taiwan recorded its first local COVID-19 case on December 21, 2020 when an over 30-year-old Taiwanese female was infected by a foreign pilot who was previously confirmed to have been infected with coronavirus in US. In addition, a hospital cluster outbreak began in mid-January, 2021 when a doctor at the hospital was infected by a COVID-19 patient in northern Taiwan. As of February 10, 2021, a total of 21 cases have been linked to this hospital cluster.

Prevention and Control Measures Against COVID-19

Prompt Preventive Measures

When COVID-19 first occurred in Wuhan in late 2019, Taiwan quickly established a response team for this mysterious illness in China on January 2, 2020 (Taiwan CDC 2020b). On January 15, 2020, the Taiwan CDC classified COVID-19 as a Category 5 communicable disease to further strengthen surveillance and containment based on the Communicable Disease Control Act. A level 3 CECC for severe pneumonia with novel pathogens was established on January 20, 2020. On January 23, 2020, the CECC was elevated to level 2 because the first case of COVID-19 was confirmed in Taiwan on January 21, 2020 (Taiwan CDC 2020b). Due to the global epidemic situation getting worse, the CECC was further upgraded to a level 1 facility on February 27, 2020 to coordinate and mobilize resources from across ministries and private stakeholders to fight against COVID-19 (Taiwan CDC 2020b). This operational infrastructure and legislation encouraged the public and medical facilities to remain aware of the COVID-19 and to adopt precautionary measures to reduce transmission risk (Taiwan CDC 2020b). Detailed timelines of measures adopted by the Taiwanese government are provided by Taiwan's Ministry of Health and Welfare and are accessible on the Web page of "Crucial Policies for Combating COVID-19" (https://covid19.mohw.gov.tw/en/sp-timeline0-206.html).

Taiwan's CECC rapidly implemented more than 100 action items, including border control from the air and sea, a requirement of a 14-days' home quarantine for inbound travelers, a ban on transit through or entry into Taiwan for foreign nationals, case identification using new data and technology, quarantine of suspicious cases, and the formulation of policies for schools and businesses to follow between January and April, 2020 (Fig. 2). During the first phase of border control between December 31, 2019, and February 11, 2020, the CECC provided many travel notices for affected areas of China according to the severity of respective COVID-19 infections. A complete entry ban was implemented on residents of China including Hong Kong and Macao on February 11, 2020.

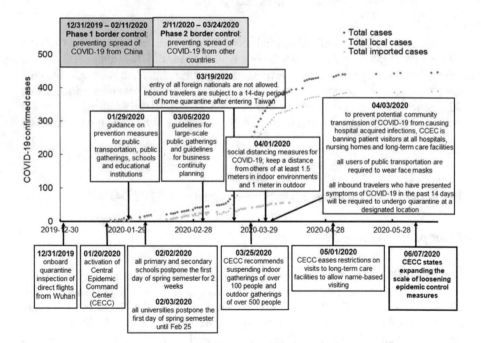

Fig. 2 Timeline of major control and preventive actions implemented by the Taiwan's CECC during the first six months since COVID-19 outbreak

Table 1 Number of people entering Taiwan during the first eight months of 2020 compared to the same month between 2015 and 2019

	Total entry persons		Mainlanders[a]		Foreigners	
	2020 monthly mean	% of the same month in past 5 years	2020 monthly mean	% of the same month in past 5 years	2020 monthly mean	% of the same month in past 5 years
January	2,262,692	115.7	88,165	31.8	586,506	125.1
February	1,136,655	53.8	5,167	1.6	329,898	73.4
March	216,002	9.8	1,297	0.5	76,423	12.8
April	22,822	1.0	476	0.2	1,917	0.4
May	22,824	1.0	581	0.2	2,445	0.5
June	32,102	1.5	761	0.3	6,325	1.3
July	41,353	1.8	906	0.3	8,682	2.0
August	47,480	2.1	1,392	0.5	13,982	2.9

Source National Immigration Agency, Ministry of Interior, Taiwan
[a]Mainlanders excluded Hong Kong and Macao residents

During the second phase of border control, the CECC issued many travel notices and raised the advisory level for other countries based on the severity of their respective COVID-19 infections from February 11 to March 21, 2020. On March 19, all foreign nationals were prohibited from entry to Taiwan, and all inbound travelers were subject to a 14-day period of home quarantine after entry. From March 24, all airline passenger transits through Taiwan were banned. Based on data from Taiwan's National Immigration Agency (Table 1), the total number of people entering Taiwan in April and May was only 1% of the corresponding numbers for the same month during the previous 5 years (2016–2019). Among these, the numbers for mainlanders and foreigners were less than 1%, respectively, of the corresponding numbers for the same month during the previous 5 years. On June 29, Taiwan relaxed entry measures for foreign nationals, Hong Kong and Macao residents to meet commercial and trade demand and humanitarian needs. The entry numbers of foreigners were increased from 1.3% in June to 2.9% in August compared to the same month between 2015 and 2019.

Preparedness and Lessons from SARS

Preparedness is crucial to reduce the effects of epidemics and avoid the spread of infections. Taiwan has made rapid progress in strengthening its public health preparation systems for infectious disease outbreaks 10 years after SARS (Inglesby et al. 2012). During the severe acute respiratory syndrome (SARS) outbreak, Taiwan reported 346 confirmed cases and 73 death, the highest number outside China and Hong Kong (Chan-Yeung and Xu 2003). Taiwan's government learned from the 2003 SARS experience and created a public health response mechanism to facilitate more effective actions for subsequent outbreaks (Wang et al. 2020a). In 2004, the National Health Command Center was established. This comprehensive platform and unified central command system quickly recognized the crisis and activated emergency management structures to address the emerging outbreak as seen from the early recognition of the COVID-19 epidemic in Taiwan.

Hospitals in Taiwan also learning from the SARS experience and have designated specific pathways to control patient flow; all patients are checked for fever, and their travel histories are confirmed (Schwartz et al. 2020; Wang et al. 2020b). A comprehensive patient diversion strategy including space partition, patient diversions, and staff subdivisions at the hospital emergency department is used to reduce the risk of infection, restrict the contaminated area (Kung et al. 2020), and protect healthcare workers (Lin et al. 2020). The CECC implemented control for nonmedical visits to the hospital during the COVID-19 outbreak on April 3, 2020. Healthcare providers are organized into discrete blocks within their working areas and into modular teams to avoid hospital-wide infection (Wang et al. 2020b). In addition, Taiwan has highly successful at care institutions for the elderly with no cluster infections in any of the 1,091 care institutions providing space 62,651 people (Yang and Huang 2020).

Rapid Risk Communication

Risk communication plays a major role for the control and prevention of restricting infectious disease outbreaks. During the COVID-19 outbreak, effective and frequent communication to the public from a trusted official could be crucial for reducing public panic and improving compliance with public health measures. Starting on January 23, 2020, a daily press briefing, an open and transparent epidemic information platform, was held to keep the public informed on the current developments of COVID-19 in Taiwan. The Minister of Health and Welfare, Chen Shih-Chung, has been the commander in charge to host news briefings almost every day to provide COVID-19 information in a language that has been easily understandable. This supported government's effort to gain public trust and build leadership in the fight against this epidemic. The daily press briefing simultaneously provided updates from relevant government departments, medical institutions, and private social media accounts.

Transparent Information for Prevention

Effective dissemination and transparency in the communication of epidemic-related information are crucial to reduce public panic. Digital technology provides easy public access to transparent information, facilitating the prevention of COVID-19 in Taiwan. Taiwan's CECC uses multiple major information transmission channels to raise the awareness of epidemic prevention and to provide clear and basic preventive knowledge. For example, Taiwan's CECC share knowledge regarding COVID-19 prevention across various social media platforms (such as LINE, Facebook, and Twitter), electronic media channels (such as YouTube), print media (such as mass transit posters), and broadcast media (such as television and radio) to regularly disseminate COVID-19 prevention information. Such efforts were also to avert public panic and educate public correct guidance on prevention and control of COVID-19 to improve personal hygiene and public awareness.

Preventing the transmission of misinformation and disinformation on social media is likely to reduce the fear and stigma during the epidemic prevention period (Lin 2020). The government's ability to respond quickly and effectively to misinformation and disinformation is as vital as the proactive publishing of information. The CECC's daily press briefings, Taiwan CDC Facebook pages, and the LINE account have become major sources of up-to-date information. Since the setup of the CECC, daily press briefings have been held to announce the latest policies and information on COVID-19 and to promptly clarify misinformation circulating on social media. The Taiwanese government has posted memes on social media as "humor over rumor" to counter disinformation. Moreover, the Taiwan CDC has been actively employing chatbots on LINE, a platform similar to WhatsApp and widely used in Taiwan, to announce key policies and statistics on a daily basis.

The Taiwan CDC began sending key messages to mobile phones through the "Public Warning System or Public Warning Cell Broadcast Service", for people who do not watch television, listen to the radio, or use social media or the Internet. The first message was sent on February 7, 2020, to those who had been to sites visited by passengers from the Diamond Princess cruise ship on January 31, advising them in regarding self-health management. The Taiwanese government has also established an infection control hotline at 1922 for inquires and the reporting of suspicious information. These multiple risk communication channels help to prevent unnecessary public panic and anxiety in Taiwan.

Social Innovation and Community Responses to COVID-19

Risk Awareness Among Communities

During an infectious disease outbreak, the perception of risk significantly affects people's actions in response to health threats. Both Taiwan and Hong Kong were severely affected by the SARS epidemic in 2003, and the public in these areas subsequently showed more risk awareness and better preparation for future outbreaks (Lei and Klopack 2020). On January 31, 2020, Taiwan's CECC convened a meeting to discuss COVID-19 prevention efforts. Participating experts advised that healthy people do not need to wear face masks and that local governments should reinforce the requirement of face masks in the following specific situations: doctor visits, patient visitations, and accompaniment of patients for doctor's appointments; presentation of respiratory infection symptoms; and for individuals with chronic diseases during outdoor activities (Taiwan CDC 2020c). However, the general population has taken precautions to strengthen environmental and personal hygiene in response to the COVID-19 outbreak. For example, most people began to wear face masks outdoors after the outbreak (Su and Han 2020). The demand for masks, disinfectants, and other anti-epidemic supplies has dramatically increased after the Lunar New Year holiday.

Religious communities have mobilized to limit mass gatherings, and some religious institutions suspended services (Taipei Times 2020). For example, organizers of large-scale public gathering events deferred or canceled their events. Dajia and Baishatun Matsu Pilgrimages, the two important annual religious processions in Taiwan, are held yearly to celebrate the birthday of the sea goddess Matsu during the third month of the Chinese lunar calendar. Because of pressure and criticism from the Taiwanese public and advice from health professionals, organizers postponed this annual religious event due to concerns regarding the COVID-19 outbreak (Chao and Chiang 2020a). As the COVID-19 pandemic has stabilized and some restrictions have been lifted in Taiwan, a scaled-down Dajia Matsu procession was rescheduled from June 11 to June 20, 2020, and temples were encouraging people to participate online and to maintain hygiene if they attend in person (Chao and Chiang 2020b). The actual procession workers were limited to 800 people, whose names are recorded and whose temperatures checked daily (Chao and Chiang 2020b). The Baishatun Matsu Pilgrimage was held from July 5 to July 13, 2020, nearly four months later than originally scheduled (Kuan and Yeh 2020). The COVID-19 precautionary measures adapted by the temple officials, including to register all participants, to instruct temple workers and worshipers wearing masks, to take people's temperature every day and to regularly sanitize participants' hands (Kuan and Yeh 2020).

Social Innovation

The "Name-Based Mask Distribution System" in Taiwan can be considered a leading example of social innovation through the application of digital technology to address COVID-19 (MOST GASE 2020). The public in Taiwan learned the potential value of mask-wearing from their SARS experience. This resulted in a shortage of medical-grade face masks during the Lunar New Year holiday period. To grant medical staff and patients priority access to medical-grade face masks, the CECC implemented a name-based rationing system for face masks distribution to the general population. Big data from the health insurance database was leveraged to implement "Name-based Mask Distribution System 1.0" on February 6, 2020. To avoid long lines of people waiting outside pharmacies, the government worked with the civil sector to launch various "mask maps" and "epidemic prevention maps" that immediately informed the public of the locations and quantities of epidemic prevention supplies in more than 6000 pharmacies in Taiwan (Tu 2020). On March 12, 2020, the updated "Name-Based Mask Distribution System 2.0" was launched where the public could order masks at either the eMask Web site or the National Health Insurance (NHI) Mobile App from a mobile phone. On April 22, 2020, the "Name-based Mask Distribution System 3.0" was introduced. Citizens can order masks by using kiosk machines at convenience stores and make cash payments onsite or can pre-purchase online and visit one of more than 10,000 convenience stores providing mask collection.

A prompt response to public opinion is key to successful epidemic prevention. "Pink masks" are another example of social innovation with a rapid online broadcast response from the Taiwan CECC (MOST GASE 2020). On April 12, 2020, the media reported that some elementary school boys did not wish to wear pink face masks for fear of being laughed at by their classmates (Liu 2020). On April 13, 2020, all male CECC officials wore pink face masks during the daily press briefing to demonstrate to young boys that no color is exclusive to boys or girls (Liu 2020). This example also provides an opportunity for education in relation to gender-related color stereotypes in Taiwanese society and to the promotion of gender equality through mass media.

To prevent the transmission of misinformation and disinformation, Taiwan established the Taiwan FactCheck Center (TFC) (https://tfc-taiwan.org.tw/) through a collaboration between two local non-profit organizations, Taiwan Media Watch and the Association for Quality Journalism, in July 2018. During the COVID-19 outbreak, the TFC identifies misinformation or disinformation regarding COVID-19 online, verifies it, and provides immediate clarification to the public on the COVID-19 section of the TFC's official Web site. In addition, an open, collaborative platform called Cofacts (https://cofacts.g0v.tw), created by Taiwanese civic tech community g0v in response to fake news, has existed in the closed messaging app LINE since December, 2016. On this platform, LINE users send suspicious links to the Cofacts account, and a bot then automatically replies if the article is already registered in its database as having been fact checked (Yang 2019). In the case of erroneous message, Cofacts provides immediate clarification to the public to reduce unnecessary panic and anxiety (Yang 2019). These non-profit fact-checking platforms are active engagements and collaborations with civil society and independent fact-checkers to provide rapid responses to misinformation regarding COVID-19 and can be also recognized as an example of social innovation that helps Taiwanese society to combat COVID-19 effectively.

Effects of COVID-19

Socioeconomic Effects

Social distancing, border control, and home quarantine have markedly reduced in economic productivity and placed a heavy burden on society. To reduce social isolation, self-blame, or even discrimination that might affect quarantined individuals, Taiwan's CECC has actively acknowledged the contribution of these people, rather than depicting them as a threat to public health, and local governments coordinated with the CECC through community-level networks to provide food and other necessities as well as emotional support (Lin and Cheng 2020).

Taiwan's export-dependent economy is vulnerable to reductions in global trade flows and recession due to weaker external demands and the disruption of global supply chains from the COVID-19 pandemic. In addition, local tourism-related businesses in Taiwan have been heavily affected by a lack of foreign visitors. Based on data from Taiwan's Directorate General of Budget Accounting and Statistics (DGBAS), Taiwan's seasonally adjusted unemployment rate increased to 4.16% in May 2020, the highest rate recorded since November 2013. On August 14, DGBAS predicted that Taiwan's gross domestic product would increase by 1.56% in 2020, a downgrade from a forecast of 2.37% made in February but much higher than the global average forecast. A major reason for this is that manufacturing activities are less affected by COVID-19 pandemic because no curfew or lockdown measures have been adopted in Taiwan, and orders have even been received from within the same industries where lockdowns elsewhere have affected operations.

Mental Health and Psychological Well-Being

A disease outbreak can trigger public fear and anxiety leading to emotional distress and panic. A Facebook questionnaire study was conducted to investigate the relationship of anxiety symptoms and preventive measures adopted for COVID-19 in Taiwan during February 14–16, 2020 (Wong et al. 2020). Anxiety symptoms were assessed using a 6-item version of the State–Trait Anxiety Inventory. From a total 3,555 completed responses, 52.1% of the respondents reported a moderate-to-severe level symptoms of anxiety between February 8 and the time of the survey, whereas 48.8% reported having had moderate-to-severe level symptoms of anxiety at the beginning of the outbreak. The results of this study indicated that an increased score for anxiety symptoms was significantly associated with an increase in preventive practices (Wong et al. 2020). Another Facebook survey study was conducted online to investigate the effects of COVID-19 on daily life and how varying levels of support may affect mental well-being in the form of sleep disturbance and suicidal thoughts in Taiwan from April 10 to 23, 2020 (Li et al. 2020). Of 1970 participants, 55.8% reported sleep disturbance, and 10.8% reported suicidal thoughts in the week prior to the study. These observations occurred at higher rates than in the study findings of Wu et al. (2016), who used the same questions to assess sleep disturbance (28.3%) and suicidal thoughts (2.1%) in Taiwan's general population. Sleep disturbance was found to be closely associated with the various factors, including increased worry regarding COVID-19, more severe effects of COVID-19 on social interactions, low perceived social support, severe academic or occupational interference due to COVID-19, low COVID-19-related support, and poor self-reported physical health (Li et al. 2020).

Collateral Benefit of Control Measures on Other Infectious Diseases

Personal hygiene, hand-washing, social distancing, and mask-wearing believe to be effective measures to limit transmission of viral respiratory diseases. In Taiwan, during the first 12 weeks of 2020, drastic decreases in influenza diagnoses in outpatient departments, in positivity rates for clinical specimens, and in confirmations of severe cases were reported compared with for the same period in 2019 (Kuo et al. 2020). Based on surveillance data on infectious diseases from the Taiwan CDC, the incidence of severe complicated influenza (Chan et al. 2020; Yang et al. 2020) and invasive pneumococcus disease (Chan et al. 2020; Tsai et al. 2020) decreased markedly after February 2020. Only one severe complicated influenza case was reported in March, whereas approximately 10–50 cases were recorded per month from January to March across the 2016–2019 period (Yang et al. 2020).

Conclusion

Because Taiwan has been excluded from a membership in the WHO, Taiwan faces challenges with vaccine availability and delay in COVID-19 vaccine delivery from international vaccine suppliers. Recently, the CECC announced that Taiwan was listed among the non-UN member states that would receive AstraZeneca COVID-19 vaccines through the WHO COVAX platform on February 4, 2021 and obtain a supply of 5 million doses of vaccine from Moderna by mid-2021 on February 10, 2021. Even though Taiwan has not been kept officially informed by the WHO about the COVID-19 pandemic, Taiwan provides a successful example of prevention of COVID-19 outbreak. By quickly integrating NHI database with immigration and customs database, NHI chip cards provide as an important digital tool to ensure public access to mask rationing system purchasing face mask and to help medical professionals tracking patient's recent travel history and medical information. The CECC daily press briefings also facilitated rapid communications, addressed public concerns while fighting misinformation about COVID-19. Using lessons learned from the SARS epidemic, both the Taiwanese government and the public have adopted proactive strategies and precautionary measures to combat COVID-19, including epidemic surveillance, tight border controls, community-based prevention and control, stockpiling of medical supplies, and health education and outreach to prevent the spread of the infection, to effectively maintain the health and safety of the Taiwanese population.

References

Chen, P. L. Y. (2020). A reachable governance to fight COVID-19: Democracy and the legacy of embedded autonomy in Taiwan. *Asian Journal of Social Science Studies, 5*(3), 18–28.

Chan, K. S., Liang, F. W., Tang, H. J., Toh, H. S., & Wen-Liang, Y. (2020). Collateral benefits on other respiratory infections during fighting COVID-19. *Medicina Clinica, 155* (6), 249–253.

Chan-Yeung, M., & Xu, R. H. (2003). SARS: Epidemiology. *Respirology, 8,* S9–S14.

Chao, L. Y., & Chiang, Y. C. (2020a, February 27). Dajia Matsu Pilgrimage to be postponed. *Focus Taiwan CNA English News.* https://focustaiwan.tw/culture/202002270004 [accessed June 17, 2020].

Chao, L. Y., & Chiang, Y. C. (2020b, June 8).Dajia Matsu Pilgrimage to take place June 11–20. *Focus Taiwan CNA English News.* https://focustaiwan.tw/culture/202006080009 [accessed June 19, 2020].

Everington, K. (2020, September 23). 18 Filipinos test positive for coronavirus after returning from Taiwan. *Taiwan News.* https://www.taiwannews.com.tw/en/news/4015132 [accessed October 5, 2020].

Gardner, L. (2020). *Update January 31: Modeling the spreading risk of 2019-nCoV.* https://sys tems.jhu.edu/research/public-health/ncov-model-2/ [accessed June 4, 2020].

Huang, I. Y. F. (2020). Fighting against COVID-19 through government initiatives and collaborative governance: Taiwan experience. *Public Administration Review, 80*(4), 665–670.

Inglesby, T., Cicero, A., Nuzzo, J., Adalja, A., Tonic, E., Rambhia, K., & Morhard, R. (2012). *Report on Taiwan's public health emergency preparedness programs 10 years after SARS.* Center for Biosecurity of UPMC.

Kuan R. P., & Yeh, J. (2020, June 16). Baishatun Matsu pilgrimage to be held from July 5–13. *Focus Taiwan CNA English News.* https://focustaiwan.tw/culture/202006160013 [accessed October 20, 2020].

Kung, C. T., Wu, K. H., Wang, C. C., Lin, M. C., Lee, C. H., & Lien, M. H. (2020). Effective strategies to prevent in-hospital infection in the emergency department during the novel coronavirus disease 2019 pandemic. *Journal of Microbiology, Immunology, and Infection.* https://doi.org/10.1016/j.jmii.2020.05.006.

Kuo, S. C., Shih, S. M., Chien, L. H., & Hsiung, C. A. (2020). Collateral benefit of COVID-19 control measures on influenza activity Taiwan. *Emerging Infectious Diseases, 26*(8), 1928–1930.

Lai, S., Bogoch, I. I., Ruktanonchai, N. W., Watts, A., Lu, X., Yang, W., Yu, H., Khan, K., & Tatem, A. J. (2020). Assessing spread risk of Wuhan novel coronavirus within and beyond China, January-April 2020: A travel network-based modelling study. *medRxiv.* https://doi.org/10.1101/2020.02.04.20020479.

Lei, M. K., & Klopack, E. T. (2020). Social and psychological consequences of the COVID-19 outbreak: The experiences of Taiwan and Hong Kong. *Psychological Trauma: Theory, Research, Practice and Policy, 12*(S1), S35–S37.

Li, D. J., Ko, N. Y., Chen, Y. L., Wang, P. W., Chang, Y. P., Yen, C. F., & Lu, W. H. (2020). COVID-19-related factors associated with sleep disturbance and suicidal thoughts among the Taiwanese public: A Facebook survey. *International Journal of Environmental Research and Public Health, 17*(12), 4479.

Lin, C. Y. (2020). Social reaction toward the 2019 novel coronavirus (COVID-19). *Social Health and Behaviour, 3*(1), 1–2.

Lin, M. W., & Cheng, Y. (2020). Policy actions to alleviate psychosocial impacts of COVID-19 pandemic: Experiences from Taiwan. *Social Health and Behaviour, 3*(2), 72–73.

Lin, C. H., Tseng, W. P., Wu, J. L., Tay, J., Cheng, M. T., Ong, H. N., et al. (2020). A double triage and telemedicine protocol to optimize infection control in an emergency department in Taiwan during the COVID-19 pandemic: Retrospective feasibility study. *Journal of Medical Internet Research, 22*(6), e20586.

Liu, L. (2020, April 13). Taiwan government officials wear 'girly' colored masks. *Taiwan News.* https://www.taiwannews.com.tw/en/news/3915248 [accessed June 19, 2020].

MOST GASE. (2020). *Interview with the "Genius Digital Minister" Audrey Tang: Civic tech and the successful fight against the epidemic*, May 26, 2020. https://gase.most.ntu.edu.tw/focus/592 [accessed June 19, 2020].

Schwartz, J., King, C. C., & Yen, M. Y. (2020). Protecting healthcare workers during the coronavirus disease 2019 (COVID-19) outbreak: Lessons from Taiwan's severe acute respiratory syndrome response. *Clinical Infectious Diseases, 71*(15), 858–860.

Su, S. F., & Han, Y. Y. (2020). How Taiwan, a non-WHO member, takes actions in response to COVID-19. *Journal of Global Health, 10*(1), 010380.

Taipei Times. (2020). Virus Outbreak: Faithful stream services as temples close, March 22, 2020. https://www.taipeitimes.com/News/taiwan/archives/2020/03/22/2003733165 [accessed June 17, 2020].

Taiwan CDC. (2020a, May 26). *Press releases: Investigation into cluster infection involving naval crews of Dunmu fleet is complete; infection found only within naval crew members aboard Panshi.* https://www.cdc.gov.tw/En/Bulletin/Detail/HrMDge9urbot2lFp1aLbWw?typeid=158 [accessed June 4, 2020].

Taiwan CDC. (2020b). Prevention and control of COVID-19 in Taiwan. Updated on June 2, 2020. https://www.cdc.gov.tw/Category/Page/0vq8rsAob_9HCi5GQ5jH1Q [accessed June 4, 2020].

Taiwan CDC. (2020c). *Domestic experts recommend that healthy students do not need to wear masks.* Press release January 31, 2020. https://www.cdc.gov.tw/En/Category/ListContent/tov1ja hKUv8RGSbvmzLwFg?uaid=fuxYlhrqkcT5drFreHzhAg [accessed June 17, 2020].

Tsai, J. R., Yang, C. J., Huang, W. L., & Chen, Y. H. (2020). Decline in invasive pneumococcus diseases while combating the COVID-19 pandemic in Taiwan. *The Kaohsiung Journal of Medical Sciences, 36*(7), 572–573.

Tu, C. C. (2020). *Lessons from Taiwan's experience with COVID-19.* Atlantic Council, April 7, 2020. https://www.atlanticcouncil.org/blogs/new-atlanticist/lessons-from-taiwans-exp erience-with-covid-19/ [accessed June 12, 2020].

Wang, C. J., Ng, C. Y., & Brook, R. H. (2020). Response to COVID-19 in Taiwan: Big data analytics, new technology, and proactive testing. *JAMA, 323*(14), 1341–1342.

Wang, C. J., Bair, H., & Yeh, C. C. (2020). How to prevent and manage hospital-based infections during coronavirus outbreaks: Five lessons from Taiwan. *Journal of Hospital Medicine, 15*(6), 370–371.

Wong, L. P., Hung, C. C., Alias, H., & Lee, T. S. H. (2020). Investigation into public anxiety symptoms and preventive measures during the COVID-19 outbreak in Taiwan. https://doi.org/ 10.21203/rs.3.rs-23432/v1.

Wu, C. Y., Lee, J. I., Lee, M. B., Liao, S. C., Chang, C. M., Chen, H. C., & Lung, F. W. (2016). Predic- tive validity of a five-item symptom checklist to screen psychiatric morbidity and suicide ideation in general population and psychiatric settings. *Journal of the Formosan Medical Association, 115*(6), 395–403.

Yang, C. J., Chen, T. C., Kuo, S. H., Hsieh, M. H., & Chen, Y. H. (2020). Severe complicated influenza declined during the prevention of COVID-19 in Taiwan. *Infection Control and Hospital Epidemiology*, 1–2. https://doi.org/10.1017/ice.2020.272.

Yang, O. (2019). Defending democracy through media literacy. *Taiwan Democracy Bulletin, 3* (6) October 09, 2019. https://bulletin.tfd.org.tw/tag/taiwan-factcheck-center/ [accessed June 22, 2020]

Yang, P., & Huang, L. K. (2020). Successful prevention of COVID-19 outbreak at elderly care institutions in Taiwan. *Journal of the Formosan Medical Association, 119*(8), 1249–1250.

Yeh, W. (2020, August 1). CORONAVIRUS/Taiwan reports 7 new COVID-19 cases, one case still a mystery (update). *Focus Taiwan CNA English News*. https://focustaiwan.tw/society/202 008010016 [accessed August 14, 2020].

Zlojutro, A., Rey, D., & Gardner, L. (2019). A decision-support framework to optimize border control for global outbreak mitigation. *Scientific Reports, 9*(1), 1–14.

Healthcare Resources in the Context of COVID-19 Outbreak—Thailand, 2020

Uma Langkulsen and Desire Tarwireyi Rwodzi

Abstract The coronavirus disease 2019 (COVID-19) has ravaged the whole world since the outbreak began in December 2019. Nations and territories have experienced varying degrees of morbidity and mortality as direct effects of COVID-19. Despite being among the very first countries outside of China to encounter the full wrath of the COVID-19 pandemic, Thailand was able to swiftly control and curb the first wave of the COVID-19 scourge. This investigation assessed the health effects of COVID-19 and the adequacy of the available healthcare resources in responding to the threats posed by the COVID-19 pandemic in Thailand. As of January 24 and February 8, 2021, Thailand had reported a cumulative total of 13,500 COVID-19 cases and 79 deaths, respectively. During the first wave, cases increased rapidly from mid-March to mid-April, reaching a peak of 188 cases per day. The second wave hit Thailand towards the end of November 2020, and cases increased rapidly to reach a second peak of 745 confirmed COVID-19 cases on 4th January 2021. More COVID-19 related deaths were reported during the first wave. The evolving COVID-19 situation in Thailand was linked to an increase in the reporting of mental health issues. However, with time, there was a gradual decline in the proportion of people reporting high levels of anxiety. Public Health resources were strategically deployed to curb further spread of COVID-19. Across the 77 provinces in Thailand, professional nurses had the highest density per 100,000 population, followed by public health technical officers. However, within each category of health workers, huge variations existed among provinces. Besides the specialized healthcare workers, there were over a million village health volunteers across the country. The Ministry of Public Health ensured that distribution of medicines and medical supplies was dependent on the numbers of COVID-19 cases reported in each province. Available resources were adequate and were deployed in areas of greatest need in a timely fashion, and this could be one of the explanations why Thailand did not have a COVID-19 burden as bad as seen in other countries.

U. Langkulsen (✉)
Faculty of Public Health, Thammasat University, Pathum Thani 12121, Thailand
e-mail: u.langkulsen@gmail.com

D. T. Rwodzi
UNAIDS Regional Support Team for Asia and the Pacific, Bangkok, Thailand

© The Author(s), under exclusive license to Springer Nature Switzerland AG 2021
R. Akhtar (ed.), *Coronavirus (COVID-19) Outbreaks, Environment and Human Behaviour*, https://doi.org/10.1007/978-3-030-68120-3_5

63

Keywords COVID-19 · Coronavirus · Thailand · Healthcare resources

Introduction

The coronavirus disease 2019 (COVID-19) outbreak started in the Hubei Province of China in December 2019. Subsequently, many cities and provinces fell victim to COVID-19, and it is believed that the heavy transportation load during the Chinese Lunar New Year period amplified the spread of the virus (Wu et al. 2020). After China, Thailand is the second country in which the COVID-19 infection occurred since early January 2020 (Ratanarat et al. 2020). Suggesting early international spread, Thailand recorded its first confirmed COVID-19 exported case on January 8, 2020, and the patient was a woman in her early 60 s who had flown directly to Thailand from Wuhan, with five family members, as part of a tour group of 16 (Okada et al. 2020). The first known case of COVID-19 community transmission was reported on January 23, 2020, and was a 51-year-old male taxi driver who had contracted the virus, potentially from Chinese tourists (Pongpirul et al. 2020).

In the very early days of the outbreak, Thailand established thermal screening points at Suvarnabhumi Airport, Don Mueang, Phuket and Chiang Mai airports targeting travelers from Wuhan since January 3, 2020 (Okada et al. 2020). COVID-19 was declared a global pandemic by the World Health Organization on March 20, 2020. In Thailand, the ratio between local transmission and importation was equal to 1:4.25, suggesting that importation contributed far more cases of COVID-19 than local transmission (Joob and Wiwanitkit 2020d). Despite being one of the first countries to be hit by the COVID-19, Thailand managed to put the scourge under control without too many casualties. This paper seeks to assess the health effects of COVID-19 and the adequacy of the available healthcare resources in responding to the threats posed by the COVID-19 pandemic in Thailand.

Methods

Study Area

With a total of 77 provinces, Thailand is divided into 12 health area regions as shown in Fig. 1.

Data Collection

We gathered COVID-19 statistics on cases and deaths from the Digital Government Development Agency (DGA) and Department of Disease Control, Ministry of Public

Fig. 1 Study area

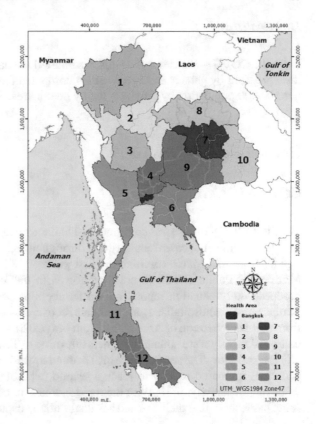

Health of Thailand from the onset of the pandemic up to January 24 and February 8, 2021, respectively. We also gathered data on health worker establishments from the Strategy and Planning Division, under the Office of Permanent Secretary in the Ministry of Public Health. This included the numbers of health workers by province and region for the fiscal year 2018. From the Drug and Medical Supply Information Center, we gathered information on ventilators as well as medicines and medical supplies distributed to provinces and regions during the COVID-19 response initiatives. The Department of Mental Health under the Ministry of Public Health conducted an online anxiety test about COVID-19. The objectives were to screen the general population for anxiety related to COVID-19, and to help reduce the anxiety. We also conducted a review of recent publications on the COVID-19 response in Thailand.

Data Analysis

From the COVID-19 cases and deaths, we charted the data against a time scale and disaggregated by gender and age. We also computed the distribution of ventilators per 100,000 population by province and by health area.

Health Effects of COVID-19

COVID-19 Cases

The first COVID-19 case was confirmed in Thailand on January 12, 2020, as shown in Fig. 2. The number of cases remained low throughout February and started increasing rapidly from mid-March reaching the first peak of 188 cases in a single day. The Ministry of Public Health in collaboration with the World Health Organization produced daily situation reports on the numbers of confirmed COVID-19 cases, deaths, numbers hospitalized, and numbers recovered. Daily cases declined steadily from mid-March to end of April 2020. From May 2020, very few cases were reported daily, and mostly were among foreigners returning to Thailand without any community transmission. The second wave hit Thailand towards the end of November 2020, and cases increased rapidly to reach a second peak of 745 confirmed COVID-19 cases on 4th January 2021. With the rapid increase in cases, demographic data for some cases including their age and sex could not be captured.

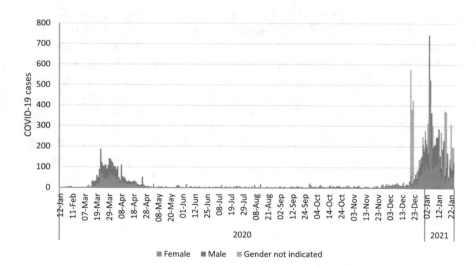

Fig. 2 Number of COVID-19 confirmed cases by sex in Thailand during January 12, 2020–January 24, 2021. *Source* Digital Government Development Agency of Thailand

As of January 24, 2021, Thailand had reported a cumulative total of 13,500 COVID-19 cases. Cases increased rapidly from mid-March to mid-April as shown in Fig. 3.

Disaggregating the COVID-19 cases by age group, 18% of cumulative cases were young people aged between 20 and 29 years as shown in Fig. 4. This was followed

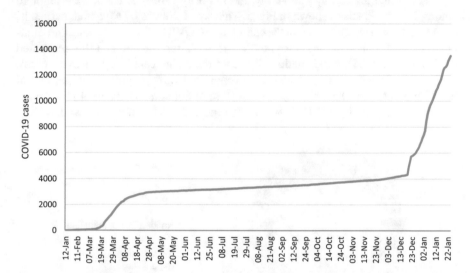

Fig. 3 Number of COVID-19 confirmed cases in Thailand during January 12, 2020–January 24, 2021. *Source* Digital Government Development Agency of Thailand

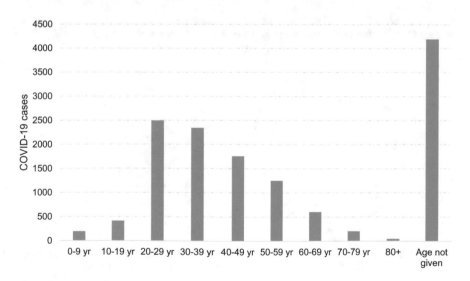

Fig. 4 Number of COVID-19 confirmed cases by age group in Thailand during January 12, 2020–January 24, 2021. *Source* Digital Government Development Agency of Thailand

by the 30–39-year age group which accounted for 17% of the reported cases. The number of reported cases continued to decrease with increasing age, and only 6% of the total cases were reported among the elderly aged 60 years and above. However, age was not indicated for 4,192 confirmed COVID-19 cases accounting for 31% of the cumulative cases.

When disaggregated by nationality, the majority (56%) of cases were reported among Thai nationals as shown in Fig. 5. This was followed by Myanmar nationals, accounting for at least a quarter of the cumulative COVID-19 cases. Among the cumulative cases, 90 nationalities across the world were represented and this accounted for 7% of the total. However, nationality could not be indicated for 12% of the cases.

The data for reported cases was also disaggregated by province of onset. Out of the 77 provinces in Thailand, 2 provinces (Samut Sakhon and Bangkok) accounted for half of the cumulative COVID-19 cases reported in Thailand as shown in Fig. 6.

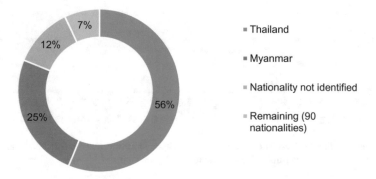

Fig. 5 Number of COVID-19 confirmed cases by nationality in Thailand during January 12, 2020–January 24, 2021. *Source* Digital Government Development Agency of Thailand

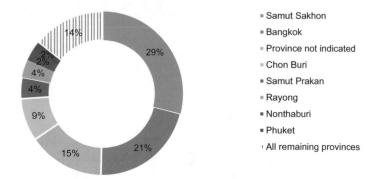

Fig. 6 Number of COVID-19 confirmed cases by province of onset in Thailand during January 12, 2020–January 24, 2021. *Source* Digital Government Development Agency of Thailand

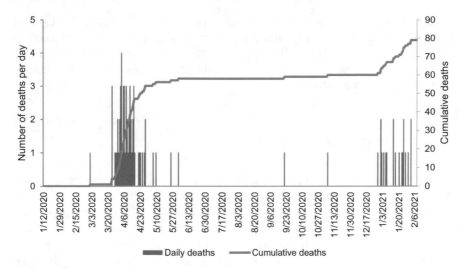

Fig. 7 Number of COVID-19 deaths in Thailand during January 12, 2020–February 8, 2021. *Source* Department of Disease Control, Ministry of Public Health

Mortality

As of February 8, 2021, Thailand had reported a cumulative total of 79 COVID-19 deaths as shown in Fig. 7. Majority of COVID-19 deaths in Thailand were reported during the first wave. The highest number of daily deaths reported stands at 4, and this was recorded in April 2020 during the first wave. The initial mortality rates for patients in the hospital were estimated to be 11–15%, but more recent data suggested a case fatality rate of 2–3% (Wu et al. 2020). Without validated treatments for COVID-19, the main strategies are symptomatic and supportive care, such as keeping vital signs, maintaining oxygen saturation and blood pressure, and treating complications, such as secondary infections or organs failure (Wu et al. 2020).

Some of the COVID-19 patients frequently needed prolonged hospitalization and stay in the ICU and thus developed typical nosocomial complications such as ventilator associated pneumonia and sepsis (Ratanarat et al. 2020).

COVID-19 Presentation in Patients

COVID-19 tended to present itself in different ways. A case study in Thailand suggested that one 40-year-old patient tested positive for COVID-19 after initially presenting with an orthopedic problem, fracture (Joob and Wiwanitkit 2020a). In a separate case, a healthcare worker contracted COVID-19 from a patient who initially presented with dengue (Joob and Wiwanitkit 2020c). With time, it became clearer that some patients with COVID-19 do not develop fever or respiratory symptoms at the time of presentation. Thailand's experience with COVID-19 highlighted the importance of strict respiratory infection control regardless of whether one has clear respiratory symptoms. There is also a possibility that a patient with COVID-19 might initially present with a skin rash that can be misdiagnosed as another common disease (Joob and Wiwanitkit 2020b).

Mental Health and COVID-19

Results from a survey conducted from March 24–May 21, 2020, showed a gradual decline in the proportion of people reporting high levels of anxiety and an increase in the proportion of those reporting low anxiety levels. The majority of the population (more than 70%), however, reported moderate anxiety levels as shown in Fig. 8.

The Department of Mental Health also conducted an Online Happy Family Survey for Thai during the COVID-19 outbreak. Out of a total of 1,500 participants who responded, the majority (77%) were females. In addition, 37% of the participants were from Bangkok and its surrounding areas, while 63% were from other provinces. Almost half (48.9%) of the participants indicated that they were very happy staying at home with their families during the COVID-19 outbreak. Majority of participants (54.1%) acknowledged that they were moderately stressed while 4.2% reported their families to be extremely stressed. Almost two thirds (65.5%) of the participants reported they supported their families and worked together as a team in dealing with problems during the COVID-19 period. More than three in four participants (76%) felt very confident with the ways that their families dealt with the COVID-19 crises.

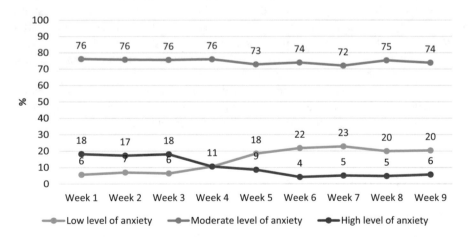

Fig. 8 Anxiety level during COVID-19 outbreak in Thailand. *Source* Department of Mental Health, Ministry of Public Health

Health System Capacity

Overall, the median number of ventilators per 100,000 population by health area ranged from 2 to 7 as shown in Fig. 9. However, there were huge variations among provinces within each health area. Largest variations were noted in health area 12 in which availability of ventilators ranged from 0 to 30 ventilators per 100,000 population.

Essential Healthcare Services

Distribution of medicines and medical supplies was dependent on the numbers of COVID-19 cases reported in each province. From the Government Pharmaceutical Organization under the Ministry of Public Health, supplies were distributed according to caseload at health area level, before they were further distributed by province and ultimately to hospitals managing COVID-19 cases as shown in Fig. 10. Patients under investigation (PUIs) for COVID-19 received N95 masks and five pieces of personal protective equipment (PPE) per patient per day. Confirmed COVID-19 cases received 15 pieces of N95 masks and PPE per patient per day. Medical supplies that were distributed include alcohol gel, masks, drugs, surgical gowns and personal protective equipment. Transportation of medical supplies was facilitated by Thailand Post Distribution Co., Ltd.

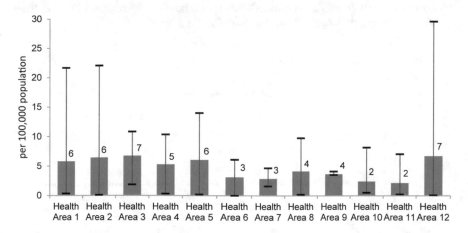

Fig. 9 Number of ventilators per 100,000 population in Thailand. *Source* Office of Permanent Secretary, Ministry of Public Health

Fig. 10 Flow diagram for
the distribution of medicine
and medical supplies in
Thailand

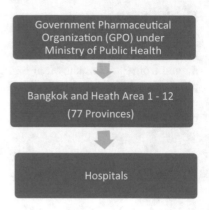

Health Workforce

Across the 77 provinces and health areas, professional nurses had the highest density per 100,000 population, followed by public health technical officers as shown in Fig. 11. Within each category of health workers, huge variations exist among provinces. For example, there were only 66 professional nurses per 100,000 population in Bangkok Province compared to 351 nurses per 100,000 population in Sing Buri Province. A similar trend was also observed for other categories of health workers. Medical Scientists, psychologists and statisticians were the rarest health workers across the provinces. Besides the specialized healthcare workers, there are over a million village health volunteers across the country, in addition to more than 15,000 public health volunteers in Bangkok (Bello 2020).

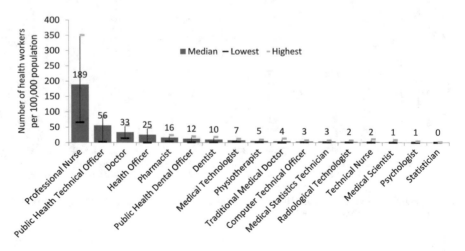

Fig. 11 Number of health personnel in the Ministry of Public Health in Thailand. *Source* Office of Permanent Secretary, Ministry of Public Health

COVID-19 Vaccination in Thailand

It is expected that Thailand would start receiving vaccinations in June 2021 at a rate of 5 million doses a month using shots from Thailand's Siam Bioessence. Following the transfer of production technology from AstraZeneca, Thailand is poised to become a production hub in ASEAN (Praphornkul 2021). However, the government of Thailand is not relying on a sole candidate to meet its COVID-19 vaccine requirements, and has contingency plans to secure the vaccines from other producers if the arrangement with AstraZeneca cannot deliver required doses timely (Rattiwan 2021). With emphasis on equity, the COVID-19 vaccination program will be rolled out in 3 phases, prioritizing healthcare workers, the elderly and those seriously ill (Arayasukawat 2021).

Discussion and Conclusions

Successful control of the COVID-19 pandemic is not only influenced by the numbers of available health workers, but also requires that the health workers be knowledgeable and practice the recommended infection prevention and control measure. In a knowledge, attitudes and practices (KAP) study conducted among health personnel in Nang Rong district, a rural region of Thailand in Buriram Province, there was a significant association between total COVID-19 knowledge score and type of medical personnel (Apaijitt and Wiwanitkit 2020). Overall knowledge about COVID-19 was low among health workers, despite Thailand having implemented several public health policies to curb the spread of COVID-19. Some physicians had lower COVID-19 knowledge score than non-physicians (Apaijitt and Wiwanitkit 2020). This was clear testimony that health personnel themselves needed some capacity building for them to be able to disseminate correct information to the generality of the population. In the case of Thailand, village health volunteers (VHVs) also played a crucial role in flattening the spread of COVID-19 at the community level. Their capacity was strengthened to monitor people's movement in and out of their villages and to conduct home visits disseminating information about COVID-19. In conclusion, available resources were adequate and were deployed in areas of greatest need in a timely fashion. This could be one of the explanations why Thailand did not have a COVID-19 burden as bad as seen in other countries.

Acknowledgements We would like to acknowledge and express our gratitude to Associate Professor Dr. Pannee Cheewinsiriwat from Department of Geography, Faculty of Arts, Chulalongkorn University, who made the health area map.

Declarations of interest We declare no competing interests.

References

Apaijitt, P. & Wiwanitkit, V. (2020). "Knowledge of coronavirus disease 2019 (COVID-19) by medical personnel in a rural area of Thailand." *Infection Control & Hospital Epidemiology, 41*(10), 1243–1244. https://doi.org/10.1017/ice.2020.159.

Arayasukawat, P. (2021). "Vaccination program will be divided into 3 phases." Retrieved 13 February 2021, from https://thainews.prd.go.th/en/news/detail/TCATG210202095723365.

Bello, W. (2020). "How Thailand Contained COVID-19." Retrieved 18 July, 2020, from https://fpif.org/how-thailand-contained-covid-19/.

Joob, B., & Wiwanitkit, V. (2020a). Carpal fracture and COVID-19 infection: observation from Thailand. *Indian Journal of Orthopaedics, 54*(3), 393.

Joob, B., & Wiwanitkit, V. (2020b). COVID-19 can present with a rash and be mistaken for Dengue. *Journal of the American Academy of Dermatology, 82*(5), e177.

Joob, B., & Wiwanitkit, V. (2020c). COVID-19 in medical personnel: observation from Thailand. *Journal of Hospital Infection, 104*(4), 453.

Joob, B. & Wiwanitkit, V. (2020d). "Outbreak of COVID-19 in Thailand: Time Serial Analysis on Imported and Local Transmission Cases." *International Journal of Preventive Medicine, 11*(43). https://doi.org/10.4103/ijpvm.IJPVM_98_20.

Okada, P., Buathong, R., Phuygun, S., Thanadachakul, T., Parnmen, S., Wongboot, W. et al. (2020). "Early transmission patterns of coronavirus disease 2019 (COVID-19) in travellers from Wuhan to Thailand, January 2020." *Euro Surveill, 25*(8).

Pongpirul, W. A., Pongpirul, K., Ratnarathon, A. C., & Prasithsirikul, W. (2020). Journey of a Thai Taxi Driver and Novel Coronavirus. *New England Journal of Medicine, 382*(11), 1067–1068.

Praphornkul, P. (2021). DPM confirms Thais will start receiving vaccinations in June at a rate of 5 million doses a month. Retrieved 13 February 2021, from https://thainews.prd.go.th/en/news/detail/TCATG210205110857576.

Rattiwan, S. (2021). Thailand has contingency plans to secure COVID-19 vaccines. Retrieved 13 February 2021, from https://thainews.prd.go.th/en/news/detail/TCATG210208143858436.

Ratanarat, R., Sivakorn, C., Viarasilpa, T. & Schultz, M. J. (2020). "Critical care management of patients with COVID-19: early experience in Thailand." *The American Journal of Tropical Medicine and Hygiene, 103*(1), 48–54. https://doi.org/10.4269/ajtmh.20-0442.

Wu, Y. C., Chen, C. S., & Chan, Y. J. (2020). The outbreak of COVID-19: An overview. *J Chin Med Assoc, 83*(3), 217–220.

How Malaysia Counters Coronavirus Disease (COVID-19): Challenges and Recommendations

Nasrin Aghamohammadi, Logaraj Ramakreshnan, and Chng Saun Fong

Abstract The arrival of COVID-19 to Malaysia was identified in January 2020 and the biggest cluster was associated with a Tablighi Jamaat religious gathering organized in Sri Petaling, Kuala Lumpur, in the early March. A number of challenges and recommendations on successful public health outbreak response tactics were addressed by the Malaysian government to combat the spread of the disease according to the national situation. As of July 3, 2020, web-based real-time tracking of the disease demonstrated that at least 8,643 confirmed cases and 121 deaths were reported in Malaysia, with at least 97.6% (8,437) of infected patients fully recovered from the disease. Meanwhile, Kuala Lumpur (2,422, 28.0%), Selangor (2,006, 23.2%), Negeri Sembilan (994, 11.5%), Johor (686, 7.9%), and Sarawak (565, 6.5%) were the top five states that recorded the highest number of COVID-19 cases in Malaysia until June 30, 2020. As of February 4, 2021, a total of 226,912 infections and 809 deaths were reported with Selangor accounted for more than 30% of the total cases in Malaysia. This pandemic highlighted the crucial role of a multilateral partnership led by the Malaysian government in handling an outbreak of highly infectious disease. The enforcement of the Movement Control Order (MCO) is the main reason for the chain of infection to be halted and controlled over a short period. The MCO is also proven to be an effective strategy to identify the clusters in which the virus spread has originated or affected the mass population in Malaysia.

N. Aghamohammadi (✉) · L. Ramakreshnan · C. S. Fong
Centre for Occupational and Environmental Health, Department of Social and Preventive Medicine, Faculty of Medicine, University of Malaya, 50603 Kuala Lumpur, Malaysia
e-mail: nasrin@ummc.edu.my

L. Ramakreshnan
e-mail: logarajramakreshnan@gmail.com

C. S. Fong
e-mail: fongcs92@gmail.com

N. Aghamohammadi
Centre for Epidemiology and Evidence-Based Practice, Department of Social and Preventive Medicine, Faculty of Medicine, University of Malaya, 50603 Kuala Lumpur, Malaysia

L. Ramakreshnan · C. S. Fong
Institute for Advanced Studies, University of Malaya, 50603 Kuala Lumpur, Malaysia

© The Author(s), under exclusive license to Springer Nature Switzerland AG 2021
R. Akhtar (ed.), *Coronavirus (COVID-19) Outbreaks, Environment and Human Behaviour*, https://doi.org/10.1007/978-3-030-68120-3_6

Keywords COVID-19 · Malaysia · Movement control order · Pandemic ·
Standard operating procedures

Introduction

The end of the year 2019 recorded the emergence of one of the deadliest pneumonia
outbreaks that threatened the humankind, known as coronavirus disease 2019 or
COVID-19. The infectious disease was caused by a novel betacoronavirus called
severe acute respiratory syndrome coronavirus 2 (SARS-CoV-2) (Xie and Chen
2020). It was first identified in December 2019 in Wuhan, Hubei, China, before
started to spread exponentially at the beginning of 2020 globally. Deeply concerned
by the unprecedented swift global spread and severity of the outbreak, the World
Health Organization (WHO) declared COVID-19 as a public health emergency
of international concern (Mahase 2020). As of February 4, 2021, the Worldometer
updates indicated that more than 104,928,503 confirmed cases and 2,278,995 deaths
have been reported across more than 221 countries with seventy-six million people
have recovered from the infections. Phylogenetic analyses indicated that the virus
has an 89% nucleotide identity with bat coronavirus (SARS-like-CoVZXC21) as the
potential animal reservoir (Chan et al. 2020). Meanwhile, Malayan pangolins were
outlined as the possible intermediate hosts due to their approximately 85.5–92.4%
genome similarity to SARS-CoV-2 (Lam et al. 2020). Even though the mode of transmission was unclear at the early stages of the pandemic, it was discovered that the
virus was mainly transmitted via droplet or contact transmission (Van Doremalen
et al. 2020). On top of that, new transmission modes were suggested when Guan
et al. (2020) detected the virus in the gastrointestinal tract specimens, saliva, urine,
and even in esophageal erosion and bleeding sites of infected peptic ulcer patients.
In a clinical study involving 4,021 infected patients, Yang et al. (2020) estimated
the reproductive number (R_0) and incubation period of the virus to be 3.77(95%
CI, 3.51–4.05) and 4.75 days (IR: 3.0–7.2 days), respectively. Besides, the clinical
presentation of COVID-19 much resembled viral pneumonia with mild cases usually
self-limiting and recovery happened within two weeks (Wu and McGoogan 2020).
In the occasion of severe infections, symptoms progressed into acute respiratory
distress syndrome (ARDS) and septic shock, which eventually ended in multiple
organ failure. Despite numerous preventive measures in various countries, it remains
a challenging task to contain and control the spread of novel COVID-19 with high R_0,
long incubation period, and limited treatment options. In line with this, this chapter
aims to elaborate on the challenges and recommendations on successful public health
outbreak response tactics by the Malaysian government to combat the spread of the
disease according to the national situation.

Epidemiology of COVID-19 in Malaysia

The arrival of COVID-19 to Malaysia was identified in January 2020, especially when positive cases were detected on travelers from China following the disease outbreak in Hubei, China (NST 2020a). On an important note, localized clusters emerged in the middle of March with the biggest cluster associated with a Tablighi Jamaat religious gathering organized in Sri Petaling, Kuala Lumpur, in early March (Aljazeera 2020; Barker 2020). Consequently, this had caused an explosive increment in the number of local cases and exportation of cases to neighboring territories. By the end of March 2020, Malaysia grew as one of the severely hit tropical countries in Southeast Asia recording more than 2,000 active cases in almost every state and the federal territory of the country (TST 2020). As of July 3, 2020, Web-based real-time tracking of the disease demonstrated that at least 8,643 confirmed cases and 121 deaths were reported in Malaysia, with at least 97.6% (8,437) of infected patients fully recovered from the disease. As shown in Fig. 1, Kuala Lumpur (2,422, 28.0%), Selangor (2,006, 23.2%), Negeri Sembilan (994, 11.5%), Johor (686, 7.9%), and Sarawak (565, 6.5%) were the top five states that recorded highest number of COVID-19 cases in Malaysia until June 30, 2020. As of February 4, 2021, a total of 226,912 confirmed cases and 809 deaths were reported in Malaysia with Selangor experiencing the worst outbreak of COVID-19 which accounted for more than 30% of the total cases in Malaysia.

Recognizing the arising health emergency in the country, Malaysian government initiated preparations to detect and monitor infections, stockpile equipment, treat and isolate the infected patients as well as device and publish preventive measures to the general public as early as January 6, 2020. A nationwide Movement Control Order (MCO) was initiated on March 16, 2020, to combat the spread of the disease via

Fig. 1 Distribution map of COVID-19 confirmed cases by states in Malaysia by June 30, 2020 (Modified from https://www.outbreak.my/#states)

social distancing and interstate travel restrictions (Sham 2020; Sukumaran 2020). In the next couple of days, a federal gazette was set to restrict travels to and from COVID-19-prone areas with heavy penalties for any form of breaching (Malaysiakini 2020). The MCO was enhanced and extended up to May 23, 2020, until the rate of new cases becomes relatively low. During this time, a gradual easing of restrictions under a Conditional Movement Control Order (CMCO) was initiated at the beginning of May to allow business operation with strict adherence to standard operating procedures (SOPs) (Latif 2020). On June 10, 2020, the CMCO was replaced with the Recovery Movement Control Order (RMCO) by lifting a host of restrictions that were enforced during the previous stage. However, forewarning that people must still abide by the given SOPs for the RMCO to be successful was issued to enable a complete lifting of restrictions by August 31, 2020 (Cindi Loo 2020).

The Challenges of COVID-19 in Malaysia

This section elaborates the challenges during the COVID-19 outbreak in Malaysia. Thereafter, recommendations and suggestions based on the challenges will be presented as learned lessons to prepare for the upcoming pandemics in the country.

Limitations of Healthcare Infrastructure, Utilities, and Manpower Capacity

One of the greatest challenges that the country was facing was inadequate healthcare infrastructure, utilities, and manpower capacity to tackle the COVID-19 pandemic. Malaysia has insufficient hospital beds to manage the peak outbreak, and this necessitates the additional health infrastructure capacity to be increased with immediate urgency (Salim et al. 2020). Even though the country has four manufacturers of face masks with higher quality, the country was still facing inadequate supply for the facemasks during the peak of the pandemics (Steven 2020). Moreover, minimal support was given by the government to these manufacturers to sustain their operations. Most of the masks produced by Malaysia were exported to high-income developed countries with only little being supplied to the country's healthcare institutions which subsequently raised the price of masks. By the middle of April, the Ministry of Health warned that supply chains of personal protective equipment (PPE) were running low with only two weeks of stocks left (The Star 2020a). On the other hand, limited manpower capacity in terms of healthcare providers and healthcare frontliners made them work tirelessly to provide treatment and support in a time of unprecedented demand for patient care (Bernama 2020a). As a result, the healthcare system was severely strained due to the increasing number of infections by the virus to support medical services and essential needs of infected patients within the country.

Management of COVID-19 Spread via Inadequate Standard Operating Procedures (SOPs)

Managing the outbreak of the new pandemic of unknown origin amid limited SOPs and protocols in terms of physical or social distancing measures, treatment and isolation procedures, and disease prevention measures was another greatest challenge in Malaysia. Consequently, many public sectors such as employment sectors, education sectors, tourism sectors, and many more faced tremendous challenges in keeping their activities on the track during the pandemic (Latif 2020). Of note, the management of COVID-19 spread must deal with the information scarcity on its transmission routes and associated prevention techniques during the earliest periods of the outbreak (Salim et al. 2020). With no concrete knowledge on the transmission routes of the novel coronavirus, these sectors have to either come up with their own SOPs in alignment with WHO's guidelines to continue their operations during the outbreak or shut down their operations until further notice was given by the government. On top of that, the WHO guidelines for risk assessment among the healthcare workers (HCWs) were adapted and treated as a base guideline for treating COVID-19 infected patients and healthcare facilities in the country (WHO 2020a).

Management and Prevention of Pseudo-information Spread on COVID-19

The spread of wrong information on COVID-19 is another challenge that needed an efficient mechanism to tackle the issue to prevent public panicking. In Malaysia, some people have been arrested for allegedly spreading false information about the COVID-19 pandemic. As of May 17, 2020, police and the Malaysian Communications and Multimedia Commission (MCMC) have opened 265 investigations on fake news and rumors associated with COVID-19 (VOA News 2020). A total of 30 people have been charged, 11 were served with a warning notice, and 18 others pleaded guilty (Zolkepli 2020).

Economic Challenges Due to Shutdown of Business Operations During COVID-19 Pandemic

COVID-19 exerted a tremendous impact on the country's economy that necessitated proper management of the impact on the public's livelihoods while concurrently managing the economic shock and its consequences (Aruna 2020). Stocks on Malaysia's stock exchange of Bursa Malaysia tumbled during the outbreak as investors sold securities due to the expected economic impact caused by the virus. With China as Malaysia's largest trading partner, the country's economy was directly impacted and economic experts have warned the prolonged virus outbreak could hit the country gross domestic product (GDP) hard. Malaysia also largely relied on tourism (Zainuddin and Shaharuddin 2020). These subsequently forced the states to shift their focus to the Southeast Asian market due to the decline of Mainland Chinese tourists. Regardless of the large losses incurred by tourism businesses, several Malaysians have voiced their concerns over the spread of the virus and urging a ban on travelers from China to the country. Due to the cancelation of the Ramadan bazaars, many Malaysian traders have used online platforms such as WhatsApp and Facebook to sell their products. In response, the government introduced a Special Fund for Tourism to help small- and medium-sized businesses affected by COVID-19. Moreover, the government launched an economic stimulus package known as the 'caring package worth RM250 billion for the welfare of the people and to support businesses, including small and medium enterprises (SMEs) (The Star 2020b). The government also announced an allocation to the Ministry of Health for the purchase of equipment and to hire contract personnel, especially nurses to help overcome the COVID-19 pandemic.

Movement Control Order (MCO) and Social Distancing Without a Proper SOP During the First Phase of MCO

The WHO had identified social distancing as a necessary action to flatten the curve and combat further transmission of the virus during the beginning of the outbreak. Among the ASEAN countries, quick actions were made by the Malaysian government in terms of MCOs within the country (Sukumaran 2020). Although MCO was implemented on March 18 in the country, specific locations were subjected to a stricter order known as Enhanced MCO (EMCO) after March 27 if a large cluster was detected within the area (Sham 2020). During this period, all businesses that supply basic needs such as supermarkets and food delivery services were subjected to strict operation hours (8 am to 8 pm) while industries and other businesses were either closed or allowed to operate with strict adherence to SOPs (Bernama 2020b). Later on, Conditional MCO (CMCO) was implemented in May 2020 as a relaxation of regulations regarding the previous MCO which was followed by Recovery MCO (RMCO) on June 2020 where a range of businesses and activities was allowed to resume operations, including eateries, cinemas, weddings, and religious gatherings (Cindy Loo 2020). Nonetheless, it should be noted that the Malaysians seem to be not prepared for the MCO without clear SOPs issues at the beginning stage of the MCO (Latif 2020). Further control orders were then instituted requiring borders to close with the suspension of all overseas travel and arrivals placed an immense burden on the locals and travelers to face difficulties in arranging their trips and basic needs while waiting for the ending of MCOs. Besides, the reliance on Malaysia's workforce in Singapore stumbled when the country closed its borders for cross-country travels that resulted in confusion and chaos at border control checkpoints.

Closure of Education Institutions and Postponement of Public Examinations

According to UNICEF monitoring, 134 countries are currently implementing nation-wide closures of education institutions, impacting about 98.5% of the world's student population (Rozaidee 2020). The COVID-19 pandemic impacted the education system of Malaysia via closure of academic institutions such as schools, colleges, and universities as well as the suspension of all extracurricular activities to contain the spread of the infections. On 15 April, the Education Minister of Malaysia announced that the Ujian Penilaian Sekolah Rendah (UPSR) and Pentaksiran Tingkatan Tiga (PT3) examinations for standard six and form three students have been canceled for 2020 in light of the COVID-19 pandemic (Mohsen 2020). He also announced that all other major school examinations including the Sijil Pelajaran Malaysia (SPM) and Sijil Tinggi Persekolahan Malaysia (STPM) would be postponed to 2021 and August 2020, respectively.

Unavailability of a Vaccine for SARS-CoV-2 and Vaccine Development Approaches

Similar to the other countries, the unavailability of a vaccine that can work against SARS-CoV-2 is one of the challenges that weakened the efficient prevention of the spread of COVID-19 in Malaysia. However, the country attempted to join the global race in creating a vaccine for the novel coronavirus that had threatened humanity worldwide. In the attempt, the Institute for Medical Research Malaysia (IMR) collaborated with the Malaysian Vaccines and Pharmaceutical Sdn Bhd (MVP) and University of Malaya Tropical Infectious Diseases Research and Education Centre (TIDREC) to begin testing existing local vaccines for infectious bronchitis viruses (avian coronaviruses) against the novel virus (NST 2020b). The trials were focused on IBV as previous studies indicated that these viruses in poultry have high genetic similarity with the human coronavirus.

Recommendations to Contain COVID-19 Pandemic in Malaysia

As a countermeasure for the COVID-19 pandemic which has severely impacted Malaysia, the following recommendations were discussed in the next sections.

Panic Management During COVID-19 Pandemic

The Malaysian government has done a good job in ensuring peace and order while tackling the COVID-19 pandemic. A survey conducted among 4,850 participants across different regions of Malaysia revealed a general willingness for participants to make behavioural changes in the face of the COVID-19 pandemic (Azlan, Hamzah, Sern, Ayub, & Mohamad 2020). The approaches of the Malaysian government in the current pandemic can be set as a guideline for future reference. In this section, the recommendation of panic management by the Malaysian government was elaborated.

Timely Updates on COVID-19 and MCO-Related Matters

The timely updates given by the ministries especially by Ministry of Health (MOH) and Ministry of Defense Malaysia (MinDef) on COVID-19 and MCO-related matters provided reassurance to the public that the government had a clear vision of handling the pandemic. MOH has also produced various infographics (available online: http://covid-19.moh.gov.my/infografik) and videos (available online: http://covid-19.moh.gov.my/video) on COVID-19-related matter for public awareness and education. Besides MOH, the MinDef has also provided the public with consistent updates on MCO-related matters. For example, MinDef provided updates on the supply of essential needs to prevent panic buying. By joining forces with the Ministry of Domestic Trade and Consumer Affairs (KPDNHEP), Ministry of Agriculture and Food Industry, and the Ministry of International Trade and Industry (MITI), the supply of food, daily essentials, and healthcare products (including surgical masks) was warranted to be sufficient across Malaysia despite the ongoing pandemic. Besides ensuring the availability of supplies, quality assurance and market prices were also well maintained via the Maximum Price Control Scheme to prevent exploitation of the supply market.

Management of Pseudo-information

The management and dissemination of authentic information to the public via social media and other platforms were essential for panic management among the public. In Malaysia, the difficulty in understanding scientific knowledge has led to a proliferation of pseudoscientific messages in social media, many of which contain half-truths and have a strong bias (Rampal 2020). Therefore, the Malaysian Communications and Multimedia Commission (MCMC) has played an active role to oversee and prevent the spread of false information online. One of the approaches was via a one-stop information hub at Sebenarnya.my Web site (URL: https://sebenarnya.my/) for information checking and to report any pseudo-information. MCMC with the collaboration of police has ensured that false information disseminated by an irresponsible individual was addressed immediately and stern actions were taken against them. In short, the initiatives from MCMC have provided a positive impact on panic management in Malaysia during the COVID-19 pandemic.

Movement Control Order (MCO) to Prevent the Outbreak of COVID-19

First and foremost, the prevention of the outspread of the virus was of utmost importance. In response to the outbreak, the government of Malaysia has announced an MCO following the Prevention and Control of Infectious Diseases Act 1988 and the Police Act 1967 beginning on March 18, 2020. The main purpose of MCO was to enforce effective social distancing to prevent the outspread of SARS-CoV-2 among the general population. Meanwhile, the MCO also serves as a strategy to identify the clusters in which the virus spread has originated or affected. Through the four phases of MCOs and close contact tracing of patients that were positive COVID-19, the chain of infection was able to be stopped and controlled over a short period as shown in Fig. 2 that depicts the distribution of COVID-19 cases in Malaysia from April to June 2020.

The effectiveness of MCO was reflected in the drastic drop in positive COVID-19 cases. During the third and fourth phases, the active COVID-19 cases had dropped by about 70% (from 3,003 to 1,311) cases as compared to the 4,314 reported cases in the first and second phases of MCO (Ministry of Health Malaysia 2020b). In short, the enforcement of MCO was one of the major contributors to flattening the COVID-19 infection curve.

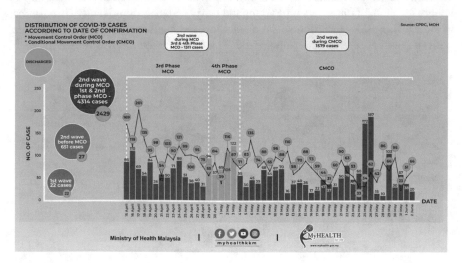

Fig. 2 Distribution of COVID-19 cases in Malaysia. *Source* Ministry of Health Malaysia (2020)

Embracing the New Norm of Social Distancing, Good Self-hygiene, and Habit of Wearing a Face Mask

Besides practicing self-isolation within the MCO period, the public communities were also advised to practice preventive measures especially those that travel out of their homes to buy groceries and to attend urgent matters. In particular, the preventive measures suggested by the Ministry of Health (MOH) Malaysia were to maintain at least 1-meter distance from one another, wash hands frequently with water and soap, and have the habit of wearing a mask when visiting crowded places (Ministry of Health Malaysia 2020c). These measures were constantly reminded through various platforms such as Facebook, Twitter, Instagram, radio stations, TV advertisements, and other platforms. Figure 3 shows an example of the preventive measures in infographic format created by MOH for dissemination to the public audience.

The social distancing of 1 m was proposed by WHO (WHO 2020b). To date, the transmission of COVID-19 was found to be mainly through virus-laden fluid particles such as droplets and aerosols formed in the respiratory tract of an infected person (Asadi et al. 2020; Bourouiba 2020; CDC 2020; Jones & Brosseau 2015). The human-to-human transmission of COVID-19 occurs when the virus-laden droplets were ejected from the mouth and nose during breathing, talking, coughing, and sneezing (Bai et al. 2020). Under many circumstances, larger droplets that were expelled from an infected person would directly affect the recipient upon close contact. Meanwhile, smaller virus-laden droplets could be easily transmitted via aerosolized droplet nuclei (Jones & Brosseau 2015). While there is still a lack of evidence, the possibility of airborne transmission of COVID-19 could not be rejected entirely. Small- and medium-sized droplets can be suspended in the ambient atmosphere for a longer time, thus prolonging the travel distance significantly (Bourouiba et al. 2014). Besides ensuring social distancing, the habit to frequently wash hands with soap and water was also emphasized. The transmission of virus infections such

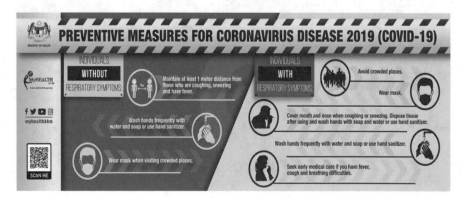

Fig. 3 Infographic on preventive measures for coronavirus diseases 2019. *Source* Ministry of Health Malaysia (2020)

as COVID-19 can occur via physical contact with droplets deposited on a surface and subsequent transfer to the recipient's respiratory system commonly via hands (Nicas and Jones 2009). The transmission via indirect contact with contaminated surfaces was highly probable as SARS-CoV-2 was found to be able to survive on many types of surfaces for hours (Van Doremalen et al. 2020). According to WHO, good hand hygiene remains the most effective method to prevent the outspread of the coronavirus (WHO 2015). Essentially, soap molecules have a polar ionic hydrophilic side and a nonpolar hydrophobic side that bonds with oils and lipids (Mittal et al. 2020). Therefore, soaps are effective to disassemble the lipid envelope of the coronavirus, thus neutralizing it (Mousavizadeh and Ghasemi 2020). The physical agitation of the hands and water further ensures that the virus is thoroughly washed away from the hand surface. On the other hand, the wearing of face mask has also become a new norm to the public as a defense against airborne pathogens. In general, face masks provide a two-way inward and outward protection. The inward protection was carried out by filtering droplets containing a virus that would then be inhaled by a non-infected person while the outward protection was carried out by trapping virus-laden droplets expelled by an infected person (van der Sande et al. 2008). In the COVID-19 pandemic, the outward protection has become an equally important matter given that the SARS-CoV-2 might be transmitted from asymptomatic patients as well as minimally symptomatic patients in the early stage of infection (Bai et al. 2020; Ye et al. 2020; Zou et al. 2020). In summary, as an implication of the measures taken, the MOH is able to flatten the curve of COVID-19 by ensuring adequate facilities to accommodate the COVID-19 patients in the country (Leong 2020). As of April 9, 2020, the COVID-19 fatality rate of 1.58% in Malaysia was among the lowest in the world (TSV 2020). At the same time, Malaysia's recovery rate of 38% was also higher than the global average of around 21.7% (TSV 2020).

Disinfection to Prevent an Outbreak of COVID-19

Disinfection is important to prevent the outbreak of disease especially in high-risk areas where COVID-19 cases are found to be prevalent. According to scientific guidelines, disinfection would be able to reduce the risk associated with the breach of host barriers, human-to-human transmission, and transmission of environmental pathogens (Centers for Disease Control and Prevention 2013). The choice of disinfectants plays a critical role in the decontamination process. Some of the identified disinfectants' active ingredients for infection prevention and control are summarized in Table 1.

According to the SOP of disinfection in Malaysia, the Malaysia Fire and Rescue Department uses eight types of chemicals in its public sanitation and disinfection operations at COVID-19 high-risk areas (Mohd Asri 2020). The choice of the chemicals, which included ethanol, sodium hypochlorite, and benzalkonium chloride, was made with advice from the MOH. As of April 30, 2020, the Malaysia Fire and Rescue Department had conducted a total of 117 public sanitation operations, which includes 10 red zones and 31 yellow zones in various states (Mohd Asri 2020). On

Table 1 Disinfectants' active ingredients and working concentrations against COVID-19

Active ingredient	Contact time (minutes)
Hydrogen peroxide (0.5%)	1
Benzalkonium chloride (0.05%)	10
Chloroxylenol (0.12%)	10
Ethyl alcohol (70%)	10
Iodine in iodophor (50 ppm)	10
Isopropanol (50%)	10
Povidone-iodine (1% iodine)	1
Sodium hypochlorite (0.05–0.5%)	5
Sodium chlorite (0.23%)	10

Source Malaysian Pharmaceutical Society 2020; National Environment Agency of Singapore 2020

a side note, there was a recent hype to introduce automated disinfection delivered in a particular confined chamber to reduce the transmission of COVID-19. Technology industries claimed these innovations to be effective in providing disinfection in high-risk areas such as hospitals, airports, train stations, bus stations, supermarkets, factories, schools, and other crowded areas. The disinfection procedure was initiated by infrared or motion sensors embedded in the device when individuals pass through the disinfection chamber (Ministry of Health Malaysia 2020a). As interesting as it seems, the WHO does not recommend spraying the external part of the body using chemicals such as alcohol or chlorine because regardless of the chemicals (disinfectants) used and the mode of delivery of disinfectants to the body surface, this disinfection method does not kill the virus inside the body (WHO 2020b). On top of that, spraying such substances can be harmful to the mucous membrane of individuals (WHO 2020b) . In short, there are mixed opinions on the use of automated disinfection chambers in high-risk areas across the globe. Hence, the use of the disinfection chamber in reducing the COVID-19 transmission may not be recommended given the lack of scientific evidence and unclear risk–benefit profile.

Challenges and Recommendations in Healthcare Infrastructure Management

During the COVID-19 pandemic, the healthcare sector was bombarded with various challenges such as the shortage in personal protective equipment (PPE) for the frontliners and the upsurge of the intensive care unit (ICU) occupants and difficulties in conducting relevant clinical research. On top of that, the lack of staff has also put immense stress on the healthcare sector. The shortcomings and recommendations related to healthcare management matters were elaborated in this section.

Infection Prevention for Healthcare Workers

According to WHO, healthcare workers providing direct care to COVID-19 patients should be equipped with PPE which includes medical masks, gowns, gloves, and eye protection with goggles or face shields to prevent the infection outbreak among the healthcare groups (WHO 2020c). Besides, for procedures that may generate aerosols such as tracheal intubation, noninvasive ventilation, tracheostomy, cardiopulmonary resuscitation, bag-mask ventilation, and bronchoscopy, the masks should be N95 or FFP2-equivalent respirators and gowns or aprons should be fluid resistant. During the early stage of the pandemic, the shortage of PPE such as medical masks and respirators occurs over the globe. As a recommendation, close attention to the logistics supply chain is necessary to ensure the continuous supply of PPE for frontliners (Qiu et al. 2020; Wonget al. 2020; Xie et al. 2020). In this matter, the coordination among the healthcare institute and another relevant ministry such as the Ministry of Domestic Trade and Consumer Affairs as well as the other industries was important to ensure the optimum operation of healthcare facilities. Under drastic measures, consideration should be given to reuse in between patients. Besides that, masks that are beyond manufacturer's expiry date (Centers for Disease Control and Prevention, 2020) and non-N95 reusable masks with high-efficiency particulate air (HEPA) filters should be considered as well (Hui et al. 2017). Besides PPE, surface decontamination was equally important to prevent infection outbreak as SARS-CoV-2 can persist on inert surfaces for up to 72 h (Setti et al. 2020). Environmental contamination of SARS-CoV-2 was detected on furniture and equipment within a patient's room and toilet (Ong et al. 2020). The proper disposal of soiled objects was also necessary as SARS-CoV-2 might be transmitted from the fecal route (Ong et al. 2020; Wang et al. 2020; Young et al. 2020). Last but not least, visits to the ICU should be restricted or banned to prevent further transmission (Gomersall et al. 2006; Liao et al. 2020). As a recommendation, videoconferencing via mobile phones or other interfaces can be used for communication between family members and patients or healthcare workers. Attention must also be given to decontaminate healthcare workers' mobile phones because they might be contaminated with viral pathogens as well (Pillet et al. 2016).

The risk stratification in one of the public hospitals known as the University Malaya Medical Centre (UMMC) guideline divided the risk into four main categories such as high, moderate, low and no risk where else the WHO guidelines stratify the HCW into two categories which are high and low risk only. The management of exposed healthcare workers in the WHO guidelines was also based on the two risk categories compared to the UMMC guidelines which provide specific management for HCW based on their stratified risk (UMMC COVID-19 Task Force 2020; WHO, 2020c). In conclusion, the WHO guidelines for risk assessment for HCW exposed to COVID-19 patients serve as a base guideline for healthcare facilities and organizations throughout the world(WHO 2020a). The UMMC guidelines have been tailored based on multiple available guidelines to ensure that the safety of the frontliner at UMMC is of utmost priority and proper risk assessment would enable the organization to improve on gaps that are still present in their efforts to prevent the transmission of COVID-19 among HCWs and directly providing a safe and healthy working environment.

Infrastructure of Intensive Care Unit (ICU)

Other patients and healthcare workers to be protected, and critically ill patients with suspected or confirmed COVID-19 should preferably be admitted to an airborne infection isolation room (AIIR) (Centers for Disease Control and Prevention 2020). Alternatively, if AIIRs are unavailable, WHO recommended that patients are placed in adequately ventilated single rooms with the doors closed (WHO 2020d). In the event where single ICU rooms are unavailable, cohorting of cases in shared rooms with beds spaced apart is recommended instead (WHO 2020d).

Preservation of Intensive Care Unit (ICU) Capacity

The hike in the number of critically ill COVID-19 patients can occur over a short period which would result in a sudden increase in demand for critical care beds(Einav et al. 2014; Grasselli, Pesenti, & Cecconi 2020; Remuzzi and Remuzzi 2020). The preservation of ICU capacity is important to ensure the provision of optimum health-care services in tackling the global pandemic. One of the crucial steps to preserve ICU capacity is via the prevention of COVID-19 outbreak among the public community (Fisher and Wilder-Smith 2020). Although difficult, it is achievable through strategic planning. For example, Malaysia proclaimed that MCO has successfully enabled Malaysia to preserve the healthcare system capacity (CodeBlue 2020). In the event where future COVID-19 cases are more than the current hospitals can accommodate, there should be contingency plans to turn other hospitals into temporary isolation and quarantine centers. On the other hand, if the ICUs become overwhelmed by patients with and without COVID-19 (Grasselli et al. 2020; Qiu et al. 2020; Remuzzi and Remuzzi 2020; Xie et al. 2020; Yang et al. 2020) critical care triage that prioritizes patients for intensive care and rations scarce resources will be required (Christian et al. 2014; Rosenbaum 2020). To reduce the strain on ICUs, elective surgeries should be postponed, and recovering ICU patients with COVID-19 should be discharged to other de-escalation wards. On a side note, telemedicine should be encouraged during the COVID-19 pandemic to reduce the inflow of patients who do not urgently require intensive care.

Staffing Issues in Intensive Care Unit (ICU)

Findings revealed that the increase in patient mortality in ICUs is associated with high ICU workload with low staffing count (Lee et al. 2017). Besides experiencing physical fatigue, healthcare workers in ICUs are also vulnerable to mental health problems because of the constant fear of being infected and the demanding workload during outbreaks like COVID-19 (Xiang et al. 2020). Therefore, it is recommended that staff from other ICUs or even non-ICU areas is relocated to provide additional technical support (Qiu et al. 2020). On top of that, other reassuring measures such as infection prevention measures; limitation of shift hours and provision of rest areas; and mental health support (Adams and Walls 2020; Xiang et al. 2020) are important to prevent over-exhaustion of staff in ICUs.

Management of Workforce: The Frontliners and Essential Services Provider

Just like elsewhere around the globe, the COVID-19 pandemic has taken Malaysia by storm. Albeit the importance of social distancing and self-isolation at home, a large workforce is needed to assist and bring control during the pandemic in Malaysia. The staffing arrangement of frontliners for medical and homeland security is elaborated in this section.

Medical Frontliners

As mentioned briefly under the challenges in healthcare infrastructure management, the lacking of trained workers imposes shortcomings in healthcare services. As a countermeasure in Malaysia, the MOH recalled retired doctors and nurses as a reinforcement to the current frontlines. Around 3,000 retired nurses have registered to come back to serve during the COVID-19 outbreak (Julia Chan 2020). The reinforcement of doctors and nurses could allow a re-coordination of manpower especially for some MOH staff that had been working nonstop since the disease outbreak. The veterans were allocated to respective departments according to their expertise, while existing nurses undergo post-basic training before being allocated to hospitals and clinics (Julia Chan 2020). As of April 23, the number of healthcare workers who contracted COVID-19 was 325 including three deaths. However, none of these healthcare workers contracted the deadly virus by attending to COVID-19 patients at healthcare facilities (Salim 2020).

Homeland Security Frontliners

Since March 18, 2020, the enforcement of MCO is carried out to curb infection outbreak in Malaysia. By Sect. 5 of the Prevention and Control of Infectious Diseases Act 1988, the Health Ministry, as the main authority in charge, has to call forth the aid from the Royal Malaysia Police to enforce the said law and regulations in responding to COVID-19 developments. Due to the escalation in COVID-19 situation, the army from Malaysian Armed Forces joined the Royal Malaysia Police beginning on March 22, 2020 (The Star 2020c).

Enhancing Logistical Support and the Supply Chain of PPE and Other Necessities During COVID-19 Pandemic

The logistical support for PPE in Malaysia and around the world was overwhelmed during the 2020 pandemic (UNICEF 2020; United Nations 2020). In facing the crisis of supply chain disruption for PPE, the Malaysian Plastics Manufacturers Association (MPMA) and the Federation of Malaysian Fashion, Textile, and Apparel (FMFTA) have contributed to addressing the PPE shortage issue (Razak 2020). Aside from the two organizations, some Bursa Malaysia listed company also stepped in by diverting their production and distribution capabilities into manufacturing medical products and equipment, such as face masks and shields, hand sanitizers and disinfectant products, as well as medical ventilators (Zainul 2020b). Besides large organizations taking lead in covering the gap of PPE shortage, there are several movements initiated at the local level to produce face masks using ready-made and recycled materials. For example, a group of residents living at Armanee Terrace Condominium in Petaling Jaya has been sewing fabric face masks using new and unused cotton fabric such as batik cloth, bedsheets, and towels to be given to nurses and hospital support staff (Jade Chan 2020). On April 27 2020, the director-general of MOH assures that the PPE supply within Malaysia is sufficient while close monitoring on the PPE supply is continued to ensure a steady supply (Zainul 2020a).

Embracing the Post-pandemic as a New Norm and the Way Forward

Malaysia will be ready to roll out its COVID-19 exit strategy once the number of daily new cases drops below the double-digit trend. The CMCO which was enforced since May 4 2020, was a soft landing approach imposed by the government before the country could enter its exit strategy for COVID-19 (Yusof and Babulal 2020). Ideally, six criteria must be met before the Malaysian government can decide to lift the MCO which is expected to end on June 9, 2020 (Povera & Naz Harun 2020). The most important of the six are border control and movement control. Insufficient border control would risk the introduction of virus from overseas. As for movement control, apart from maintaining appropriate social distance, e-learning and working from home should be encouraged at all times. The third criterion is to improve the country's healthcare system in terms of laboratory capability with a sufficient number of wards and intensive care units. The fourth criterion is to protect high-risk groups such as the elderly, the disabled, those undergoing chemotherapy, and those with comorbidities. As for the fifth, the public must practice the new normal, which is to ensure social distancing, always washing their hands, and avoiding gatherings. The last is to work together with communities to implement preventive measures. In a nutshell, the COVID-19 pandemic has utterly changed the norm of daily routine which has been practiced since a couple of decades ago. Despite the successful control over the coronavirus outbreak in Malaysia, the understanding of the subject matter remains scarce. There is still no vaccine that can assure the prevention of another outbreak that may occur soon. Until then, the public must learn to embrace the new norm as a new way of living for at least one to two years down the road.

Conclusion

In summary, this pandemic highlighted the crucial role of a multilateral partnership led by the Malaysian government in handling an outbreak of highly infectious disease. The enforcement of MCO is the main reason that the chain of infection was able to be halted and controlled over a short period. The MCO is also proven to be an effective strategy to identify the clusters in which the virus spread has originated or affected the mass population in Malaysia. During these critical times, the usage of the various media platform for dissemination of preventive measures to the public audience was found to be effective in controlling the spread of disease. The timely updates on the COVID-19 with facts and data are key factors in panic management among the public. Besides that, the strong cross-sector partnership among governmental bodies and other relevant industries was crucial while facing various challenges such as the shortage in personal protective equipment (PPE) for the frontliners and the upsurge of intensive care unit (ICU) occupants. The commitment from retirees to rejoin the frontliners for medical and homeland security helps to solve the issue with staff shortage during the pandemic. Lastly, the pandemic could not be controlled if it was not for the cooperation given by the public to the various action plans imposed by the government of Malaysia.In conclusion, it was also learned that the health care in Malaysia needs to be always prepared and updated in terms of healthcare capacity and facilities, information management and dissemination as well as operating procedures during global health emergencies to combat the spreading of infections in the country.

References

Adams, J. G., & Walls, R. M. (2020). Supporting the health care workforce during the COVID-19 global epidemic. *JAMA—Journal of the American Medical Association, 323,* 1439–1440. https:// doi.org/10.1001/jama.2020.3972.

Aljazeera. (2020). Made in Malaysia: How mosque event spread virus to SE Asia. Retrieved from https://www.aljazeera.com/news/2020/03/malaysia-mosque-event-virus-hotspot-se-asia-200318021302367.html.

Aruna, P. (2020). Coronavirus fears hit Bursa Malaysia. Retrieved from https://www.thestar.com.my/business/business-news/2020/01/29/coronavirus-fears-hit-bursa-malaysia.

Asadi, S., Bouvier, N., Wexler, A. S., & Ristenpart, W. D. (2020). The coronavirus pandemic and aerosols: Does COVID-19 transmit via expiratory particles? *Aerosol Science and Technology, 54,* 635–638. https://doi.org/10.1080/02786826.2020.1749229.

Azlan, A. A., Hamzah, M. R., Sern, T. J., Ayub, S. H., & Mohamad, E. (2020). Public knowledge, attitudes and practices towards COVID-19: A cross-sectional study in Malaysia. *PLoS ONE, 15*(5), e0233668. https://doi.org/10.1371/journal.pone.0233668.

Bai, Y., Yao, L., Wei, T., Tian, F., Jin, D. Y., Chen, L., et al. (2020). Presumed asymptomatic carrier transmission of COVID-19. *JAMA—Journal of the American Medical Association, 323,* 1406–1407. https://doi.org/10.1001/jama.2020.2565.

Barker, A. (2020). Coronavirus COVID-19 cases spiked across Asia after a mass gathering in Malaysia. This is how it caught the countries by surprise. Retrieved from https://www.abc.net.au/news/2020-03-19/coronavirus-spread-from-malaysian-event-to-multiple-countries/12066092.

Bernama. (2020a). Mercy Malaysia launches Covid-19 fund. Retrieved from https://www.thesundaily.my/local/mercy-malaysia-launches-covid-19-fund-BA2147388.

Bernama. (2020b). MCO: New business hours 8 Am-8 Pm, starting April 1—Ismail Sabri. Retrieved from https://www.bernama.com/en/general/news_covid-19.php?id=1826628.

Bourouiba, L. (2020). Turbulent gas clouds and respiratory pathogen emissions: Potential implications for reducing transmission of COVID-19. *JAMA—Journal of the American Medical Association, 323,* E1–E2. https://doi.org/10.1001/jama.2020.4756.

Bourouiba, L., Dehandschoewercker, E., & Bush, J. W. M. (2014). Violent expiratory events: On coughing and sneezing. *Journal of Fluid Mechanics, 745,* 537–563. https://doi.org/10.1017/jfm.2014.88.

CDC. (2020). Standard operating procedure (SOP) for triage of suspected COVID-19 patients in non-US healthcare settings : Early identication and prevention of transmission during triage background / purpose. Retrieved June 6, 2020, from Centres for Disease Control and Prevention website: https://www.cdc.gov/coronavirus/2019-ncov/hcp/non-us-settings/sop-triage-prevent-transmission.html.

Centers for Disease Control and Prevention. (2013). Guideline for Disinfection and Sterilization in Healthcare Facilities, 2008; Miscellaneous Inactivating Agents. CDC Website, (May), 9–13.

Centers for Disease Control and Prevention. (2020). Strategies for Optimizing the Supply of N95 Respirators: COVID-19 I CDC. Retrieved June 6, 2020, from https://www.cdc.gov/coronavirus/2019-ncov/hcp/respirators-strategy/index.html.

Chan, J. (2020). Condo residents turn face mask makers I The Star. Retrieved June 5, 2020, from The Star website: https://www.thestar.com.my/metro/metro-news/2020/05/08/condo-residents-turn-face-mask-makers.

Chan, J. F. W., Kok, K. H., Zhu, Z., Chu, H., To, K. K. W., Yuan, S., et al. (2020). Genomic characterization of the 2019 novel human-pathogenic coronavirus isolated from a patient with atypical pneumonia after visiting Wuhan. *Emerging Microbes and Infections, 9*(1), 221–236.

Chan, J. (2020). Covid-19: 3,000 retired nurses to return to service I Malaysia I. Retrieved June 5, 2020, from Malay Mail website: https://www.malaymail.com/news/malaysia/2020/03/25/covid-19-3000-retired-nurses-to-return-to-service/1850174.

Christian, M. D., Sprung, C. L., King, M. A., Dichter, J. R., Kissoon, N., Devereaux, A. V., et al. (2014). Triage: Care of the critically ill and injured during pandemics and disasters: CHEST consensus statement. *Chest, 146*(4 Suppl), e61S–e74S. https://doi.org/10.1378/chest.14-0736.

Cindi, L. (2020). CMCO ends June 9, Recovery MCO from June 10 to Aug 31 (Updated). The Sun Daily. Retrieved https://web.archive.org/web/20200607110612/https://www.thesundaily.my/local/cmco-ends-june-9-recovery-mco-from-june-10-to-aug-31-updated-EM2538754.

Code Blue. (2020). MOH: Covid-19 Can't Be Wiped Out, Only Contained I CodeBlue. Retrieved June 5, 2020, from https://codeblue.galencentre.org/2020/04/27/moh-covid-19-cant-be-wiped-out-only-contained/.

COVID-19 Outbreak Live Updates. Retrieved from https://www.outbreak.my/#states on July 3, 2020.

Einav, S., Hick, J. L., Hanfling, D., Erstad, B. L., Toner, E. S., Branson, R. D., et al. (2014). Surge capacity logistics: Care of the critically ill and injured during pandemics and disasters: Chest consensus statement. *Chest, 146*(4 Suppl), e17S–e43S. https://doi.org/10.1378/chest.14-0734.

Fisher, D., & Wilder-Smith, A. (2020). April 4). The global community needs to swiftly ramp up the response to contain COVID-19. *The Lancet, 395,* 1109–1110. https://doi.org/10.1016/S0140-6736(20)30679-6.

Gomersall, C. D., Tai, D. Y. H., Loo, S., Derrick, J. L., Goh, M. S., Buckley, T. A., et al. (2006). Expanding ICU facilities in an epidemic: Recommendations based on experience from the SARS epidemic in Hong Kong and Singapore. *Intensive Care Medicine, 32*(7), 1004–1013. https://doi.org/10.1007/s00134-006-0134-5.

Grasselli, G., Pesenti, A., & Cecconi, M. (2020). Critical care utilization for the COVID-19 outbreak in Lombardy, Italy: Early experience and forecast during an emergency response. *JAMA—Journal of the American Medical Association, 323,* 1545–1546. https://doi.org/10.1001/jama.2020.4031.

Guan, W. J., Ni, Z. Y., Hu, Y., Liang, W. H., Ou, C. Q., He, J. X., et al. (2020). Clinical characteristics of coronavirus disease 2019 in China. *New England Journal of Medicine, 382*(18), 1708–1720.

Hui, C. Y. T., Leung, C. C. H., & Gomersall, C. D. (2017). Performance of a novel non-fit-tested HEPA filtering face mask. *Infection Control and Hospital Epidemiology, 38,* 1260–1261. https://doi.org/10.1017/ice.2017.105.

Jones, R. M., & Brosseau, L. M. (2015). Aerosol transmission of infectious disease. *Journal of Occupational and Environmental Medicine, 57*(5), 501–508. https://doi.org/10.1097/JOM.0000000000000448.

Lam, T. T. Y., Jia, N., Zhang, Y. W., Shum, M. H. H., Jiang, J. F., Zhu, H. C., et al. (2020). Identifying SARS-CoV-2-related coronaviruses in Malayan pangolins. *Nature*, pp. 1–4.

Latif, R. (2020). Malaysia defends easing of coronavirus curbs as new infections jump. Retrieved https://www.reuters.com/article/us-health-coronavirus-malaysia/malaysia-defends-easing-of-coronavirus-curbs-as-new-infections-jump-idUSKBN22E09R.

Lee, A., Cheung, Y. S. L., Joynt, G. M., Leung, C. C. H., Wong, W. T., & Gomersall, C. D. (2017). Are high nurse workload/staffing ratios associated with decreased survival in critically ill patients? A cohort study. *Annals of Intensive Care, 7*(1). https://doi.org/10.1186/s13613-017-0269-2.

Leong, I. (2020). MCO: Malaysia succeeds in flattening curve of COVID-19—Dr Noor Hisham. Retrieved June 4, 2020, from Astro Awani website: http://english.astroawani.com/malaysia-news/mco-malaysia-succeeds-flattening-curve-covid-19-dr-noor-hisham-240616.

Liao, X., Wang, B., & Kang, Y. (2020). Novel coronavirus infection during the 2019–2020 epidemic: Preparing intensive care units—the experience in Sichuan Province, China. *Intensive Care Medicine, 46*(2), 357–360. https://doi.org/10.1007/s00134-020-05954-2.

Mahase, E. (2020). China coronavirus: WHO declares international emergency as death toll exceeds 200. *BMJ: British Medical Journal (Online), 368.*

Malaysiakini. (2020). All states gazetted as Covid-19-hit area, ban on inter-state travelling. Retrieved from https://www.malaysiakini.com/news/515211.

Malaysian Pharmaceutical Society. (2020). Covid -19 pandemic guidance document version 2. Retrieved from https://t.me/cprckkm.

Ministry of Health Malaysia. (2020a). Disinfection box/ chamber/ tunnel /booth / partition/ gate on the transmission of COVID-19.

Ministry of Health Malaysia. (2020b). Home | COVID-19 MALAYSIA. Retrieved May 8, 2020, from http://covid-19.moh.gov.my/.

Ministry of Health Malaysia. (2020c). Infografik | COVID-19 MALAYSIA. Retrieved June 4, 2020, from http://covid-19.moh.gov.my/infografik.

Mittal, R., Ni, R., & Seo, J.-H. (2020). The Flow Physics of COVID-19. *Journal of Fluid Mechanics, 894,* F2. https://doi.org/10.1017/jfm.2020.330.

Mohd Asri, M. A. (2020). BERNAMA—COVID-19 : Eight types of chemicals used in public sanitation, disinfection ops by Fire and Rescue Dept - DG. Retrieved June 4, 2020, from https://www.bernama.com/en/general/news_covid-19.php?id=1838197.

Mohsen, A. (2020). Covid-19: UPSR, PT3 cancelled, SPM, STPM postponed. Retrieved from https://www.thesundaily.my/home/upsr-pt3-cancelled-spm-stpm-postponed-XD2261473.

Mousavizadeh, L., & Ghasemi, S. (2020, March 31). Genotype and phenotype of COVID-19: Their roles in pathogenesis. Journal of Microbiology, Immunology and Infection. https://doi.org/10.1016/j.jmii.2020.03.022.

National Environment Agency of Singapore. (2020). Interim list of household products and active ingredients for disinfection of the COVID-19 Virus. 19, pp. 1–7. Retrieved from https://www.nea.gov.sg/our-services/public-cleanliness/environmental-cleaning-guidelines/guidelines/interim-list-of-household-products-and-active-ingredients-for-disinfection-of-covid-19.

NST (News Straits Times). (2020a). [Breaking] 3 coronavirus cases confirmed in Johor Baru. Retrieved from https://www.nst.com.my/news/nation/2020/01/559563/breaking-3-coronavirus-cases-confirmed-johor-baru.

NST (2020b). IMR begins testing to develop vaccines for Covid-19. Retrieved from https://www.nst.com.my/news/nation/2020/03/578079/imr-begins-testing-develop-vaccines-covid-19.

Nicas, M., & Jones, R. M. (2009). Relative contributions of four exposure pathways to influenza infection risk. Risk Analysis, 29(9), 1292–1303. https://doi.org/10.1111/j.1539-6924.2009.01253.x.

Ong, S. W. X., Tan, Y. K., Chia, P. Y., Lee, T. H., Ng, O. T., Wong, M. S. Y., et al. (2020). Air, surface environmental, and personal protective equipment contamination by severe acute respiratory syndrome coronavirus 2 (SARS-CoV-2) from a symptomatic patient. JAMA—Journal of the American Medical Association, 323, 1610–1612. https://doi.org/10.1001/jama.2020.3227.

Pillet, S., Berthelot, P., Gagneux-Brunon, A., Mory, O., Gay, C., Viallon, A., et al. (2016). Contamination of healthcare workers' mobile phones by epidemic viruses. Clinical Microbiology & Infection, 22(5), 456.e1–456.e6. https://doi.org/10.1016/j.cmi.2015.12.008.

Povera, A., & Naz Harun, H. (2020). Dr Noor Hisham: Six criteria must be met before MCO is lifted. Retrieved June 5, 2020, from New Straits Times website: https://www.nst.com.my/news/nation/2020/04/586369/dr-noor-hisham-six-criteria-must-be-met-mco-lifted.

Qiu, H., Tong, Z., Ma, P., Hu, M., Peng, Z., Wu, W., et al. (2020). Intensive care during the coronavirus epidemic. Intensive Care Medicine, 46, 576–578. https://doi.org/10.1007/s00134-020-05966-y.

Rampal, S. (2020). Covid-19: So much information, what do we believe? I The Star. Retrieved June 5, 2020, from The Star website: https://www.thestar.com.my/lifestyle/living/2020/03/17/covid-19-so-much-information-what-do-we-believe.

Razak, A. (2020). Covid-19: MPMA, FMFTA gear up to address PPE shortage. New Straits Times. Retrieved from https://www.nst.com.my/news/nation/2020/04/586005/covid-19-mpma-fmfta-gear-address-ppe-shortage.

Remuzzi, A., & Remuzzi, G. (2020). April 11). COVID-19 and Italy: What next? The Lancet, 395, 1225–1228. https://doi.org/10.1016/S0140-6736(20)30627-9.

Rosenbaum, L. (2020). Facing covid-19 in Italy—Ethics, logistics, and therapeutics on the epidemic's front line. The New England Journal of Medicine, 382(20), 1873–1875. https://doi.org/10.1056/NEJMp2005492.

Rozaidee, A. (2020). MOE: All sports and co-curricular activities to be postponed due to coVID-19". Retrieved from https://says.com/my/news/moe-all-sports-and-co-curricular-activities-in-march-to-be-postponed-due-to-covid-19.

Salim, S. (2020). COVID-19: Malaysia healthcare worker infections more than doubled to 325 in three weeks. Theedgemarkets. Retrieved from https://www.theedgemarkets.com/article/covid19-malaysia-healthcare-worker-infections-more-doubled-325-three-weeks.

Salim, N., Chan, W. H., Mansor, S., Bazin, N. E. N., Amaran, S., Faudzi, A. A. M., et al. (2020). COVID-19 epidemic in Malaysia: Impact of lock-down on infection dynamics. medRxiv.

Sham, N. S. (2020). COVID-19: PKPD dikuat kuasa di dua kawasan di Simpang Renggam. Retrieved from http://www.astroawani.com/berita-malaysia/covid-19-pkpd-dikuat-kuasa-di-dua-kawasan-di-simpang-renggam-235454.

Setti, L., Passarini, F., De Gennaro, G., Barbieri, P., Perrone, M. G., Borelli, M., et al. (2020, April 23). Airborne transmission route of covid-19: Why 2 meters/6 feet of inter-personal distance could not be enough. *International Journal of Environmental Research and Public Health, 17*, 2932. https://doi.org/10.3390/ijerph17082932.

Steven, C. K. W. (2020). The facts behind the shortage of face masks. Daily Express. Retrieved from http://www.dailyexpress.com.my/read/3629/the-facts-behind-the-shortage-of-face-masks/.

Sukumaran, T. (2020 March 16). Coronavirus: Malaysia in partial lockdown from March 18 to limit outbreak". South China Morning Post. Retrieved fromhttps://www.scmp.com/week-asia/health-environment/article/3075456/coronavirus-malaysias-prime-minister-muhyiddin-yassin.

The Star. (2020a). Covid-19: Need more PPE, current supply can only last for two weeks, says Health DG. Retrieved from https://web.archive.org/web/20200421142611/https://www.thestar.com.my/news/nation/2020/04/13/covid-19-need-ppe-donations-current-supply-can-only-last-two-weeks-says-health-dg.

The Star. (2020b). Muhyiddin unveils RM10bil stimulus package for SMEs (updated). Retrieved from https://www.thestar.com.my/news/nation/2020/04/06/muhyiddin-unveils-rm10bil-special-stimulus-package.

The Star. (2020c). Covid-19: Mindef and Armed Forces FAQ | The Star. Retrieved June 5, 2020, from https://www.thestar.com.my/news/nation/2020/03/22/covid-19-mindef-and-armed-forces-faq.

TST (The Strait Times). (2020). Coronavirus: Malaysia records eight deaths; 153 new cases bring total to 1,183. Retrieved from https://www.straitstimes.com/asia/se-asia/malaysia-records-fourth-coronavirus-death.

TSV, T. S. T. (2020). Covid-19: Malaysia's death rate among the lowest in the world. Retrieved June 4, 2020, from https://www.thestartv.com/v/covid-19-malaysia-s-death-rate-among-the-lowest-in-the-world.

UNICEF. (2020). COVID-19 impact assessment and outlook on personal protective equipment | UNICEF Supply Division. Retrieved June 5, 2020, from https://www.unicef.org/supply/stories/covid-19-impact-assessment-and-outlook-personal-protective-equipment.

United Nations. (2020). Supply Chain and COVID-19: UN rushes to move vital equipment to frontlines | United Nations.

UMMC COVID 19 Task Force. (2020). Guidelines for surveillance of healthcare workers.

van der Sande, M., Teunis, P., & Sabel, R. (2008). Professional and home-made face masks reduce exposure to respiratory infections among the general population. *PLoS One, 3*(7). https://doi.org/10.1371/journal.pone.0002618.

Van Doremalen, N., Bushmaker, T., Morris, D. H., Holbrook, M. G., Gamble, A., Williamson, B. N., et al. (2020). Aerosol and surface stability of SARS-CoV-2 as compared with SARS-CoV-1. *New England Journal of Medicine, 382*, 1564–1567. https://doi.org/10.1056/NEJMc2004973.

VOA News. (2020). Malaysia Arrests Thousands Amid Coronavirus Lockdown. Retrieved from https://www.voanews.com/science-health/coronavirus-outbreak/malaysia-arrests-thousands-amid-coronavirus-lockdown.

Wang, W., Xu, Y., Gao, R., Lu, R., Han, K., Wu, G., et al. (2020). Detection of SARS-CoV-2 in different types of clinical specimens. *JAMA—Journal of the American Medical Association, 323*, 1843–1844. https://doi.org/10.1001/jama.2020.3786.

WHO. (2015). The evidence for clean hands. WHO. Retrieved from http://www.who.int/gpsc/country_work/en/.

WHO. (2020a). Risk assessment and management of exposure of health care workers in the context of COVID-19: Interim guidance, 19 March 2020.

WHO. (2020b). Coronavirus disease (COVID-19) advice for the public. Retrieved June 4, 2020, from Coronavirus disease 2019 website: https://www.who.int/emergencies/diseases/novel-coronavirus-2019/advice-for-public.

WHO. (2020c). Rational use of personal protective equipment for coronavirus disease 2019 (COVID-19). WHO, 2019 (February), 1–7. Retrieved from https://www.who.int/csr/resources/publications/putontakeoff.

WHO. (2020d). Infection prevention and control during health care when COVID-19 is suspected. WHO, (i), 1–5. Retrieved from https://www.who.int/publications/i/item/10665-331495.

Wong, J. E. L., Leo, Y. S., & Tan, C. C. (2020). COVID-19 in Singapore—Current experience: Critical global issues that require attention and action. *JAMA—Journal of the American Medical Association, 323*, 1243–1244. https://doi.org/10.1001/jama.2020.2467.

Wu, Z., & McGoogan, J. M. (2020). Characteristics of and important lessons from the coronavirus disease 2019 (COVID-19) outbreak in China: Summary of a report of 72 314 cases from the Chinese Center for Disease Control and Prevention. *JAMA, 323*(13), 1239–1242.

Xiang, Y. T., Yang, Y., Li, W., Zhang, L., Zhang, Q., Cheung, T., et al. (2020). Timely mental health care for the 2019 novel coronavirus outbreak is urgently needed. *The Lancet Psychiatry, 7*, 228–229. https://doi.org/10.1016/S2215-0366(20)30046-8.

Xie, J., Tong, Z., Guan, X., Du, B., Qiu, H., & Slutsky, A. S. (2020). Critical care crisis and some recommendations during the COVID-19 epidemic in China. *Intensive Care Medicine, 46*(5), 837–840. https://doi.org/10.1007/s00134-020-05979-7.

Xie, M., & Chen, Q. (2020). Insight into 2019 novel coronavirus—an updated intrim review and lessons from SARS-CoV and MERS-CoV. *International Journal of Infectious Diseases, 94*, 119–124.

Yang, X., Yu, Y., Xu, J., Shu, H., Xia, J., Liu, H., et al. (2020a). Clinical course and outcomes of critically ill patients with SARS-CoV-2 pneumonia in Wuhan, China: A single-centered, retrospective, observational study. *The Lancet Respiratory Medicine, 8*(5), 475–481. https://doi.org/10.1016/S2213-2600(20)30079-5.

Yang, Y., Lu, Q., Liu, M., Wang, Y., Zhang, A., Jalali, N., et al. (2020). Epidemiological and clinical features of the 2019 novel coronavirus outbreak in China. MedRxiv.

Ye, F., Xu, S., Rong, Z., Xu, R., Liu, X., Deng, P., et al. (2020). Delivery of infection from asymptomatic carriers of COVID-19 in a familial cluster. *International Journal of Infectious Diseases, 94*, 133–138. https://doi.org/10.1016/j.ijid.2020.03.042.

Young, B. E., Ong, S. W. X., Kalimuddin, S., Low, J. G., Tan, S. Y., Loh, J., et al. (2020). Epidemiologic features and clinical course of patients infected with SARS-CoV-2 in Singapore. *JAMA—Journal of the American Medical Association, 323*(15), 1488–1494. https://doi.org/10.1001/jama.2020.3204.

Yusof, T. A., & Babulal, V. (2020). Exit strategy after Covid-19 cases dropped below double-digit figure. Retrieved June 5, 2020, from New Straits Times website: https://www.nst.com.my/news/nation/2020/06/597402/exit-strategy-after-covid-19-cases-dropped-below-double-digit-figure.

Zainuddin, A., & Shaharuddin, H. (2020). Prolonged Covid-19 may hit Malaysia's GDP hard. The Malaysian Reserve. Retrieved fromhttps://themalaysianreserve.com/2020/02/20/prolonged-covid-19-may-hit-malaysias-gdp-hard/.

Zainul, E. (2020a). Health ministry no longer facing PPE shortage | The Edge Markets. Retrieved June 5, 2020, from The Edge Markets website: https://www.theedgemarkets.com/article/health-ministry-no-longer-facing-ppe-shortage.

Zainul, E. (2020b). These Malaysian companies jump onto the pandemic bandwagon, how many will make it? | The Edge Markets. Retrieved June 5, 2020, from The Edge Markets website: https://www.theedgemarkets.com/article/these-malaysian-companies-jump-pandemic-bandwagon-how-many-will-make-it.

Zolkepli, F. (2020). Covid-19: Over 200 fake news cases have been recorded so far, says Ismail Sabri. Retrieved from https://www.thestar.com.my/news/nation/2020/05/17/covid-19-over-200-fake-news-cases-have-been-recorded-so-far-says-ismail-sabri.

Zou, L., Ruan, F., Huang, M., Liang, L., Huang, H., Hong, Z., et al. (2020). SARS-CoV-2 viral load in upper respiratory specimens of infected patients. *New England Journal of Medicine, 382*, 1177–1179. https://doi.org/10.1056/NEJMc2001737.

COVID-19 in Indonesia: Geo-Ecology and People's Behaviour

Budi Haryanto, Triarko Nurlambang, and Silvia R. Dewi

Abstract Indonesia has been struggling to response the COVID-19 since the first reported COVID-19 cases in Indonesia on 2 March 2020. It is about three months after intensively scary news of pandemic occurrence among neighbour countries. All efforts had been conducted to prevent and control the spread of disease, included how to implement effectively the message of using mask, social distancing, stay at home, frequently washing hands (health protocol for COVID-19) by local wisdom, in regards that Indonesia has more than 30 races and ethnicities and spread out in more than 3000 inhabited islands. People's behaviour, based on its local wisdom, are dominantly influence the people's response to the spread of COVID-19. In some provinces, the people's behaviour influences the spread of COVID-19 effectively. However, in some other provinces, it may not influence the increasing incidence of the disease. Thus, the significantly decreasing of the COVID-19 incidence is still under expectation in the near future. Many efforts had been conducted by the Government of Indonesia, from central command in the beginning to delegate autonomy of command to province and city or district. Understanding the geo-ecology and characteristics of population dynamic are important to describe the mechanism of the spread of COVID-19 for more focus in control interventions.

Keywords Covid-19 · Indonesia · People's behaviour · Local wisdom · Geo-ecology

B. Haryanto (✉) · S. R. Dewi
Department of Environmental Health, Faculty of Public Health, Universitas Indonesia, Depok, Indonesia
e-mail: bharyant@cbn.net.id

B. Haryanto · T. Nurlambang
Research Center for Climate Change, I-SER, Universitas Indonesia, Depok, Indonesia

T. Nurlambang
Department of Geography, Faculty of Mathematics and Basic Sciences, Universitas Indonesia, Depok, Indonesia

© The Author(s), under exclusive license to Springer Nature Switzerland AG 2021
R. Akhtar (ed.), *Coronavirus (COVID-19) Outbreaks, Environment and Human Behaviour*, https://doi.org/10.1007/978-3-030-68120-3_7

Introduction

COVID-19 cases have been developing rapidly in China since January 2020. Although any COVID-19 cases have not been found in Indonesia during that time, Indonesian societies seemed to start taking more consideration in travelling, especially in terms of going to other countries. This action developed as the Indonesian Government quickly issued an order of flights restriction to China and any other countries (BNPB, Universitas Indonesia 2020a). The Central Statistics Agency of West Java recorded the number of international air passengers decreased 74.40% from January to August 2020 (PemerintahProvinsiJawa Barat 2020). Furthermore, the mobilization of inter-city decreased due to societal behaviour and anxiousness of being exposed to COVID-19. From January to August 2020, the total number of trains in West Java only reached up to 6.64 millions passengers (39.13% decrease compared to the same period last year) (*PemerintahProvinsiJawa Barat* 2020). Seeing the decrease in mobilization and low demand in tourism sectors, in the early cases of COVID-19, Indonesian Government misprioritized the taken actions. Although Indonesian Government has stricken international flights, Indonesian Government is said to fail in pressing possible transmissions of COVID-19 as the Government rather promoted domestic flights through a program of "*Ayo Liburan*" (Let's Go Vacation). Through this program, the Government provided incentives on airfare and lodging in tourist destination areas (BNPB, Universitas Indonesia 2020a).

Indonesia confirmed the first COVID-19 cases on 2 March 2020. When the first cases appeared in Depok, West Java, the President, and Ministry of Health Republic of Indonesia urged the societies not to panic in facing COVID-19 as the Country is ready in tackling the virus (KementerianKomunikasidanInformatika 2020). As a result of these first cases, the phenomenon of panic buying showed up among the societies. Panic buying is a situation in which many people suddenly buy as much food, fuel, etc., as they can because they are worried about something bad that may happen (Cambridge Dictionary n.d). Panic buying appears as the consumers perceive the scarcity of goods (Shadiqi et al. 2020). The feeling of insecurity which increased as supply chain interruption is frequently observed during a disaster proceeds a behaviour of coping the feeling of insecurity through saving basic needs by purchasing as many as possible (Shou et al. 2013; Arafat et al. 2020). Further, in facing the uncertain situation, panic buying is a result of societies' willingness to feel safe and continue living normally despite the instability of situations (Bacon and Corr 2020; Arafat et al. 2020). This behaviour can be mitigated by providing accurate and comprehensive information related to current accidents (Shadiqi et al. 2020). Social media also contribute in intensifying the insecurity and stress felt by the societies related to COVID-19 (Garfin et al. 2020). With the strong social interaction, Indonesian societies perform especially between neighbourhoods, and societies tend to adopt a worrying opinion rather than a calming opinion (Shadiqi et al. 2020). The COVID-19 pandemic has spurred many analyses from various scientific disciplines.

These include the fundamental analysis from microbiology, epidemiology, mathematical modelling from various disciplines, as well as from economy, sociology, and public policy. However, a geographic perspective on the pandemic is still lacking.

The question of how a pandemic spreads and intensifies in a particular area is inherently geographic one, and it can be understood through behavioural geography and health geography. In practical terms, a geographic analysis is concerned with three principal elements: point, distance, and area. All three elements form a spatial pattern or organization to a given phenomenon, including the spread of a disease such as COVID-19. The movements of people which determine the dynamics of COVID-19 contagion can be explained in terms of how it occupies a specific temporal and spatial dimension. The temporal dimension of a spatial behaviour can be examined using different temporal units. This includes how events occur in the spaces of a day, be it by dividing days into periods of morning-noon and evenings, or by its hours. Weekly analysis is also a possibility and especially useful for examining differences between routines in weekdays and weekends. A longer yearly analysis divided into months, seasons, or even multiple years is also possible. The differences between these time dimensions are described in the frequency of movement from a certain origin to destination.

As the first cases of COVID-19 were found in Depok, Indonesia, societies' responses towards the first cases were fast, especially in Jabodetabek (Jakarta, Bogor, Depok, Tangerang, Bekasi) area. In the first weeks, panic buying happened not only in Depok, but also in Jakarta, Bogor, Tangerang, and Bekasi considering the high mobility of the societies to travel among these cities. Masks, hand sanitizers, vitamins, and supplements are the primary targets of the buyers. Besides, rice, oil, and various instant foods are also hunted by the buyers in each supermarket. With the high demand from the societies, the increase in price is seen in several items. Several well-known brands of hand sanitizers have increased by 81.12% on several e-commerce platforms in Indonesia including *Shopee, Bukalapak,* and *Tokopedia* (Machmud and Minghat 2020). The excessive buying and behaviour of piling up have reduced accessibility for several people. For instance, a study among Bali societies revealed that 86.78% of the respondents have difficulty getting masks and hand sanitizers (Yuniti et al. 2020). The difficulty in finding masks and hand sanitizers is supported by the price of these necessities which undergo an increase in price up to 100% in retail shops (Matompo 2020).

COVID-19 first cases have naturally changed societies' behaviour to pay more attention to their health with various preventive measures each individual perceived. A study by Yuniti et al. (2020) stated that 83.06% respondents of Bali societies eat healthy food in maintaining health to prevent from COVID-19. Fuhrer, 42.56% respondents also consume multivitamins in increasing their immune system. With incomplete information spread among the societies, another alternative believed to be useful in becoming a potential cure for COVID-19 is herbs/herbal medicine (*Jamu*). *Jamu* is claimed to be capable of dealing with COVID-19 mainly in maintaining and increasing the human immune system (Hartanti et al. 2020). From this information, Indonesian societies started seeking herbs for daily consumption in tackling COVID-19 even before COVID-19 entered Indonesia. With the high demand of *Jamu* as well as high interest from the societies in implementing their knowledge to make an efficacious herbal medicine, Indonesia National Agency of Drug and Food Control as an agent supervising the quality of herbal medicine spread across Indonesia has received a high number of traditional medicine licensing from January to July 2020. The licensing is given to 178 traditional medicine and 149 local health supplements with efficacy on maintaining immune systems (BadanPengawa-sObatdanMakananRepublik Indonesia 2020). Besides herbal medicine, sunbathing is believed to boost the immune system in preventing COVID-19. However, incomplete information regarding sunbathing has led to various misinformation. With the news spreading around the societies and lack of double-checking have directed some societies to sunbathe around 10 am–4 pm for 15 min, the time of UV rays may have harmful effects on the skin, eye, and immune system (WHO 2003) .

COVID-19 Cases in Indonesia

After the first cases of COVID-19 in Depok, West Java, the cases of COVID-19 keep going up significantly. With the total cases of 1,528 and 114 new daily cases on 31 March 2020, Indonesian Government issued government regulation 21/2000 regarding large-scale social restrictions/LSSR to accelerate handling COVID-19. The implementation of the issued law varies in each province. For instance, DKI Jakarta started implementing LSSR on 10 April 2020, whereas West Java started implementing LSSR on 6 May 2020. Similar to the law issuance, each region in Indonesia has their own characteristics in handling COVID-19 and in implementing health protocols according to how local societies perceive the accidents of COVID-19 and interfere with local culture. The comparison of COVID-19 cases increase is shown in Fig. 1.

Based on the graphic above, DKI Jakarta, East Java, and South Sulawesi are the top three provinces with high increase of COVID-19 cases. In East Java, there were 52 clusters found on 10 May 2020 (BNPB, Universitas Indonesia 2020b, p. 146). The biggest and first cluster was a hajj health workers training centre who came from 38 districts across East Java. The second cluster was a boarding school, and the third cluster was an industry. The high increase of COVID-19 cases in East Java is associated with the low awareness of the societies to implement health protocols. Further, no strict sanctions to those who did not implement health protocols have contributed to the uncontrolled transmission.

Unlike DKI Jakarta, East Java, and South Sulawesi, Tegal City, a city in Central Java, succeeds in handling COVID-19 as it has the lowest number of COVID-19 cases (BNPB, Universitas Indonesia 2020b, p. 109) (Fig. 2).

When a society of Tegal City was stated positive of COVID-19, territorial quarantine became a key success of handling COVID-19 (BNPB, Universitas Indonesia 2020b, p. 116). From that incident, the Mayor of Tegal City closed access to Tegal using movable concrete barriers. Seeing the success that Tegal City experienced in handling the first case, the Government of Tegal City proposed an implementation of

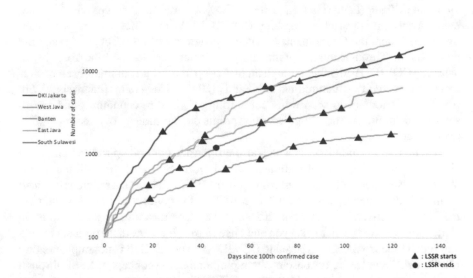

Fig. 1 Comparison of COVID-19 increase in five provinces in Indonesia

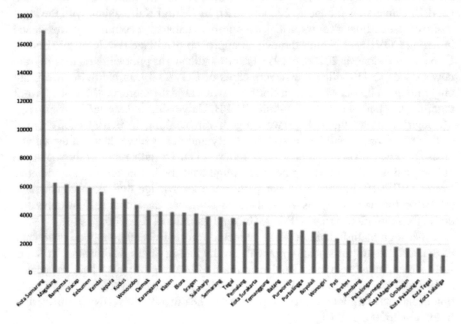

Fig. 2 Spread of COVID-19 cases in central java (Pemerintah Provinsi Jawa Tengah 2020)

LSSR to the Ministry of Health. As of 20 June 2020, only two active cases of COVID-19 found in Tegal (Pemerintah Kota Tegal 2020). Central Java, as a province, implements a cultural approach in handling COVID-19. With a thick culture performed by the societies, the Government initiated a program of "Jogo Tonggo" (keep the neighbours). Through this program, the government encouraged the societies in handling COVID-19 together through mutual cooperation and community empowerment systematically, structured and thoroughly (BNPB, Universitas Indonesia 2020b, p.121). With four main scopes of "Joko Tonggo" (health, economics and social, safety, and entertainment), "Jogo Tonggo" points out the strategy of grass root rather than top-down strategy.

Several regions in Indonesia which also emphasize cultural approach in tackling COVID-19 are Bali, Central Kalimantan, and North Sulawesi. In Bali, the use of masks is associated with a concept coming from Hindu Religion, a religion practiced by most Bali societies, called "Tat Twam Asi" or the concept of I am you, you are me (BNPB, Universitas Indonesia 2020b, p. 12). The meaning implied within the context of mask usage that when you and I use masks, it means that I protect you and you protect me from being exposed to COVID-19. Though Bali is not implementing LSSR, Bali societies are unconsciously implementing the concept of LSSR through spiritual approach. Further, when cases of COVID-19 were found in a traditional market in Bali, the merchants obeyed the orders not to sell in the market for a while (BNPB, Universitas Indonesia 2020b, p. 12). In Central Kalimantan, with the local motto of "Isen Mulang" or persistent, a program of enhancing community's resilience in facing COVID-19 is built through an approach of village development (BNPB, Universitas Indonesia 2020b, p. 191). From 14 villages as pilot villages, the program now extends to 119 villages with zero cases of COVID-19. In North Sulawesi, not only through religion, societies in North Sulawesi hold the motto of "Mapalus" (work together) as a part of their local culture (BNPB, Universitas Indonesia 2020b, p. 410). Through this motto, the societies work together in breaking the transmission chain of COVID-19 through building posts voluntarily and doing surveillance in the village posts (BNPB, Universitas Indonesia 2020b, p. 411). The adherence of the society to public figures and religious figures has a huge contribution in keeping the societies from COVID-19. Further, with comprehensive knowledge that the societies have related to the danger of this virus, the societies have their own self-awareness in doing the regulations issued (BNPB, Universitas Indonesia 2020b, p. 414).

Other regions in Indonesia which reflected efforts in keeping a good obedience of the societies towards health protocols are including South Kalimantan. In South Kalimantan, in order to make implementation of health protocols a habit especially in the prone areas, the local task force of COVID-19 initiated a program of two traditional markets as a pilot project of implementing health protocols (BNPB, Universitas Indonesia 2020b, p. 184).

The obedience level of the societies is reflected from the number of COVID-19 cases in the areas where the number of COVID-19 keeps going up. For example, Surabaya, a city in East Java, as a region where people are still lacking in its awareness and weak law implementation has COVID-19 cases of 15,845 compared to

Tegal City with cases of 453 in 30 October 2020. The transmission of COVID-19 is associated with how societies perform recommended prevention actions. In performing a healthy behaviour, one's behaviour is influenced by internal factors such as knowledge, perception, emotions, and motivation. According to a research done by Yanti et al. (2020), societies who have a good knowledge tend to perform a good behaviour (93.3%) in preventing COVID-19 transmission.Not only internal factors, one's behaviour is also influenced by external factors such as physical and non-physical environment (Yanti et al. 2020). Economic conditions, as a non-physical environment, also contribute to the success of COVID-19 programs. The implementation of how economics affects the COVID-19 transmission can be seen in East Kalimantan. In East Kalimantan, lower class societies tend to violate the restrictions during COVID-19 due to their urgency to make money in fulfilling their basic needs such as eating (BNPB, Universitas Indonesia 2020a, p. 237).

The high number of COVID-19 in Indonesia is caused by various reasons. The response of the societies in this COVID-19 pandemic takes a big role in the spreading of COVID-19. When the policy does not accommodate what societies want, although it is for the sake of common good, some societies tend to ignore the policy. For example, in preventing the spread of COVID-19, as LSSR coincided with Eid Al-Fitr, the Indonesian Government anticipated the mobility of the societies through restriction of going to other cities declared in the regulation of minister of transportation number 25/2020. Even though the regulation has not been issued, the societies have gotten prior information. The prior information was used by the societies to start travelling to other cities before the issuance of the regulation. The mobilization of the societies during IdulFitr is associated with effective communication built between the Government and its societies even before COVID-19 cases were found. In West Sulawesi, for instance, self-awareness has been built for 90% due to the early communication system that the Government has provided since February 2020 (BNPB, Universitas Indonesia 2020c p. 233). The communication has been proven to be effective seeing that the situation in the celebration of Eid-Al Fitr was quiet as societies preferred to stay at home and did not do any mobilization.

In keeping the individual's health from COVID-19, some societies responded very strictly as a result of their high concern of being infected. Some people were exaggerating by using excessive personal protective equipment in a not high-risk area. This goes along with a research done by Sapp (2002) saying that incomplete knowledge perceived by an individual may lead to irrational behaviour. Whereas, for going to the supermarket, the usage of masks is enough for personal protection. Besides the overuse of PPE, to avoid themselves from COVID-19, some societies refused the burial of infected dead people in their area. The case of burial refusal happened in several cities in Indonesia such as Makassar and Gowa (South Sulawesi) and Semarang (Central Java). The response of the societies may be caused by the lack of information related to the COVID-19 transmission itself and the adjusted protocol of burial in this pandemic.

As the Indonesian Government declared a new normal starting on 1 June 2020, prevention efforts of COVID-19 cases seemed to be forgotten by some individuals. As a result, several regions in Indonesia experienced an increase in COVID-19 cases, such as in Southeast Sulawesi. On 5 July 2020, Southeast Sulawesi the number of COVID-19 cases intensified to 482 cases (BNPB, Universitas Indonesia 2020c, p. 256). The increase in COVID-19 cases is also reflected from the increase of mobilization during the new normal era as seen in the increase of domestic flights. In January to August 2020, the total number of domestic flight passengers in West Java was as many as 145.68 thousands (80.09% decrease compared to the same period last year of 731.62 thousand passengers) (PemerintahProvinsiJawa Barat 2020). From this number, the increase in passengers is seen as relaxation of LSSR is implemented. In August 2020, domestic flight passengers were 6,372 and increased to 125.96% compared to July 2020.

Not only in the airline sector, but the increase also happens in shopping centres. Although the number of visits has increased, almost all formal sectors have tightened the health protocol in restricting the spread of COVID-19. The limitation of the visitors, the obligation in wearing masks, body temperature checking, and a set social distancing system have been done by the formal sectors. However, health protocol in informal sectors may not be as tightened as it is in informal sectors. Since the declaration of new normal, activities in traditional markets are now not as restricted as they used to be during LSSR. During this new normal, although the obligation to wear masks and do social distancing still exists in traditional markets, the implementation is far from what is expected to be. Both merchants and market visitors often ignore the mask. This condition has contributed to the increase of COVID-19 cases. As an example, in 4 June 2020, a transmission cluster of COVID-19 in a traditional market of PasarBesar, Central Kalimantan, kept getting higher and reached up to 50 new cases were found that day (BNPB, Universitas Indonesia 2020b, p. 200). When the government of Central Kalimantan shifted the LSSR into Humanist Village Social Restrictions (HVSR), a program similar to LSSR with the difference in targeting red zone areas such as traditional markets, the implementation of HSVR was far from effective due to the disobedient of the merchants and market visitors (BNPB, Universitas Indonesia 2020b, p. 209). As a result, new clusters coming from the traditional market of PasarBesar were found. Not only in traditional markets, within a neighbourhood, masks usage often neglected. For instance, in East Java, housewives who shopped to itinerant greengrocers at the same time and made a crowd ignored using masks (BNPB, Universitas Indonesia 2020b, p. 146). Further, the habit of not staying at home and gathering in coffee shops without implementing any health protocols is also seen in East Java. In preventing the spread of COVID-19, the implementation of health protocols should be the foundation of activities done by the societies in this new normal era.

Like other coronaviruses, health facilities are mainly hospitals and clinics treatment is a place prone to transmission. One study reported approximately 40% of cases of infection contracted in health care facilities (Del Rio and Malani 2020). Health care workers, including attending doctors, nurses, and midwives direct patients, are groups prone to virus transmission. Compared to children, patients with 60 years of age or older is at increased risk of infection. Compared to the elderly, the infection in children is mildly symptomatic and even without symptoms (Velavan & Meyer, 2020). Those who travel a lot often hang out in groups and live in the area densely populated population at high risk of contracting or transmitting the virus (Velavan and Meyer 2020). In the context of epidemics and response, special protection measures are necessary given to population groups prone to transmission.

Strategic Policy for Countermeasures for COVID-19

Government response through various sectors, Local Government, and society not yet uniform in translating the policies to tackle COVID-19. Commitment, understanding the clinical course and disease epidemic, and the influence of the media mass and social influences the public response to epidemics and policies epidemic response. The COVID-19 epidemic that continues to grow until now with the negative socio-economic impact it causes presents the Government with a strategic policy dilemma choice countermeasures. The government naturally wants results-capable policies reduce the epidemic, but is able to do it within the limited resources, technology, and sociocultural norms, and limit the negative socio-economic impacts. As a large and broad country that adopts a partially decentralized system size of the development sector, the imbalance between planning and implementing the policy is very possible. Implementation of national policies, including guard policies social distance, chaining through two levels of managerial bureaucracy: (a) Central: strategy national countermeasures in the form of program planning and budgeting involving Local Government and other related stakeholders; (b) Area: strategy for implementing prevention programs and services. Strategy implementation of countermeasures in the regions is an elaboration of the national strategy adapting to the local capacity and socioculture. Strategy implementation may differ from one region to another. Central government, Local government, and service units need to unite, unite, coordinate carry out and closely monitor prevention programs and services.

The COVID-19 pandemic has spurred many analyses from various scientific disciplines. These include the fundamental analysis from microbiology, epidemiology, mathematical modelling from various disciplines, as well as from economy, sociology, and public policy. However, a geographic perspective on the pandemic is still lacking.

The question of how a pandemic spreads and intensifies in a particular area is inherently geographic one, and it can be understood through behavioural geography and health geography. In practical terms, a geographic analysis is concerned with three principal elements: point, distance, and area. All three elements form a spatial pattern or organization to a given phenomenon, including the spread of a disease such as COVID-19. The movements of people which determine the dynamics of COVID-19 contagion can be explained in terms of how it occupies a specific temporal and spatial dimension (Hafner 2020). The temporal dimension of a spatial behaviour can be examined using different temporal units. This includes how events occur in the spaces of a day, be it by dividing days into periods of morning-noon and evenings, or by its hours. Weekly analysis is also a possibility and especially useful for examining differences between routines in weekdays and weekends. A longer yearly analysis divided into months, seasons, or even multiple years is also possible. The differences between these time dimensions are described in the frequency of movement from a certain origin to destination.

One analysis of movement frequency could reveal patterns of decrease in movements correlated with increasing distance, as proposed in Tobler First Law of Geography (Waters 2017), and this information is usually paired with data on reasoning for route choices and time of travel. These factors decide how one will eventually decide on a route or a location, and they are linked to how a location is perceived. A location perceived as dangerous or unpleasant induces topophobia, while one that is perceived as safe and is liked elicits topophilia (Tuan 2001). The advantage of taking into account these factors is that it enables the differentiation of movement decisions based on behavioural goals with those determined by a yet unknown or unpredicted exogenous force.

Behavioural geography has significantly grown since the 1970s, alongside the emergence of radical and post-modern thoughts in the discipline. Parts of the discourse had pushed for a more humanistic approach, while others preferred a more analytical approach. Some focus of the analytical approach includes the study of behaviour, expectation, risk and uncertainty, habit formation, learning process, decision-making, and more specifically on place-related decisions, cognitive maps, and other processes related to spatial knowledge (Golledge et al. 2000 and Kitchin 2001). This dynamic growth occurred due to the increasingly open geographic discipline, which resulted in more intensive cross-disciplinary research. This increasingly cross-disciplinary interaction opens geography to the epistemological concepts, theories, methods, models, and even data procured by other discipline. A shift of focus from research that describes forms to those more concerned with the formative process took hold. Models of spatial behaviour start to emerge from observation of small-scale individual behaviour and larger societal patterns. These models were further developed on the idea that individual spatial decisions and patterns could not be analysed in isolation, as they are highly dependence on factors emerging from other individuals, crowds, or communities.

A potentially useful approach in investigating the COVID-19 pandemic is through the Lund time geography approach initiated by Thomas Haggerstand (Ellegard 2018). This approach focuses on three variables: space, time, and human behaviour. By assessing those three variables, a behavioural model could be formed and aggregated to community or group level. A behavioural model applying those principles to show how COVID-19 infection may occur in office workers in working days is presented below: (Fig. 3)

Though the model is originally constructed to describe the behaviour of an office worker, it can also be extrapolated to other roles within a household, such as adults who are homemakers, traders, as well as children or young adults in the family. The model can also be used to describe behavioural patterns of people living in apartments, dorms, or other types of homes. This model can then be aggregated into the community level and categorized based on domicile, social and economic conditions, reasoning for commute, as well as its destination, transport, and patterns across different days. A comparative analysis can also be conducted, for example, one

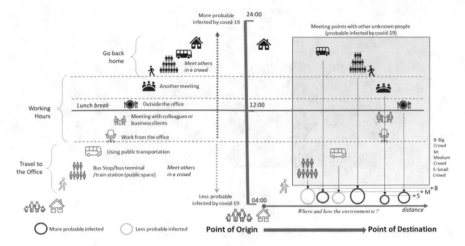

Fig. 3 Office worker's daily behaviour (space and time)

could analyse differences between pre-pandemic behaviour and behaviour during the pandemic, which may open new insights in understanding the spread of COVID-19.

The Lund time geography approach was later developed to investigate the situation or characteristics in points of meeting or crowding, which would require a geo-ecological approach. This means that comprehensive investigation of the interconnectedness between behaviour and situation in spaces where crowds occur is highly relevant. As an example, the above model describes movement for a family member who is a worker. Within their movement dynamics, there exists points of crowding when they are waiting for public transportation (an open area), within the public transportation vehicle (an enclosed space, possibly with air conditioner as means of air circulation), in the office and during meetings (enclosed space with air conditioners), and when they return home with public transports (enclosed space) as well as when they arrive at pick up points for the transport.

One's behaviour can be determined by their perception and understanding of a place. This behaviour can also shift if there is a change in the function or value within the environment of the place in question, which will in turn drive a change in its cognitive and affective value. A change in decisions regarding the place will then occur as a consequence (Fig. 4).

To conduct such an analysis, a more detailed picture of the points of meeting or crowding must be acquired, as well as how the place is perceived. The points may be construed as having a spatial dimension or are located within a setting. A temporal dimension may also be added, for example, one could assess whether the points occur in certain times during the day or within a week, and how its density fluctuates. In the interest of COVID-19 risk assessment, points of crowding are important as they are vectors for transmission, especially when its density is the highest. An even greater risk is the fact that crowds may consist of health individuals as well as those who have little knowledge of their COVID-19 status, or those who are unknowingly carriers.

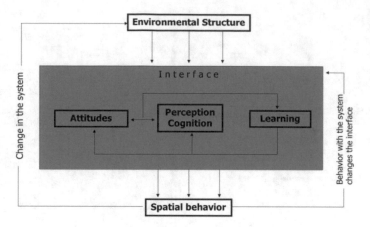

Fig. 4 Environment and human behaviour (*Source* adapted from Golledge et al. 2000)

It has been more than four months since the first COVID-19 case emerged in Indonesia. The enactment of the LSSR (large-scale social restriction, PSBB) changed the value or function of many spaces. This change follows local policies on restrictions. These measures, however, are more recommendations rather than strict regulation, which requires a high degree of understanding and initiative from the public.

There are five recommendations, sorted here from the most to the least known: wearing masks in public (97% out of total respondents), to stay at home (94%), physical distancing (93%), washing hands using soap (91%), and a recommendation for people to refrain from travels (87%) (Kamil and Riski 2020). Between the five recommendations, the only one having no obvious geographic/spatial dimension is hand washing recommendation. Meanwhile, understanding of the transmission of COVID-19 through droplets carrying the virus is widely understood, requiring a safe amount of space between people and possible infection vectors.

Figure 5 describes the transmission of COVID-19 via droplets emitted from talking, sneezing, and coughing. Masks are shown to highly reduce transmission, as well as keeping distances between individuals. Three types of masks are shown to be able to reduce droplet spread, home-made tea cloth masks, surgical masks, and N95 masks.

As such, to reduce transmission and cut off infection chains, points of meeting and crowding must be given utmost attention. More detailed data on the setting of those places and crowds, as well as their behaviour relevant to COVID-19 prevention and regulation adherence, is needed. Categorizing those places in terms of its risk potential will give us the following conditions: a) outdoor places with natural crowding potential (e.g. train stations or bus stops), and b) indoor places with inadequate air circulation (e.g. offices, meeting rooms, markets, malls, conference halls, places of worship), where more of COVID-19 risk(Salas and Mariano 2020) (Fig. 6).

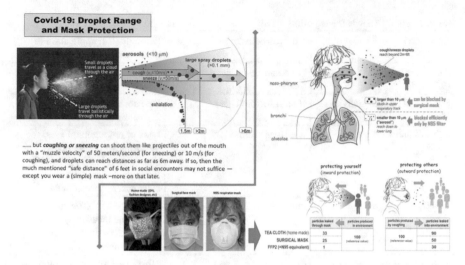

Fig. 5 Droplet range and mask protection (*Source* adapted from Sui Huang 2020)

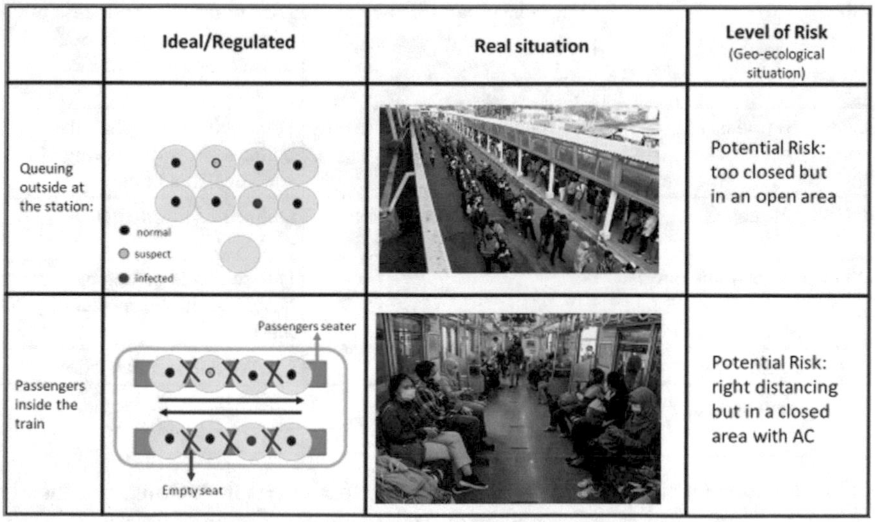

	Ideal/Regulated	Real situation	Level of Risk (Geo-ecological situation)
Queuing outside at the station:			Potential Risk: too closed but in an open area
Passengers inside the train			Potential Risk: right distancing but in a closed area with AC

Fig. 6 Crowd during LSRR working-day in greater jakarta at outside station and inside train

Jakarta and its neighbouring cities have become Indonesia's COVID-19 epicentre and are where LSSR policies are most enforced. The LSSR policies are intended to reduce transmission of the disease by preventing crowds to form and face-to-face meetings from occurring. A survey conducted regarding public understanding of the LSSR policies (Table 1) shows that respondents tend to understand what behaviours are prohibited by the LSSR, but only less than 50% understood which behaviours are allowed (Kamil and Riski 2020).

Table 1 Public knowledge regarding LSSR

Restricted	Allowed activities	Difficulties in application in regional governments
Activities in public facilities (92%)	Food and beverages sector (43%)	Lack of socialization regarding LSSR
Social and cultural activities (86%)	Health sector (18%)	Late social aid and mistargeting
Schools and workplaces (84%)	Logistics sector (10%)	Lack of enforcement from the government; contradictory regulations; ineffective public communication; public disregard for restrictions
Transportation (84%)	Public services sector (8%)	Public resistance towards quarantine measures; lack of consensus of the meaning of LSSR policies
Security and defence (35%)	Energy sector: water, electricity, oil, and gas (4%)	Most of population income are low, and therefore, they have to find jobs outside of home
Others:	Financial sector (3%)	Lack of restraint and sense of crisis in the public; LSSR; resistance towards LSSR policies
Eid travels (*mudik*)	Communication sector (3%)	Difficulties in adopting the online aspects of working or studying from home
Sporting events	Other more critical sectors (10%)	Online misinformation
Other cross regional travels		Difficulties of getting care in hospitals; limited availability of masks
Physical distancing		Lack of a tiered evaluation

Source Kamil and Riski (2020)

The Greater Jakarta area (Jabodetabek) has experienced fluctuating, but mostly increasing, numbers of cases ever since the pandemic emerged in March 2020. Early in the adoption of quarantine measures or LSSR, there are generally little issue in public adherence, at least until 3-4 weeks after its introduction. Very real behaviour change could be observed in Jakarta and its neighbouring areas. There was very little panic buying, and they are mostly isolated to cases early in the pandemic. Restrictions placed on various public services, especially in transportation, were enacted. Quarantine seems to have also created another issue entirely, namely its negative impact to mental health (Love 2020). The logical consequence of the restriction was a massive reduction in travels and reduction of density in public places. In other words, activities have moved from public places to private residences. In the initial month of the LSSR measures, we see a significant reduction in the COVID-19 test positive rate in Jakarta and its surrounding region. Data from Health Unit Jakarta Local Authority shows that reproductive number on March 2020 was 4 and it was drop until 1 on mid of April 2020 (Sari 2020).

A direct impact of this policy is that it negatively impacts business on several sectors, has caused a large decrease in activity and even some closure. However, online-based services see an increase in business. Many businesses adapt their activities into a more online model. People with middle-lower income bear the brunt of the economic impact, especially those that gain income through face-to-face businesses and activities, and with little savings. Many lost their jobs, and some face eviction.

After over a month of LSSR, the month of Ramadhan shows massive movements, especially nearing Eid Al-Fitr. Many in Jakarta came from other regions, and the Eid is a moment where many of them return to their hometown. This return is commonly known as "mudik". Even though the government has recommended people in Jakarta to not travel outside, mudik especially, this restriction proved to be very difficult to enforce. Even travel operators only allowed 50% of its capacity and have refunded many tickets. The government has also enforced screening measures for those travelling in and out of Jakarta. However, these measures did not suffice in preventing the spread of crowds outward.

The public desire to return was enough that it seemed to overcome LSSR restrictions. An estimate of 3.7 million Jabodetabek citizens travelled outside of the region, most likely to go on a mudik. Roughly 38% were bound for Central Java, 33% to West Java, 21% to East Java, Yogyakarta, and Bali (Elyazar, et al. 2020). This massive migration caused many pockets of epidemics in mudik destinations, namely in Bandung and its surrounding area in West Java, and in Central Java, with Solo and Semarang showed increase in case numbers. These new outbreak clusters originate in train stations, airports, bus terminals, as well as markets. A clear pattern is that the spread of COVID-19 was concentrated in transportation hubs.

The diffusion of COVID-19 outside of Jakarta enforces the epicentre or egocentric nature of the spread of the disease. There are clear records of the disease emerging in points of concentration in both open and enclosed spaces, in cities which became mudik destination from those originating from previous epicentres in Jabodetabek and other large cities in the country. Since the third month of June, or a month after Eid Al-Fitri, the condition has unequivocally become more severe, with July showing around 83% (103.000) of cases emerging in Java and Bali.

As LSSR restrictions are relaxed, the number of cases continue to rise. The baseline condition of vulnerable areas has become increasingly worrying, as infections become more and more common (Kantor Berita Antara, 24 Juni 2020). The intensity of human movement has also increased, approaching levels before the COVID-19 pandemic. Though testing capacity has increased since the early days of the pandemic in March, the national positivity rate (the number of positive cases over total number of tests) remained at a high level, at 10.5% by mid-July 2020. This number has increased well beyond the normal rate of 5% (Umasugi, 2020). It is obvious that the pandemic has yet to reach its peak in Jabodetabek, at least at the time of this writing (Fig. 7).

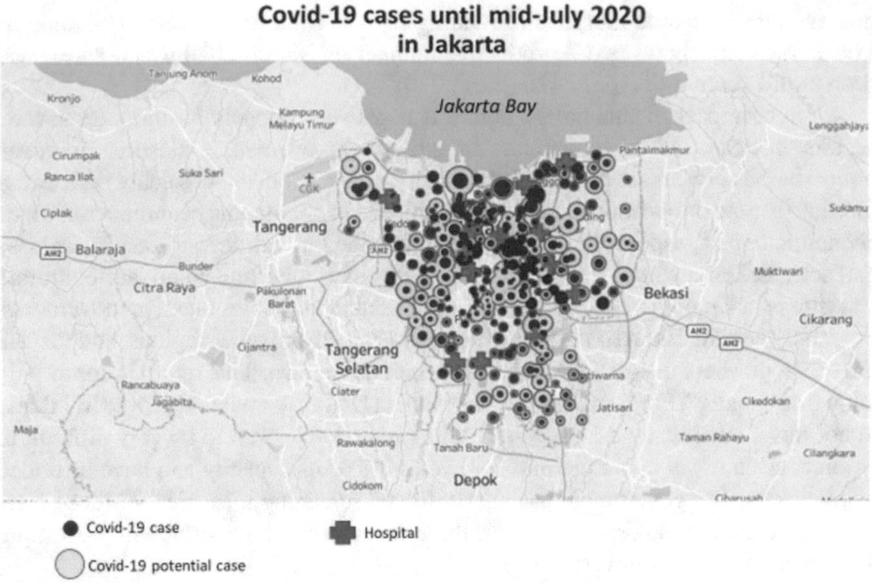

Fig. 7 COVID-19 cases until mid-July 2020 in Jakarta *Source* Sicovid-Dept. Geography UI, 2020

The numbers of positive cases in Jabodetabek show that there is a lack of adherences to regional and national restriction policies. Recommendations and restrictions seem to carry little effect in shutting down infection from spreading. The public has generally shown a lack of concern and discipline in adhering to the restrictions. However, a great portion of the blame also lies in unproductive behaviour from the government and its officials, as they also show a lack of leadership in the pandemic in that they produce many contradictory policies and many government personnel do not themselves adhere to COVID-19 protocol, such as not wearing a mask properly, which in combination communicates a lack of seriousness in the handling of the pandemic and a failure to produce a sense of crisis. What follows from these are the re-emerging of COVID-19 hot spots, such as markets and offices.

Conclusion

Indonesia is still in struggling to response the increasing of COVID-19 since March 2020. All efforts had been conducted to prevent and control the spread of disease include implementing health protocol for COVID-19 through people's behaviour based on its local wisdom. Commitment, understanding the clinical course and disease epidemic, and the influence of the media mass and social influences the public response to epidemics and policies epidemic response are the key actions taken into account by the government. Understanding the geo-ecology and characteristics of population dynamic are important to describe the mechanism of the spread of COVID-19 for more focus in control interventions.

References

Antara. (2020). DKI Jakarta KembaliMasuk Zona Merah Covid-19?IniFaktanya, edisi 24 Juni 2020, Kantor Berita Antara, Jakarta.

Arafat, Y. S. M., Alradie-Mohamed, A., Kar, K. S., et al. (2020). Does COVID-19 pandemic affect sexual behaviour? A cross-sectional, cross-national online survey. *Psychiatry Research, 289*, e113050. https://doi.org/10.1016/j.psychres.2020.113050.

Bacon, M. A., Corr, P. J. (2020). Coronavirus (COVID-19) in the United Kingdom: A personality-based. *British Journal of Health Psychology*, 25(4). https://doi.org/10.1111/bjhp.12423.

BadanPengawasObatdanMakananRepublik Indonesia. (2020, August 10). Demand Tinggi SelamaPandemi, Badan POM KawalPengembanganObat Herbal Berkualitas. *Badan POM*. Retrieved from: https://www.pom.go.id/new/view/more/berita/19178/Demand-Tinggi-Selama-Pandemi–Badan-POM-Kawal-Pengembangan-Obat-Herbal-Berkualitas.html.

Berskala Besar Pada Masa Transisi, Pemda DKI Jakarta, Jakarta.

BNPB, Universitas Indonesia. (2020a). *Pengalaman Indonesia dalamMenanganiWabah COVID-19 di 17 ProvinsiPeriodeJanuari - Juli 2020*. Jakarta: PenerbitBadan Nasional PenanggulanganBencana.

BNPB, Universitas Indonesia. (2020b). *Pengalaman Indonesia dalamMenanganiWabah COVID-19 di 17 ProvinsiPeriodeMaret-Juli 2020*. Jakarta: PenerbitBadan Nasional PenanggulanganBencana.

BNPB, Universitas Indonesia. (2020c). *Pengalaman Indonesia dalamMenanganiWabah COVID-19 di 17 ProvinsidanPembelajarandariMancanegaraPeriodeMaret-Juli 2020. Jakarta*: Penerbit-Badan Nasional Penanggulangan Bencana.

Cambridge Dictionary (n.d). Panic Buying. *Cambridge Dictionary*. Retrieved from: https://dictionary.cambridge.org/dictionary/english/panic-buying.

Del Rio, C., & Malani, P. L. (2020). 2019 Novel corona virus – Important information for clinicians. *Journal of American Medical Association, 323*(11), 1039–1040.

Ellegard, K. (2018). Thinking time geography: Concepts, methods and applications, Routledge, Oxon.

Elyazar, I., et.al. (2020). Mobilitas Penduduk Meluaskan Zona Merah Covid-19; Larangan Mudik Jolowi Harus Diperkuat, Eijkman-Oxford Clinical Research Unit (EOCRU), https://theconversation.com/mobilitas-penduduk-meluaskan-zona-merah-covid-19-larangan-mudik-dari-jokowi-harus-diperkuat-pengawasannya-136681 downloaded on August 2 2020.

Garfin, D. R., Silver, R. C., & Holman, E. A. (2020). The novel coronavirus (COVID-2019) outbreak: Amplification of public health consequences by media exposure. *Health Psychology, 39*(5), 355–357. https://doi.org/10.1037/hea0000875.

Golledge, R., et al. (2000). Cognitive maps, spatial abilities, and human wayfinding. *Geographical Review of Japan. 73*(Ser. B)(2), 93–104.

Hafner, C. M. (2020). The spread of the Covid-19 pandemic in time and space. *International Journal on Environmental Research and Public Health, 17*(11), 3827.

Hartanti, D., Dhiani, B. A., Charisma, S. L., & Wahyuningrum, R. (2020). The Potential roles of Jamu for COVID-19: A learn from the traditional Chinese medicine. *Pharmaceutical Sciences & Research, 7*(4), 12–22. https://doi.org/10.7454/psr.v7i4.1083.

Huang, S. (2020). COVID-19: Why we should all wear masks—there is new scientific rationale, https://medium.com/@Cancerwarrior/covid-19-why-we-should-all-wear-masks-there-is-new-scientific-rationale-280e08ceee71, downloaded on July 31 2020.

Kamil, I., & Oktaviani, R. (2020). *Kajian damak penerapan PSBB terhadap pemutusan rantai penyebaran Covid-19 di Indonesia.* Padang: Universitas Andalas.

Kementerian Komunikasidan Informatika. (2020, March 2). Kasus Covid-19 Pertama, Masyarakat-JanganPanik. *Indonesia.go.id Portal Informasi Indonesia.* Retrieved from: https://indonesia.go.id/narasi/indonesia-dalam-angka/ekonomi/kasus-covid-19-pertama-masyarakat-jangan-panik.

Kitchin, R. (2001). Cognitive maps, International Encyclopedia of the Social & Behavioural Sciences, Elsevier Science Ltd.

Love, T. (2020). Effective CPTED during Covid-19, https://designoutcrime.org/index.php/free-resources/effective-cpted-during-covid-19 downloaded on August 2 2020.

Machmud, A., Minghat, A. D. B. (2020). The price dynamics of hand sanitizers for COVID-19 in Indonesia: Exponential and cobweb forms. *Indonesian Journal of Science and Technology, 5*(2). https://doi.org/10.17509/ijost.v5i2.24431.

Matompo, O. S. (2020). Legal Protection of Online Business Transaction (E-Commerce) During the COVID-19 Pandemic in Indonesia. *JurnalIlmuHukum, 4*(1), 146–154. Retrieved from: http://journal.umpo.ac.id/index.php/LS/article/view/2660/1445.

Pemerintah Kota Tegal. (2020). Data Pantauan COVID-19 Kota Tegal. *CoronaTegal Kota.* Retrieved from: https://corona.tegalkota.go.id/?page=beranda.

PemerintahProvinsiJawa Barat. (2020, October 3). TransportasiUdaraMenggeliat, ModaLainnya-masihTurun. *PemerintahProvinsiJawa Barat.* Retrieved from: https://jabarprov.go.id/index.php/news/39498/2020/10/03/Transportasi-Udara-Menggeliat-Moda-Lainnya-masih-Turun.

PemerintahProvinsiJawa Tengah. (2020). StatistikKasus COVID-19 JawaTengah. *Tanggap COVID-19 ProvinsiJawa Tengah.* Retrieved from: https://corona.jatengprov.go.id/data.

Salas, J. & Zafra, M. (2020). An analysis of three Covid-19 outbreaks: How they happened and how they can be avoided, Science and Tech magazine, 12 June 2020 edn.

Sari, N. (2020). Tren Kasus Positif Covid-19 Selama 2 Periode PSBB di Jakarta, GrafikMasihNaik-Turun, Kompas.com, downloaded on August 12020.

Sapp, S. G. (2002). Incomplete knowledge and attitude-behaviour inconsistency. *Social Behaviour and Personality an International Journal, 30*(1), 37–44. https://doi.org/10.2224/sbp.2002.30.1.37.

Shadiqi, M. A., Hariarti, R., Hasan, K. F. A., I'anah, N., & Al Istiqomah, W. (2020). Panic buying padapandemi COVID-19: Telaahliteraturdariperspektifpsikologi. *Jurnal Psikologi Sosial.* Retrieved from: http://jps.ui.ac.id/index.php/jps/article/view/221.

Shou, B., Xiong, H., & Shen, Z. M. (2013).Consumer panic buying and quota policy under supply disruptions. In Working paper. Hong Kong. Retrieved from: http://personal.cb.cityu.edu.hk/biyishou/files/MSOMPanic.pdf.

Tuan, Y. F. (2001). *Place and space; The perspective of experience.* Minneapolis: University of Minnesota Press.

Umasugi, R. A. (2020). Dinkes DKI: Positivity Rate 10,5 Persenuntuk Sehari Kemarin, Secara Mingguan Angkannya 5,5 Persen. Kompas.com, downloaded on 05 August 2020.

Velavan, T. P., & Meyer, C. G. (2020). The COVID-19 epidemic. *Tropical Medicine and International Health, 25*(3), 278–280.

Waters, N. (2017). Tobler First Law of Geography, https://theconversation.com/mobilitas-penduduk-meluaskan-zona-merah-covid-19-larangan-mudik-dari-jokowi-harus-diperkuat-pengawasannya-136681, downloaded on July 31 2020.

WHO. (2003, September 13). Radiation: Sun protection. *World Health Organization.* Retrieved from: https://www.who.int/news-room/q-a-detail/sun-protection.

Yuniti, I. G. A. D., Sasmita, N., Komara, L. L., Purba, J. H., & Pandawani, N. P. (2020). The impact of Covid-19 on community life in the province of Bali, Indonesia. *International Journal of Psychosocial Rehabilitation, 24*(10), 1918–1929. https://doi.org/10.37200/ijpr/v24i10/pr300214.

Yanti, B., Mulyadi, E., Wahiduddin, W., Novika, R. G. H., Arina, Y. M. D., Martani, N. S., et al. (2020). Community knowledge, attitudes, and behaviour towards social distancing policy as a means of preventing transmission of COVID-19 in Indonesia. *JurnalAdministrasiKesehatan Indonesia, 8*(1), 4–14. https://doi.org/10.20473/jaki.v8i2.2020.4-14.

Does Socioeconomic Status Matter? A Study on the Spatial Patterns of COVID-19 in Hong Kong and South Korea

Keumseok Koh, Sangdon Park, Yuen Yu Chan, and Tsz Him Cheung

Abstract While Hong Kong and South Korea have effectively controlled and managed the COVID-19, they have also suffered from multiple waves of infections with distinctive clusters of the pandemics. While people with lower socioeconomic status (SES) had been reported more vulnerable in Hong Kong and Korea, we found no significant correlations between the COVID-19 infection and the SES at the local level. Instead, we found a positive association between the COVID-19 infection and annual tax payment per capita at the district level in Korea. The findings imply the potentially vulnerable areas of COVID-19 may vary by country.

Keywords Hong Kong · South Korea · Socioeconomic status (SES) · COVID-19 · Spatial patterns

Introduction

Hong Kong and South Korea have been frequently mentioned among model areas for effective COVID-19 control and management during the pandemic period. As of August 31, 2020, the World Health Organization (WHO) (2020) reported that Hong Kong and South Korea recorded a total of 4,811 confirmed cases (64.2 per 100,000 persons) and 20,182 cases (39.1 per 100,000 persons). Of the total cases, the fatality rates were 1.9% (90 deaths) in Hong Kong and 1.6% (324 deaths) in Korea.

The governments of Hong Kong and South Korea have implemented similar interventions to control the COVID-19. First, both governments have actively embraced

K. Koh (✉)
Department of Geography, Faculty of Social Sciences, The University of Hong Kong, Hong Kong, China
e-mail: peterkoh@hku.hk

S. Park · Y. Y. Chan
The University of Hong Kong, Hong Kong, China

T. H. Cheung
Division of Social Science, The Hong Kong University of Science and Technology, Hong Kong, China

© The Author(s), under exclusive license to Springer Nature Switzerland AG 2021
R. Akhtar (ed.), *Coronavirus (COVID-19) Outbreaks, Environment and Human Behaviour*, https://doi.org/10.1007/978-3-030-68120-3_8

advanced technologies. For example, they were among the first countries which used smartphones, wearable devices, and location-tracking apps for managing people under self-quarantine. Despite the privacy issues, they also have been releasing the details of the contact tracing of the patients confirmed within the past 14 days to the public, by collecting information from patient interviews, daily transaction records, and other big data. Second, both governments successfully acquired rapid and accurate COVID-19 testing kits in early February 2020 for early detection. Finally, strict yet flexible social distancing measures—e.g. meal gathering size, business hours, types of food outlets allowed—have been also effective to control the spread of the pandemic.

At the same time, people in Hong Kong and Korea have stayed alert and vigilant to fight the virus. Since the onset of the virus, people in Hong Kong and Korea have been proactively wearing face masks, to cooperatively follow the government guidelines and to willingly accept inconveniences from the social distancing policies. Recent outbreaks of the 2002–2004 Severe Acute Respiratory Syndrome (SARS) in Hong Kong and the 2015 Middle East Respiratory Syndrome (MERS) in South Korea may have made people more aware of the COVID-19 outbreak.

Although Hong Kong and Korea have been less impacted by the pandemic compared with other countries across the world, it would be of interest for researchers, policymakers, and health professionals to examine if there are any notable patterns and characteristics in the COVID-19 infection in Hong Kong and South Korea. This study examines the timelines and the spatial patterns of the COVID-19 using local area-level data from January 2020 to July 31, 2020 in Hong Kong and Korea. The overarching goal of this study is to investigate if there is any association between the COVID-19 infection and socioeconomic status by using area-level information.

COVID-19 in Hong Kong

Timeline

Hong Kong's COVID-19 infection from January 2020 to July 31, 2020, can be divided into three back-to-back waves as described below (HK Centre of Health Protection 2020).

First Wave (January 2020 to March 2020)

The first wave of COVID-19 in Hong Kong can roughly be defined as the period between January 2020 and March 2020. During this period, Hong Kong had a relatively flatter epidemic curve compared to other major cities across the world.

On January 23, 2020, two imported cases—i.e. a 39-year old male with Wuhan residence in China and a 56-year old Hong Kong male who had been in Wuhan in early January—were reported as the first confirmed cases of COVID-19 in Hong Kong (SCMP 2020a). Until March 31, 2020, the city accumulated 715 confirmed COVID-19 cases, 94 of which were asymptomatic infections and four of which led to deaths. Occasionally, there have been clusters of confirmed cases that have mostly been characterized by social clusters such as the 'Bar and Band' cluster in Lan Kwan Fong and Tsim Sha Tsui, and the 'Hotpot Family' cluster. The relative success of spread control has been widely accredited to intense surveillance and social distancing control measures implemented by the local government and rapid response and high hygienic awareness among the public.

Second Wave (April 2020 to June 2020)

There is no wide consensus regarding the date of the onset of the second wave of COVID-19 in Hong Kong. However, The Centre for Health Protection in mid-March 2020 reported a 'second wave' of COVID-19 infections marked by imported cases (RTHK 2020a). The majority of confirmed cases in this period were imported—i.e., residents, domestic workers, migrant workers, and foreign students returned to Hong Kong from Europe, North America, and Southeast Asia. There were also occasional local clusters of infections in public housing residences with untraceable roots such as LukChuen House at Lek Yuen Estate, Sha Tin, in June 2020, where seven infections were reported. From April 1, 2020 to June 30, 2020, there were 491 new confirmed cases of COVID-19, 379 of which were imported cases.

On June 16, 2020, Hong Kong government decided to ease gradually its social distancing restrictions since there were no major confirmed local cases observed in May and June 2020: For instance, in June 2020, the government allowed to reopen two major theme parks like Ocean Park Hong Kong and Hong Kong Disneyland and to set the Hong Kong Book Fair hosted in July 2020 (Nikkei Asia 2020; SCMP 2020b).

Third Wave (July 2020 to Present)

Beginning from July 2020, the third wave of COVID-19 has swept Hong Kong with an exponential increase in confirmed infections, a large majority of which are local infections. Following a period of 21 days of no local infections, on July 5, 2020, a cook at a restaurant in Ping Shek Estate, a public housing estate, was diagnosed with untraceable roots. The number of local infections experienced a rapid upfold and cases were reported daily; for instance, 41 local infections reported on 13 July, even reaching 83 new local infections reported on 19 July, which was the highest since March 2020. In total, from July 5, 2020 to July 31, 2020, there were 2014 confirmed cases reported in Hong Kong, 828 of which were local cases, and 952 of which were close contacts of local cases. A large majority of the local infections during the third wave can be traced to the virus spreading during social gatherings, events, or population clusters, such as celebratory banquets, elderly homes, and workplaces.

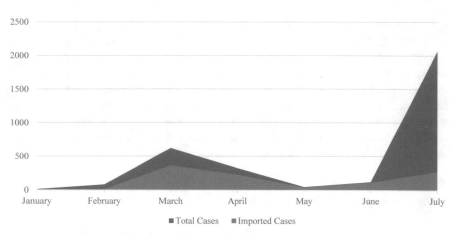

Fig. 1 Number of reported COVID-19 cases in Hong Kong 2020

Table 1 Number of reported COVID-19 cases in Hong Kong 2020

	January	February	March	April	May	June	July	Total
Total Cases	13	82	620	323	47	121	2067	3273
Imported Cases	11	11	367	227	42	110	277	1045

Source HKCHP. The authors
Note Date collected until July 31, 2020

Figure 1 and Table 1 illustrate and summarize the timeline of the total confirmed and reported cases and imported cases of COVID-19 in Hong Kong from January 2020 to July 2020. It is seen that a temporary peak was reached in March 2020, yet from July onwards, namely the onset of the third wave, confirmed numbers have skyrocketed along with the proportion of local cases. In addition, as of July 31, 2020, there have been a total of 27 deaths related to COVID-19 in Hong Kong, only seven of which were from the first and second wave (Worldometers 2020).

Geographical Patterns by District

All 18 districts in Hong Kong have reported confirmed cases. This phenomenon can be attributed to the fact that Hong Kong is a compact city with a high population density. In addition, almost all districts are highly accessible by the local mass transit system or other types of public transportation. Inter-district traveling is commonplace for commuting to work or attending social gatherings, which can lead to infections easily spreading across Hong Kong.

Table 2 Confirmed patients by region

Regions	No. of patients	Total population
Hong Kong Island	543	1,253,417
Kowloon	1,088	2,241,347
New Territories	990	3,840,620
Outside HK	32	N.A.
N.A.	620	N.A.
Total confirmed patients	3,273	7,335,384

Source HKCHP. The authors
Note Date collected until July 31, 2020

For a more macro analysis, Table 2 demonstrates the confirmed cases of three major regions in Hong Kong, namely Hong Kong Island, Kowloon, and New Territories. We observe that Kowloon and New Territories have almost double the cases compared to Hong Kong Island.

Table 3 summarizes the further classification of the confirmed cases by the source of contact. We can see there is a difference between the compositions of confirmed cases between the three regions as demonstrated in Table 3. Firstly, close to half

Table 3 Confirmed patients in Hong Kong's three regions by source of contact

Source	Confirmed cases in HKI	% of total in HKI	Confirmed cases cases in KL	% of total in KL	Confirmed cases cases in NT	% of total in NT
Close contact of local case	142	26.15%	504	46.32%	386	38.99%
Local	90	16.57%	419	38.51%	324	32.73%
Imported	216	39.78%	136	12.50%	215	21.72%
Possibly local	49	9.02%	14	1.29%	38	3.84%
Close contact of imported case	8	1.47%	10	0.92%	8	0.81%
Close contact of possibly local	38	7.00%	5	0.46%	19	1.92%
N.A.	0	0.00%	0	0.00%	0	0.00%
Total	543	100.00%	1088	100.00%	990	100.00%

Source HKCHP. The authors
Note Date collected until July 31, 2020. HKI: Hong Kong Island, KL: Kowloon, NT: New Territories

(40%) of cases reported with Hong Kong Island residency were imported cases, instead of local infections, yet imported cases only represent 13 and 22% of cases in Kowloon and New Territories, respectively. In these two territories, the majority of cases are local cases or close contact of local cases, which implies that local clusters could have emerged in these two regions.

Table 4 lists the confirmed cases at the district level, a lower geographic yet an important administrative unit in Hong Kong.

Table 4 Confirmed patients by district

	Districts	No. of patients
HKI	Central & Western	142
	Wan Chai	160
	Eastern	150
	Southern	91
	Subtotal	543
KL	Wong Tai Sin	407
	Kwun Tong	216
	Sham Shui Po	118
	YauTsimMong	174
	Kowloon City	173
	Subtotal	1088
NT	Islands	54
	Kwai Tsing	128
	North	75
	Sai Kung	137
	Sha Tin	194
	Tai Po	56
	Tsuen Wan	68
	Tuen Mun	177
	Yuen Long	101
	Subtotal	990
Other	Outside HK	32
	N.A.	620
Total		3273

Source HKCHP. The authors
Note Date collected until July 31, 2020

Wong Tai Sin

Wong Tai Sin district, located in Kowloon, has a population of 420,200 (HK Census 2019). It is mainly a residential area housing several public housing estates, the Wong Tai Sin temple, and several local schools. The district has been relatively unscathed during the first two waves of COVID-19 in Hong Kong, with only 14 cases before July 7, 2020. Yet, the district has experienced a large increase in cases during the third wave. A large majority of cases in Wong Tai Sin during the third wave have been related to elderly homes, and this includes residents of elderly homes, the staff of the elderly homes, and also their family members. Elderly homes in the district affected by COVID-19 include Kong Tai Care for the Aged Centre and Helping Hand Hong Kong Bank Foundation Lok Fu Care Home. In addition, a large number of cases happened within Tsz Wan Shan, a residential area within Wong Tai Sin district that contains several major public housing estates. As of July 31, 2020, 220 cases have been recorded in Tsz Wan Shan. The area has been described by media as an 'Epidemic Zone', and the government has provided free testing to public housing residents in the area (HK01 2020). It is also worth noting that Wong Tai Sin has the lowest average household income among the 18 districts in Hong Kong (HK Census 2019) and also a largely elderly population that live in elderly homes or public housing estates, which may lead to the fact that its median age of confirmed patients, 57 years-old, is the lowest among the three most heavily affected districts.

Kwun Tong

The second most heavily-hit district is Kwun Tong, which borders Wong Tai Sin to the North and is also located in Kowloon. It has a population of 693,900 (HK Census 2019) and is the district with the highest population density in Hong Kong. The area contains major residential areas such as Sau Mau Ping and Lam Tin and is also a major industrial area. It also houses major shopping centers such as APM and Megabox. The median age of infected patients in Kwun Tong is 46. Similar to Wong Tai Sin, the area has been relatively unscathed during the first two waves of COVID-19, with only 51 cases until June 30, 2020. During the third wave, a large number of cases have been residents living in public housing estates, including Choi Fook Estate, Ping Shek Estate, Kai Yip Estate, and Choi Ha Estate, all of which have been included in the government's free testing scheme (RTHK 2020b). The district also houses a restaurant named Bun Kee, where the infected cook who marked the onset of the third wave of COVID-19 worked. There has since been a cluster surrounding this restaurant, but most do not reside in the Kwun Tong district. Another restaurant causing infections has been the Windsor restaurant located in Sau Mau Ping, which contained up to five infected employees who reside in the Kwun Tong district.

Sha Tin

The third most-affected district in Hong Kong is Sha Tin, a new town located in the New Territories region with a population of 692,500 (HK Census 2020). The region has a range of housing properties from public housing estates, residences under the Home Ownership Scheme, private housing properties, and villages. It also houses commercial shopping centers with high activity such as the New Town Plaza and the industrial area of Fo Tan. The median age of infected patients of Sha Tin is 47. Among the three most-affected districts in Hong Kong, it has been relatively most affected by COVID-19 during the first two waves, with 64 cases before July 5, 2020, more than half of which (58%) were imported cases. Sha Tin also experienced a mini-outbreak during the second wave; towards the end of May, there was an outbreak of COVID-19 cases at LukChuen House, Lek Yuen Estate, forming a cluster of seven confirmed residents, which had untraceable roots. During the third wave, there were 130 cases, all of which were local cases or close contacts of local cases. During the third wave, public housing estates were also heavily affected, the most prominent being Ming Chuen House of Shui Chuen O Estate in early July, with 11 residents confirmed with COVID-19, which authorities thought to be linked to spread in elevator facilities (hkcnews, 2020).

Spatial Analysis with a Crude Incidence Rate

As shown in Fig. 2, the geographic pattern of COVID-19 in Hong Kong was further analyzed with the crude incidence rate, the confirmed COVID-19 cases per 100,000 population. A quantile classification was used to categorize the areas with the lowest (1st quantile) to highest (4th quantile) incidence rates.

While Wong Tai Sin was still among the areas with the highest COVID-19 incidence rates (4th quantile), Kwun Tong and Sha Tin were classified the 3rd and the 2nd quantile groups, respectively. In contrast, Kowloon City, YauTsimMong, Central & Western, and Wanchai were newly identified areas with the highest COVID-19 incidence rates in Hong Kong.

Clusters

Population Clusters

Population clusters appear to be the most prevalent form of the cluster; this is seen in two forms: population clusters in public housing estates and elderly homes. Compared to private housing buildings and apartments, public housing estates across Hong Kong including Lek Yuen Estate, Cheung Hong Estate, Choi Fook Estate, Ping Shek Estate, and Shan King Estate recorded a large number of confirmed cases. As shown in Table 5, several elderly homes were the clusters of infection in Hong Kong.

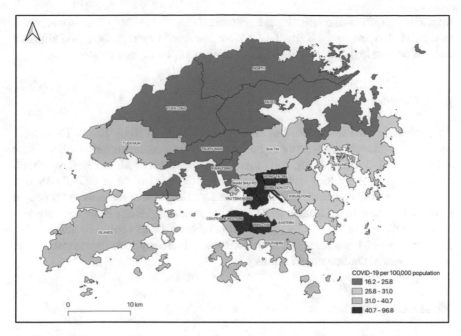

Fig. 2 Spatial pattern of reported COVID-19 cases by district in Hong Kong 2020. *Source* HKCHP. The authors. *Note* The authors created. A quintile classification

Table 5 Confirmed patients aged 65 or above and their elderly home residence

Total cases		534
Total cases living in elderly homes	Name of elderly home (District)	53
	Kong Tai Care for the Aged Centre (Wong Tai Sin)	31
	Helping Hand Hong Kong Bank Foundation Lok Fu Care Home (Wong Tai Sin)	1
	TWGHs Jockey Club Sunshine Complex for the Elderly (Southern)	1
	Salvation Army Lung Hang Residence for Senior Citizens (Sha Tin)	1
	Cornwall Elderly's Home Golden Branch (Tuen Mun)	17
	King Fuk Home for the Aged (Sham Shui Po)	2

Source HKCHP. The authors; Note. Date collected until July 31, 2020

Approximately 10% of confirmed elderly patients reside in elderly homes with more than half belonging to the Kong Tai Care for the Aged Centre cluster, and almost a third belonging to the Cornwall Elderly's Home Golden Branch cluster.

Work Clusters

Transmission in the workplace has also been prevalent in Hong Kong. There have been small outbreaks, consisting of no more than ten people, among retail workers in Sasa and Marks & Spencer, and other institutions such as Mount Kelley School. However, the most serious work cluster during the third wave of COVID-19 was a direct marketing company, Star Globalin MongKok, where there were 59 people COVID-19 patients as of August 2020. Relatively speaking, Hong Kong has not experienced as many work clusters as other comparable cities; this could be attributed to the fact that work-from-home practices have been adopted by government institutions and commercial entities since February.

Social Gathering Clusters

Along with population clusters, social gathering clusters, where people gather in a certain place for social interaction or some other form of recreational activity, have also been a prevalent form of transmission. The first cluster in Hong Kong was the Hotpot cluster, which emerged after a 19-people extended family gathering in Kwun Tong on the second day of Chinese New Year. Since then 13 people have been related to the cluster, two of which are colleagues of the infected family members (Now News 2020).

Another social gathering cluster is the Bar and Band cluster, which involves a music band that travelled to different bars to perform during the first wave of COVID-19. The involved bars are Insomnia in Lan Kwai Fong, Dusk Till Dawn and Centre Stage in Wan Chai and All Night Long in Tsim Sha Tsui (SCMP 2020c). The cluster involved 24 infected patients including 12 musicians, five bar staff members, and seven customers (Now News 2020). It is believed the same musicians could have transmitted the disease to different bar customers and staff while performing.

Next, the only religion cluster is the North Point Buddhist Temple cluster, which surrounds the Fook Wai Cheung She, a Buddhist temple established in 1958. Cases related to the temple started being reported in mid-February 2020, and since then 19 cases have accumulated including its master and multiple worshippers and related family members, many of whom are elderly-aged (SCMP 2020c).

There have also been two notable clusters related to local restaurants. The first notable cluster is the Bun Kee Restaurant cluster, involving the Hong Kong-style restaurant located in Ping Shek Estate, Kwun Tong. The restaurant marked the first COVID-19 case of the third wave, when a cook in the restaurant was infected locally. Since then and until July 31 2020, 16 cases have emerged including restaurant staff, customers, and related family members. The second is the Sun Fat Restaurant &

Taxi Drivers cluster, which is a Hong Kong-style restaurant located in Jordan. The restaurant is frequently visited by local taxi drivers and is known as a 'canteen' for taxi drivers. The first reported case related to this cluster was on July 7, 2020, when an employee of the restaurant was reported with COVID-19. Since then and until August 9, 2020, there have been 11 infections related to this cluster including taxi drivers, staff members of the restaurant, and other customers (Now News 2020).

COVID-19 in South Korea

Until early March, the largest infection of COVID-19 outside China had taken place in South Korea. While there are still tens of new cases recorded daily until July 31, 2020, South Korea has been considered as a model country of managing and controlling COVID-19 due to several effective responsive measures such as rapid test kits, the use of big data, and mobile technology (Salmon 2020).

General Overview

South Korea's remarkable economic growth from the 1960s brought substantial changes to its health care system. The country first implemented 'social health insurance' from the 1970s, made the transition to universal health care in 1989 and adopted a single-payer system by 2004 (Kwon et al. 2015). Statistics from OECD (2017a, b) suggest that South Korea possesses a well-established medical system even among other high-income countries, with 12.3 hospital beds per 1,000 people and 16.6 consultations annually per capita which excel all the other OECD countries. However, South Korea failed to respond effectively in the 2015 MERS outbreak, and several policy changes were made afterward to improve the government's preventative measures, as well as preparedness towards contagious diseases (Ariadne Labs 2020). Such lessons from the previous outbreak, along with an established health care system, enabled effective diagnosis and containment of COVID-19.

The first outbreak of COVID-19 had taken place on 20th of January 2020, imported from Wuhan, China. In response to the first confirmed case and soaring infection occurrences in China, an emergency response committee was established by the Korea Center for Disease Control and Prevention (KCDC). The figure remained minimal until 19th of February where President Moon even claimed the outbreak will 'disappear soon' (Choe, 2020). However, such early declaration of 'victory' was overturned by a super-spreader in DaeGu Metropolitan City (DGMC), also known as Patient No.31, who brought a rapid surge of confirmed cases across the whole country through Shincheonji church gatherings.

According to the KCDC data (2020), the biggest cluster is from 'Shincheonji' church, partaking 36% of the total cases, followed by mass infections with 27%, entries from overseas with 17%, and infection from contacts with 9%. 'Mass infection', the second significant cluster in Korea, refers to the transmissions that have taken place among a group of people, such as religious ceremonies, religious events, sports group gatherings and workers tightly packed in a closed space (Ariadne Labs 2020). For example, in Seoul Metropolitan City, In Cheon Metropolitan City, and GyeongGi Province, which consist of Korea's capital area, the infections among party-goers in Itaewon, Seoul (a district known for nightlife and multicultural environment) resulted in more than 200 confirmed cases. In addition, there were also cases of mass infections in elderly homes, workplaces, and churches across different parts of the country.

Despite such mass transmissions, South Korea successfully contained a further drastic surge of confirmed cases. The Korean case is distinct from other countries in ways that it embodies immediate responses, real-time dissemination of information to the public, close public–private sector coordination, and national organization of preventative efforts (Cha and Kim 2020). Recently, the daily average of the COVID-19 incidence remains about 100+ in August 2020.

Geographical Patterns of COVID-19 Spreading in Korea

Table 6 lists the total numbers of confirmed COVID-19 cases in Korea across the Large Autonomous Governments (LAG) collected by the Korea Centers for Disease Control and Prevention (KCDC) (2020). Figure 3 illustrates the spatial patterns of COVID-19 at the LAG level in Korea.

The incidence of COVID-19 infections at the LAG level in Korea is concentrated on two pairs of metropolitan cities and their neighboring provinces: DaeGu MC (DGMC) and GyeongBukProvince (GBP) are the ones and Seoul Metropolitan City

Table 6 Total confirmed cases by Large Autonomous Governments in Korea

Cities/Provinces	Cases	Cities/Provinces	Cases
DaeGuMC (DGMC)	6940	BuSan MC (BSMC)	130
GyeongBukP (GBP)	1356	ChungBuk P (CBP)	56
Seoul MC (SMC)	1600	GangWon P (GWP)	53
GyeongGiP (GGP)	1546	SeJong MC (SJMC)	45
InCheon MC (ICMC)	303	UlSan MC (USMC)	34
GwangJu MC (GJMC)	179	JeonBuk P (JBP)	18
ChungNam P (CNP)	160	JeonNam P (JNP)	17
DaeJeon MC (DJMC)	147	JeJu P (JJP)	11
		Total	14,336

Source KCDC. The authors

Note Data as of July 31, 2020; MC = Metropolitan City; P = Province

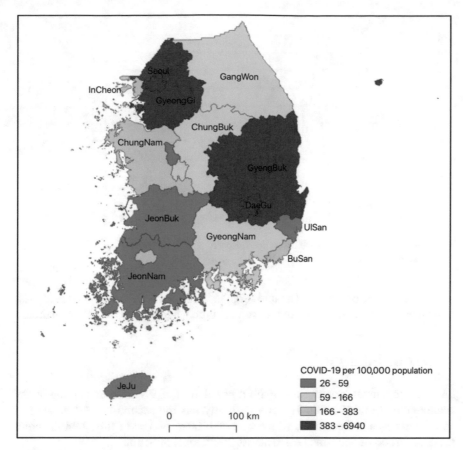

Fig. 3 Spatial patterns of COVID-19 spreads at the Large Autonomous Government level in Korea.
Note Authors created. A quintile classification was used

(SMC) and GyeongGi Province (GGP) are the others. DGMC and GBP yielded a
total of 8,342 out of 14,336 confirmed cases, while SMC and GGP yielded 3,155 as
of 31st of July 2020. DGMC and GBP, which are adjacent city and province in the
south-eastern part of Korea, were swept by the infection that started by a pseudo-
Christian cult called 'Shincheonji Church' in late February 2020. Figure 4 visualizes
the daily trend of the confirmed COVID-19 cases in Korea heavily impacted by a
drastic surge of the Shincheonji-based infections in late February.

About 44.5% of the nation's population live in SMC and GGP—the nation's
capital area, which may lead to the higher number of cases in these two LAGs. Except
for these two pairs, the remaining parts of the country have been experiencing less
severe occurrence of transmission. The confirmed cases in ICMC, another metropolis
in the nation's capital area, were largely influenced by its geographical proximity to
SMC and GGP. GJMC, a city located in the southwestern part of the country with a

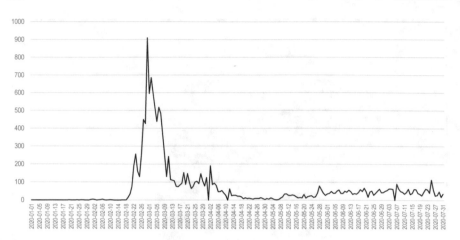

Fig. 4 Daily Confirmed Covid-19 cases, South Korea.
Note Authors created. *Data Source* Our World in Data

total of 204 cases, experienced a rapid surge in the figure from early July due to mass infections that took place in badminton club, church, and private education institutes.

Seoul Metropolitan City

SMC is the capital city of Korea with a population of 9.7 million, serving as the economic and cultural hub. The city currently has the second-highest number of confirmed cases in the country with a total of 1600 as of 31st of July. Table 7 shows the distribution of confirmed cases across 25 districts in Seoul.

The districts with the highest figures of confirmed cases are located in the southern parts of the city, including Gangnam-gu (Small Autonomous District), Songpa-gu, Gwanak-gu, and Gangseo-gu with 98,98, 144, and 103 cases, respectively. These are also the areas that possess a larger population than the most of other districts, where Gangnam-gu also serves as Central Business District (CBD) of the city. In contrast, districts that are located on the upper half of the city possess a relatively lower prevalence of confirmed cases with exceptions to Eunpyeong-gu and Dobong-gu where large clusters were formed based on infections that took place in elderly homes for both districts.

Table 8 illustrates clusters that are responsible for the infections, as well as the number of confirmed cases for each of these clusters. The biggest cluster, apart from oversea entries, is infections from Itaewon nightclubs. There are also several 'work-related' clusters partaking high proportion of confirmed cases including infections from office buildings in Gangnam-gu and Gwanak-gu, 'Rich-way' Sales, Coupang logistics center, and call center in Guro-gu. Rich-way and the call center are 2nd and 3rd largest clusters, with a total of 122 and 98 cases of infections.

Table 7 Distribution of confirmed cases in Seoul by district (as of July 31, 2020)

District (gu)	Cases	District (gu)	Cases
Jongno-gu	28	Mapo-gu	52
Jung-gu	20	Yangcheon-gu	71
Yongsan-gu	54	Gangseo-gu	103
Seongdong-gu	57	Guro-gu	91
Gwangjin-gu	22	Geumcheon-gu	39
Dongdaemun-gu	44	Yeongdeungpo-gu	72
Jungnang-gu	44	Dongjak-gu	70
Seongbuk-gu	42	Gwanak-gu	144
Gangbuk-gu	28	Seocho-gu	63
Dobong-gu	64	Gangnam-gu	98
Nowon-gu	50	Songpa-gu	98
Eunpyeong-gu	63	Gangdong-gu	49
Seodaemun-gu	41	Others	93
		Total	**1600**

Data Source Korea Centers for Disease Control and Prevention (KCDC)

Table 8 Clusters of confirmed cases in Seoul (as of July 31, 2020)

Cluster	Cases
Oversea Entry	343
Church in Songpa-gu	18
Elderly Home in Gangseo-gu	24
Office in Gangnam-gu	29
K' Building Office, u in Gangnam-gu	7
Private gathering in Gangnam-gu	13
Rich-Way' Sales Company	122
Coupang' Logistics Center in Bu-cheonsi	24
Office in Gwanak-gu	14
Itaewon Nightclub	139
Call Center in Guro-gu	98
Church in Gwanak-gu	25
Contacts from Other Cities and Provinces	55
Others	525
Under Investigation	164
Total	**1600**

Note Authors Created
Data Source KCDC

It is known that in both clusters, a large number of employees were working in a closed space, without sufficient preventative measures such as social distancing and mask-wearing (Kim 2020). Another major cluster was formed based on churches in Songpa-gu and Gwanak-gu with 18 and 25 cases, respectively. For these churches, the transmission was fostered during weekly masses and 'camping trip' that was organized by Wang-sung church in Gwanak-gu. Other clusters include elderly homes and private gatherings.

Overall, Seoul was heavily impacted by infections from workplaces, especially from sales and customer service industries where a large number of people gathered in confined spaces. Religious gatherings were another important cluster where weekly masses and church trips were still pushed ahead despite the looming threats of infection. Furthermore, higher figures of infections are found in the southern parts of the city where there are not only larger populations but also offices and service industries that turned out to be the major source of infection. Containing transmissions in Seoul possesses immeasurable significance since the city is also closely connected to ICMC and GGP, where people often commute through extensive public transportation networks. For instance, most of the major clusters in Seoul have also exerted their influences by forming other key clusters in ICMC and GGP.

Gyeonggi-Do Province (GGP)

GGP is a large province that surrounds Seoul with a population of 14 million. As shown in Table 9, a high number of confirmed cases were detected in Bucheon-si (Small Autonomous City), Seongnam-si, and Pyeongtaek-si with 187, 187, and 148 cases, respectively. The distribution suggests that areas that are in a commutable distance from Seoul, which include Bucheon-si and Seongnam-si, were subject to a higher number of cases compared to cities that are located further away from the capital. For instance, not a single case of infection was reported in Yeoncheon, an area that shares a border with North Korea, up until now while cities and counties that are located at the further east of the province, including Gapyeong, Yangpyeong, and Yeoju, have at most two confirmed cases. Pyeongtaek-si is an outlier in a way that it possesses one of the highest confirmed cases in the province despite its distant location from the capital. The phenomenon could be explained by the presence of United States forces stationed in the province. Camp Humphreys, the biggest U.S Army Garrison in the country, is located in Pyeongtaek-si, and 70% of the confirmed cases were related to U.S soldiers and their families who entered Korea from foreign origins (MBC News 2020).

Table 10 shows the clusters in GGP where the call center in SMP resulted in 54 confirmed cases. Churches were also clusters of transmissions in GGP, as shown by the Shincheonji cult Gwancheon headquarter as well as the influx of worshippers from DGMC. Other clusters are still unidentified, which make it difficult to further discuss the transmission status in the region. Such a problem further exacerbates in DGMC and GBP where the rapid surge in the confirmed cases made it impossible to trace clusters and transmission routes.

Table 9 Distribution of confirmed cases in GGP by city (as of July 31, 2020)

City/County	Cases	City/County	Cases
Gapyeong	2	Anseong	9
Goyang	76	Anyang	72
Gwacheon	12	Yangju	14
Gwnagmyung	35	Yangpyeong	1
Gwangju	39	Yeoju	2
Guri	13	Yeoncheon	0
Gunpo	80	Osan	10
Gimpo	42	Yongin	125
Namyangju	53	Uiwang	11
Dongducheon	6	Uijeongbu	72
Bucheon	187	Yicheon	17
Seongnam	187	Paju	24
Suwon	113	Pyeongtaek	148
Siheung	36	Pocheon	38
Ansan	58	Hanam	13
		Hwaseong	51
		Total	1546

Note Authors Created
Data Source KCDC

Table 10 Clusters of confirmed cases in GGP (as of July 31, 2020)

Clusters	Cases
Oversea Entry	460
ShincheonjiCult_Daegu	15
ShincheonjiCult_Gwacheon	22
Influx from Daegu &Gyeongbuk Province	44
Zumba ClassCheongnang Sejong	4
Call CenterSeoul	54
Unidentified	947
Total	**1546**

Note Authors Created; Data Source: KCDC

GyeongBuk Province (GBP) and DaeGu Metropolitan City (DGMC)

Confirmed cases from GBP and DGMC consist of more than half of the total cases in Korea. As briefly mentioned above, the 'super-spreader' from Shincheonji church contributed to such a situation by attending numerous events and places without acknowledging her infection status. Table 11 shows the distribution of infections in GBP according to cities and counties. Government data on clusters have yet to be

Table 11 Distribution of confirmed cases in GBP (as of July 31, 2020)

City/County	Cases	City/County	Cases
Gyeongsan	639	Gunwi	6
Gyeongju	56	Bonghwa	71
Gumi	78	Seongju	22
Gimcheon	19	Yeongduk	2
Munkyung	3	Yeongyang	2
Sangju	16	Yechun	43
Andong	55	Ulleung	0
Yeongju	6	Uljin	1
Yeongchun	38	Uisung	43
Pohang	54	Cheongdo	142
Goryeong	10	Cheongsong	2
		Chilgok	51
		Total	**1359**

Note Authors Created
Data Source KCDC

published, while only the total number of confirmed cases and its trend over time are available for DGMC at the current stage. There are total of 6,940 confirmed cases in DGMC and 1,401 in GBP as of 31st of July.

Spatial Analysis with a Crude Incidence Rate

As shown in Fig. 5, the geographic pattern of COVID-19 in Korea was further analyzed with the confirmed COVID-19 cases per 100,000 population. A quintile classification was used to categorize the areas with the lowest (1st quantile) to highest (4th quantile) incidence rates.

As shown in Fig. 5a, Seoul has 10 out of 16 local governments with the highest COVID-19 incidence rates (4th quantile). Most areas with the lowest incidence rates (1st and 2nd quantiles) were located in GGP. Figure 5b reveals that the areas nearby Daegu (denoted by a star), a major hot spot of COVID-19 from Shincheonji church, have higher COVID-19 incidence rates in GBP. The majority of localities in GBP and several areas in the capital area were identified as problematic areas with the highest COVID-19 incidence rates in Fig. 5c.

Major Clusters of COVID-19 Infections in Korea

Apart from the clusters that were mentioned above, religion and work-related infections served as critical transmission points across nearly all cities and provinces.

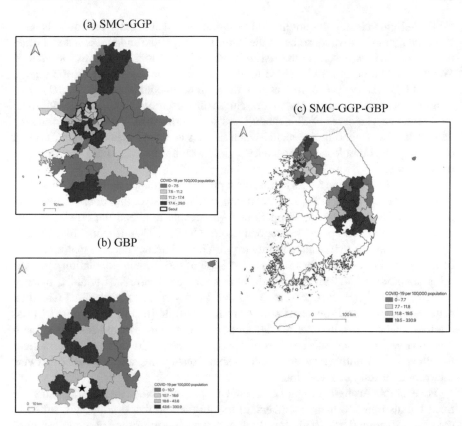

Fig. 5 Spatial patterns of reported COVID-19 cases by localities in selected areas in Korea 2020. *Source* The authors. *Note* The authors created. A quantile classification

Religion Clusters

The first cluster of the COVID-19 spread was a pseudo-Christian cult called Shincheonji church contributed to the rapid surge of infections from late February to early April. Their religious gatherings in DMC, the city with a population of 2.6 million located in the southern part of the country, served as the major source of the city's confirmed cases surged from 443 on 24th of February to 6,211 by 19th of March (Daegu Metropolitan City 2020). Apparently, the cult refused to follow the safety guideline set by the authority by banning masks during gatherings since it is 'disrespectful', arranging close physical proximity between members and inactively coordinating with the authority's attempt to trace the individuals who were involved into the cult (Rashid, 2020). A super-spreader, the 'Patient No. 31st', infected many others who were present at the gatherings, ultimately triggering the giant leap, between February and March and finally causing many confirmed cases in other major metropolitan cities including SMC, BuSan, InCheon, DaeJeon, and UlSan.

The religion cluster was not limited to so-called cults but extended towards other Protestant churches where some of them did not suspend Sunday masses and other religious activities that raise the risk of potential infection. The figures show that Wang-sung church in SMC yielded total of 25 confirmed cases during the second wave of COVID-19 spread. In Incheon, there were 56 confirmed cases solely from religious gatherings, while On-cheon church in Busan also resulted in additional 32 infections. The cases of Shincheonji and other Protestant churches illustrate the danger of mass gathering, especially when there is no safety measure such as social distancing or compulsory mask-wearing that would otherwise limit the range of transmission.

One may question the forces behind such phenomenon, especially given that these churches are not some cults, like the one mentioned above, that demand unconditional devotion from the worshippers or force people to attend the on-site masses despite all the risks imposed by the pandemic (Quadri 2020). It seems like Korea's strong Christian culture could possibly explain how churches became major clusters of infection. Protestants constitute around 30% of the Korean population but their religious passion and zeal overwhelm others who are Catholic, Buddhist, or have no religion at all (Hazzan 2016). Korea can be described as a 'zealous Christian state' where Korean Protestants still share huge similarities to Evangelical Christianity (Steger 2020). In other words, attending mass regularly and continuing with missionary works constitute an important part of people's faith that they are unwilling to follow the preventive measures, such as social distancing, which would otherwise interrupt their religious devotion.

For example, Wang-sung church in Seoul not only continued their weekend on-site mass but also pushed their 'membership training' camp that eventually resulted in the mass infection (Park 2020). It is thought that such significance of 'church-going' among Korean Christians as 'an important part of your faith and devotion [where] unless you're really sick on your death bed...you go to church' (Steger 2020). Therefore, the evangelical mindset and unwillingness to follow the government's safety guideline led several Protestant churches to become major clusters across many cities and provinces. In other words, Korea's prevalence of religion-related transmission illustrates how one's religious faith could 'overcome' the fear of epidemic and pursue choices that ignore the significance of 'social distancing' and many other preventive rules. It is obvious that a religious congregation serves as an optimum dispersal hub for COVID-19 as people are arranged in proximity for a long time under the enclosed environment. Accordingly, similar mass infections were found in religious gathering spots of Malaysia, Pakistan, and India where a huge number of worshippers from different regions were assembled in a small closed space. Therefore, there is a need for stricter social distancing requirements, as well as potential suspensions of religious gathering as an uncontrolled religious congregation imposes enormous risks.

Work Clusters

Work-related transmission is another prevalent cluster that could be easily found across different cities and provinces. It is important to note that the type of 'work' did not range from different industries but was focused on manual labour sites, simple service, and sales industries where strict adherence towards social distancing could have been more difficult than other occupations. In Seoul, the outbreak in 'customer call service center' in Guro-gu, SMC resulted in 98 confirmed cases which have extended its influence on Incheon and Gyeonggi-do Province as well as some of their employees were commuting to Seoul.

Another key infection started from a company called 'Rich-Way', a pyramid scheme organization that targeted elders aged over 60. There were a total of 122 people who were infected from the organization's sales seminar and educations sessions, mainly composed of older patients. In GGP, an outbreak in Coupang logistics center played a crucial role in lifting the figure, with approximately 150 confirmed cases. Considering the intensity of physical labour and a high number of labourers in restricted space, proper social distancing and mask usage were not properly implemented in the logistics center (Kim 2020).

Association Between Areal COVID-19 Incidence and Socioeconomic Status

A handful of studies in the USA reported that African American and Hispanic people suffer from disproportionately higher rates of COVID-19 infection compared to Whites and Asians (Blundell et al. 2020; Oronce et al. 2020). It implies that lower socioeconomic status may unequally result in negative health outcomes in relation to COVID-19, as the pandemic situation continues more than a half year. However, it is unclear that East Asian areas have a similar association between COVID-19 incidence and socioeconomic status. There is no study, to our knowledge, to examine the association between COVID-19 incidence and socioeconomic status at the local level in East Asia.

Due to the limited availability of data, we performed a set of univariate regression analyses. The geographical units of analysis are small autonomous localities in Korea and districts in Hong Kong. Two localities in Korea were excluded from the analysis because there are no confirmed cases reported. The COVID-19 incidence rates at the local level were estimated the total confirmed cases per 100,000 population. To capture the socioeconomic status at the local level, we used the median income per capita at each district in Hong Kong released by HK Census Department (2020). Instead of income, we used the total amount of annual tax payment per capita at each locality in Korea since there is no data available regarding personal income. However, the annual tax is also a good indicator of socioeconomic status because it is calculated based on one's personal income and wealth. The data were log-transformed to normalize the values and to account for skewness. Figure 6 and Table 12 summarize the results of the analysis.

Figure 6a and Table 12a shows a statistically significant positive association between the total annual tax payment and the COVID-19 cases in SMC and GGP— if the tax payment increases 1%, the chance to be infected to COVID-19 increases 0.28%. The finding is interesting because it is contradictory to the findings of the previous studies in the U.S. In GBP shown in Fig. 6b and Table 12b, we found a

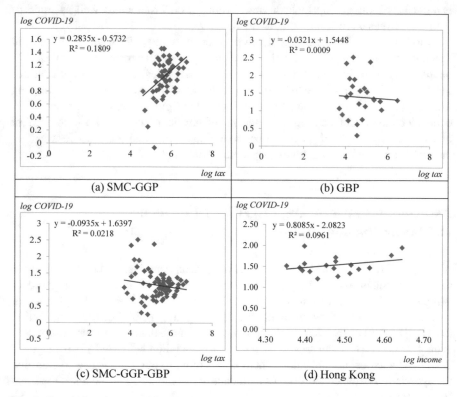

Fig. 6 Correlations between COVID-19 and tax/income. *Note* The authors created

Table 12 Results of a set of univariate regression models

Area\ Values	Sample Size	Coefficients	P-value
(a) SMC-GGP	55[*]	0.2835	0.0012
(b) GBP	22[#]	-0.0321	0.8143
(c) SMC-GGP-GBP	77	-0.0935	0.1999
(d) Hong Kong	18	0.8085	0.2106

Note [*]Yeoncheon-gun in SMC-GGP and [#]Ulleung-gun in GBP were excluded due to zero case of the COVID-19 infection

negative correlation between the total annual tax payment, which was found statistically insignificant. When analysing both the capital area and GBP shown in Fig. 6c and Table 12c, we found another negative correlation between the total annual tax payment, but it was statistically insignificant. In Fig. 6d and Table 12d, we observed a positive correlation between income and the COVID-19. However, the association was found statistically insignificant.

The findings imply that the association between COVID-19 incidence and socioeconomic status may be different in the U.S. and East Asia (Finch and Finch 2020). In Hong Kong and Korea, the responses to COVID-19 have been more active, cooperative, and participatory. Despite several sharp increases in COVID-19 infection, two Asian entities have been managing well the pandemic, compared to the other part of the world. Accordingly, the two governments could relax their social distancing measures to allow more daily activities to their people. There is no doubt that more people with a lower socioeconomic status, may expose to higher risks from disadvantaged living conditions in public housing, elderly homes, and labour-intensive and impossible to work-from-home occupations (e.g. logistics, retail workers). However, relaxed social distancing measures may cause added risks of COVID-19 infection to the public through social gatherings and extra-economic activities. Most importantly, the finding may imply that active government interventions of the pandemic coupled with the public's awareness and vigilance may prevent an unequal burden of COVID-19 potentially concentrating on people with lower socioeconomic status.

The findings have limitations. First, there would be potential confounders to influence the COVID-19 infection, since the results are from a univariate regression. Future studies need to follow up with more independent variables and study areas. Second, the area-level analysis may preclude us from making any inferences about individual-level associations between socioeconomic status and COVID-19 infections due to 'ecological fallacy'. Governments and related health agencies may need to share individual-level data with researchers for further studies. Despite these limitations, our findings may be informative for policymakers considering additional interventions to mitigate the effects of COVID-19 on people with lower socioeconomic status.

Implications from the COVID-19 Spreads in Hong Kong and Korea

In this study, we found several implications from COVID-19 spreads in Hong Kong and Korea. First, in Hong Kong and Korea, similar work clusters exist. More confirmed cases were identified in the labour-intensive workplace such as sales industries both in Hong Kong and Korea. While most individuals with lower socioeconomic status are responsible for the labour-intensive works, most of them may not enjoy the proper hygienic protection equipment due to the negligence of corporate responsibility (e.g., avoidance of obligations by subcontract hiring or outsourcing).

In addition, the labour-intensive work normally renders individuals less chance to work from home. Due to the nature of the work that has to be carried out in the workplace, it is inevitable for workers to face many people. The frequent human contacts of workers oppose the principle of social distancing. The workers have had no capacity but to suffer from a higher risk of infection because of the livelihoods. On the contrary, for individuals with higher socioeconomic status who work in professional sectors, very few confirmed cases have been observed both in Hong Kong and Korea. Workers in professional sectors are mostly individuals with higher socioeconomic status. While health welfares and work-from-home policies are normally provided and implemented in professional sectors, workers can benefit from them and protect themselves from the virus.

Second, it is interesting to note that the spread of COVID-19 in Hong Kong and Korea is highly driven by their unique cultures, while the cultures are related to the socioeconomic status of individuals. In Hong Kong, the culture of living is special compared to other countries in the world. Given the residences are highly dense in Hong Kong, public housing estate is the most notable type of residences. A public housing estate in Hong Kong is only provided to individuals with low incomes and social disadvantages, where most flats are densely packed on the same floor. Also, the hygienic measures in public housing estates are relatively looser than those in private housings. In turn, the transmissibility of the virus in the public housing estates would be relatively higher. Therefore, a high number of confirmed cases occurred in public housing estates across Hong Kong. Also, due to the unaffordability of the Hong Kong properties, the economically inactive individuals (mostly the elderly) may have to live in subdivided flats or elderly homes, in which the hygienic standards are usually unsatisfactory. Therefore, clusters of the confirmed cases were identified in elderly homes. In South Korea, cultural behaviours of individuals are also related to their socioeconomic status. Driven by a strong religious culture, certain religious practices are deeply rooted among South Korean populations. Worshippers of different religions (e.g. churches) in South Korea continued their religious assembles out of 'faith' (e.g. the idea that the virus can be overcome by belief), especially for older generations who are comparatively less educated. The individuals with less education or the elderly would be vulnerable to some extreme ideas from faith communities, creating a bandwagon effect among them. Even though the South Korean government announced the guidelines of social distancing, religious groups, especially the conservative ones, were not willing to adhere to public health advice. One potential reason could stem from the political regime of the current South Korean government, which is more liberal and democratic, contradicting the conservative or even far-right nature of the South Korean churches nowadays. It can be observed that young populations in South Korea continue to condemn the unreasonably zealous behaviours of religious groups, urging the senior religious leaders to reason.

Third, despite some population clusters like workplaces, social gatherings, and residences, there may be no or less negative correlation between socioeconomic status and the COVID-19 at the local level in Hong Kong and Korea. Interestingly, we found a positive association meaning relatively affluent areas may have higher risks of the COVID-19 incidence, especially in the capital area of Korea. Under a well-managed pandemic situation like in Korea, economically more developed areas may have a higher risk of the COVID-19 infection due to increased daily activities. Active government interventions and the public's cooperation may be also important to address the pandemic as well as its potential uneven impacts on population.

Addendum

Since December 2020, both Hong Kong and South Korea have experienced new waves of COVID-19. As of February 10, 2020 the total number of confirmed cases and deaths were 10,710 and 189 in Hong Kong and 82,434 and 1,496 in South Korea, respectively (The Government of Hong Kong Special Administrative Region 2021; KCDC 2020). Since January, 'ambush-style' temporary (12 to 48 hours long) lockdown orders have been imposed for mandatory COVID-19 test to several neighbourhoods in Hong Kong (Cheng et al. 2021). In Korea, the first COVID-19 test for a confirmed patient's pet was implemented in Seoul on February 10, 2021 (TBS 2021). COVID-19 vaccination is expected to start from the end of February in both regions. While both countries plan to use vaccines from multiple makers, the first batch of vaccines will be provided by the Pfizer-BioNTech in Hong Kong and by AstraZeneca in South Korea (The Government of Hong Kong Special Administrative Region 2020; BBC News Korea 2021).

Concluding Remarks

Hong Kong and South Korea, being the model areas for effective COVID-19 control and management, have shown that the socioeconomic status of population may be an important factor of the COVID-19 infection. It is noted that the individuals with lower socioeconomic status should be addressed and supported during the pandemic, even though the overall COVID-19 outbreak control is said to be satisfactory in both Hong Kong and South Korea. While work clusters and culture clusters have been significantly associated with the socioeconomic status of the population, more evidence-based researches about other social subgroups such as individuals working in 'shutdown sectors', a household with children and individuals that are less educated the knowledge about COVID-19 urgently demand scholarly attention.

References

Ariadne Labs. (2020). Emerging COVID-19 success story: South Korea learned the lessons of MERS. Exemplars in Global Health.

BBC News Korea. 한국도 아스트라제네카 백신 사용승인... 65세 이상도 포둑. https://www.bbc.com/korean/news-56007507. Retrieved 11 February 2021.

Blundell, R., Costa Dias, M., Joyce, R., & Xu, X. (2020). COVID-19 and inequalities. *Fiscal Studies, 41*(2), 291–319.

Cha, V., & Kim, D. (2020, March 27). A Timeline of South Korea's Response to COVID-19. Retrieved from Center for Strategic & International Studies: https://www.csis.org/analysis/timeline-south-koreas-response-covid-19.

Cheng, S., Ho, K., & Grundy, T. (2021). Hong Kong imposes first lockdown in Covid-hit Jordan – 10,000 residents affected, 3,000 personnel deployed. Hong Kong Free Press. https://hongkongfp.com/2021/01/23/just-in-hong-kong-imposes-first-lockdown-in-covid-hit-jordan-10000-residents-affected-3000-personnel-deployed/. Retrieved 27 January 2021.

Choe, S.-H. (2020, February 27). South Korean Leader Said Coronavirus Would 'Disappear.' It Was a Costly Error. Retrieved from The New York Times: https://www.nytimes.com/2020/02/27/world/asia/coronavirus-south-korea.html.

Daegu Metropolitan City. (2020, July 7th). Daily Update COVID-19. Retrieved from Daegu Metropolitan City COVID-19 Response Webpage: http://www.daegu.go.kr/dgcontent/index.do?menu_id=00936642&menu_link=/icms/bbs/selectBoardArticle.do&bbsId=BBS_02112.

Finch, W. H., & Hernández Finch, M. E. (2020). Poverty and Covid-19: Rates of incidence and deaths in the United States during the first 10 weeks of the pandemic. *Frontiers in Sociology, 5,* 47.

Hazzan, D. (2016, April 7). Christianity and Korea. Retrieved from The Diplomat: https://thediplomat.com/2016/04/christianity-and-korea/.

HK01. (2020). 【新冠肺炎】慈雲山「全區檢測」只包公屋 有疫廈至今未收樽仔. Available from https://www.hk01.com/18%E5%8D%80%E6%96%B0%E8%81%9E/506833/%E6%96%B0%E5%86%A0%E8%82%BA%E7%82%8E-%E6%85%88%E9%9B%B2%E5%B1%B1-%E5%85%A8%E5%8D%80%E6%AA%A2%E6%B8%AC-%E5%8F%AA%E5%8C%85%E5%85%AC%E5%B1%8B-%E6%9C%89%E7%96%AB%E5%BB%88%E8%87%B3%E4%BB%8A%E6%9C%AA%E6%94%B6%E6%A8%BD%E4%BB%94. Accessed 5 August 2020.

Hkcnews. (2020). 【第三波疫情】公屋再爆疫 水泉澳邨明泉樓共12人確診 Available from https://www.hkcnews.com/article/31834/%E5%85%AC%E5%B1%8B-%E6%AD%A6%E6%BC%A2%E8%82%BA%E7%82%8E-%E6%B0%B4%E6%B3%89%E6%BE%B3%E9%82%A8-31834/%E3%80%90%E7%AC%AC%E4%B8%89%E6%B3%A2%E7%96%AB%E6%83%85%E3%80%91%E5%85%AC%E5%B1%8B%E5%86%8D%E7%88%86%E7%96%AB-%E6%B0%B4%E6%B3%89%E6%BE%B3%E9%82%A8%E6%98%8E%E6%B3%89%E6%A8%93%E5%85%B112%E4%BA%BA%E7%A2%BA%E8%A8%BA. Accessed 20 July 2020.

Hong Kong Special Administrative Region Census and Statistics Department. (2019). Population and household statistics analysed by district council district. Available from https://www.statistics.gov.hk/pub/B11303012019AN19B0100.pdf. Accessed 20 July 2020.

Hong Kong Special Administrative Region Census and Statistics Department. (2020). Hong Kong in Figures (Latest Figures). Available from https://www.censtatd.gov.hk/hkstat/hkif/index.jsp. Accessed July 10 2020.

Kim, Y. (2020, July 8). 정은경 "쿠팡부천물류센터서방역수칙제대로안지켜져". Retrieved from Yonhap News: https://www.yna.co.kr/view/AKR20200708147900530?input=1195m.

Korea Centers for Disease Control and Prevention. (2020). Coronavirus Disease-19, Republic of Korea. Retrieved from Korea Centers for Disease Control and Prevention (질병관리본부): http://ncov.mohw.go.kr/en/.

Kwon, S., Lee, T., & Kim, C. (2015). Republic of Korea Health System Review. Manila: World Health Organization, Regional Office for the Western Pacific. Retrieved from http://www.searo.who.int/entity/asia_pacific_observatory/publications/hits/hit_korea/en.

MBC News. (2020, July 31). 경기도, '한미연합훈련취소' 통일부에건의…"코로나19 확산우려". Retrieved from MBC News: https://news.naver.com/main/read.nhn?mode=LSD&mid=sec&sid1=102&oid=214&aid=0001055527.

Nikkei Asia. (2020). Hong Kong Disneyland reopens amid uncertainty over expansion plans. Nikkei Asia. Available from https://asia.nikkei.com/Business/Travel-Leisure/Hong-Kong-Disneyland-reopens-amid-uncertainty-over-expansion-plans. Accessed 20 October 2020.

Now. (2020). 【持續更新‧附群組圖】香港新型冠狀病毒肺炎確診個案追蹤. Available from https://news.now.com/home/local/player?newsId=379456. Accessed 18 August 2020.

Organisation for Economic Co-operation and Development (OECD). (2017a). *Doctors' consultations—yearly, total per capita, 2017, South Korea [data set]*. Paris: OECD Data.

Organisation for Economic Co-operation and Development (OECD). (2017b). *Health equipment – hospital beds per capita, 2017, South Korea [data set*. Paris: OECD Data.

Oronce, C. I. A., Scannell, C. A., Kawachi, I., & Tsugawa, Y. (2020). Association between state-level income inequality and COVID-19 cases and mortality in the USA. *Journal of General Internal Medicine, 35*(9), 2791–2793.

Park, J.-w. (2020, June 26). 서울관악구왕성교회확진자 12명…MT 8·성가대 3·예배 1명(종합). Retrieved from Chosun Biz: https://biz.chosun.com/site/data/html_dir/2020/06/26/2020062602419.html.

Quadri, S. A. (2020). COVID-19 and religious congregations: Implications for spread of novel pathogens. *International Journal of Infectious Diseases, 96*, 219–221.

Rashid, R. (2020, March 9). Being Called a Cult Is One Thing, Being Blamed for an Epidemic Is Quite Another. Retrieved from The New York Times: https://www.nytimes.com/2020/03/09/opinion/coronavirus-south-korea-church.html.

RTHK. (2020a). HK hit by 'second wave' of virus infections Available from https://news.rthk.hk/rthk/en/component/k2/1514791-20200316.htm. Accessed 14 July 2020.

RTHK. (2020b). 政府明起擴大社區檢測計劃 涵蓋46樓宇逾8萬居民. Available from https://news.rthk.hk/rthk/ch/component/k2/1542204-20200806.htm. Accessed 8 August 2020.

Salmon, A. (2020, June 15). Inside Korea's low-cost, high-tech Covid-19 strategy. Retrieved from Asia Times: https://asiatimes.com/2020/06/the-secrets-behind-south-koreas-covid-19-success/.

South China Morning Post. (2020a). China coronavirus: death toll almost doubles in one day as Hong Kong reports its first two cases. South China Morning Post. Available from https://www.scmp.com/news/hong-kong/health-environment/article/3047193/china-coronavirus-first-case-confirmed-hong-kong. Accessed 7 July, 2020.

South China Morning Post. (2020b). Coronavirus: Hong Kong Book Fair to tentatively go ahead in December after being postponed amid city's third wave of infections. South China Morning Post. Available from https://www.scmp.com/news/hong-kong/hong-kong-economy/article/3098801/coronavirus-hong-kong-book-fair-tentatively-go. Accessed 19 October 2020.

South China Morning Post. (2020c). Coronavirus: Hong Kong Buddhist temple linked to Covid-19 cluster 'sincerely sorry'. Available from https://www.scmp.com/news/hong-kong/health-environment/article/3074640/coronavirus-hong-kong-buddhist-temple-linked. Accessed 20 August 2020.

Steger, I. (2020, February 26). How religion is playing a role in the spread of coronavirus in Korea. Retrieved from Quartz: https://qz.com/1808390/religion-is-at-the-heart-of-koreas-coronavirus-outbreak/.

TBS. (2021). 의심 증상에 반려동물 코로나19 검사, 서울서 ㅅ 실시. http://tbs.seoul.kr/news/newsView.do?typ_800=7&idx_800=3422823&seq_800=20415400. Retrieved 10 February 2021.

The Government of Hong Kong Special Administrative Region. (2020). Press releases: government announces latest development of COVID-19 vaccine procurement. https://www.info.gov.hk/gia/general/202012/12/P2020121200031.htm. Retrieved 10 February 2021.

The Government of Hong Kong Special Administrative Region. (2021). Coronavirus disease in HK. https://www.coronavirus.gov.hk/eng/index.html. Retrieved 11 February 2021.

To, K. K. W., & Yuen, M. D. (2020). Responding to COVID-19 in Hong Kong. Available from https://www.hkmj.org/abstracts/v26n3/164.htm. Accessed 25 August 2020.

Worldometers. (2020). China, Hong Kong SAR Coronavirus Cases. Available from https://www.worldometers.info/coronavirus/country/china-hong-kong-sar/. Accessed 1 August 2020.

Lives of Migrants During COVID-19 Pandemic: A Study of Mobile Vendors in Mumbai

S. S. Sripriya

Abstract During COVID-19 pandemic, in India where a large chunk of labourers and workers in the informal sector resides, migrants from urban clusters returned to their rural areas in some of the states, thus creating many possible vulnerabilities. In the absence of a fixed place for vending activity, mobile vendors, who are the poorest of the poor in urban India, face challenges to do their economic activity in local trains and also take care of their families with their meagre earnings. The high vulnerability of mobile vendors was reported in a recent study conducted among a sample of 150 (Male: 60; Female: 90) mobile vendors in local trains of Mumbai. During a pandemic like COVID-19, it is important to understand changes in the livelihood options of this section of urban poor. The precarious nature of problems reported in the media and write-ups as reviewed in this paper have brought the fact that they lost livelihood options during the pandemic and subsequently moved out of the city. To make the city to be more inclusive for the urban poor, therefore, there is a need to provide suitable opportunities through which, not only can we enhance their livelihood skills, but also make them to be the contributors to the overall development of the society.

Keywords Informal sector · Migrants · Mobile vendors · COVID-19 · Urban poor · Policy measures

Introduction

During COVID-19 pandemic, the developing economies like India which has a large chunk of labourers and workers in the informal sector migrants from urban clusters returned to their rural areas in some of the states, thus creating many possible vulnerabilities. Most of these migrants work as labourers in the sectors like agriculture, brick kilns, construction sites, domestic-related services like maids, watchman, drivers, industrial non-skilled workers and small tiny roadside petty businesses like

S. S. Sripriya (✉)
Tata Institute of Social Sciences, Mumbai, India
e-mail: sivaraju.sripriya@gmail.com

tea shops, dhabas, small eateries, hotels and restaurants. The top five destination states for migrants as per 2011 census include Maharashtra, Delhi, Uttar Pradesh, Gujarat and Haryana, as they attract over half of the inter-state migrants. Further, the data also showed that the states like Uttar Pradesh, Bihar, Rajasthan and Madhya Pradesh are the top origin states of migrants. From the state of Bihar alone, there are 7.5 million migrants and a majority of them cited 'search for work or employment as their reason for migration'. Rural to urban migrants are mainly concentrated in 53 million-plus urban agglomerations (with one million and more) that comprises 140 million out of 377 million urban population of the country equivalent to 43% of total urban population as per 2011 census. Eight megacities reported about two-fifth of coronavirus positive cases (Bhagat et al. 2020). Out of 482 million workers in India, about 194 million are permanent and semi-permanent migrant workers. Also, there are about 15 million short-term migrant workers of temporary and circulatory nature. There are conspicuous migration corridors within the country—Bihar to Delhi, Bihar to Haryana and Punjab, Uttar Pradesh to Maharashtra, Odisha to Gujarat, Odisha to Andhra Pradesh and Rajasthan to Gujarat (Bhagat et. al 2020). The National Commission for Enterprises in the Unorganised Sector (NCEUS) reports that around 92% of India's workforce with informal employment is substantially drawn from migrant labour (NCEUS, 2007). About 30% of migrant workers are working as casual workers, and they are quite vulnerable to the vagaries of the labour market and lack social protection. The skewed development of most of these states is an important contributory factor for the rural push as a major reason for rural–urban migration. Given their low socio-economic status, it is interesting to critically analyse the extent and nature of their vulnerability during the COVID-19 pandemic, and the paper is an attempt in this direction.

The Migrants: Most Vulnerable Sections

The cities become a magnet for disadvantaged groups looking for possibilities unavailable in their villages and provinces. Migrants from rural areas move to the major cities with the hope of getting employment and thereby to earn a living of their own. There is a section of migrants who become vendors in the absence of not getting a regular job in the place of destination. These migrants who are largely experiencing poverty try to adjust in the urban centres with their meagre financial supports. Given their marginalization and helplessness to get any regular employment in any governmental or non-governmental organizations, they turn towards carrying out other low-paid livelihood activities like street or mobile vending. The existence of vendors in all the countries, irrespective of its level of development, is highly noticed as they fill a 'gap in the retail chain' (Bhowmik 2010:17). It is documented in the literature on administrative policies about how mobile vending brings changes in the economy, power relations and juxtapositions in the use of space. Since they are 'lacking a voice and effective means of competing in the sphere of the state, a sphere that ideally seeks to offer benefit to the greatest number of people', the vendors are at a disadvantage when reference is made on their inclusion in the policy framework (Bhowmik 2010). Street vending activity reduces the expenses related to renting and utilities, mostly suitable for low-skilled employment, and fits with a new ideology of independence and entrepreneurship. Gore (1968) is of the view that the informal sector is simply a school of urban skills from which most rural immigrants eventually graduate to formal-sector jobs (Gore 1968).

It has been noted that the levels of informal employment have been increasing over time, while the levels of formal employment which have been in the decades post-1999–2000 have been declining (Sharma 2012: 29). Informal employment comprises more than 90% of the total employment in India, with India having more than 450 million informally employed workers with unorganized enterprises that range from pushcart vendors to home-based enterprises related to the diamond and gem industries (Sharma 2012: 30). The proportion of persons who are informally employed is not only increasing over time, but more than 90% of the informal workforce are either self-employed workers or casual workers.

Vendors: Poorest of the Poor in Urban India

In the Indian economy, workers in the informal sector, particularly women workers who are low paid in the informal sector, are highly vulnerable amongst different groups in society, due to the nature of work that the sector is characterized by Mohapatra (2012). There are two dominant forms of vending—street vending and mobile vending. Most of the earlier studies have concentrated on street vending, and mobile vending like selling in suburban trains, buses, etc., has not yet received much attention from researchers. Studies (Bhowmik 2001) have confirmed that street vendors are the most marginalized sections of the urban poor. They provide the urban rich everyday household goods at affordable prices, and thereby, they constitute an important part of the urban economy. Martha Chen and Raveendran (2011–12) point out that street vending in India represented 4% of the total urban employment and 5% of the informal urban employment, with the percentage of men who were street vendors (5%), being 1.7 times higher than that of the women (3%). The survey coordinated by Bhowmik and Saha (2011) in ten cities across India (Bhubaneswar, Bengaluru, Delhi, Hyderabad, Imphal, Indore, Jaipur, Lucknow, Mumbai and Patna) provides some useful details about the working and living conditions of street vendors, the views of consumers and the spatial-temporality of urban street vending in India. Appadurai's (2000) description of street life in Mumbai presents the embedded complexities of street vendors' issues. Over the past 50 years, vending has expanded not only in its scope but also of its importance. The increase in hawkers is due to the increase in the urban poor. Although hawkers are viewed as a problem for urban governance, they are the solution to some of the problems of the urban poor. By providing cheaper commodities, hawkers are in effect providing subsidy to the urban poor, something which the government should have done.

Mobile vending and its resulting reproduced spaces also subvert state control, and they act in one sense as an important logistical network that makes living in the city endurable, and this is especially true for low-wage workers. But their economic illegibility and illegitimacy are clear signs of state failure in reaching its goals of modernity, and their mobile semi-illegal status contradicts the state's abstract production of the city as being competently and coherently ordered. Urban marginal spaces are sites of struggles where inhabitants with lesser power must creatively negotiate a variety of economic and social obstacles to survive. The entrepreneurial spirit and the energy they put into their selling schemes dovetail nicely with the neoliberal narrative. In Cambodia and the Philippines, it is found that women dominate vending activities (Tinker 2003). However, in India, there are very few women vendors though the country has a large vending scene, i.e. an estimated 260,000 in Delhi, 500,000 in Mumbai- (Bhowmik and Saha 2012).

According to Bhowmik (2003), vendors have had a long historical presence in Mumbai, provides essential services to most of the population and provide direct employment for over three lakh people, in addition to indirectly employing hundreds of thousands more. The total employment provided through hawking becomes larger if we consider the fact that they sustain certain industries by providing markets for their products. A lot of the goods sold by train hawkers, such as moulded plastic goods and household goods, beauty products and cosmetics, homemade sweets, snacks, jewellery and stationery, are manufactured in small-scale or home-based industries. The most distressing part is the lack of social security for all the unorganized sector workers in most developing countries. Train hawkers are part of the unorganized sector and the regulation of employment would assure them security and independence. Those that hawk in trains in Mumbai buy their stock of goods daily from wholesale markets in the city.

Living Conditions of Vendors in Mumbai

The income of the hawkers is so low that it is difficult for them to employ others to help in their business and pay towards their wages. Hawkers are supported by their wives and one or two of their children. These people do not do the selling, but they help in the other activities. The wives help their husbands in procuring the goods, and the children help in sorting out different goods at home. Family help becomes necessary because the hawker cannot do all the activities on his / her own. The low income from the trade makes it impossible to employ others for helping out. With the result, it is found that some of the children, especially girls, do not go to school so that they can help their parents.

In Mumbai, consumers prefer hawkers because they provide services at convenient places. Hence, a lot of time is saved in making purchases. They feel that hawkers near their homes and the railway stations are most convenient for them. Consumers from the middle-income groups find vendors near railway stations most convenient because they can buy their necessities while returning home from their offices as this saves them time. These people purchased vegetables, fruits and other items for home use while returning home. It should be noted that hawkers plying their trade near the railway stations are the frequent targets of the eviction staff.

The SNDT-ILO study (2000) shows that 85% of the street vendors covered suffer from ailments associated with stress. These include hyperacidity, migraine, digestive problems and lack of sleep. In Mumbai, where there are around 200,000 hawkers, the municipal corporation has granted only 14,000 licences. Moreover, the municipal corporation has stopped granting new licences for the past two decades; hence, most of these licence holders do not ply the trade at present as they are too old or they have died. The census undertaken by TISS-YUVA on hawkers in Mumbai (1998) found that only 5,653 hawkers, out of a total of 102,401 hawkers covered, had licences.

Mobile Vendors and Their Vulnerability in Mumbai

Trains come in the no hawking zones. These 'no hawking zones', street vendors claim that they are in residential areas. This puts train hawkers at a greater risk of encountering officials that may threaten them with such a huge fine taking advantage of the rule. Train hawkers fear that getting together by themselves will worsen their plight if it comes to the notice of the police. With a hand to mouth existence and the daily struggle for survival, it leaves them with little time and no courage to unite and fight. Train hawkers are a person's striving for economic self-reliance; hence, they are contributors to a new India, as business people. Train hawkers are part of the world of commerce and collectively must find a place in the chambers of commerce to contribute further to the country's economy. A policy designed to develop the informal sector tends to ignore or minimize the potential or the benefits of street merchants and underestimate the difficulties of making micro-businesses which bear the development burden of formal enterprises. The activities of the vendors are highly insecure and have unstable income, composed of long hours, poor conditions, and no legal or social protection, limited access to credit and very limited bargaining power (ILO 2004).

In the absence of a fixed place for vending activity, mobile vendors face challenges to do their economic activity in local trains and also take care of their families with their meagre earnings. It is important therefore to conduct a scientific study of local train mobile vendors, and such study is expected to provide valuable inputs to evolve suitable policies and programmes for their well-being. For creating an inclusive city, it is very important to make efforts in addressing various issues that the marginalized sections of the city dwellers like mobile vendors experience on a day-to-day basis. The present paper is based on the recently conducted study on the mobile vendors in local trains of the city of Mumbai.

Local train of Mumbai is called as the lifeline of the city. It not only supports the public of Mumbai in commuting but also the life of vendors in the train for their daily bread and butter. Local trains are crowded and chaotic during peak hours. But some people hope to make a few bucks and get back home. Mumbai Suburban Railway is to travel within the city. The network of railways has three lines, namely the Western Line, the Central Main Line and the Harbour Line. Mumbai is a linear city covered by the local train network. Most of the time of Mumbai people is spent in travelling with the locals.

Mumbai local train network is the largest and busiest in the world. It is spread over an area of 465 sq. km including that of suburbs and carries nearly 7.24 million commuters daily (Ministry of Railways 2015). The Mumbai Suburban simply called as local trains consist of suburban railway lines. It has categorized as Western, Central, Harbour, Trans-Harbour and Nerul-Uran railway lines. For the present study, the three lines, namely Western, Central and Harbour lines, were considered, and those vendors who were carrying mobile vending in those trains, with an equal proportion of 50 vendors, were chosen as respondents for the study.

According to the railway station manager at CST railway station, Mumbai, the mobile vendors come under the category of unauthorized commuters. With this apprehension, mobile vendors avoid approaching the railway station authorities even in needy situations. The Indian Railway Protection Force service is a security force established by the Railway Protection Force Act 1957. The role of RPF in relation to railway platform activities is enormous. RPF can catch and prevent these mobile vendors as a law, but there were incidents where vendors helped RPF personnel to trace the criminals. Most of the time vendors run away from the RPF. RPF, as a rule, has to seize the goods from the vendors and produce in the court. They will be bailed out after warning or minimal penalties.

Local train vendors save the time of the public by selling attractive products like ladies cosmetics, colourful nail polishes, hair bands, clutches and even the products related to different seasons like in monsoon, colourful umbrellas, wind shelter, plastic mobile covers to protect from rain, plastic folders to protect the official documents. Eatables like vada pao, fruits, snacks and other eatables are also available with the vendors. Quality of the products is not that fine, but the products are relatively cheaper compared to the market price. The profit that they make after selling the product is marginally very low. The amount largely goes for buying food to eat or for purchasing products to sell for the next day.

Socio-Economic Profile of Mobile Vendors

A study was conducted by the author of this paper in 2018 among a sample of 150 (male: 60; female: 90) mobile vendors in local trains of Mumbai. While the average age of the respondent is 30.1 years, it is relatively higher among females (33 years) when compared to males (25.7 years). Male vendors are involved in mobile vending as their main occupation in the initial years after migration and thereby are relatively at younger ages. However, in the case of females, their entry into the mobile vending is at a later stage of their life mainly after completing their familial responsibilities like bearing and rearing of children. The education status of the vendors is observed to be at 39.3% illiterate, and another 39.3% have attained only secondary level of education. Most of the sampled vendors (74.7%) reside in Chawls which have only basic amenities. On average, they earn rupees 893.2 per day. Over one-third of the female vendors (37.8%) have very low income (below Rs 300 per day) and are more financially vulnerable when compared to males as they have to take care of the entire household expenditure with their limited earnings as compared to men.

All the sampled vendors are either first- or second-generation migrants. Nearly half of them (49.3%) are from Maharashtra state. Those migrants who have come from other states indicate that nearly one-fifth (19.3%) have come from southern states like Andhra Pradesh, Telangana, Tamil Nadu, Karnataka and Kerala. Migrants from Gujarat constitute 8.7%. Their representation from northern states are: Uttar Pradesh (13.3%), Bihar (6%) and other states (3.3%). The data related to various reasons for their migration to Mumbai showed that lack of income opportunities have influenced migration. A few of them (8%) have also stated reasons such as loss in business, rise in debt, marriage (especially among females) and low level of income influenced them or their families to migrate to Mumbai. The data on reasons for choosing Mumbai for livelihood indicate that most of the first- and second-generation migrants (84.7%) have stated that they migrated to the city in search of a job as their livelihood was a major issue in their place of origin. The mean estimated sales per day are worked out to be rupees 876.5 among the total vendors. The mean sales are considerably higher among the male vendors (rupees 1020.8) as against the sales among the females (rupees 779.2). The total vendors have their daily profit to the extent of rupees 391.6 on an average. Male vendors have considerably higher levels of profit (rupees 448.3) by selling their products, and in the case of females, the daily profit is at a lower level (rupees 353.4). Majority among the total vendors (73.3%) have mentioned positively about their partnerships with other mobile vendors. Two-thirds of the total vendors (66.6%), without much variation across males and females, have partnerships mainly for lending credit to each other. Since mobile vending in local trains is against the law and declared as an illegal activity, while the vendors carry out their activities, they pay a certain amount to the station authorities like Railway Protection Force and Ticket Collectors in exchange to get their informal approval to carry out their activity. One-third of the total vendors in the study (39.3%) have accepted that they do pay a certain amount to the authorities associated with the railways. It is noticed that mobile vending activity is very challenging, and also, they face several constraints. Given the high vulnerability of mobile vendors as reflected in this study, it is very significant to make an effort to understand changes in their livelihood options and their quality of life during a pandemic like COVID-19.

Migrants and COVID-19

Migrants suffer from the double burden of being poor and migrants. A telephonic survey of more than 3000 migrants from north-central India by Jan Sahas (2020) shows that majority of the workers were the daily wage earners, and at the time of lockdown, 42% were left with no ration; one-third was stuck at destination city with no access to food, water and money, and 94% do not have worker's identity card (Jan Sahas 2020). Coronavirus outbreak leads to a loss of livelihood for those who either work on short-term contracts or those who are without any job contracts. India is likely to face the job crisis because of the COVID-19. Migrant workers and workers in the informal sector are likely to be badly hit (ILO 2020). The coronavirus outbreak is going to affect them badly leading to their further impoverishment due to loss of livelihood. It may also hugely affect their food and nutritional intake, access to health care and education of children.

The imposition of the lockdown as a measure to contain the exponential progression of the COVID-19 pandemic has hit the unskilled and semi-skilled migrant labourers the most (Dandekar and Ghai 2020). International Labour Organization (ILO) estimates are that around about 400 million workers in the informal economy are at the risk of falling deeper into poverty during the crisis. When the Government of India (GOI) announced the sudden 'lockdown' in March to contain the spread of the pandemic, migrant informal workers were mired in a survival crisis, through income loss, hunger, destitution and persecution from authorities policing containment and fearful communities maintaining 'social distance' (Sengupta and Jha 2020). The initial government response to prevent migrant movement towards their homes was informed by the fear that they would carry the contagious coronavirus to their places of origin. Despite facing problems, the migrants when they reached their villages, they were placed under home quarantine or put in quarantine centres under deplorable conditions. In some instances, the return of migrant workers leads to social tensions. Soon after the announcement of the extension of the lockdown, in Mumbai, a false rumour about a special train sent thousands of migrants to a suburban train station, creating a riot situation (Miglani and Jain 2020). An assessment by the International Monetary Fund (IMF) shows that overall support by the central and state governments through various cash and kind transfers and other measures, such as healthcare infrastructure, testing facilities and tax relief, was only about 0.2% of India's GDP (IMF Policy Tracker). The emergency created by the COVID-19 lockdown and the uncertainty about the future trajectory of the pandemic requires immediate and near-universal cash transfers to the poor. Activists and volunteers have raised alarm about the abysmal conditions in the temporary shelters where migrants are confined (Burman 2020). The high proportion of infection and registered deaths from coronavirus in crowded informal settlements, like Dharavi in Mumbai, where 800,000 people are packed in 2.5 km, underscores the welfare deficit in the shape of affordable housing, sanitation and health care for migrants and the urban poor in destination areas. Social distancing and self-quarantine are strategies in places where the poor live in tiny, crowded tenements, queue up for water and share minimal public sanitation facilities. At present, the state needs to intensify and expand the coverage of emergency relief, income, food transfers and free health services, including COVID-19 detection tests and the cost of hospitalization (Sengupta and Jha 2020). Millions of migrant workers are anticipated to be left unemployed in India due to the lockdown and subsequent fear of recession. The risk is particularly higher for those who are working in unorganized sectors, and those who do not have writer contracts or those whose contracts are at the verge of completion (Khanna 2020).

The ILO has estimated that up to 25 million people might become unemployed worldwide due to the impact of COVID-19. According to the early estimates, tens of millions of migrant workers were left unemployed in India by the end of March 2020 due to the lockdown. According to the United Nation's World Food Programme (WFP), an estimated 265 million people could be pushed to the brink of starvation by the end of the year 2020 (Dahir 2020).

The rural migrants with the existing vulnerabilities in various socio-economic and health aspects become further vulnerable during pandemics and disasters. For instance, Stranded Workers Action Networks (SWAN) observed that during 21 days of COVID-19 lockdown, 70% of workers had less than 200 rupees left with them, 98% of them had not received any cash from the government, and 89% of them had not received any pay from their employers. Further, more than 85% of workers had to pay for their journey to their place of origin from their own pockets. SWAN also stated that 90% of migrant workers in various states did not get paid by their employers, 96% did not receive ration from the government, and 70% migrants did not get cooked food (The Hindu 2020). In another estimate, it was reported that around 60 thousand people have returned from urban centres to rural areas in just a few days of the lockdown (Economic Times 2020). According to The Centre for Monitoring Indian Economy, enrolment under MGNREGA (provides 100 days of guaranteed wage employment in form of unskilled manual work) surged by 40 to 50% during the lockdown. Also, it was reported that 90 million of small traders and labourers have lost their jobs and were looking for livelihood options.

COVID-19 Situation in Mumbai

A medical survey of nearly 7000 people in Mumbai has found that one in six or about 16% of residents in the city had contracted the coronavirus. In slum areas, where lakhs of people live in cramped spaces usually sharing toilets, the number was a whopping 57% (Roy 2020). The city is home to around 1.2 crore people, of which some 65% live in slums. A further 60 lakh people are estimated to live in peripheral districts. "Antibodies stay for 3–6 months in your body. Then, they weaken. This will not create herd immunity. Unlocking is happening now, and if local trains start, then the second wave of coronavirus can come", said Dr. Avinash Bhondve, President of the Maharashtra Medical Association. An aggressive, proactive approach, called 'chasing the virus' that included tracking, tracing, testing and treating by the public health administration helped flatten the curve in Dharavi within two months (Sahu and Dobe 2020).

Ward-wise data of COVID-19 cases in Mumbai:

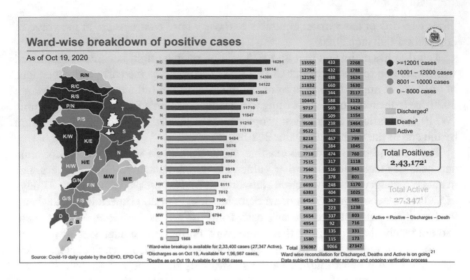

Source https://stopcoronavirus.mcgm.gov.in/assets/docs/Dashboard.pdf

The Well-Being of Migrants During Post-COVID-19: Some Suggestions

To mitigate the effect of the lockdown on the vulnerable groups, the Government of India on 26 March 2020 announced Rs. 1.70-lakh-crore package under the Pradhan Mantri Gareeb Kalyan Yojana. The Ministry of Home Affairs issued an advisory for health actions at the place of congregation of migrant workers (Government of India, 2020). The Government of India also talked about the mental health of these migrant workers and issued guidelines.

A positive step made by some of the state governments to help the return migrant labourers though welcomed, the results are yet to be assessed and its impacts are yet to be critically analysed. Some of the schemes mainly refer to: Jharkhand government's scheme of sending 1684 workers to work in Border Road Organization (BRO); the Uttar Pradesh government's initiatives through associations like FICCI, LAGHU Udyog Bandhu, Indian Industry Association and NADRECO for providing jobs to 11 lakhs skilled and semi-skilled workers of the state; Bihar government's skill development mission disbursement of rupees 1 thousand migrants by using app-based systems and Chandigarh state's programme titled Rajiv Gandhi Kisaan Nyay Yojana through which farmers would get 13 thousand acres in one year to cover the rice, maize and sugarcane farmers a minimum income support scheme (The Hindu 2020).

Given the high vulnerability of migrants, especially mobile vendors who are the poorest of the poor in cities like Mumbai, their whereabouts especially could not be ascertained by the author of the paper. It was found that most of them could not be contacted as they already moved out of the city, and also, their mobile numbers were not reachable. The precarious nature of problems reported in the media and write-ups as reviewed in this paper have brought in the fact that they lost livelihood options due to the lockdown and subsequently moved out of the city as they could not afford to stay in the city.

The lessons to be learned in this regard are that there is a need to maintain a registration system through which the government can maintain the basic details of mobile vendors. Such data creation and management of vendors will provide them with an identity through which they can have access to obtain identity cards, licences for carrying out their economic activities, voter id, ration card and other such documents for their day-to-day living. Role of the media in highlighting various issues that the migrants face is to be acknowledged, and they need to be encouraged to take up a proactive role in highlighting issues of this section of vulnerable urban dwellers. Mobile vending reflects on the social entrepreneurship nature among the vendors. Given the encouragement in connecting them with suitable entrepreneurial skills, professional training and connecting them to local markets, it will greatly help in achieving successful social entrepreneurs in our society.

Urban planning needs to be evolved more comprehensively by incorporating inclusiveness through fulfilling the fundamental rights of all the urban dwellers irrespective of their status, revising the existing developing plans, modifying the existing land use regulations and various government programmes and schemes, especially for the urban poor. Most of the street vendors are migrants, especially from the poorer states, and there is a need to emphasize on the creation of employment opportunities through industrialization and agriculture development in their home states. Such measures will not only develop the concerned states but also minimize the flow of migration of poor people to major cities like Mumbai for their livelihood opportunities. Also, there is a need to identify the most vulnerable sections in urban areas like mobile vendors and to connect them to other employment opportunities that are less hazardous and provide them with adequate earnings for their living.

Given the demographic dividend as reflected in the young age group of mobile vendors and also their potential to carry out their livelihood activity in such harsh situations, there is a need to harness this section of the population and provide suitable opportunities through which not only can we enhance their livelihood skills but also to make them to be the contributors to the overall development of the society.

References

Ajay Dandekar, R. G. (2020). Migration and Reverse Migration. *Economic and Political Weekly*, 28–31.

Appadurai, A. (2000). Spectral housing and urban cleansing: Notes on millennial Mumbai. *Public Culture, 12*(3), 627–651.

Bhagat, R., et al. (2020). *The COVID-19, Migration and livelihood in India a background paper for policy makers international institute for population sciences, Mumbai the COVID-19.* ResearchGate: Migration and Livelihood in India.

Bhowmik, S. (2001). Hawkers and the urban informal sector: A study of street vending in seven cities. *Prepared for National Alliance of Street Vendors in India (NASVI).* Available at: https://wiego.org/sites/wiego.org/files/publications/files/Bhowmik-Hawkers-URBAN-INFORMAL-SECTOR.pdf.

Bhowmik, S. K. (2003, April 19). National policy for street vendors. *Economic and Political Weekly,* 1543–46.

Bhowmik, S. (2010). Legal protection for street vendor. *Economic and Political Weekly,* 12–15.

Bhowmik, S. (2010). Street vendors in the global urban economy. Routledge.

Bhowmik, S. K., & Saha, D. (2012). Street vending in ten cities in India. Delhi National Association of Street Vendors of India.

Bhowmik, S. K., & Saha, D. (2011). *Financial accessibility of the street vendors in India: Cases of inclusion and exclusion.* Mumbai: Tata Institute of Social Sciences.

BMC (2020) *Corona virus ward wise data.* https://stopcoronavirus.mcgm.gov.in/assets/docs/Das hboard.pdf.

Burman, S. (2020, April 28). Delhi police report on migrant camps: Fans not working, bad food. Retrieved from https://indianexpress.com/article/india/delhi-police-report-on-migrant-camps-fans-not-working-bad-food-6382213/.

Chen, M. A., & Raveendran, G. (2012). Urban employment in India: Recent trends and patterns. *Margin: The Journal of Applied Economic Research, 6*(2), 159–179.

Dahir, A. L. (2020, April 22). Instead of coronavirus, the hunger will kill us. A Global Food Crisis Looms. *The Economic Times.* https://www.nytimes.com/2020/04/22/world/africa/coronavirus-hunger-crisis.html.

Dandekar, A., & Ghai, R. (2020). Migration and reverse migration in the age of COVID-19. *Economic and Political Weekly, 55*(19), 28–31.

Economic Times, https://economictimes.indiatimes.com/news/politics-and-nation/mp-cm-shivraj-singh-hints-at-extending-lockdown-says-lives-of-people-more-important/videoshow/750331 00.cms.

Gore, M. S. (1968). *Urbanization and family change.* Popular Prakashan.

Government of India. (2020). https://www.dgmhup.gov.in/documents/Advisory_Quarantine_Migr ant_Workers.pdf.

ILO. (2004). The informal economy and workers in Nepal. international labour organization (ILO), Series 1, Kathmandu, Nepal.

International Labour Organization. (2020). ILO Monitor 2nd edition: COVID-19 and the world of work, Available at: https://www.ilo.org/global/topics/coronavirus/impactsand-responses.

Khanna, A. (2020). Impact of migration of labour force due to global COVID-19 pandemic with reference to India. *Sage,* 181–191.

Miglan, S, & Jain, R, (2020, April 14). Update 3-India extends world's biggest lockdown, ignites protest by migrant workers, Retrieved from https://br.reuters.com/article/asia/idUSL3N2C2ILI.

Mohapatra, K. (2012). Women workers in informal sector in India: Understanding the occupational vulnerability. *International Journal of Humanities and Social Science, 2*(21), 197–207.

Monalisha Sahu, M. D. (2020). *How the largest slum in India flattened the COVID curve?* SEEJPH: A Case Study.

NCEUS. (2007). *Report on conditions of work and promotion of livelihoods in the unorganised sector, national commission for enterprises in the unorganised sector.* New Delhi: Govt. of India.

Railways, M. O. (2015). Retrieved from Indian Railways.

Roy, D. D. (2020, July). *Mumbai survey finds 57% have had COVID-19 in slums, 16% in other areas.* Retrieved from https://www.ndtv.com/mumbai-news/coronavirus-mumbai-sero-survey-finds-57-have-had-COVID-19-in-slums-16-in-other-areas-2270412.

Sahas, J. (2020). *Voices of the invisible citizens: A rapid assessment on the impact of COVID-19 lockdown on internal migrant workers.* April, New Delhi.

Sharma, K. (2012). Role of women in informal sector in India. *Journal of Humanities and Social Science, 4*(1), 29–36.

Sohini Sengupta, M. K. (2020). Social policy, COVID-19 and impoverished migrants: Challenges and prospects in Locked Down India. *Sage*, 152–172.

The Hindu. (2020, April). *Data | 96% migrant workers did not get rations from the government, 90% did not receive wages during lockdown: survey.* Retrieved from https://www.thehindu.com/data/data-96-migrant-workers-did-not-get-rations-from-the-government-90-did-not-receive-wages-during-lockdown-survey/article31384413.ece.

The Hindu. (2020, June). *Jharkhand gives BRO approval to recruit over 11,800 workers for border area projects.* Retrieved from https://www.thehindu.com/news/national/other-states/jharkhand-gives-bro-approval-to-recruit-over-11800-workers-for-border-area-projects/article31779372.ece.

The Railway Protection Force Act. (1957). Govt of India. https://www.indiacode.nic.in/handle/123456789/1450?view_type=browse&sam_handle=123456789/1362.

Tinker, I. (2003). Street foods: Traditional microenterprise in a modernizing world. *International Journal of Politics, Culture, and Society, 16*(3), 331–349.

YUVA/TISS. (1998). 'Census survey of hawkers on BMC lands', survey conducted by Tata institute of social sciences and youth for voluntary action and unity for Brihanmumbai Municipal Corporation.

Beating Back COVID-19 in Mumbai

Hem H. Dholakia

Abstract The world is facing an unprecedented health crisis. With over 279,000 cases and 10,000 deaths (as on 27 November 2020), Mumbai remains one of the most affected cities in India due to the COVID-19 pandemic. The Municipal Corporation has mounted an aggressive response that includes tracking, testing, tracing and treatment of patients. More than 1.8 million tests have been conducted, and more than 3.9 million contacts have been traced. Nearly, 18,000 hospital beds and 70,000 beds in care centres dedicated to COVID have been established in Mumbai. With stringent social distancing measures and infrastructure for medical management, Mumbai has been able to reduce COVID-related deaths despite the increase in number of cases. The COVID-19 crisis is far from over. The pandemic has laid bare inadequacies across science, emergency response, health systems strengthening as well as policy across the globe. Yet, this provides a great opportunity for policy makers to address deep-rooted structural challenges. Lessons from this experience will help in overcoming future pandemics.

Keywords COVID-19 · Mumbai · Pandemic · Health systems · Mental health · Megacity

Introduction

With nearly 1.44 million deaths and more than 61 million recorded cases globally (as on 27 November 2020), the Coronavirus-19(COVID-19) pandemic remains an unprecedented global crisis.[1] The knowledge of the infection, its control, community engagement, health system capacity, vaccine development, antiviral therapy, border controls and stabilising the economy have all been major challenges posed by the

---

[1] Data are sources from Google News https://news.google.com/COVID19/map?hl=en-IN&mid=%2Fm%2F03rk0&gl=IN&ceid=IN%3Aen Last accessed 27 September 2020.

H. H. Dholakia (✉)
Independent Health and Environment Expert, 1A/12 Talmakiwadi, Tardeo Road, Mumbai 400007, India
e-mail: dr.dholakia@gmail.com

© The Author(s), under exclusive license to Springer Nature Switzerland AG 2021
R. Akhtar (ed.), *Coronavirus (COVID-19) Outbreaks, Environment and Human Behaviour*, https://doi.org/10.1007/978-3-030-68120-3_10

pandemic (Han et al. 2020). Urgent efforts to generate knowledge based on science, formulate effective policies, develop appropriate metrics for monitoring, to stem the spread of the virus and provide quality medical care have overwhelmed health systems worldwide. An immediate response by countries, the world over, has been partial or complete lockdown. The lockdowns have impacted business, wellbeing and daily life (Chu et al. 2020). Low-resource settings, such as India, have been disproportionately burdened by the pandemic.

The first case in India was detected on 30 January 2020 where the affected had a travel history to Wuhan, China (Vara 2020). Since January, India has recorded nearly 9.3 million cases and about 136,000 COVID-19 deaths (as on 27 November 2020) (See Footnote 1). Across the different states, the highest number of cases have been recorded in Maharashtra (~1.8 million cases), Karnataka (~881,000 cases), Andhra Pradesh (~866,000 cases) and Tamil Nadu (~778,000 cases). There is a higher prevalence of COVID-19 cases in urban areas. On average, urban districts reported 2.7 times more cases per million persons as compared with rural areas (Kawoosa 2020). This disparity may be explained by the higher population density in urban areas that facilitates virus transmission. In addition, rural areas have a lower share of COVID-19 testing (Kawoosa 2020). Some of the cities most affected include Mumbai, Delhi, Hyderabad, Bengaluru and Patna.

Mumbai is the fifth largest megacity in the world with a population of over 12.5 million people. It is also the financial capital of India, a base for many industries and multi-national companies, contributing 5% to the GDP of India. The city is densely populated with more than 55% people live in slums (Bhagat et al. 2006). On average, the population density of Mumbai is 20000 people per square kilometre. This overcrowding poses several challenges as people lack access to good sanitation, clean water and public health services (Raju and Ayeb-Karlsson 2020). From the perspective of COVID-19, the precarious conditions of living in densely populated areas precludes the possibility of social distancing and may accelerate the spread of the disease (Wasdani and Prasad 2020). Therefore, while social distancing may be the most effective way to control the spread of the pandemic, it is not a realistic option for thousands of people. As on 27 November 2020, Mumbai had recorded more than 279,000 cases and 10,700 deaths related to COVID-19.

COVID-19 Cases and Deaths in Mumbai

Central to the COVID-19 response has been the collection of meaningful data. For most countries, this implies data on confirmed cases relative to the testing that has been carried out. Through the initial phase of the outbreak, criteria for diagnosis were being understood, testing kits were not as widely available and the sensitivity and specificity of tests were poorly understood (Fang et al. 2020). Today, as laboratories have scaled up their operations, serological testing as well as real-time polymerase chain reaction (RT-PCR) remain far more accessible. Further, better recognition of clinical signs and symptoms are leading to a more accurate diagnosis.

In India, the recommended test for COVID-19 diagnosis by the Indian Council on Medical Research (ICMR) is the real-time polymerase chain reaction (RT-PCR) across all symptomatic persons (ICMR 2020). For rapid surveillance in containment areas, serological tests (as per ICMR protocol) could be used. However, recent systematic reviews show that higher quality studies assessing the diagnostic accuracy of serological surveys are needed (Bastos et al. 2020).

For Mumbai, a key step in COVID-19 surveillance and management has been the sharing of data by the Municipal Corporation of Greater Mumbai (MCGM). The institution has played a key role in testing and identification of cases and management. Through their official twitter account, MCGM has shared official data on COVID-19 starting 16th March 2020. The data reports include the total number of suspected cases, total positive cases, total deaths and the number of recoveries on a daily basis. We have compiled data over the last nine months (from 16 March 2020 to 27 November 2020) in Figs. 1, 2 and 3, respectively.

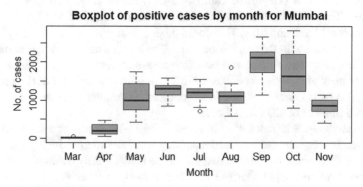

Fig. 1 Boxplot of COVID-19 positive cases in Mumbai. *Source*: Author's compilation from various MCGM reports

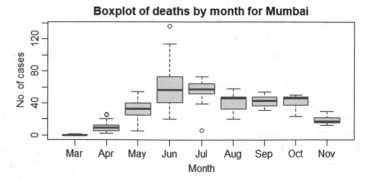

Fig. 2 Boxplot of COVID-19 deaths in Mumbai. *Source*: Author's compilation from various MCGM reports

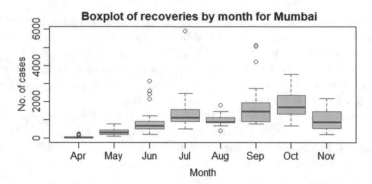

Fig. 3 Boxplot of COVID-19 recoveries in Mumbai. *Source*: Author's compilation from various MCGM reports

Figure 1 shows a boxplot of confirmed COVID-19 cases that from March to November. There has been a sharp rise in the median number of COVID positive cases between March and May. This could be due to an expansion of clinical testing as well as an increase in the number of cases. The median number of cases stabilised through June, July and August and saw an increase in September. In September, the median number of new cases per day were 2163, twice the number of the preceding months. The subsequent months of October and November have seen a decline in the median number of COVID-19 cases.

Figure 2 shows a boxplot of confirmed COVID-19 deaths that from March to November. The highest median deaths per day were recorded in July (57 deaths per day). Since then, there has been a decline in the median daily deaths in August (46 deaths per day) and September (43 deaths per day), October and November. The fact that absolute number of deaths are declining despite an increase in the number of cases is a positive sign. While several factors may have contributed to this, one factor is the responsiveness of the health system in providing appropriate medical management.

We find that the number of recoveries for people with COVID-19 is increasing over time (Fig. 3). The highest median recovery was observed in October. This is a positive sign and reflects the effectiveness of interventions that have been put in place.

A pertinent question that remains relates to the dynamics of COVID-19 and future spread of the disease. Few studies have attempted to predict the future course of the pandemic for India using mathematical, statistical and ensemble models (Sardar et al. 2020). Predictions are estimated for a short time period 15 to 20 days in advance for the month of May 2020 and indicate an increase in number of cases for the short term. Given the short time-series data that were available (March through April) to calibrate these models, predictions for longer time horizons were difficult (Sardar et al. 2020). The modelling studies also suggest that elimination of the disease is possible through a reduction in the contact rate between infected and uninfected individuals through interventions such as contact tracing and social distancing (Sarkar et al. 2020). As our understanding improves and more data are available, more studies that assess future transitions of the pandemic are required.

Distribution of Cases Within Mumbai

Whereas understanding the absolute burden of disease due to COVID-19 is important, it is equally necessary to study the distribution of disease across geographies, age groups and income classes. COVID-19 is disproportionately affecting the poor minorities due to the inequitable spread in densely populated areas (Shadmi et al. 2020). Asia's largest slum, Dharavi, has a population density of 335,000 people per square kilometre (Nutkiewicz et al. 2018). There were concerns that spread of infection in Dharavi would overwhelm public health authorities. An aggressive, proactive approach, called 'chasing the virus' that included tracking, tracing, testing and treating by the public health administration helped flatten the curve in Dharavi within two months (Sahu and Dobe 2020).

Table 1 shows the ward-wise distribution of COVID-19 cases in Mumbai as on 3 June 2020. Some wards (namely A, B, D, E, F-South, G-South) belonging to the southern and central portions of Mumbai had recorded more than 5 cases per 1000 population on average, and wards in the northern part had less than three cases. Figure 4 shows the number of days for doubling of COVID-19 cases. What is interesting to note is that wards with higher cases per 1000 population (namely A, B, D, E, F-South, G-South) showed a slower rate of spread of the infection. A more detailed study trying to understand the drivers behind these observations is required.

Mounting a Response for the COVID-19

Containing the COVID-19 pandemic necessitated a multi-pronged response that addresses prevention as well as provision of quality medical care to affected persons. Lockdowns, social distancing, surveillance, contact tracing and community education have been some of the preventive aspects. The medical side has involved ramping up availability of medical beds, protective equipment and medical devices such as ventilators as well as having adequate human resources.

The first nationwide lockdown for fourteen hours was on March 22, which was followed by 21 days lockdown starting from March 24. Since then, there have been several extensions of the lockdown (Phases 2, 3 and 4). With each phase of the lockdown, there has been a change in the rules that govern the plying of vehicles on the roads, running of shops and businesses, gathering in public places as well as opening of schools, colleges and educational institutions. India also enforced a closure of its international borders, thereby giving the government time to prepare a

Table 1 Distribution of COVID-19 cases across different wards in Mumbai

Ward	Population	Total COVID-19 positive cases	Cases per 1000 population
A	185,014	1075	5.8
B	127,290	568	4.5
C	166,161	476	2.9
D	346,866	1921	5.5
E	393,286	2738	7.0
F-North	529,034	2770	5.2
F-South	360,972	2269	6.3
G-North	599,039	3258	5.4
G-South	377,749	2239	5.9
H-East	557,239	2476	4.4
H-West	307,581	1102	3.6
K-East	823,885	2571	3.1
K-West	748,688	2559	3.4
L	902,225	2847	3.2
M-East	807,720	2124	2.6
M-West	411,893	1496	3.6
N	622,853	2127	3.4
P-North	941,366	1683	1.8
P-South	463,507	1163	2.5
R-Central	562,162	981	1.7
R-North	431,368	506	1.2
R-South	691,229	1227	1.8
S	743,783	1882	2.5
T	341,463	908	2.7

Source: Author's compilation based on MCGM reports as on June 3, 2020

coherent response for tackling the pandemic (The Lancet 2020) . This move made it the largest COVID-19 national lockdown in the world.

Under the fifth phase, the lockdown has been extended until 31 October 2020 (Banerjea 2020).

Surveillance and Contact Tracing

Surveillance is critical for the development and implementation of evidence-based interventions. Early detection helps control the outbreak at source, mitigates impacts by stopping spread of the virus, informs design of interventions and helps optimise

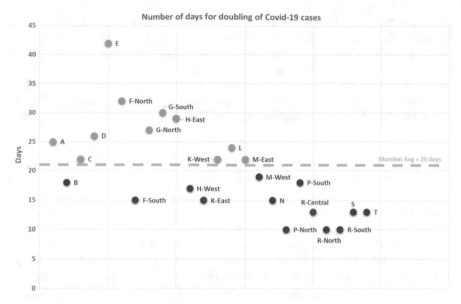

Fig. 4 Ward-wise distribution for expected number of days for doubling of COVID-19. *Source*: Author's compilation based on MCGM Reports as on June 3, 2020

resource use (WHO 2017). In the past, surveillance has played a central role in containing several outbreaks such as Nipah Virus (Plowright et al. 2019), MERS (Memish et al. 2014) as well as SARS (Heymann and Rodier 2004). In addition to surveillance, contact tracing is a common intervention for controlling disease outbreaks. It involves the identification of contacts and their isolation to stem the spread of the infection (Hellewell et al. 2020). Contact tracing of individuals, the use of cluster approach as well as technology such as AarogyaSetu app have been the key for contact tracing in Maharashtra (The Lancet 2020) . Contact tracing applications are designed to assist health officials in tracking down exposures after an infected individual is identified and often rely on Bluetooth technology.

The government deployed hundreds of doctors and healthcare workers for surveillance in Mumbai. Through several campaigns, the Mumbai Municipal Corporation invited professionals to volunteer their time for COVID-19 surveillance as well as treatment. In areas such as Dharavi, door-to-door screening was conducted with particular attention to houses or workspaces in the vicinity of positive cases (Golechha 2020). In addition to RT-PCR tests for symptomatic people, use of pulse oximeters and thermal scanners were deployed for mass screening among vulnerable groups such as the elderly. People who tested positive were isolated in quarantine facilities. To ensure physical distancing during the lockdown, Maharashtra has used drones (The Lancet 2020) .

As on 3 October 2020, there were 1.1 million tests conducted in Mumbai. Of these, more than 200,000 positive cases were detected and more than 3.1 million contacts have been traced. Of the contacts traced, 40% have been identified as high risk.[2] One innovative step has been the use of artificial intelligence (AI)-based voice testing for the detection of COVID (ETGovernment 2020). The application asks the suspected patient to speak, and analysis of the voice indicates the presence of COVID. If suspected, a sample will be taken to conduct the RT-PCR test (ETGovernment 2020). Given the ubiquitous penetration of smartphones across India (Agrawal 2018), this would help in increasing the number of people tested.

Community Education and Campaigns

One challenge in controlling the pandemic is the spread of misinformation, fear and stigma. Education of people on topics such as COVID symptoms, mode of spread and treatment by providing actionable information empowers them to take decisions to protect health (WHO 2020). Empowerment education is important for behaviour and social change in the context of COVID-19.

Concerted efforts and campaigns for education have been launched by the Mumbai Municipal Corporation. A key campaign 'My Family, My Responsibility' launched on 15 September 2020 contains a series of simple and implementable steps to prevent spread of the infection.[3] These steps outline measures for housing societies, shopping malls, offices as well as travel in public or private vehicles. Some of the measures include the use of masks, maintaining a safe distance, frequent hand washing, use of COVID-19 tracing applications and sanitisation of surfaces as well as limiting the number of people who can congregate in a single place at any given time. Several NGOs have supported this through designing knowledge collaterals in the local language on social distancing, sanitisation and limiting congregations. They have also spearheaded COVID education in the communities where they provide health and education services. COVID information as part of the caller tune (when phoning a person) jingles on the radio, and use of hoardings have been other media to spread awareness. Several film stars and prominent personalities have contributed to infor-mation campaigns run by the Municipal Corporation. These efforts have helped in creating mass awareness about the disease.

Health Infrastructure

Health infrastructure has played a key role in the management of COVID-19. At the core of infrastructure has been making more than 18,000 hospital beds available

---

[2]Data from the Mumbai Municipal Corporation available at https://stopcoronavirus.mcgm.gov.in/key-updates-trends.

[3]Public information available from the official Twitter Account of the Mumbai Municipal Corporation.

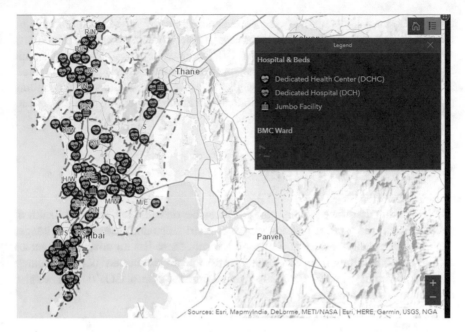

Fig. 5 Health infrastructure for addressing COVID-19 in Mumbai. *Source*: Available at https://sto pcoronavirus.mcgm.gov.in/IoM-treatment-facilities

for COVID-19 patients. These include 15,000 beds at dedicated COVID hospitals and health centres, 2000 intensive care unit beds, 9000 beds with oxygen capacity and 1100 ventilator beds (See Footnote 2). Emphasis has been laid on setting up of institutional quarantine facilities. There are a total of 70,000 beds in COVID Care Centres (See Footnote 2). Figure 5 shows the distribution of these centres across Mumbai.

In addition to infrastructure development, a multi-pronged strategy that includes—decentralised hospital bed management system; setting up of 24/7 war rooms across all 24 wards; deployment of 10 ambulances and 30 phone lines in each War Room dedicated for COVID-19 patients and a dashboard with real-time data—has been adopted.

As soon as an individual is diagnosed to be COVID positive, a copy of the laboratory report is shared with the Mumbai Municipal Corporation. Public health experts posted within the War Room contact the person and discuss and determine the conditions for quarantine. A detailed history of all their contacts within the last seven days is made. These contacts are isolated at home and followed up on a regular basis. For the elderly, people with co-morbidities or symptoms, admission to dedicated COVID hospitals is preferred. Young or asymptomatic persons are quarantined and monitored at COVID Care Centres (CCC). Medical management of all persons is undertaken, and at the end of 10 days of quarantine, people from the CCC are send for home isolation for another 14 days. In some cases, where home quarantine is advised, the War Room follows up every day to understand any change in symptoms (e.g., temperature, cough, sense of smell, shortness of breath or falling oxygen saturation). An Interactive Voice Recording System (IVR) collects key data on patients. In this manner, people can be grouped and the appropriate medical care can be provided. This leads to resource optimisation as well as improved health outcomes. The falling number of COVID deaths, despite an increase in number of cases, partially reflects the success of robust health infrastructure.

Other Impacts of COVID-19

COVID-19 has had far-reaching consequences beyond just the physical health of people. Social distancing, lockdowns, travel restrictions and economic challenges have impacted daily society and life. As a result, there has an adverse impact on mental health resulting from emotional distress and an increased risk of psychiatric illness (Pfefferbaum and North 2020). The measures to control COVID-19 have also resulted in environmental and economic impacts.

Mental Health

COVID-19 has thrown the world and lives of people into a state of uncertainty. Uncertain prognoses, resource shortages, financial losses, conflicting messages on infection prevention, lack of vaccines and curtailment of personal freedom are some drivers of worsening mental health (Pfefferbaum and North 2020). Mental health challenges have been observed across COVID-19 patients, medical caregivers as well as society at large (Chen et al. 2020).

In India, studies have documented an increased incidence of depression, anxiety, stress (Verma and Mishra 2020) as well as cases of self-harm and suicide during the pandemic (Dsouza et al. 2020). For instance, the Indian Psychiatry Society surveyed 1871 persons online and found that more than two-fifths of the people were experiencing mental disorders due to the lockdown (Grover et al. 2020a). These findings have been corroborated by other cross-sectional surveys that document sleep difficulties, paranoia about acquiring COVID-19 infection and distress-related information on social media (Roy et al. 2020). In the context of Mumbai, there are few studies that look at mental health issues in light of COVID-19. Someshwar et al. (2020) studied the health-related quality of life in Mumbai and found that social distancing negatively impacted physical and psychological domains.

There has been a clarion call for the need to intensify awareness around mental health and institute systems to expand mental health services across India. Though there is a slow expansion in telecommunication and online services, majority of health centres providing advanced care have completely stopped on account of the pandemic (Grover et al. 2020b). Further, a focus on treating COVID-19 patients has led to the collapse of regular mental health services (Grover et al. 2020b). Yet, meeting this demand remains an uphill task. Education and training of psychosocial issues for health system leaders; addressing needs for special populations and evidence-based community health interventions remains the need of the hour (Pfefferbaum and North 2020). Recently, on 7 September 2020, the Mumbai Municipal Corporation set up a helpline for emotional wellness in partnership with civil society organisations. This is a positive step to address mental health in Mumbai. Going forward, mental health should be incorporated as one of the key pillars of the overall pandemic management programme across India.

Environment

The imposition of restrictions on account of COVID-19 has had several indirect impacts. A prominent impact among these is a dramatic reduction in air pollution levels across Indian cities. For instance, in Delhi, data across 34 monitoring stations show a 39% reduction in fine particulate matter ($PM_{2.5}$) and 52% reduction in nitrogen oxides (between 24 March 2020 and 3 May 2020) as compared with the same time period during the previous year (Mahato et al. 2020). This is significant as Indian cities are some of the most polluted in the world. Air pollution is among the largest environmental risks and is responsible for a large disease burden (Dholakia and Garg 2017). It is estimated that air pollution contributed to 1.24 million deaths in India in 2017, which were 12.5% of the total deaths (Balakrishnan et al. 2019).

Particulate matter in India has contributions from vehicles, residential energy use, industrial activity and background dust (Guo et al. 2019). Lockdowns have meant curtailment of transport, construction as well as industrial activities. As a result, emissions from passenger and commercial vehicles, air planes, construction-related activity as well as industrial emissions have declined (Sharma et al. 2020). Further, the reduction of primary emissions implies lower pollutant formation through secondary chemical reactions in the atmosphere (Purohit et al. 2019). Mumbai recorded a 76% decline in the nitrogen oxide level concentration as well as a 54% decline in particulate matter during the lockdown compared to the same period in 2019 (Borwankar 2020). Compared to other cities—Kolkata, Delhi and Ahmedabad—Mumbai saw the highest gains in air quality during the lockdown period. These findings suggest that policies, such as a shift to public transport or reducing energy demand across sectors, as outlined in India's National Clean Air Plan (NCAP) will be beneficial in cutting pollution, if implemented stringently.

Conclusion

The COVID-19 crisis is far from over. As the number of cases continues to grow, measures to curb the pandemic remain in place. The world over while most countries are doing a reasonable job in managing the health impacts, they should also focus on policies to address the economic and social challenges that are likely to ensue. Pharmaceutical companies have invested heavily in vaccine development. However, there remains uncertainty with respect to the large-scale roll-out of COVID-19 vaccine.

The city of Mumbai which has been severely impacted by the pandemic has played a leading role in pro-actively managing the crisis. Yet, as the number of cases increase, there remains the need to be vigilant and step up health interventions. The pandemic has laid bare inadequacies across science, emergency response, health systems strengthening as well as policy across the globe. Yet, this provides a great opportunity for policy makers to address deep-rooted structural challenges. Lessons from this experience will help in overcoming future pandemics. COVID-19 has paved the way for the world to deal with low-probability, high-impact events. It also serves as a stark reminder to safeguard fragile ecosystems of the world.

Addendum

At the time of publication, Mumbai had carried out more than 3 million COVID-19 tests. Of these, 316487 people had tested positive (as on 18 February 2021). The recovery rate was close to 94 percent. However, over the last few days, there has been a surge in the number of positive cases. Recognizing this, the Municipal Corporation has bolstered its efforts for testing and treatment. Maharashtra state is bringing in stringent measures for social distancing and isolation to prevent the spread. A positive development has been the commencement of the vaccination drive for COVID-19. However, there are concerns about newer strains from UK, South Africa and Brazil entering India. The efficacy of the vaccine against these newer strains remains to be ascertained.

References

Agrawal, R. (2018). *India connected: how the smartphone is transforming the world's largest democracy*. New York: Oxford University Press.

Balakrishnan, K., Dey, S., Tarun Gupta, R. S., Dhaliwal, M. B., Cohen, A. J., Stanaway, J. D., et al. (2019). The impact of air pollution on deaths, disease burden, and life expectancy across the states of India: The global burden of disease study 2017. *the Lancet Planetary Health, 3*(1), e26-39. https://doi.org/10.1016/S2542-5196(18)30261-4.

Banerjea, A. (2020). 'Unlock 5: Lockdown extended till 31 Oct, cinema halls can operate at 50% capacity'. *Mint*, 1 October 2020. https://www.livemint.com/news/india/unlock-5-lockdown-ext ended-to-31-oct-cinema-halls-can-operate-at-50-capacity-11601472472816.html.

Bastos, M. L., Tavaziva, G., Abidi, S. K., Campbell, J. R., Haraoui, L.-P., Johnston, J. C., & Lan, Z. et al. (2020). 'Diagnostic accuracy of serological tests for COVID-19: Systematic review and meta-analysis'. *BMJ 370*(July). https://doi.org/10.1136/bmj.m2516.

Bhagat, R. B., Guha, M., & Chattopadhya, A. (2006). Mumbai after 26/7 Deluge. *Population and Environment, 27,* 337–349.

Borwankar, V. (2020). 'Air pollution levels in Mumbai drop by 76% post lockdown'. *The TImes of India,* 13 May 2020. https://timesofindia.indiatimes.com/city/mumbai/air-pollution-levels-in-mumbai-drop-by-76-post-lockdown/articleshow/75708175.cms#:~:text=MUMBAI%3A%20The%20lockdown%20has%20a,air%20among%20five%20other%20cities.&text=During%20the%20lockdown%2C%20SAFAR%20monitored,%2C%20Chennai%2C%20Pune%20and%20Kolkata.

Chen, Q., Liang, M., Li, Y., Guo, J., Fei, D., Wang, L., et al. (2020). Mental health care for medical staff in China during the COVID-19 outbreak. *The Lancet Psychiatry, 7*(4), e15-16. https://doi.org/10.1016/S2215-0366(20)30078-X.

Chu, D. K., Akl, E. A., Duda, S., Solo, K., Yaacoub, S., Schünemann, H. J., et al. (2020). Physical distancing, face masks, and eye protection to prevent person-to-person transmission of SARS-CoV-2 and COVID-19: A systematic review and meta-analysis. *The Lancet, 395*(10242), 1973–1987. https://doi.org/10.1016/S0140-6736(20)31142-9.

Dholakia, H. H., & Garg, A. (2017). 'Climate change, air pollution and human health in Delhi, India'. In *Climate change and air pollution.* Springer Climate Book Series. Delhi: Springer. https://doi.org/10.1007/978-3-319-61346-8_17.

Dsouza, D. D., Quadros, S., Hyderabadwala, Z. J., & Mamun, M. A. (2020). Aggregated COVID-19 suicide incidences in India: Fear of COVID-19 infection is the prominent causative factor. *Psychiatry Research, 290*(August), 113145. https://doi.org/10.1016/j.psychres.2020.113145.

ETGovernment. (2020). 'Mumbai: BMC to use ai-based voice analysis test to detect COVID-19'. *The Economic Times,* 10 August 2020. https://government.economictimes.indiatimes.com/news/technology/mumbai-bmc-to-use-ai-based-voice-analysis-test-to-detect-Covid-19/77456454.

Fang, Y., Zhang, H., Xie, J., Lin, M., Ying, L., Pang, P., & Ji, W. (2020). Sensitivity of chest CT for COVID-19: Comparison to RT-PCR. *Radiology, 296*(2), E115–E117. https://doi.org/10.1148/radiol.2020200432.

Golechha, M. (2020). 'COVID-19 Containment in Asia's largest urban slum dharavi-Mumbai, India: Lessons for policymakers globally'. *Journal of Urban Health,* August. https://doi.org/10.1007/s11524-020-00474-2.

Grover, S., Mehra, A., Sahoo, S., Avasthi, A., Tripathi, A., D'Souza, A., et al. (2020a). State of mental health services in various training centers in India during the lockdown and COVID-19 pandemic. *Indian Journal of Psychiatry, 62*(4), 363. https://doi.org/10.4103/psychiatry.IndianJPsychiatry_567_20.

Grover, S., Sahoo, S., Mehra, A., Avasthi, A., Tripathi, A., Subramanyan, A., et al. (2020b). Psychological impact of COVID-19 lockdown: An online survey from India. *Indian Journal of Psychiatry, 62*(4), 354. https://doi.org/10.4103/psychiatry.IndianJPsychiatry_427_20.

Guo, H., Kota, S. H., Sahu, S. K., & Zhang, H. (2019). Contributions of local and regional sources to PM2.5 and its health effects in North India. *Atmospheric Environment, 214*(October), 116867. https://doi.org/10.1016/j.atmosenv.2019.116867.

Han, E., Tan, M. M. J., Turk, E., Sridhar, D., Leung, G. M., Shibuya, K. & Asgari, N., et al. (2020, Sep). 'Lessons learnt from easing COVID-19 Restrictions: An analysis of countries and regions in Asia Pacific and Europe'. *The Lancet.* https://doi.org/10.1016/S0140-6736(20)32007-9.

Hellewell, J., Abbott, S., Gimma, A., Bosse, N. I., Jarvis, C. I., Russell, T. W., et al. (2020). Feasibility of controlling COVID-19 outbreaks by isolation of cases and contacts. *The Lancet Global Health, 8*(4), e488–e496. https://doi.org/10.1016/S2214-109X(20)30074-7.

Heymann, D., & Rodier, G. (2004). Global surveillance, national surveillance, and SARS. *Emerging Infectious Diseases, 10*(2), 173–175.

ICMR. (2020). 'Advisory on strategy for COVID-19 testing in India'. National task force on COVID-19. New Delhi: Indian council on medical research. https://www.icmr.gov.in/pdf/Covid/strategy/Testing_Strategy_v6_04092020.pdf.

Kawoosa, V. M. (2020). 'What data tells us about testing in Urban and rural India'. Hindustan times. 24 September 2020. https://www.hindustantimes.com/india-news/what-data-tells-us-about-testing-in-urban-and-rural-india/story-woXp8xYNrpARnkeesAdKXI.html.

Lancet, T. (2020). India under COVID-19 lockdown. *Lancet (London, England), 395*(10233), 1315. https://doi.org/10.1016/S0140-6736(20)30938-7.

Mahato, S., Pal, S., & Ghosh, K. G. (2020). Effect of lockdown amid COVID-19 pandemic on air quality of the megacity Delhi, India. *The Science of the Total Environment, 730*(August), 139086. https://doi.org/10.1016/j.scitotenv.2020.139086.

Memish, Z. A., Almasri, M., Turkestani, A., Al-Shangiti, A. M., & Yezli, S. (2014). Etiology of severe community-acquired pneumonia during the 2013 Hajj—Part of the MERS-CoV surveillance program. *International Journal of Infectious Diseases, 25*(August), 186–190. https://doi.org/10.1016/j.ijid.2014.06.003.

Nutkiewicz, A., Jain, R. K., & Bardhan, R. (2018). Energy modeling of urban informal settlement redevelopment: exploring design parameters for optimal thermal comfort in Dharavi, Mumbai, India. *Applied Energy, 231*(December), 433–445. https://doi.org/10.1016/j.apenergy.2018.09.002.

Pfefferbaum, B., & North, C. S. (2020). Mental health and the COVID-19 pandemic. *New England Journal of Medicine, 383*(6), 510–512. https://doi.org/10.1056/NEJMp2008017.

Plowright, R. K., Becker, D. J., Crowley, D. E., Washburne, A. D., Huang, T., Nameer, P. O., et al. (2019). Prioritizing surveillance of Nipah virus in India. *PLOS Neglected Tropical Diseases, 13*(6), e0007393. https://doi.org/10.1371/journal.pntd.0007393.

Purohit, P., Amann, M., Kiesewetter, G., Rafaj, P., Chaturvedi, V., Dholakia, H. H., et al. (2019). Mitigation pathways towards national ambient air quality standards in India. *Environment International, 133*(December), 105147. https://doi.org/10.1016/j.envint.2019.105147.

Raju, E., & Ayeb-Karlsson, S. (2020). 'COVID-19: How do you self-isolate in a refugee camp?' *International Journal of Public Health*, May, 1–3. https://doi.org/10.1007/s00038-020-01381-8.

Roy, D., Tripathy, S., Kar, S. K., Sharma, N., Verma, S. K., & Kaushal, V. (2020). Study of knowledge, attitude, anxiety and perceived mental healthcare need in Indian population during COVID-19 pandemic. *Asian Journal of Psychiatry, 51*(June), 102083. https://doi.org/10.1016/j.ajp.2020.102083.

Sahu, M., & Dobe, M. (2020). 'How the largest slum in India flattened the COVID curve? A case study'. *South Eastern European Journal of Public Health (SEEJPH)*, August. https://www.seejph.com/index.php/seejph/article/view/3614.

Sardar, T., Nadim, S. S., Rana, S., & Chattopadhyay, J. (2020). Assessment of lockdown effect in some states and overall India: A predictive mathematical study on COVID-19 outbreak. *Chaos, Solitons, and Fractals, 139*(October), 110078. https://doi.org/10.1016/j.chaos.2020.110078.

Sarkar, K., Khajanchi, S., & Nieto, J. J. (2020). Modeling and forecasting the COVID-19 pandemic in India. *Chaos, Solitons, and Fractals, 139*(October), 110049. https://doi.org/10.1016/j.chaos.2020.110049.

Shadmi, E., Chen, Y., Dourado, I., Faran-Perach, I., Furler, J., Hangoma, P., et al. (2020). Health equity and COVID-19: Global perspectives. *International Journal for Equity in Health, 19*(1), 104. https://doi.org/10.1186/s12939-020-01218-z.

Sharma, S., Zhang, M., Anshika, J. G., Zhang, H., & Kota, S. H. (2020). Effect of restricted emissions during COVID-19 on air quality in India. *Science of the Total Environment, 728*(August), 138878. https://doi.org/10.1016/j.scitotenv.2020.138878.

Someshwar, H., Sarvaiya, P., Desai, S., Gogri, P., Someshwar, J., Mehendale, P. & Bhatt, G. (2020). 'Does social distancing during the lock down due to COVID-19 outbreak in Mumbai affect quality of life?' *International Journal of Clinical and Biomedical Research*, April, 1–4. https://doi.org/10.31878/ijcbr.2020.62.01.

Vara, V. (2020). 'Coronavirus in India: How the COVID-19 could impact the fast-growing economy'. Pharmaceutical technology. 30 April 2020. https://www.pharmaceutical-technology. com/features/coronavirus-affected-countries-india-measures-impact-pharma-economy/.

Verma, S., & Mishra, A. (2020). Depression, anxiety, and stress and socio-demographic correlates among general Indian public during COVID-19. *International Journal of Social Psychiatry, 66*(8), 756–762. https://doi.org/10.1177/0020764020934508.

Wasdani, K. P., & Prasad, A. (2020). The impossibility of social distancing among the urban poor: The case of an Indian slum in the times of COVID-19. *Local Environment, 25*(5), 414–418. https://doi.org/10.1080/13549839.2020.1754375.

WHO. (2017). 'WHO guidance for surveillance during an influenza pandemic'. Geneva: World Health Organisation. https://www.who.int/influenza/preparedness/pandemic/WHO_Guidance_ for_surveillance_during_an_influenza_pandemic_082017.pdf.

WHO. (2020). 'Key messages and actions for COVID-19 prevention and control in schools'. Geneva: World Health Organisation. https://www.who.int/docs/default-source/coronaviruse/ key-messages-and-actions-for-Covid-19-prevention-and-control-in-schools-march-2020.pdf? sfvrsn=baf81d52_4.

COVID-19: The Australian Experience

Kevin McCracken

Abstract This chapter examines various aspects of the COVID-19 pandemic in Australia from January to late October 2020. It shows how, by comparison with most other developed nations, the country fared well in terms of both cases and deaths suffered in the disease outbreak. The chapter conveys both an epidemiological perspective on the pandemic and related social, economic, political and behavioural dimensions. Mistakes made and painful lessons learnt from dealing with the outbreak are described. The pandemic still has a considerable way to go in Australia before any return to 'normal' life can be anticipated, the chapter concluding with a warning of the dangers of complacency towards the still-present viral threat. The scattered outbreaks of COVID-19 in eastern Australia in early March, 2021 are a relevant reminder of this.

Keywords COVID-19 · Pandemic · Lockdown · Quarantine · Long COVID · Social disadvantage · Complacency · Face masks

Introduction

In late December 2019/early January 2020, Australians, like people in most countries around the world, learnt from media reports of a novel respiratory disease identified in the city of Wuhan, Hubei Province, China. The actual first detected case of the infection, a SARS-like coronavirus, was reportedly earlier than this, on 17 November 2019, Chinese authorities allegedly initially keeping the discovery under wraps for a month or so.

Since then, the COVID-19 outbreak, as it has become known, has rapidly spread around the globe, taking on truly pandemic proportions. At the time of writing (late October 2020), there have been almost 44 million confirmed cases of the disease globally and 1.2 million deaths. A global infectious disease event like this came as

K. McCracken (✉)
Geography and Planning, Macquarie School of Social Sciences, Macquarie University, Sydney, NSW, Australia
e-mail: kevin.mccracken@mq.edu.au

© The Author(s), under exclusive license to Springer Nature Switzerland AG 2021
R. Akhtar (ed.), *Coronavirus (COVID-19) Outbreaks, Environment and Human Behaviour*, https://doi.org/10.1007/978-3-030-68120-3_11

no surprise to most international medical, public health and epidemiological professionals. Most, however, had expected a virulent influenza pandemic as the likely
prospect.

The first officially confirmed case of COVID-19 in Australia was identified on
25 January 2020, a man returning from China. In the following few weeks, the
reported case numbers slowly expanded, but public and government concern about
the outbreak was limited. To many, the infection was simply seen to be like 'a little
flu'. That assessment, however, has now been well and truly revised as over 900
persons have now officially lost their lives to the disease and more than 27,000
people have been confirmed as positive cases.

Both of these numbers, deaths and cases, in fact probably understate the true
toll, as in the early stages of the pandemic testing for the disease was relatively
limited and numerous positive cases quite likely slipped through undetected and
some deaths mistakenly deemed as other causes (e.g. pneumonia, dementia, diabetes,
etc.). Meanwhile, recent serological testing suggests further undercount of true case
numbers.

Despite the numerical uncertainty, the COVID-19 outbreak indisputably counts
as one of the country's greatest health crises ever; although by comparison with most
countries, Australia's population has survived well. In terms of deaths, the outbreak
falls a good deal short of the 1919 Spanish Flu toll (the usual comparative pandemic
yardstick) of around 15,000 deaths Australia-wide. However, the economic, social
and political effects of COVID-19 mortality and morbidity are without comparison.
Also, despite the Prime Minister's statement at the onset of the epidemic that "We are
all in this together" that homily proved well off the mark as the outbreak progressed
and multiple cleavages in Australian society were revealed. Also shown were the
various advantages and disadvantages of the Australian federal system of governance
for handling a crisis like COVID-19. The fact of being an "island" nation has also
been an important variable, making it easier than for most to shut off the bulk of
international travel and other physical linkages with the wider world which would
have led to greater illness and mortality.

The Temporal Course of the Pandemic in Australia

In typical epidemic fashion, COVID-19 began slowly in Australia. From the first
formal positive case, identified on 25 January, cumulative-confirmed cases crawled
to 25 by the end of February. New cases continued at a slow pace in the first week and
a half of March, but then picked up speed, improved COVID testing levels pushing
up detection numbers. The unwanted distinction of being the first day to record over
100 new cases occurred on March 18th, but this in turn was quickly overshadowed
as the month progressed by days with totals several times that. On two days, March
24 and 28, over 400 new cases were diagnosed. Fortunately, however, these alarming
figures were to prove the peak of the 'first wave' of the pandemic and daily case
numbers fell away ('flattening the curve') over April, to reach low double figures by
the end of the month.

Deaths at this time were relatively limited, the first of the pandemic occurring on March 1, the victim an elderly man who had fallen ill on a cruise ship in Japanese waters and then evacuated back to Australia. Another 30 or so deaths occurred over the course of that month, all sad events but nothing to suggest the large (and in many cases avoidable) losses that lay ahead later in the pandemic.

After a slow start to dealing with the developing pandemic, the Federal and state/territory governments moved into action from mid-March with various regulations (see section below) designed to limit the incidence and spread of COVID-19 in Australia. While many of these regulations were far from perfect in design and implementation, they nonetheless probably need to be given credit for the bulk of the decline in positive cases. Government modelling that suggested in the worst-case projection that the virus could claim up to 150,000 deaths was also significant in starting to communicate the potential seriousness of the emerging pandemic to the public.

With a large fall in positive cases over April and low death numbers, there was growing community confidence that the pandemic was being beaten. Experienced infectious disease/public health experts though cautioned against complacency, warning the pandemic could very easily take off calamitously if government restrictions were lifted too soon. In essence, two (related) crises confronted authorities—a 'health' crisis and an 'economic' one. Although saving lives was seen as the key concern, the need to kick-start the badly damaged economy was also recognized as vital. It was also recognized that long-term economic hardship and uncertainty in the community would in turn produce further health problems (e.g. mental health, domestic violence, etc.). With these dual considerations in mind, easing restrictions was started in some states and territories at the end of April and early May.

May and most of June encouragingly saw confirmed case numbers remain low. July, however, brought change, with positive diagnoses starting a surge to hitherto unseen heights (see Fig. 1 and Table 1). This 'second wave' peaked at just under 700 cases on August 5. Reported daily numbers then fell away over the rest of the month, on August 31 registering below 100 cases for the first time in two months.

As shown by Fig. 1, this second surge of the pandemic was very different from the first one in terms of magnitude. Almost 18,000 national new cases were registered over July–August. It was also very different geographically, being exceptionally heavily focused on the southern state of Victoria with around 95% of those 18,000 July–August new cases being located in that single state. This concentration on Victoria was the result of a small number of returning overseas travellers carrying the infection into the state in conjunction with infection leakages in the poorly managed hotel quarantine system that had been established for travellers by the state government. Weaknesses in the state's contact tracing scheme and isolation of community cases added to difficulties in fighting the outbreak. Breaches to the hotel quarantine rapidly saw community transmission of the virus become established. Genomic sequence testing has confirmed these case links, as well as ties to NSW cases tested at this time (e.g. the large Crossroads Hotel cluster in south western Sydney).

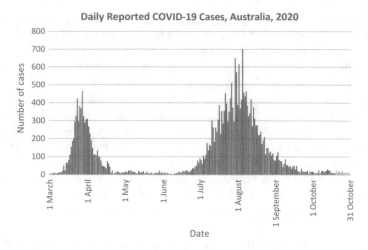

Fig. 1 Data Source: Australian government, department of health. https://www.health.gov.au/
news/health-alerts/novel-coronavirus-2019-ncov-health-alert/coronavirus-covid-19-current-situat
ion-and-case-numbers

Table 1 Australia: Cumulative COVID-19 cases, deaths and source of infection, March—October,
2020

Date (up to)	Cumulative Total Cases (a) No.	Cumulative Deaths (b) No.	Overseas Acquired Infections (b) (%)
1 March	20	1	...
1 April	4854	34^	66.2^
1 May	6948	93	63.5
1 June	7370	103	62.2
1 July	8095	104	59.4
1 August	17053	200	29.6
1 September	25782	657	20.0
1 October	27082	888	19.5
31 October	27590	907	20.1

^5 April figures
Data Sources: Australian government, department of health
(a)   https://www.health.gov.au/news/health-alerts/novel-coronavirus-2019-ncov-health-alert/cor
onavirus-covid-19-current-situation-and-case-numbers
(b)   https://www.health.gov.au/resources/collections/coronavirus-covid-19-at-a-glance-infrograp
hic-collection

Demographic and Epidemiological Toll

Compared with most other countries, Australia has fared well in the fight against COVID-19. Caution of course needs to be exhibited in the fight though as complacency towards the COVID threat could very easily rapidly see a new major outbreak emerge. For the moment (late October 2020), however, daily reported new confirmed cases have stabilized in low double figures and will hopefully fall away even further.

Revisiting the course of the pandemic shows a disease that affected persons of all demographic and social circumstances—young and old, rich and poor, males and females, different employment groups, etc. Relative vulnerabilities to the disease however were not equal. Occupational type, as will be seen, for example, was an important differentiator. Multiple social disadvantage was central in other cases, for example in the second wave outbreak in Victoria.

Disadvantage also contributed to COVID-19 cases (and some deaths) in special population groups such as homeless people, indigenous persons, and people with disabilities. The pandemic hit these groups, but fortunately not to the dire extent feared at the start of the outbreak. Numerous homeless persons' lives, for example, were undoubtedly saved by more than 5000 "rough sleepers" being moved off the streets into hotel and motel accommodation (Homelessness Australia 2020). Indigenous-led actions to keep communities safe from COVID-19 (e.g. cancelling non-essential travel to remote communities) were also important (Cousins 2020), at the end of June fewer than 60 indigenous cases having been recorded across the country. COVID-19-related indigenous mental health wellbeing needs have also been considered (Dudgeon et al. 2020). Alongside these positives, there have been lessons however to be learnt in relation to these groups. For instance, an already running Royal Commission on persons with disability unveiled a severe lack of data on such persons. Indeed, it was conceded to the Commission by a senior federal health official that persons with disabilities were totally overlooked in the federal government's first pandemic plan (Australia, Disability Royal Commission 2020).

Many people denied being at any risk from the virus, but this belief was soon proved to be ill founded. A substantial proportion of young adults in particular did not feel personally threatened and engaged in 'at-risk' group activities, flouting social distancing regulations. Official testing, however, clearly put the lie to this view, many in their 20 s and 30 s falling ill with the virus (Fig. 2). Although the severity of the infection was generally lower for younger ages, this was not invariably so. The median age of cumulative-confirmed cases (currently 37 years) has fallen during the pandemic, showing again all ages can be affected.

Just reporting diagnosed case numbers can give the impression that episodes of the disease are generally over and can be managed within the regular 14-day quarantine period for diagnosed cases. This, however, underplays on-going health issues that often occur after COVID-19 has supposedly been cured. The so-called long COVID with persisting fatigue, brain fog, lung inflammation and other conditions can plague victims of the virus for months (Nabavi 2020). In some cases, more serious heart, lung and neurological problems also occur as part of a 'post-viral syndrome'. How

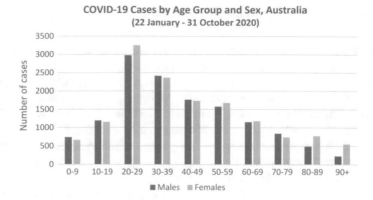

Fig. 2 Data Source: Australian government, department of health, *Coronavirus (COVID-19) current situation and case numbers.* https://www.health.gov.au/news/health-alerts/novel-cor onavirus-2019-ncov-health-alert/coronavirus-covid-19-current-situation-and-case-numbers#at-a-glance

common and lengthy these long-term issues are is being pursued by researchers at a number of leading Australian medical institutions. 'Long haulers' (as they have come to be known) experiencing these after effects can find it impossible to fully resume 'normal' daily living activities.

As noted in the Introduction, true COVID-19 case numbers may be higher than the official ones published. Researchers from the Australian National University have developed a blood antibody test that indicates previous exposure to the virus, and this research suggests many thousands of people have been unknowingly infected with the virus and will need to be factored into on-going COVID-19 burden of disease considerations.

Sadly, more than 900 Australians who fell ill with the virus did not survive, succumbing to its virulence. The nature of the infection made very old persons particularly vulnerable due to their frequently weakened immune systems and co-morbidities (e.g. hypertension, dementia, diabetes, etc.). This is illustrated by a national median age of deaths from the virus in the mid-80s years. Over 680 of these elderly deaths were of nursing home residents (mainly in Victoria) and, in a disturbing number of cases, avoidable. Standards of care and infection control were poor in many such establishments and brought unnecessary bereavement to many families. Detailed information on the condition of the nursing home sector can be found in a recently concluded Royal Commission investigation into aged care and COVID-19 (Australia, Royal Commission into Aged Care 2020). A snapshot of cases and deaths in residential aged care facilities is presented in Table 2. To date, female deaths slightly out number those of males, in large measure reflecting the greater number of 'old-old' women and their associated mortality risk (Fig. 3).

A not always recognised side of the pressure put on the healthcare system by COVID-19 has been negative flow-on effects for other areas of public health. These can easily get lost sight of under coronavirus pressures. Elective surgery operations

Table 2 Australia: cumulative COVID-19 outbreaks in residential aged care facilities, (to 30 October 2020)

Location	Total Cases No.	Resident Total No.	Staff Total No.	Resident Deaths No.
Australia (Total)	4256	2027	2229	678
Victoria (Selected Facilities)				
Facility A	232	101	131	30
Facility B	189	103	86	38
Facility C	188	94	94	45
Facility D	134	49	85	17
Facility E	125	79	46	22

Data Source: Australian government, department of health (30 October 2020)
https://www.health.gov.au/resources/collections/covid-19-outbreaks-in-australian-residential-aged-care-facilities

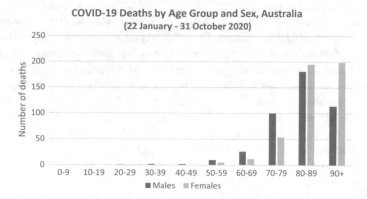

Fig. 3 Data Source: Australian government, department of health, *Coronavirus (COVID-19) current situation and case numbers.* https://www.health.gov.au/news/health-alerts/novel-coronavirus-2019-ncov-health-alert/coronavirus-covid-19-current-situation-and-case-numbers#at-a-glance

for example were cut back early in the pandemic in anticipation of a massive surge in serious COVID-19 cases, blowing out already lengthy waiting times for operations. Breast cancer has been another highly affected area, a just released Australian Institute of Health and Welfare report (AIHW 2020) stating that screening mammograms nationally fell from over 74,000 in April 2018 to around just 1100 in April 2020. Such declines heighten the dangers of delayed diagnosis and avoidable death.

In similar fashion, waiting times for assistance at drug and alcohol centres length-ened substantially, likewise, mental health (Wand and Peisah 2020) and domestic violence (Boxall et al. 2020) services. Additionally, many GPs anecdotally reported that patient numbers in their surgeries had fallen significantly due to patient fears of contracting the coronavirus when seeing a doctor.

There was one important positive aspect however that flowed from the outbreak, namely a large reduction in seasonal flu deaths. In 2019, more than 800 older Australians died from influenza, whereas in 2020, only 28 died from it in the first nine months of the year. COVID-19 self-isolation and social distancing protocols and a higher influenza vaccination rate all likely contributed to this welcome result.

Government Responses to the Pandemic

Australia has a three-tiered federal system of government comprising an overarching Commonwealth level, eight states and territories, and more than 500 local councils. The Commonwealth (Federal) Government is responsible for national issues that affect all Australians; the State and Territory Governments are responsible for issues affecting people in that state or territory; and Local Councils are responsible for issues that affect local communities.

Most government action to fight the COVID-19 outbreak was undertaken by the federal and state/territory levels. Initially, this worked well with the two levels of government combining productively. The political partisanship that frequently causes problems in Australia was briefly eschewed and collaborative efforts pursued under the aegis of a national cabinet comprising the Prime Minister and all state and territory premiers and chief ministers. Advice to the national cabinet was provided by the Australian Health Protection Principal Committee (AHPPC), a body established in 2006 for leading national public health emergency management and disease control. This general collaborative goodwill, however, did not last the course of the pandemic as policy differences between the eight states and territories surfaced, along with differences between the Prime Minister and some premiers. State and territory border closures and lockdown rules generally underlaid these conflicts.

As mentioned above, there was initially comparatively little sign of urgency from the Commonwealth Government in responding to the pandemic. That gradually changed however as cases in China and other north-east Asian countries swelled. In light of the apparent origins of the outbreak in China and rapidly growing numbers of cases there, a ban on foreign nationals arriving from that country was introduced on January 31. Similar actions against arrivals from Iran and South Korea were imposed a month or so later in recognition of surging cases in those two nations. A reluc-tance to impose a similar ban on arrivals from COVID-19 hot-spot Italy coming to Australia for the Australian FI Grand Prix caused considerable debate, many linking the reluctance to concerns not to compromise the Grand Prix's economic importance to Victoria. Eventually, however, the race was cancelled at the last minute and arrivals from Italy blocked.

From mid-March, however, government actions to confront the pandemic ramped up. Restrictions across virtually all aspects of Australian life and livelihood were enacted—border closures, travel, the permissible size of outdoor and indoor gatherings, social distancing, infection testing, quarantining of overseas arrivals and diagnosed cases, closure of deemed non-essential businesses, workplace staffing, schooling, etc. Perhaps not surprisingly some of the government pronouncements confused as much as they illuminated. Hairdressing, for example, was ruled as an essential service. Mixed messaging extended to the very top. The Prime Minister for instance, moments after announcing planned strict social distancing regulations, declared his intention to go and watch his favourite rugby league team the next weekend. Whether or not to close schools was another example, the Prime Minister arguing to keep them open while state authorities argued for closure.

In a pandemic of a highly contagious disease such as COVID-19 community testing to identify infected persons, tracing close contacts of diagnosed cases, and requiring those people to self-isolate for the infectious period is vital. At the beginning, the testing regime was poor. For weeks, returning overseas air travellers walked out of Sydney and other Australian airports without any temperature or microbiological testing for the virus. These travellers were then permitted to freely disperse to their homes across Australia, running the risk of spreading the infection to others. Restrictive testing criteria were another initial weakness, people with apparent COVID-19 symptoms but not having recently been overseas being excluded. Asymptomatic cases were likewise largely bypassed in the early stages of the pandemic.

Like in a number of other countries, a smart phone-based tracing app (*COVIDSafe*) was introduced by the Federal Government. This app also gave access to information on restrictions in different areas and the location of testing clinics/centres. Unfortunately, however, the digital tracing side of the app was not as successful as had been hoped. Communication difficulties between different models of smartphone caused difficulties, while public unease about personal data security deterred many from downloading the app. Authorities, however, still argue it is of value and an important part of the high testing levels strategy emphasised by government as an essential part of defeating the pandemic. QR code technology is also being increasing advocated for contact tracing.

Early in the pandemic, the Federal Government advised Australians who were overseas and wanted to return to Australia to do so as soon as possible by commercial means. Limits on overseas returnees were subsequently introduced, capped at 4000 per week, later lifted to 6000. Sydney handled the largest number of such residents (2450 per week), at times to the voiced chagrin of NSW government authorities. Cutbacks on flights, consequent limits on seats available and expensive fares however made it difficult for many to return. The exact number still stranded overseas is uncertain, but in mid-October was put at more than 30000. Amongst the stranded, there are reported feelings of having been deserted by their country. Realistically, given the current rate of returning numbers, many such people will not succeed in returning until well into 2021. An expanded rescue plan to start bringing back around 5000 people from London, New Delhi, and Johannesburg and quarantine them in a camp near Darwin was announced just after writing the above. Conveying all people who want to return back home, however, will still be a lengthy process.

Lockdown

To stop the spread of COVID-19, the Federal and state/territory governments progressively introduced the measures highlighted in the previous section. These restrictions came to be termed "lockdown " and varied in severity over time and between states (See Box 1). Businesses and industries were differentiated to be essential or non-essential, with the latter category to cease operations during the pandemic's course. Not surprisingly, this ruling, along with tight restrictions on movement and social distancing requirements, cut a rapid brutal swathe through the national economy. Virtually overnight, thousands of businesses ceased operations and tens of thousands of workers found themselves unemployed. The Australian Bureau of Statistics estimated almost 600,000 workers (55% females) lost their jobs in April due to COVID-19-related restrictions, and a further 227,700 in May. In June, unemployment officially reached a two-decade high of 7.4%. To help keep people in jobs, in late March, the Commonwealth government established JobKeeper, a A$130 billion wage subsidy programme. Without this initiative, the real unemployment rate would have been over 13%. Not all employers however were eligible for this wage assistance, significant numbers (e.g. universities) being excluded for not meeting the government's business revenue criteria for inclusion. The non-inclusion of universities in JobKeeper was a surprise as the tertiary education sector with its normal enormous earnings from overseas fee-paying students was, along with tourism, the area most severely hit by the ban on travel to Australia. It has been estimated that around 21,000 jobs could be lost in the tertiary education sector due to this loss of overseas fee income.

A major income support programme for unemployed persons (JobSeeker) was established around the same time. As well, a temporary Coronavirus Supplement payment scheme was introduced. Without these support programmes, hundreds of thousands of Australians would have slid into poverty and homelessness. Once these schemes come to an end, however, many people will fall into dire income and housing circumstances again.

In the second Victorian wave, a strong socio-geographic patterning developed, cases being concentrated in Melbourne's more disadvantaged areas. Residents in these areas tend to have low incomes, limited education and employment skills, often poor English capability and constrained housing affordability. Cramped, overcrowded housing was a further problem. Additionally, a significant proportion of residents had traumatic refugee backgrounds. Conveying easily understood public health messages to non-English speaking background (NESB) people proved a difficulty early on, but was gradually improved upon. Taken together this multiple disadvantage made for particular vulnerability to COVID-19 and not surprisingly that transpired. Numerous such areas were hit hard by the infection and residents made subject to restrictions. What was to prove to be one of the most controversial government actions in this regard was the sudden hard lockdown on 4 July of nine public housing tower blocks in the suburbs of Flemington and North Melbourne. At the lockdown of the blocks, there was a strong police presence, making many residents uneasy. Adding further stress for some of the residents was temporarily not being able to return to their apartments and in effect being made homeless.

Hard lockdown was used again by the Victorian government later in the month, with extension to all of Metropolitan Melbourne in the light of alarmingly high new case numbers. Slightly less restrictive lockdown regulations were applied to regional Victoria. While obviously difficult for Victorian residents, these lockdowns proved successful in helping bring down new locally acquired case numbers, daily totals now (late October) being in low single figures. October 26 was a landmark, with no new cases recorded across the state, the first time since June. Several other zero case days have subsequently been recorded. An important step in reining in the outbreak was making the use of fitted face marks mandatory when leaving home. This was a contentious move and resisted by many people, but generally endorsed by medical experts. The NSW government by contrast 'urged' the use of masks, but refused to make them mandatory in face of wide public opposition.

The high economic costs of lengthy lockdown and Premier Andrews' caution in opening up the state were the subject of heated political and business debate throughout September and October, but most public health specialists supported the Premier's gradual easing approach, aware of the dangers of a premature lifting of restrictions. A planned 4-step "roadmap" back to normal conditions has been used to guide the progressive easing. (See https://www.coronavirus.vic.gov.au/coronavirus-covid-19-restrictions-roadmaps). Buoyed by the encouraging decline in new cases, the Premier finally felt confident in the last week of October to remove a substantial range of lockdown restrictions, putting both Melbourne and Regional Victoria in so-called Step 3. Neighbouring NSW has waited to see the impact of the relaxed restrictions before making any changes.

Perhaps not surprisingly many people found the lockdown regulations annoying and frequently ignored them, reports of attempted illegal crossings of state borders were not uncommon. Both in Melbourne and Sydney, beautiful weather days on several occasions attracted thousands of people to beaches, totally ignoring social distancing regulations. Likewise, police were called to numerous house parties breaching permitted attendee numbers. Many hotels were also charged and fined for allowing excessive numbers of patrons. Refusing to be tested for COVID-19 and not meeting self-isolation requirements were other fairly common breaches. Anti-maskers were also active and organized several "human rights" anti-masking rallies. Other rallies were held protesting against the lengthy hard lockdown. Police frequently levied fines on violators, but to members of the population observing the restrictions, these fines often appeared paltry and inconsistent. At other times, police actions seemed to be unnecessarily heavy-handed.

Box 1:

Details of current COVID-19 restrictions can be found at:

Australian Government, *healthdirect*. COVID-19 Restriction Checker
https://www.healthdirect.gov.au/covid19-restriction-checker

Victoria's restriction levels
https://www.dhhs.vic.gov.au/victorias-restriction-levels-covid-19

NSW Public Health Orders and restrictions
https://www.health.nsw.gov.au/Infectious/covid-19/Pages/public-health-ord
ers.aspx

Restrictions in Queensland
https://www.qld.gov.au/health/conditions/health-alerts/coronavirus-covid-19/
current-status/public-health-directions.

Mistakes and Painful Lessons

When working to deal with a disease threat of the scale of COVID-19, inevitably numerous mistakes will be made. The Australian outbreak was no exception in this regard and offers a variety of painful lessons for pandemic preparedness, planning and management. Sadly, thousands of people suffered serious COVID illness episodes due to mistakes that in hindsight could and should have been avoided. Worse, many unnecessarily lost their lives.

The following briefly discusses three of the major such instances—the *Ruby Princess* cruise ship disembarkation error, the Newmarch House nursing home COVID-19 outbreak and the Melbourne hotel quarantine debacle.

(a) *The Ruby Princess*: On March 19, the *Ruby Princess*, a cruise ship in the Carnival line, returned from a cruise around New Zealand and its 2700 passengers allowed to disembark at Circular Quay in Sydney without COVID-19 testing, despite several sick passengers on board. Around 900 passengers and crew subsequently went on to become COVID positive. Additionally, 28 deaths were linked to the ship, eight in the USA. In Australia, *Ruby Princess* cases were involved in the spread of the infection to NW Tasmania (Melick 2020) and Western Australia (on a Qantas flight). Responsibility for the catastrophic disembarkation error was ultimately attributed to the NSW State Department of Health, after earlier argument that it was the Federal Australian Border Force's fault. For a detailed examination of the *Ruby Princess* saga, see the *Report* of the Special Commission of Inquiry into the Ruby Princess (August 2020). In the *Report,* the Commissioner labels some NSW Health actions as "inexplicable" and "inexcusable".

(b) *Newmarch House Nursing Home Outbreak*: Newmarch House is an Angli-
     care residential aged care home in western Sydney, registered for 102 places.
     Between 11 April and 15 June 2020, the home experienced a severe prolonged
     COVID-19 outbreak, with 71 cases diagnosed in residents and staff and 19
     residents dying. As the outbreak progressed, there was intense debate as to
     whether infected residents should remain at Newmarch (so-called Hospital in
     the Home) or be moved to hospital for care. Infected residents continued to
     be kept at Newmarch, but many relatives of healthy patients felt this was just
     a conduit for their loved ones to contract the infection. Management of the
     nursing home reported receiving conflicting advice on this issue from NSW
     and Federal Government officials, and the Anglicare CEO reported that "in
     hindsight" this would have been treated differently. Loss of staff to quarantine
     and difficulties obtaining replacements also caused difficulties. Shortages of
     PPE supplies and the expertise to properly use them were further problems.
     Families felt communication linkages with their relatives in the home were
     also unsatisfactory, creating a sense of being disconnected (Gilbert and Lilly
     2020).

(c) *Melbourne Hotel Quarantine debacle*: At the end of March, the national cabinet
     announced that from then on all returning international travellers would need
     to undergo their required 14-day quarantine in a hotel. This system generally
     worked well around the country with the exception of Victoria where it became
     a public health disaster (Coate 2020). The hotel programme there (Operation
     Soteria) was put together in a very pressured short time with lines of command
     and responsibility and day-to-day operation details unclear. Private security
     guards with little training in infection control and on low wages were engaged
     by Victorian authorities. Some recruiting in fact was reported as having been
     done via WhatsApp. Unwise personal interaction between security guards and
     quarantined returnees is also alleged to have occurred. Who actually decided
     upon using private security guards became a matter of angry dispute, as did
     claims that the Australian Defence Force offered (or did not offer) personnel to
     assist the quarantine operation. In short, the situation was the perfect recipe for
     disaster. And so it transpired. Outbreaks of the virus occurred amongst staff
     and security guards at two downtown hotels (Rydges on Swanston and the
     Stamford Plaza) and from there seeded Victoria's second COVID-19 wave.

Risky Work

Over the course of the pandemic, COVID-19 infection struck persons in all sections
of the workforce. Some workers though, because of the nature of their employment,
proved to be at special risk of contracting the virus. Foremost amongst these not
surprisingly were aged care workers, hospital and medical clinic staff and other
frontline health professionals.

Across the residential aged care sector, over 2200 staff fell ill with the infection, 200 more in fact than diagnosed resident cases (see Table 2). This led to staff shortages at many establishments over the months of the pandemic. Care staff in such facilities are generally poorly paid, and many were working shifts at several facilities to survive financially. This practice, however, multiplied the risk environments to which they were exposed. Likewise, staying at home from work in self-isolation to check out possible COVID-19 symptoms was simply not a feasible financial option for many workers, continuing to go to work and potentially spreading infection. Provision of personal protective equipment (PPE) and training in its use was also inadequate at many establishments, further raising staff risk levels.

Healthcare workers in the hospital system (particularly in Victoria) were exposed to many of the same risks of contracting COVID-19 as the aged care employees discussed above. On 4 August, *The Guardian* newspaper reported there were 911 healthcare workers with active infections of COVID-19 in Victoria, alarmingly up by 101 persons from the day before. Nurses were particularly hit hard. The State's Chief Health Officer was reported as saying nurses were more represented in these healthcare workers numbers than doctors, probably due to "the closeness of inter-action that nurses are engaged in with their care provision". Representatives of the nursing workforce expressed concern that their members were not being provided with sufficient personal protection (PPE) and training against COVID-19 infection.

Outside health and aged care, high risk of infection also existed within a number of other employment groups. Abattoir workers in Victoria, for example, stood out as particularly vulnerable, being linked to several hundred COVID-19 cases. As with aged and healthcare workers, there were a mix of factors behind these high numbers. By the labour-intensive nature of meat processing work, widespread transmission of the disease was virtually inevitable with social distancing being very difficult to maintain. Adding to the mix, a high proportion of workers in meat processing have low levels of education and inadequate appreciation of the importance of rigorously following anti-COVID personal hygiene requirements. Unwittingly, many workers took the infection home from work, spreading transmission into the community. Financial needs to support their families in turn made it impossible for some workers to take time off work when they were feeling ill.

There is only space here in this chapter to discuss these three high at-risk groups. Many others, however, could have been used as further examples—e.g. public transport workers, police officers, paramedics, social workers, hospitality workers, disability support persons, etc. For all, the pandemic brought special risks.

Political Tussles

With a Commonwealth government, six states, and two territories and the partisan nature of Australian politics, it was inevitable that political issues would play an important role in the management of the pandemic in Australia. On the one hand, federalism offers a political structure with the attraction of allowing actions most appropriate to the specific state/territory circumstances, but conversely suffers the disadvantage of the states/territories perhaps 'going their own way' at the expense of some necessary national level actions. Tensions between the different states and territories and others between the Commonwealth and one or more of the states/territories thus are likely and duly appeared.

As noted earlier in the chapter, in the initial months of the pandemic, the political confrontationalism to which Australians are accustomed was put to one side and politically collaborative policies pursued under the leadership of the new national cabinet. The (Liberal) Prime Minister, Scott Morrison, chaired the cabinet, but Labour Party Premiers/Chief Ministers held a 5–4 majority on the body. To his credit, the Prime Minister, known prior to the pandemic as an aggressive Liberal Party 'warrior', handled the cabinet in an inclusive, bipartisan fashion, to the surprise of many familiar with his previous political persona. Later, in the pandemic, policies became more contested.

The second wave in Victoria saw a lot of the previous consensus lost. Personal criticism of Victorian Labour Premier Dan Andrews' ability to bring the surge in infection under control and successfully kick-start the economy became common in the media. The catastrophic disarray of the Melbourne hotel quarantine scheme was held by the Prime Minister to lie in the state authorities' poor planning, resourcing and implementing. Claims rolled back and forth about whether Victoria had been offered Australian Defence Force staff to help out with the scheme. Conscious of the costs of the Melbourne lockdown to the national economy, the Prime Minister pressured Premier Andrews to ease the state lockdown restrictions and bring forward plans to re-open the state border. Added tensions flared up in mid-October between the Premier and the Federal Government over travellers from New Zealand being allowed to fly on from Sydney to Melbourne allegedly without the Premier's knowledge. Similar pressure to re-open the Queensland border was placed on Labour Premier Annastacia Palaszczuk. In the eyes of many critics, the Premier's delaying a re-opening was due more to a forthcoming state election than concerns over the threat of facilitating the virus' spread into the state, this of course being denied by the Premier. Cynics meanwhile argued that the Prime Minister's criticisms seemed to be largely confined to states/territories governed by the opposition Labour Party. There is some substance to this claim, though the heated Commonwealth (Liberal/National) government versus NSW (Liberal/National) government dispute over responsibility for the *Ruby Princess* debacle stands as an exception.

The Queensland border re-opening issue also led to political jousting between that state and its southern neighbour, New South Wales. NSW Liberal Premier Gladys Berejiklian argued that Queensland's requirement for re-opening—28 consecutive zero-infection days in NSW—was unrealistic and may never be met. Adding fuel to the fire, the Queensland Deputy Premier publicly stated that NSW had unfortunately given up trying to totally eradicate the virus, further cooling relations between the two states. Further tension was subsequently generated by NSW's threat to send bills to the other states and territories for the hotel quarantine costs of their residents that had been incurred by NSW.

In Victoria, intra-government differences also caused tensions over responsibility for the Melbourne hotel quarantine debacle, blame being shunted around the jobs, police and health ministers. This issue eventually claimed the scalp of the Health Minister (Jenny Mikakos) with her resigning from the Andrews ministry and government after the Premier stated at the quarantine inquiry that she was accountable for running the programme.

Political tensions also extended to the international sphere when Australia called for an independent international inquiry into the COVID-19 outbreak and the World Health Organization to be given 'weapons inspector' powers to investigate its causes. China took offence at these calls, seeing them as blaming China for the start of the pandemic, further souring already poor relations between the two countries.

Looking Back: Living Through the Pandemic

The 2020 COVID-19 pandemic will go down in Australia, like most other countries, as an unforgettable period in the nation's history. For a small proportion of the population, it will be remembered as the event that took the lives of loved (mainly elderly) family members and friends. For others, it was a time of seriously debilitating attack on their own health. For some women, the stresses of lockdown, COVID-19 threat and financial pressures translated into rising physical and psychological violence in the home. Meanwhile for others, the pandemic will be remembered as a brief, generally light encounter with a new disease. For yet others, life basically just went on as normal healthwise.

Alongside these differing health experiences, the country went through the worst economic crisis it has ever experienced. Tens of thousands of people who before the pandemic hit had had good employment and prospering businesses suddenly saw themselves without work or re-employment prospects. Attempting to keep the population's head above water, the Federal Government made available previously unenvisaged sums of support through economic programmes such as JobKeeper, JobSeeker, etc. The government debt levels incurred by these relief measures will dominate the federal budget for many years ahead.

For many, the most unedifying memory of the pandemic will be the aggressive widespread panic buying of food early in the outbreak, supermarket aisles being stripped of toilet paper, tissues, pasta, tinned foods, etc., within hours of the initial announcement of lockdown restrictions. Human nature being what it sometimes is, there were unfortunately also those who tried to use the pandemic as an avenue for illicit financial gain, using the web and other means to spruik fake vaccines, cures, protective equipment and other items (Broadhurst et al. 2020; ACCC, *Scamwatch*-https://www.scamwatch.gov.au/.). In an alert posted on 30 September, the Australian Competition and Consumer Commission (ACCC) reported having received over 4560 scam reports mentioning the coronavirus since the outbreak began, with losses in excess of $5 million. Racism also reared its head in the early stages of the pandemic, the linkages of the outbreak with China unfortunately seeing some Chinese and other Asian-Australians subjected to racial vilification and physical abuse.

For school children, the pandemic at its height was a time of adjustment—to online learning at home, to cancellation and restrictions on school and community sport, to not being able to physically meet friends each school day, to lockdown when they were not permitted to leave their homes, to living in households in which financial and emotional security suddenly became jeopardised, etc. University students faced similar upheavals, the pandemic driving a move to online teaching that will probably forever change the face of tertiary teaching in Australia. The casual social interaction between students that is a central part of the university experience was also a casualty, a victim of social distancing and limitations on group gatherings.

For older persons, the pandemic carried frightening prospects, knowing the risk they were at of contracting the disease due to pre-existing health problems and aged immune systems. Many could remember their parents talking of the deadly Spanish Flu pandemic in Australia in 1919. Fear of COVID-19 saw many lock themselves away in their homes and not see their children and grandchildren for many weeks. For the elderly resident in nursing homes, the prospect of infection was a particular worry, especially for those in establishments in which a number of COVID-19 deaths had occurred. Final in-person goodbyes to elderly family members facing imminent death often were not possible.

The Future?

What COVID-19 holds for Australia in the future is obviously impossible to tell with certainty. Total eradication of the virus though is unlikely, stabilization at low case and death figures probably the best that can be achieved in such a geographically large and diverse country. Eradication at the global scale is even more unlikely, so realistically Australians, like all other nationalities, will likely face a world in which COVID-19, as with influenza, is a regular (i.e. endemic) part of the epidemiological landscape. If so, sensible social distancing, rigorous personal hygiene and other anti-COVID behaviours will continue to be essential and improved. It will also clearly be vital to maintain careful screening for COVID-19 infection in those entering Australia.

If effective and safe vaccines and/or anti-COVID-19 therapies become available, 'normal' living with COVID-19 could become a prospect. Various medical establishments in Australia (e.g. University of Queensland, Doherty Institute, Walter and Eliza Hall Institute, etc.) are engaged in such research in collaboration with overseas laboratories. Funding assistance of $66 million for this research was announced by the Federal Minister for Health at mid-year. Reasonably, confident hopes for success are widely held, but anti-viral vaccines and therapies are notoriously difficult to produce.

What may be the greatest threat to conquering the threat of COVID-19 could in fact prove to be complacency towards the virus. There have been numerous worrying signs of this during 2020, the bewildering reluctance of most people to wear protective masks on crowded public transport or in shopping centres, etc., being only the visually most obvious.

Time will tell. For now, the COVID-19 story, both in Australia and globally, has a considerable distance to travel.

Note

For simplicity, the term 'pandemic' is used through the chapter, although in some places 'epidemic' would be epidemiologically more appropriate.

Addendum

October 31, 2020

- Australia records first day of zero cases of community COVID-19 transmission since early June

November 11, 2020

- Federal Minister for Health reports Australia "'on track' to deliver a coronavirus vaccine by March next year"

November 13, 2020

- *Australian COVID-19 Vaccination Policy* published by Federal Department of Health

Mid-Late November, 2020

- South Australia records new large COVID-19 cluster; state government imposes short, hard lockdown based on untruthful information from COVID-positive pizza shop worker
- political controversy following NSW Premier's failure to self-isolate while waiting for results of COVID-19 test

Mid-Late December, 2020

- all State border restrictions lifted, then many reimposed a few days later in response to new, rapidly emerging Sydney Northern Beaches cluster

January 2021

- NSW government makes masks mandatory on public transport and in public indoor settings

- Governmental concern at diagnosis of UK (B.1.1.7) and South African (501Y.V2) highly transmissible mutant COVID-19 variants in overseas arrivals
- Over 35000 Australian expatriots still stranded overseas
- Debate over some non-specialists' advocacy of hydroxychloroquine and ivermectin drugs to treat COVID-19

February 1, 2021

- Recorded cumulative totals: Australia (cases—28807, deaths—909); World (cases—107 million, deaths—2.3 million)

Late February/Early March, 2021

- Phased roll-out of COVID-19 vaccination commenced, initially to identified priority groups (e.g. quarantine and border workers, front-line health care workers, aged care and disability staff and residents, elderly adults, etc.) See https://www.health.gov.au/resources/publications/australias-covid-19-vac cine-national-roll-out-strategy for details.

References

Australia, Disability Royal Commission. (2020). Royal commission into violence, abuse, neglect and exploitation of people with disability: Public hearing 5: Experiences of people with disability during the ongoing COVID-19 Pandemic. https://disability.royalcommission.gov.au/public-hea rings/public-hearing-5.

Australian Government, Department of Health. (2020). Coronavirus (COVID-19) epidemi-ology reports, Australia, 2020. https://www1.health.gov.au/internet/main/publishing.nsf/Con tent/novel_coronavirus_2019_ncov_weekly_epidemiology_reports_australia_2020.htm.

Australian Institute of Health and Welfare (AIHW). (2020). Cancer screening and COVID-19 in Australia, (Web Report), Canberra: 8 October. https://www.aihw.gov.au/reports/cancer-screen ing/cancer-screening-and-covid-19-in-australia/.

Australia, Minister for health (Hon. G. Hunt). (2020). $66 million for coronavirus-related research, *Media Release*, 2 June, 2020.

Australia, Royal Commission into Aged Care Quality and Safety. (2020), Aged care and COVID-19: A special report, commonwealth of Australia, Canberra. https://agedcare.royalcommission. gov.au/publications/aged-care-and-covid-19-special-report.

Boxall, H., Morgan, A. & Brown, R. (2020). *The prevalence of domestic violence among women during the COVID-19 pandemic, Statistical Bulletin* no. 28. Canberra: Australian institute of criminology. https://www.aic.gov.au/publications/sb/sb28.

Broadhurst, R., Ball, M. & Jiang, C. J. (2020). Availability of COVID-19 related products on Tor darknet markets, *Statistical bulletin*, no. 24. Canberra: Australian institute of criminology. https:// www.aic.gov.au/publications/sb/sb24.

Coate, J. (2020). COVID-19 hotel quarantine inquiry, Government of victoria, Melbourne. https:// www.quarantineinquiry.vic.gov.au/.

Cousins, S. (2020). Indigenous Australians avert an outbreak—for now, *Foreign policy*, 19 May, pp. 7. https://foreignpolicy.com/2020/05/19/indigenous-australians-avert-coronavirus-outbreak-for-now-aboriginal/.

Dudgeon, P., Derry, K. L. & Wright, M. (2020). A national COVID-19 pandemic issues paper on mental health and wellbeing for aboriginal and torres strait Islander peoples. Transforming Indigenous mental health and wellbeing grant, the university of Western Australia Poche Centre for Indigenous Health. https://apo.org.au/node/306661.

Gilbert, L. & Lilly, A. (2020). Newmarch house COVID-19 Outbreak [April-June 2020], Independent review—Final report, Australian government, department of health. https://www.health.gov.au/resources/publications/newmarch-house-covid-19-outbreak-independent-review.

Homelessness Australia. (2020). Federal budget submission, August 2020. https://www.homelessnessaustralia.org.au/news/blog/federal-budget-submission-august-2020.

Melick, G. (2020). Independent review of the response to the north-west Tasmania COVID-19 outbreak, department of premier and Cabinet, government of Tasmania. https://www.dpac.tas.gov.au/independent_review_of_the_response_to_the_north-west_tasmania_covid-19_outbreak.

Nabavi, N. (2020). Long covid: How to define it and how to manage it, *British Medical Journal*, 370m, 3489. https://doi.org/10.1136/bmj.m3489.

Special Commission of Inquiry into the Ruby Princess. (2020). Report of the special commission of inquiry into the ruby princess, New South Wales, Australia. https://www.rubyprincessinquiry.nsw.gov.au/report/.

Wand, A. P. F. & Peisah, C. (2020). COVID-19 and suicide in older adults. *Medical Journal of Australia,213*(7), 335–335. https://www.mja.com.au/journal/2020/213/7/covid-19-and-suicide-older-adults.

Europe, Russia and Africa

COVID-19 in France: The Revenge of the Countryside

Isabelle Roussel

Abstract To manage the crisis linked to the COVID-19 pandemic, France, very poorly prepared, had to resort to very strict lookdown measures so that the health system can ensure the care of the many patients. The virus is still circulating and establishing a comprehensive review of this pandemic which has killed more than 90,000 people and is still surging. The territory and the population have been very unequally affected by the virus which has reached, as for any epidemic, the most densely populated areas. This is how, for the first time in decades, metropolises are losing their appeal and are calling for new land use planning more favourable to rural areas.

Keywords Air pollution and COVID-19 · The course of pandemic · Countryside attractivity during the pandemic · COVID-19 in France · Lockdown and inequalities · Lockdown and climate

Introduction

With the COVID-19 (corona virus disease) epidemic, the urbanized and hyper connected humanity at the start of the twenty-first century has found itself extremely vulnerable: The virus quickly spread to France in February, a few days after making a deadly entry into Italy. This extremely rapid mode of dissemination took advantage of the extreme mobility of the inhabitants of the planet (aircrafts, high-speed railways). The fight against the coronavirus has therefore led to a radical lockdown on the inhabitants: It was necessary to brutally immobilize a world which for decades had revered the principle of mobility. The only possible option for the management of the pandemic because France was unpreparedness to face this shock was lockdown. In the absence of effective tests and, in the presence of people carrying the virus but asymptomatic, it was necessary to rely on the capacities of the hospital since, like China, severe forms of the disease required the use of ventilation and resuscitation.

I. Roussel (✉)
University of Lille, Lille, France
e-mail: appa.irou@gmail.com

© The Author(s), under exclusive license to Springer Nature Switzerland AG 2021
R. Akhtar (ed.), *Coronavirus (COVID-19) Outbreaks, Environment and Human Behaviour*, https://doi.org/10.1007/978-3-030-68120-3_12

The 7000 available beds were filled at the time of the peak of the pandemic with an uneven geographical distribution since three regions (Ile-de-France, Grand Est et Auvergne Rhône-Alpes) totalled three-quarters of the people treated.

The map (Fig. 1) showing the increase in the number of deaths over the same period between 2020 and 2019 shows the country's heterogeneity in the face of the pandemic. It is these three regions which, during the various epidemics that France has faced in history, have always been the most affected. Added to this regional disparity is a strong difference between the neighboring towns and countryside. Population density and air pollution explain these differences, but the lethality of the disease, according to initial results, mainly affected elderly men with co-morbid traits (diabetes, overweight, heart problems). The effects of lockdown are of a different nature and relate to the quality of life of populations, the environment and the economy. The decline in motorized mobility during the lockdown may have given the illusion of controlling green house gases (GHGs), but the sudden and temporary nature of the drop in emissions cannot foreshadow better management of climate change which would impose more painful and more lasting restrictions. The fact of having, during the crisis, put health concerns on the front of the stage "whatever the

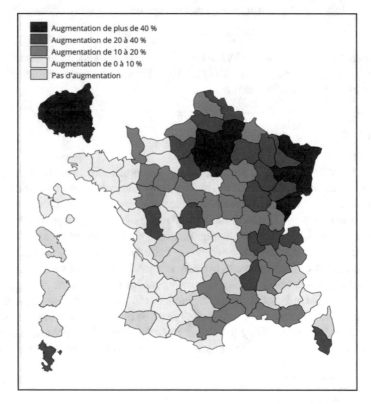

Fig. 1 Evolution of cumulative deaths from March 1 to April 30, 2020, compared to cumulative deaths from March 1 to April 30, 2019, by department. Source INSEE

cost" represents a major event whose economic consequences are still to come and are not the subject of this report.

Many imprecisions surrounding the nature and effects of the virus make predictions about the end of the pandemic still uncertain.

The Course of the Epidemic in France

This period of crisis has been marked by uncertainty and even denial. Despite the alerts given by China and Italy, the COVID-19 was assumed to be to a banal winter flu that the quality of the French healthcare system could easily contain; thus, many caregivers, for lack of precautions, have been affected by the disease. It then took several days between realizing the danger and equipping the caregivers with protective equipment (masks, gloves, over-gowns).

* **From the end of February, several clusters have developed** in connection with travel and the arrival of tourists from abroad and particularly from China: The epidemic started around Roissy airport and in the nearby air base in which 193 French citizens from Wuhan were repatriated contaminating several soldiers in this air base. Another cluster broke out around Mulhouse where a gathering of evangelicals took place between February 17 and 24. The pastor, organizer of the meeting, estimates that almost all of the faithful present were infected, or about 2,000 out of 2,500 people. However, according to medical authorities, the coronavirus was already present in Alsace before the evangelical gathering. Faced with these first clusters, the health authorities were helpless. A nurse from the Strasbourg University Hospitals University was at the origin of the contamination of 250 nursing colleagues.

* **During this time of uncertainty**, during the month of February, the virus spread with impunity. These initial outbreaks made it possible to launch the alert and generate strong reactions, but it was too late to treat the epidemic, as Germany did, by observing the instructions issued by the WHO: tests, masks, isolation of contaminated persons. In France, references to past experiences tended to downplay the epidemic even though, subsequently, the medical profession reacted, denial and ignorance remained present. The healthcare system was taken by surprise thus showing France's unpreparedness unlike neighboring Germany which, being better prepared, had 8,144 deaths as of May 21, 2020, i.e., 97 per million inhabitants against 431 in France.

The lack of stocks of masks has led to a denial of the effectiveness of this form of protection. It is true that France but like most other European countries, unlike Asian cultures, has not adopted the wearing of masks when pollution peaks.[1] But, the discovery of the number of healthy carriers has transformed each healthy person into a possible vector. It became imperative to generalize the use of the mask, but France had very low stocks; their tailoring mobilized the ingenuity of the apprentice seamstresses.

---

[1] The effectiveness of masks in the event of pollution peaks much lower than those observed in China is disputed.

The French strategy to fight the virus has been organized and focused around hospital capacities.

* **Statements are accelerating**.

On March 5, with five dead, the rate of death began to accelerate.

On March 10, an independent scientific council, bringing together several personalities with recognized scientific specialism in different fields, was created. The relationship between this "ad hoc" structure and official institutions has not always been very clear; this scrambling contributed to disorientation the population who did not know who to believe.

On March 11, WHO declared a global pandemic.

On March 12, during a speech, the President declared the closure of all school premises because the role of children in the spread of the virus is uncertain. However, the municipal elections are maintained on March 15.

On March 14, contaminations doubled in 24 h with 91 deaths, 148 on March 16.

On March 15, the first round of municipal elections contributed to the spread of the virus not so much within the polling stations which were the subject of numerous health precautions but rather during the electoral campaign which gave rise to many meetings. Several candidates and politicians were contaminated by the virus. The maintenance of the elections was felt to contradict the recommendations disseminated to *advise against* large public gatherings (Fig. 2).

*On March 16, a new speech by the President of the Republic announced "the war against the virus" and the implementation of lockdown**. This, quite strict decision, was the only possible solution not to saturate reception capacities of the hospital if the slope of the curve for the number of hospitalized patients maintained the same growth rate (see Fig. 3). France is characterized by centralized health management with controlled lockdown. To go out if necessary, it was imperative to personally authorize yourself to go out and to bring a duly completed certificate.

March 27, confinement was extended until April 15 and then be until May 11, when the curve of the epidemic's spread begins to flatten (see Fig. 3).

On March 30, the threshold of 3,000 hospital deaths was crossed. There are concerns about the number of ventilators, as the capacity of 7,000 resuscitation beds is at risk of being full despite military aid in Alsace.

Fig. 2 Evolution of the COVID-19 epidemic in France on July 24: new cases, new deaths and new admissions to intensive care units. *Source Public Health France*

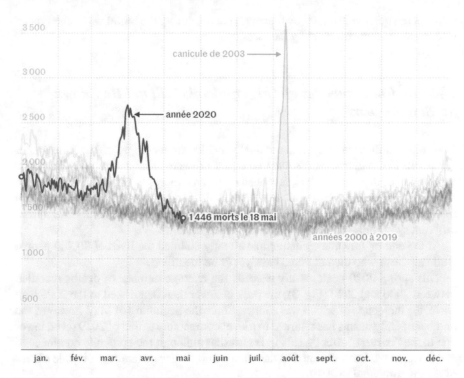

Fig. 3 Daily number of deaths in France between 2000 and 2020 source INSEE

Since March 26, patients in intensive care in saturated hospitals have been transported to hospitals with places available in western France or Germany. In this regard, helicopters and TGVs (high-speed trains) were requisitioned.

Gradually, the number of deaths decreased, and, at the end of April, the balance between hospitalized patients and deaths has freed up beds.

* The "deconfinement" took place gradually from May 11. The circulation of the virus being slowed down, the management of "deconfinement," which was very authoritarian at the beginning, has evolved by empowering civil society. The medical power is organized to try to intervene in the "clusters" as soon as they are formed. The availability of tests has improved although their reliability is poor, and it is unrealistic to test the entire population at regular intervals. However the disease is better known, the clinical symptoms are quickly detected and managed, while at the start of the epidemic, the medical advice given was to stay at home as much as possible.

Figure 2 showing the health situation report on July 24 indicates the changes in the management of the crisis. More intensive screening has allowed new cases to be identified, but these, at this date, do not translate into an overload of intensive care beds or an increase in the number of deaths which remain at a very low level. But will the sense of individual responsibility be sufficiently developed in regions that were spared but affected later by tourists? Indeed, during the initial wave, the French were very unevenly affected by the virus.

The Full Consequences of This Pandemic Will not Be Known for Several years

The virus is still circulating, and the final statistics are incomplete: The full number of patients is difficult to assess due to the lack of widespread and reliable tests. The very notion of disease is to be questioned between « disease» and « illness» because of the large but unknown number of healthy carriers who carry the virus and can transmit it but without being sick. Is identified as sick the one who suffers from different symptoms or the one who carries the virus?

At the end of July, the epidemic had already claimed the lives of 30,209 people while all those who died at home have yet to be identified.

This spring 2020 peak greatly exceeds the average number of deaths recorded between 2000 and 2020 (Fig. 3); the peak is lower than that linked to the 2003 heat wave, the incidence of which was shorter. Since the beginning of May, however, the number of daily deaths has returned to a level comparable to that of 2019 (even lower due to the "harvest" effect). This indicates that although the epidemic continues to claim victims, it has weakened to the point of no longer generating a visible excess of deaths compared to 2019.

This peak in mortality is clearly different from the recurrent epidemics of winter flu that have struck France in previous years (Fig. 3). Thus, according to the first results published by INSERM (Guillot and Klat 2020), between January and May 2020, 287,000 deaths from all causes (COVID-19 or not) were recorded against 270,000 deaths from all causes, recorded over the same period in 2019. The excess of deaths in 2020 is therefore around 17,000 deaths. Finally, it should be noted the "indirect" effects of containment measures, which, by profoundly modifying the lifestyle of the French, may have led to a drop in deaths due to causes other than COVID-19, for example road traffic accidents or the reduction of air pollution. But, on the other hand, a deficit of care provided during this period when all the medical personnel were mobilized by the Covid can result in faster deaths later.

Geographic and Socio-Economic Inequalities of Affected People, Hospitalized and Cured

If the pandemic was managed, centrally, by the French government, it has not affected all of France and the French in the same way. The virus has spread sparing many areas of France; it is true that the virus, the full characteristics of which are still unknown, affects individuals in a very unequal manner, mainly depending on the population density and the age of the individuals.

The Unequal Spread of the Virus

In 1960, the geographer A. Siegfried showed what were the main routes of dissemination of epidemics which, most often, radiate from an Asian focus to spread along the trajectory of the caravans of Central Asia then, over for years bacteria have followed the main routes used by maritime traffic, and, today, airlines represent the fastest route of diffusion. The lockdown period made possible to stop the progression of the virus towards the West and to spare many regions.

* **Ile-de-France** is the region which was the most impacted (+ 64% compared to expected mortality), followed by Grand-Est (+ 37%) and in particular the departments of Bas-Rhin, Haut-Rhin and the Moselle. The Auvergne-Rhône-Alpes, Bourgogne-Franche-Comté and Hauts-de-France regions also have a major excess of deaths (between + 15% and + 20% compared to the expected mortality).

The heavy toll paid to the virus by the departments located on the eastern border and the Paris region is etched in memories of inhabitants. The death of the mayor of the city of Saint-Louis, a town close to the Swiss border, which occurred some time after his election in the first round of municipal elections, has become a symbol of the questioning about the relevance of the organization of municipal elections. The city of Mulhouse also remembers an exceptional influx of coffins, with 630 cremations counted in 2 months, compared to 220 during a period of severe flu.

The geographical differences between the east and the west of France appear strongly on the maps[2] (Figs. 1 and 4): With an equivalent number of inhabitants, the Vosges department had six times as many hospitalized patients and ten times more deaths than that of Haute-Vienne.

* **The Rhone corridor**, located between Marseille and Lyon, has also been particularly affected by respecting the ancestral route of penetration of the virus through the port of Marseille. In this town, a doctor, Professor Raoult, claims to have treated the disease better thanks to an antimalarial drug, chloroquine, which he promoted. This hypothesis that brought hope has never been scientifically verified and is now discredited.[3]

Even if the death toll by municipality has not yet stabilized, the differences are also marked between the cities and the countryside. Population density is an essential factor in the human-to-human spread of the virus via the respiratory tract; this is why most of the world's major cities have excess mortality: New York + 364%, Madrid + 175, London + 130, Île-de-France + 98, Stockholm + 90, Istanbul + 30."

---

[2] https://theconversation.com/geographie-de-la-pandemie-de-covid-19-en-france-et-en-allemagne-premiers-enseignements-139367.

[3] https://www.ncbi.nlm.nih.gov/pmc/articles/PMC7132364/.

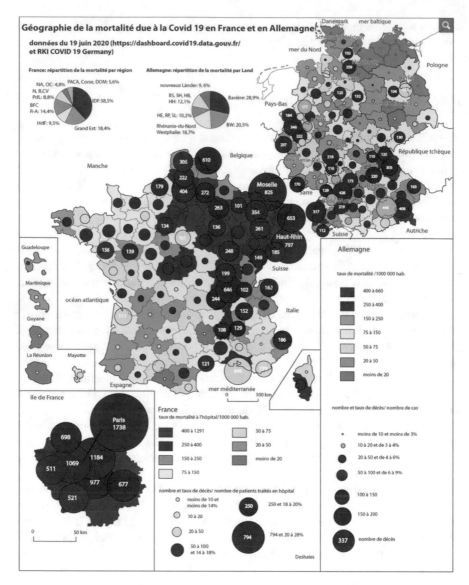

Fig. 4 Distribution of the number of cases treated in hospital by department in France and by Germany as of May 21, 2020, *according to Mr. Deshaies*

Even in Germany, the contrast between the more densely populated southern Germany and the northern great plains is also very marked.

The Environmental Factors of the Spread of the Virus

The virus is spread by droplets expelled by infected people when they speak or cough. Without a mask, these droplets carrying the virus can travel up to 3.7 m; with a mask, this distance is reduced to a few centimeters in the worst case. The so-called general public masks are especially useful for protecting those around a person infected with the virus.

The Italian environmental medicine society (2020) and a Chinese study (Hao et al. 2020), noting that the areas most affected by this health crisis also corresponded to areas with a high level of air pollution, showed the geographical correlation between the levels of air pollution and the incidence of illness (rate of hospitalizations or death). Some scientists then hypothesized that air pollution (especially small particles) could promote the maintenance and transport of the SARS-CoV-2 virus in the air, thereby promoting the spread of the disease. Thus, based on established knowledge,[4] exposure to air pollution seems be an increased risk factor for COVID-19. However, no study has yet succeeded in conclusively proving (the correlation does not necessarily imply causation) the veracity of this mode of transmission, and if it exists, it is certainly not the majority transmission mode (Bontempi 2020). The preponderant role of droplets and fomites (contaminated objects or surfaces) in the transmission of COVID-19 has been confirmed by the WHO. This discussion joins that formulated in an open letter, signed by 239 researchers, addressed to the World Health Organization (WHO) and relayed by the New York Times on July 4.[5] They underline the risk of transmission of the virus by micro-droplets (the cutoff would be at 5 microns) which could spread up to several meters.

The possibilities of contamination are higher in confined spaces because the humidity level in the air is high, and therefore, the more aerosols can contribute to the spread of the virus, for example in restaurants or hospitals while it is more quickly dispersed in the open air. Air conditioning can make contamination worse.

The role of air pollution on the inflammation, development and exacerbation of chronic pathologies (cardiovascular, respiratory) is well established. Populations exposed to this cocktail of pollutants on a daily basis, even at low levels of concentrations, see their respiratory and immune systems in particular become more vulnerable to other stresses, such as Severe Acute Respiratory Syndromes—Coronavirus (SARS-CoV). There is therefore a biological plausibility to a relationship between exposure to air pollution and greater susceptibility to the virus, as well as the occurrence of severe forms of COVID-19 disease.

Indeed, it is evident that the most polluted areas also correspond to the most densely populated areas, those which stir up multiple movements and interactions between populations, the main cause of the spread of the coronavirus.

---

[4]COVID-19 as a factor influencing air pollution is an emerging topic, rapidly evolving. https://www.ncbi.nlm.nih.gov/pmc/articles/PMC7144597/.

[5]https://www.nytimes.com/2020/07/04/health/239-experts-with-one-big-claim-the-coronavirus-is-airborne.html.

The Inequality of the Health Care System

As a factor explaining the spatial differences in mortality is still, for the moment, difficult to specify. In the most affected areas, with equivalent case densities, Hauts-de-Seine has a much lower death rate than that of Paris. The same goes for the Bouches-du-Rhône which, with a comparable density of cases, has a mortality rate that reaches 77% of that of the Rhône department and 55% of that of Meurthe-et-Moselle. But the measurement of the density of cases is itself irregular because the patients are more or less declared according to the severity of the symptoms. Should we deduce from this the differences linked to the quality of the hospital system and the treatments provided therein? For the moment, the realization of such an assessment is too early, it is therefore impossible to analyze what relates to the characteristics of the population and what follows from the qualities and defects of the health system and the effectiveness of the treatments used.

A Major Inequality: Age and Secondly Sex

While the winter flu mainly strikes people over 80, COVID-19 primarily affects people over 65. 18% of those who died (between March 1 and July 1, 2020[6]) were between 65 and 75 years old. This finding surprised the very large generation of "baby boomers" who suddenly felt vulnerable as the steadily rising life expectancy gave them hope for a long and healthy old age. Among all excess deaths from all causes, during the same period, over 93% were over 65 years of age and 0.6% were aged 15–64 (+ 7% compared to expected mortality in this age group). In this last age group, it is people aged 45–64 who have been affected. Conversely, mortality among children under 15 decreased by 14% over the period at the national level (Fig. 5).

Since deaths mainly affect the elderly, the decline in life expectancy in France due to COVID-19 seems likely to be modest (Guillot and Klat 2020). For men, this would be a decrease of 0.2 years, or a life expectancy of 79.5 years in 2020 against 79.7 years in 2019. As regards women, life expectancy is predicted to drop by 0.1 years, or 85.5 years in 2020 against 85.6 years in 2019. The greater loss of life expectancy among men reflects their greater vulnerability to this disease and the importance of co-morbidity factors that they develop.

This virus most spares children and adolescents but while being asymptomatic, can they transmit the disease? It would seem that the answer is negative but calls for points of vigilance since family meetings seem to be at the origin of new "clusters." We must wait until the start of the school year to better understand the contamination by children (ECDC 2020). People living in elder facilities were very heavily infected by caregivers or by those who came to visit them, and they were penalized by their poor ability to withstand strong resuscitation.

---

[6]Source Public Health France https://www.santepubliquefrance.fr/.

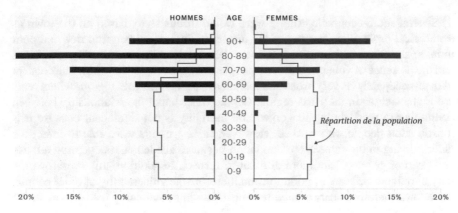

Fig. 5 Age distribution of people who died from coronavirus (COVID-19) in French hospitals between March 1 and July 19, 2020, by age group and sex. (*Source Public Health France*)

This generational inequality in the face of the disease may have caused some interrogation. Why impose lockdown which severely penalizes the least vulnerable young people to spare the older ones who more easily endure this period which does not significantly disrupt their daily rhythm of life? The constraint experienced during confinement by young people explains the feeling of liberation experienced at the time of deconfinement and the refusal to continue to endure physical distancing.

The statistics currently available allow us to formulate hypotheses, which are still weakly supported, on vulnerability to the Covid of poverty. Statistics concerning the ethnic origin of the patients are not available and the count of deaths according to their place of residence or their socio-economic level is not yet available.

What Are the Social Inequalities in the Face of Illness?

Thanks to statistics issued by INSEE, researchers (Brandily et al. 2020) have been able to highlight a link between poverty and observed excess mortality. This study, unable to count precisely the number of deaths linked to Covid, concerns the excess mortality, calculated for the municipalities included in urban areas, between April 2019 and April 2020. Before April, the entire population was surprised by the disease which, during confinement, affected the population more selectively.

In heavily infected regions such as Ile-de-France and Grand Est, the pandemic caused a 50% increase in mortality in most municipalities (compared to areas with low infection), but this increase actually reached 88% for municipalities in the poorest quartile. (i.e., municipalities where the median household income is below the 25th percentile of the national income distribution). On the other hand, the comparison of rich and poor municipalities in regions with a low infection rate does not reveal any significant difference in the evolution of mortality.

Several socio-economic criteria were tested in this study based on the town as a statistical basis. Sixty percent of the mortality differential between rich and poor municipalities in highly exposed areas can be explained by poor housing conditions and the presence of vulnerable elderly people in over-occupied housing. The concept of over-occupancy[7] is very broad and includes many dwellings in regions where rents are high, such as in the Paris region. On the other hand, the relationship between continuing to work during lockdown and mortality is not significant because it is mostly adult people who worked. However, these workers were able to infect the elderly living in the same accommodation because, inside houses, wearing a mask and "barrier gestures" are more difficult to respect. In addition, this overcrowding may have increased indoor pollution and therefore the vulnerability of older people. Lockdown therefore bears a large responsibility in accentuating inequalities in the face of disease.

Lockdown and Its Environmental and Socio-Economic Consequences

It is the brutal ban on travel that has caused the worst consequences of the pandemic on the environment. The French quickly complied choosing rapidly the best place of confinement.

Lockdown Had a Direct Impact on Mobility

Thanks to data from Facebook and mobile phones, the variations in the mobility of people between March 14 and March 17 (date of the imposed confinement) compared to the previous week could be mapped. (Fig. 6) We can see the decrease in traffic flows while the blue dots represent the municipalities whose population has increased between 5 and 35%. The Alsatian region, already severely affected, has remained very isolated from the rest of France. Nevertheless, the inhabitants of Alsatian towns took refuge in the neighboring mountains, in the Vosges.

Thirteen percent of Parisians have left Paris for a city in Île-de-France or another French region. Contrary to what one might think, this exodus has also occurred in all the major French cities such as Bordeaux, Lyon and Marseille.

As a result, unlike the metropolitan centers, a number of places usually deprived of holidaymakers at this season—mainly located between the south of Normandy and the south of the Pyrénées-Orientales—have seen their population increase significantly (blue dots on Fig. 6). The map shows a significant number of blue dots in

---

[7]Overcrowded accommodations are those that have less than "one living room, one room for each couple, one room for each other adult aged 19 or older, one room for two children if they are of the same sex or are under 7 years of age, and one room per child otherwise".

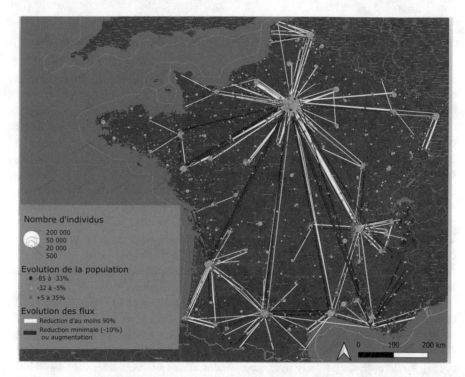

Fig. 6 Evolution of the population and long-distance mobility (over 70 km) between March 14 and 17 compared to the previous week (*Source: Facebook Data for Good, Olivier Telle*)

Normandy, Brittany and along the Atlantic coast. Although this increase is relative due to the small number of inhabitants of these small towns, they have seen the number of their inhabitants increase by more than 10%. Conversely, in the Alps, the many red dots represent the tourists leaving the closed winter sports resorts.

The increase in outgoing long-distance mobility has been accompanied by a rapid suspension of periphery-urban mobility. Once confined, the French reduced their movements by 80%. The decline in regional mobility has been most sustained in the regions most affected by the virus. The use of public transport, a factor in the spread of the virus, quickly collapsed. The reduction in travel is most strongly associated with the 25–59 age group that of the most active population.

However, a difference was observed between the well-to-do segments of the population who were able for teleworking and those who were forced to travel to continue their work. However, a partial unemployment scheme has allowed many workers to stay at home with their children who are not in school. Only the children of caregivers were accommodated in certain schools.

Heavy truck traffic has hardly decreased even though the working conditions of the truck drivers have deteriorated in the absence of catering points.

Would Confinement Be Favourable to Air Quality and the Climate?

However, logistics continued to function to supply the population, who also benefited from electricity. This was essential for households to heat their homes during the end of winter and also to supply Internet connections which had become an essential bond of sociability. Electricity consumption fell by 10–20% in the first weeks of containment. This decrease is explained by the slowdown in economic activity and by the decrease in the number of trains while overconsumption was observed in the residential sector, but the impact of these variations on CO_2 emissions is low because the electricity produced in France is almost completely carbon-free (in average, nuclear energy provides more then 70% of electricity consumed); on the other hand, power stations running on fossil fuels have been shut down in neighboring countries. The spring period was sunny, and renewable energies were able to be widely used because they are a priority on the networks. The fall in electricity prices (production greater than demand) automatically increases the cost of public support for renewable energies with a guaranteed purchase price.

Motorized transport is responsible for 29% of emissions of CO_2 (CITEPA[8]) and of a significant fraction of nitrogen oxide (NOx) emissions; in 2017, their national contribution was 63%. But, transport sector is only responsible of 14% for PM10 emissions and 17% for PM2.5. According to air quality monitoring associations,[9] during confinement, the daily average concentrations of NOx near highways has decreased by 50% in Bordeaux; 70% in Toulouse; 67% in Fort-de-France; 69% in Dijon; 62% in Rennes and 69% in Marseille, but the drop in particulate matter content is much less marked. In several French regions (Hauts-de-France, Ile-de-France, Grand Est), an increase in the levels of particulate matters was observed despite the confinement instructions. These high levels are explained by the importance of domestic heating in late spring; this season is also favourable for the application of fertilizers and slurry which results in the emissions of ammonia which react with nitrogen oxides to form particles.

Air pollution therefore results from complex atmospheric processes involving emissions from many sources of pollutants, as well as weather conditions that may or may not favour the dispersion of pollutants. However, at the beginning of spring, the cold nights generated numerous temperature inversions which blocked the dispersion of pollutants and increased the level of pollutants despite the reduction in emissions linked to transport.

* **The influence of lockdown on air pollution through the example of the agglomeration of Grenoble** (Fig. 7).

In this alpine metropolis of Grenoble, the start of confinement marks a clear decrease in nitrogen oxides in 2020 compared to the same period (1–31 mars) the

---

[8] https://www.citepa.org/wp-content/uploads/publications/secten/Citepa_Secten-2019_20_Emis sions-par-secteur.pdf.

[9] https://atmo-france.org/covid-19-focus-sur-lexposition-des-riverains-a-la-pollution-automobile-pres-des-grands-axes-avant-pendant-le-confinement-21-avril-2020/.

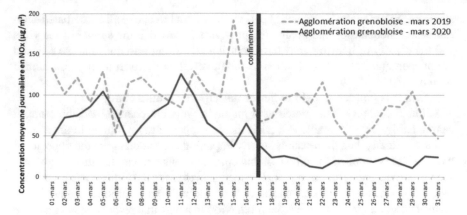

Fig. 7 Change in daily averages of nitrogen oxides NOx (mg / m³) in the agglomeration of Grenoble. Comparison between data for March 2020 and March 2019 (*Source*: *Atmo Rhône-Alpes measurement network*)

previous year. This graph highlights the importance of fluctuations in meteorology on variations in air quality.

The exposure of residents to air pollution also depends on the quality of the air inside homes. However, the confinement has changed the exposure of residents by increasing the time spent in homes and the number of residents present in homes all day. They are potentially more exposed to pollution from indoor equipment (combustion devices, furniture, coatings, paints, etc.) and from the use of cleaning and furnishing products, as well as from possible smoking. Hence, the need to ventilate his home without restriction since confinement allows doors and windows to be opened.

The reduction in noise and the silence that followed the decrease in motorized mobility had a beneficial effect on the health of the inhabitants. They were able to listen with pleasure to the singing of birds and taste the joys of nature.

According to a study by the Greenly app,[10] which measures the carbon footprint of users based on their bank statements, it has decreased by 62% during the last two weeks of march. The food budget has also fallen significantly, not so much to ensure ordinary food as the elimination of restaurant meals.

What Lessons Can We Learn from Lockdown?

The main consequence of this period of confinement is economic, but from an environmental point of view, it is possible to draw some lessons after more than four months of this crisis, which is far from over. Some have been able to detect through this pandemic an awareness of humanity's vulnerability to climate change. For others,

---

[10]https://www.greenly.earth/.

according to the analysis outlined above, confinement has above all exacerbated environmental inequalities and finally, the proven link between the spread of the virus and the density of housing, may have contributed calling into question the city and the phenomenon of metropolisation as the most effective form to organize human society.

* Can this pandemic help to better anticipate climate change?

Climate change is considered as a major irreversible change, and therefore, it cannot hardly be comparable to a crisis that is supposed to be limited in time.

Paradoxically, health measures strictly imposing confinement are the opposite of measures to be applied to combat global climate change since climate change transcends borders and is based on global cooperation to make the atmosphere breathable by all. This raises the whole question of the global organization of production to control flows and relocate industrial production in a way that respects standards. The priority given to health over the economy shows that public health measures have a much more significant effect than environmental policies on reducing emissions. "To take just one emblematic example, global air traffic was cut by two-thirds in March 2020: Even in their wildest dreams, proponents of a kerosene tax could not hope for such a result. We are well beyond the objectives of the Paris Agreement, which require an annual reduction of between 2.7% per year (to reach a maximum of +2 °C. at 2100) and 7.6% per year (to reach at +1.5 °C)"observes F.Gemenne.[11] This sudden decrease, beneficial both for health through improved air quality, and for the climate, cannot be sustained. Prevention in terms of air quality or even climate cannot be carried out abruptly by focusing only on one sector (that of city traffic or transport) because automobile traffic and mobility are essential drivers of urban life. There could be no question of banning them but of promoting alternative and active modes based on solidarity that confinement made difficult. Controlling CO_2 will not occur through lockdown, which shows trompe l'oeil [optical illusion] results with, of course, a drop in NOx and GHGs associated with urban traffic, but with, in return, significant risks to indoor air quality. It involves the implementation of large-scale strategies covering the various sectors of economic life and over the long term. Particular attention will need to be paid to the exit from containment and the possible "rebound effect." Can we consider that the refocusing of consumption on "essential needs" could be extended over time? Support for the resumption of economic activity will have to take fully into account France's commitments in terms of the climate in the short and long term, as expressed in the report of the High Council for the Climate.

The analysis of this episode can only confirm the importance of limiting pollution at the source and the sources themselves through shared and proportionate efforts with a view to improving the health of all for the common good that is air quality and climate.

---

[11] https://aoc.media/opinion/2020/07/27/habiter-la-terre-au-temps-des-pandemies-2/.

Lockdown Accentuates Environmental Inequalities

The lockdown imposed in elders facilities has generated solitude which may have been detrimental to the good health of people. The visitation ban was painfully felt, but it helped prevent the spread of COVID-19, which for older people often manifests as a severe and fatal form.

By showing a difference in the quality of life between those who can work remotely and enjoy housing that is more pleasant to live in and further away from their workplace and those who are obliged to go to their workplace daily while being restricted to public transport or the car (see above), it is therefore all urban development that is questioned.

The observed flight from urban dwellers to the countryside to avoid the virus and enjoy larger housing, with a garden underlines the unequal conditions of confinement suffered by city dwellers and rural dwellers.

In families, women experienced confinement much less well than men because their mental load was reinforced by teleworking (for women able to be connected) which had to be combined with the difficult management of the entire confined family. During this period, the number of meals taken at home increased while shopping was difficult; children had to be helped with their Internet exercises, while many home services have disappeared.

The much-maligned countryside and outskirts have taken on a certain appeal for the quality of life they offer: More spacious accommodation, better air quality, food supply provided locally (by "short circuits"). Is it, in a more lasting way, a revaluation of rural life which is taking shape and which the pandemic has accelerated? Many authors and journalists see the migration of Ile-de-France residents to their second homes during confinement as the harbinger of a future urban exodus. In 1876, 5.2% of the French population lived in Paris compared to 3.2% today (Talandier 2020).

A Questioning of the City ?

The health risks highlighted by density and urbanization call into question the doctrine advocated by town planners, namely the favour accorded to metropolisation and to the dense city and even to the densification of the urban system in general. The model of cities concentrated in metropolises has established itself in France in recent decades; it is reinforced by the concern to save energy and to fight against the artificialization of soils according to the commitment of zero net artificialization planned for 2030 (Talandier 2020). It has become a collective representation, an unsurpassable horizon of urban planning while the lessons learned from the pandemic should encourage planners to return to more reasonable densities. Controlling heat waves also opposes a strong densification of centers whose mineral character traps heat, while the presence of vegetation or cool corridors allows better ventilation without resorting to air conditioners. *"De-densification would help to thwart the well-started process of gentrification in the center of the city. Rather than defending at all costs a housing policy allowing the poorest to stay in intramural Paris, which has never really worked until now, it would be a question of making the most affluent populations want to leave Paris resulting in lower price of housing and lower density in the capital"*. (Dupuy 2020).

This movement back to the countryside at the time of confinement seems to be continuing with a more lasting movement already initiated previously and described by (Faburel 2018). Fifty-four percent of Parisians who took part in the survey carried out in May by the start-up "Paris jet'arte" declared that they wanted to leave the capital as soon as possible, against 38% before confinement. They say they seek a less stressful environment (59%) to get closer to nature (59%). They also want a simpler life, more in tune with their values (57%). They want to have access to larger and cheaper housing (52%) with a better balance between their professional and personal life (46%). In fact, metropolisation has traded proximity to living things and the autonomy of supply for the conveniences of incessant movement, permanent entertainment or even continuous connections. The fragility of the densest territories emerges as a prospect of socio-territorial reorganization and upgrading of the outskirts without rebuilding a border between towns and countryside that no longer exists. But, it is a question of giving back its place to peri-urban and rural town planning since life in the countryside encourages more ecologically sustainable behaviour: energy production (products derived from agro-forestry, village power plant, etc.), behaviour of limiting resource requirements (self-construction, re-use), consumption (sobriety, frugality, simplicity, etc.) or equipment (compost, natural phyto-purification rather than heavy equipment, etc.). (Eymard 2020). France has a whole network of medium and small towns that need to be revitalized. The COVID-19 may offer them this opportunity.

Conclusion

While at the beginning of March 2020, the number of patients requiring intensive care increased exponentially, the only solution for France was to decree the implementation of a severe lockdown which made it possible to stop the progression of the virus already strongly present in Ile-de-France and eastern France. Indeed, the country, despite being warned of the likelihood of an epidemic, was unprepared: hospital beds, medical equipment, tests and masks were in short supply. The period of confinement, between March 17 and May 11, was experienced very unequally: a moment of slowing down and a return to nature for some, a period of work in difficult conditions for others. The sudden decline in motorized mobility has resulted in a decrease in nitrogen oxides in the air, but it has created great social and economic difficulties. Since May 11, the virus has not stopped circulating, but the characteristics of the virus and the disease are better known and while waiting for a hypothetical vaccine, strategies calling for individual responsibility are put in place to avoid "clusters." While children do not appear to be affected by the virus, young people, often asymptomatic, like to get together after having undergone a long period of lock-down. They thus transmit the virus to older people who can develop serious or even fatal forms of the disease. Within days, the spread of the virus around the world highlighted the fragility of humanity, attacking societies and people who, thanks to modernity, felt invulnerable. Is it possible to sustainably assume the primacy of health over the economy, as the leaders have declared? If, the French, had the opportunity to glimpse what could mean a more sober way of life, they also saw all the dysfunc-tions that a brutal lockdown could generate in terms of inequalities and sociability without forgetting the lack of economic viability of such a device. Only European and global intergenerational solidarity will make it possible to resist the virus and the troubles linked to climate change which, beyond the crisis represented by the pandemic, threaten humanity.

ANNEX

The pandemic did not end on July 31, a second wave arose in France

While at the end of July, it could be concluded that 30,000 deaths were attributable to COVID; by mid-November, there must be at least 10,000 additional deaths who succumbed to a second wave of the epidemic which brings the number of people affected by the virus to 2 million (versus 67 million French people).

This second wave can be seen on the curve of the number of daily deaths in France over the last three years (Fig. 8). The two peaks of spring and autumn 2020 are clearly distinguishable from the winter increase, the graph of which only represents the terminal phase, and from the effects of summer heat waves (1500 deaths are attributed to the 2019 heat wave).

How the occurrence of this second wave that began on September 1 can be explain? Summer temperatures and outdoor life have been unfavourable for the virus during the summer times. These factors, combined with a period of euphoria that followed the constraints imposed during a very hard spring lockdown, have caused a general decline in the vigilance of the French and especially the youngest of them. Unlike other European countries, the rate of mobility of inhabitants quickly returned to the level reached in 2019: A week after May 11, trips to Paris reached 25% of usual trips, then 75% and 90% in September.

Thus, the months that followed the spring lockdown resulted in a premature and generalized abandonment of "barrier gestures" despite the recommendations given by the health authorities.

At the beginning of September, the coming back to classes and work caused an upsurge in contamination. But, unlike the month of March, the tests were much more available and made it possible to consider, in addition to the barrier gestures, a strategy: "test-trace-isolate". Admittedly, numerous tests were carried out and made it possible to produce incidence maps such as the one from October 21 to 7 (see Fig. 9). This strategy failed for lack of reliability of the tests, for lack of knowledge on the ways of contamination, for non-respect of the isolation of the positive people and for a general reluctance to the concept of "tracing," the application "all-anti-Covid" dedicated to this function was a fiasco.

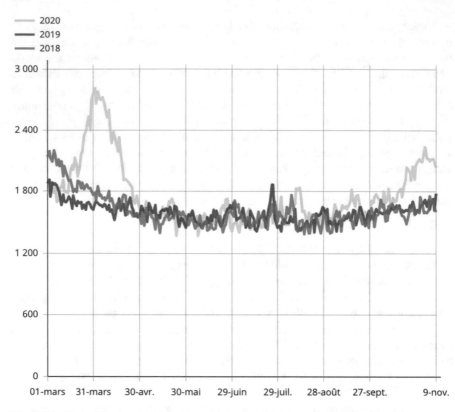

Fig. 8 Number of deaths per day in France between March 1 and November 9 during the years 2018, 2019, 2020. Source INSEE

Fig. 9 Incidence rate per 100,000 inhabitants from October 21 to 27. Source: *public health France*

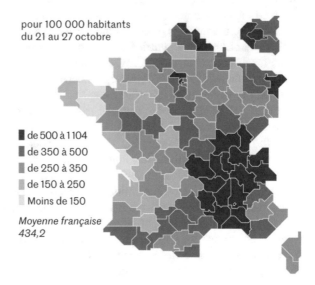

Until October 17, vigilance was encouraged but left to the care of individual responsibility; however, many clusters formed mainly in bars, gyms and family or business gatherings. So, in many cities, a 9 p.m. curfew has been imposed. But, even before knowing the results of this device, faced with the exponential rise in contaminations, the President of the Republic, for fear of the onset of winter and the risk linked to the time spent inside, imposed a new lockdown from October 31 until December 1.

This second wave and the means implemented to contain it are different from those used in the spring.

After the mixing of populations during the holidays, the whole country was affected with a very marked epicenter in the Lyon region and in the Rhône valley. This area, very touristy, had been very visited during the holidays; it also concentrates many university establishments bringing together students who, neglecting the barrier gestures, have been very affected by the virus. Nevertheless, this young population has experienced very mild forms of the disease but they have also been in contact with more vulnerable people who have been more severely affected requiring hospitalization or even intensive care.

Figure 10 shows how the number of people hospitalized. Both waves reached the same level of over 32,000 with rapid acceleration before containment measures. But, knowledge of the disease and the therapies has evolved, and, proportionally, the number of people in intensive care has fallen (4,637 instead of 7,019). Better use of corticosteroids and oxygen therapy has reduced the mortality rate of patients on mechanical ventilation.

The proportion of elderly people in hospital remains high 85.2% of them, as of November 15, were over 60 years old. Eighty-two percent of French people who died from the coronavirus were also over 65. Hence, the special vigilance that seniors must show so as not to impose too strict lockdown on the younger generations, the economic consequences of which would be disastrous.

This second lockdown, lighter than that of spring, disrupted the recovery of the economy that began this summer. The pandemic affects two particularly prosperous sectors in France, namely aeronautics and tourism, without forgetting the disruptions suffered by the world of entertainment and culture reduced to occurring in empty rooms. Preserving the lives of the elderly supposes many sacrifices, but the priority given to life over the economy is an achievement of our civilization.

On Christmas Eve, perhaps we can hope that the curve is turning definitively and that the efforts made on the vaccine will make it possible to eradicate this virus definitively while waiting for the next attack that we promise. Biologists.

Addendum

We would have liked to be able to put an end to this chapter on the pandemic but the virus was able to disclaim many rational arguments presented in this chapter; a second wave has indeed taken place, a third is announced and a fourth is looming.

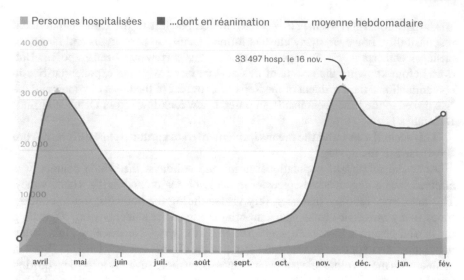

■ Personnes hospitalisées   ■ ...dont en réanimation   —— moyenne hebdomadaire

Fig. 10 Evolution of the number of people hospitalized and in intensive care between March 18, 2020 and February 1, 2021 (weekly average). *Source Santé Publique France*

3,251,160 positive cases have been confirmed since the start of the epidemic, or 5% of the French population.((4 900 cases for 100 000 hab) Recent information suggest that France is currently (4th of April, 2021) battling a peak of about 5000 COVID-19 patients in ICUs. On 2nd April, 2021, the country recorded 46,677 new cases and 304 deaths. Special medical planes dispatched COVID-19 patients from overrun Paris intensive care units to less saturated regions. The country has entered its third national lockdown on 3rd April, 2021, as it stuggles a surge in cases of COVID-19.

The death toll has doubled from the figure announced in July to reach 77,600 dead at the end of January, 2021, and the death toll is increasing every day (more than 90,000 on 1st April, 2021). 78% of covid-related deaths concern people over 75, this average tends to decrease.

The second wave, after the summer respite, began in November, as soon as the first cold days appeared. Should we see in these curves a sign of the influence of temperature? Should we see in the increase of the epidemic in October the influence of the return from vacation and the mixing of the student population which has been very affected by the virus with weak symptomatic manifestations. The curve has started to drop slightly but forecasts for the coming months are pessimistic because the different emerging variants are more contagious and perhaps more dangerous.

Science has progressed, doctors are less clueless about the disease which, proportionally kills less, and many patients are hospitalized without needing to go to intensive care. Above all, hope for a vaccine is looming even if its production and distribution appear chaotic. Only 2 million of the oldest and most vulnerable French people are vaccinated.

The spatial differences between the regions affected by the virus remains high. Overall, Western France continues to be preserved, but the new variants have not said

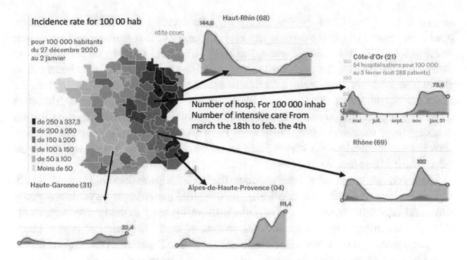

Fig. 11 Incidence rate for the whole of France at the start of 2021 and evolution in the number of people hospitalized and in intensive care between March 18, 2020 and February 1, 2021. *Source Santé Publique France 1869 / 5000*

their last word. The English variant, especially present in western France, reached 14% of all French people infected. The appearance of the other variants remains, for the moment, marginal and localized.

For the year as a whole, the entire eastern half of the country was the most affected, with excess mortality varying around 15%. The departments of the western half were more affected by the second wave but the number of patients remains very low (see example of Haute-Garonne). The "Haut Rhin" had been badly affected by the first wave while the second wave reached it weaker. On the contrary, the region of Nice (Côte d´Azur) was relatively spared during the first wave but it is currently severely affected. In "Côte d´Or" and in the Rhône, the two waves are well marked. The causes of the spread of the virus remain the same, rural areas are spared while cities are more vulnerable.

For almost a year, the French have been tired and aspire to a normal life. No one was prepared to live through such a difficult period which is probably not the last because other pandemics are likely to repeat themselves and climatic disturbances also represent a threat: droughts and floods have followed one another during these periods of lockdown.

The virus management strategy has evolved since the strict containment of the spring. Wearing a mask has become widespread as it has become compulsory in all public places from the age of 6. The French have adopted barrier gestures, offices and businesses have been fitted out to respect a certain distance between individuals. Testing has become easier and more accessible. They made it possible to exclude for 7 days those who tested positive as well as those with whom they had been in contact. At the beginning of November, all gatherings were forbidden, bars, restaurants, cinemas and theaters were closed. From October 15 in some cities and from October 31, the curfew at 8 p.m. was decreed throughout the territory. It will be lowered at 6 p.m. from the end of January. But schools are open and economic activity has resumed except in the culture and tourism sector which are badly affected despite the compensatory economic measures that have been put in place. The students have taken practically all of their lessons by videoconference since the beginning of the year and they deplore the absence of social life since gatherings are prohibited and bars closed. The population as a whole finds it difficult to accept these measures which undermine fundamental freedoms, hence the tension between the medical world which fears an overload of hospitals and which advise more restrictive measures and the aspirations of the population.

However, we must take into account the deterioration in the state of health of the French: difficulty in accessing healthcare, increase in psychiatric problems and domestic violence, overweight and an upsurge in addictive behaviour: alcohol consumption has increased. On the other hand, the number of flu, bronchitis and gastroenteritis has fallen significantly due to the health precautions taken. Epidemiological studies are in progress to validate these observations.

Faced with this contradiction, will vaccination make it possible to rapidly strengthen the population's immunization?

References

Benkimoun, S. (2020). Évolution des mobilités et diffusion du Covid-19 en France: ce que les données Facebook dévoilent, *The conversation*, 18 mai.

Bontempi, E. (2020). First data analysis about possible COVID-19 virus airborne diffusion due to air particulate matter (PM): The case of Lombardy (Italy) https://doi.org/10.1016/j.envres.2020. 109639.

Brandily, P., Brébion, C. & Briole, S. (2020). A poorly understood disease? The unequal distribution of excess mortality due to COVID-19 across French municipalities. ffhalshs-02895908ff https:// halshs.archives-ouvertes.fr/halshs-02895908.

Dupuy G., (2020). Villes denses, villes vertueuses: Un modèle indépassable? https://theconversat ion.com/villes-denses-villes-vertuelles-un-modele-indepassable-139686.

ECDC European Centre for Disease. (2020). Prevention and control COVID-19 in children and the role of school settings in COVID-19 transmission technical report 6 aug 2020 https://www.ecdc. europa.eu/en/publications-data/children-and-school-settings-covid-19-transmission.

Eymard, L. (2020). Let's stop densifying our cities! "- https://theconversation.com/canicule-et-urb anisme-arretons-de-densifier-nos-villes-142504.

Faburel, G. (2018). *Les métropoles barbares : démondialiser la ville, désurbaniser la terre, le passager clandestin*, p. 380.

France-stratégie, (2020). L'impact de la crise sanitaire (Covid-19) sur le fonctionnement du système électrique. https://www.strategie.gouv.fr/.

Guillot, M., & Klat, M. (2020). Épidémie de COVID-19, (2020): quel impact sur l'espérance de vie en France ? *The Conversation*.

Hao, X., Yan, C. & Fu, Q. (20 August 2020) Possible environmental effects on the spread of COVID-19 in *China science of the total environment* (Vol. 731, p. 139211). https://doi.org/10.1016/j.sci totenv.2020.139211.

Italian Aerosol Society. (2020). Informativa sulla relazione tra inquinamento atmosferico e diffusione del COVID-19. https://www.iasaerosol.it/attachments/article/96/Nota_Informativa_IAS.pdf.

Siegfried, A. (1960). *1960, Itinéraires de contagion, épidémies et idéologie* (p. 165). Paris: Colin.

Talandier, M., (2020). Tous au vert? Scénario rétro-prospectif d'un exode urbain, https://theconversation.com/tous-au-vert-scenario-retro-prospectif-dun-exode-urbain-137800.

Past Major Infectious Diseases and Recent COVID-19 Pandemic: Health and Social Problems in Italy

Cosimo Palagiano

Abstract In this study, the various significant infectious diseases that have affected Italy are discussed. Most of these diseases have affected not only Italy, but a large part of the world countries. Italy, as a destination for population movements for many centuries, has had to face infections from the East and the South. In recent decades, China has been the largest center of infectious disease irradiation. China is the country of origin of the latest flu pandemics, including COVID-19. Pandemics today have a greater virulence and expansion, due to the greater speed of the means of communication and the numerous points of gathering of people, such as tourist resorts, large cities and reckless behaviour regardless of precise rules.

Keywords Infectious diseases · Cholera · Plague · Flu · Covid-19 · Population movements · Social behaviours

Introduction

The pandemics that have affected Italy from past centuries to today are numerous, also because Italy, due to its geographical position, occupies the center of gravity of population movements between the East, the West and the North of Eurasian continent. Furthermore, the country is in the center of the Mediterranean Sea, in close contact with the northern regions of Africa.

Infectious diseases can be spread by bacteria and viruses.

As we know, infectious diseases are very numerous. Although each region or country has its own diseases, some can spread to most or all of the world. But many of these diseases can be avoided precisely by reacting to reclaim the environment and transform it from natural into humanized. Malaria was defeated in the swamps of temperate areas with the sprinkling of DDT, which in hot countries is more difficult to fight, as the vector, the Anopheles mosquito, can live well and multiply in that particular climate. For the treatment of the disease a certain prophylaxis was the

C. Palagiano (✉)
Sapienza University of Rome, Rome, Italy
e-mail: cosimo.palagiano@uniroma1.it

© The Author(s), under exclusive license to Springer Nature Switzerland AG 2021
R. Akhtar (ed.), *Coronavirus (COVID-19) Outbreaks, Environment and Human Behaviour*, https://doi.org/10.1007/978-3-030-68120-3_13

administration of quinine. Quinine is still present in a well-known tonic drink today, which the British introduced when they administered India, to defend themselves against malaria.

Tuberculosis

In the first decades of the last century tuberculosis was the leading cause of death in Italy as well as in other countries of our continent, because tuberculosis made the sick very weak and a remedy was not possible to find. From 1890 onwards, conferences, congresses and political debates were numerous and fruitful, so much so as to allow the launch of a specific insurance legislation which, since 1927, affected about half of the population in our country. The First World War caused a serious halt in the virtuous trend relating to the mortality and morbidity of the tuberculosis epidemic, while between 1900 and 1914 a significant reduction in the epidemic had been observed. Bologna, and some Tuscan and Lombard cities took initiatives both in the curative (sanatoriums) and preventive (marine and mountain colonies) fields. Around 1930, despite this social-health and welfare growth, tuberculosis in Italy still continued to be an important problem. In Europe, Germany was the country that, more than any other, was able to lead the fight against tuberculosis on a social, scientific and legislative level. As early as 1883, a health insurance law had been enacted in that country and a network of sanatoriums was established.

For this reason, sanatoriums were also built in Italy to treat the sick. One of the most important sanatoriums, if not the most important, was the Carlo Forlanini, founded on 10 June 1934.

Tuberculosis is acquired almost exclusively by air through the inhalation of airborne particles (droplet nuclei) containing *Mycobacterium tuberculosis*. These particles are spread by people with active pulmonary or laryngeal tuberculosis whose sputum contains a significant number of microorganisms. The treatment of tuberculosis makes use of numerous drugs in combination with each other for adequate periods of time. Vaccines are also studied to enhance the immune defenses. Nonetheless, tuberculosis today kills 1.5 million people worldwide every year and an average of one person every seven hours in Europe.

Spanish Flu

But the greatest epidemic of the twentieth century was the so-called Spanish flu, which spread as a pandemic from 1918, at the end of the First World War, until December 1920. The long duration of this epidemic must make us reflect on the fact that the flu cannot be affected by high summer temperatures. Indeed, the high summer temperatures, as will happen for COVID-19, facilitate the gathering of people in public places, such as beaches, theaters, restaurants, etc. and therefore a greater spread of diseases. The Spaniard caused 500 million cases and over 100,000 deaths, out of a world population of less than 2 billion people. In a Colorado mountain town, Gunnison, the authorities raised barricades along the streets, closed the train station, ordered citizens, who were then 1390, to stay indoors, quarantined anyone who wanted to enter the city, and jailed those who disobeyed the peremptory order.

I wanted to compare the tuberculosis infection and that of the so-called Spanish one, not only because in large part these epidemics occurred simultaneously, but because this disease is now closer in effects both to the corona virus that haunts us today, and to the remedies that the health authorities have tried to devise, such as forced isolation, the spasmodic search for prevention with a vaccine and an unlikely pharmaceutical remedy. Another affinity that binds infectious diseases is the immune system, which, if strong, is able to mitigate the consequences of the disease and even avoid them. The immune system is very weak in the elderly, especially if they are carriers of other diseases. Also those who live in degraded and poor areas, as in the case with tuberculosis have a weak immune system. However, COVID-19 is also beginning to affect young and healthy people.

The remedy of lockdown, although appropriate, but launched on a large scale to defend ourselves from COVID-19, is contrary to the need for living beings to socially share spaces. Furthermore, since the success achieved in defeating evil is about to be achieved in China, even countries with liberal democracy must follow the same very strict rules in confined spaces. Spaces in nature are the scene of struggles for survival between species. In the case of infectious diseases, pathogens and their vectors must be included. In this sense goes the herd immunity, which allows a "natural" immunity. As the British Prime Minister Boris Johnson wanted to implement in the first place, who then gave up on starting the project, aware of the damage it would produce.

Cholera

The bacterium *vibrio* of cholera spreads in unhealthy areas rich in organic waste. We all remember the spread of cholera, especially in southern Italy in 1973. Above all mussels, known by different names in various regions of Italy were accused. Cholera mainly affected the mussel farms in the Mar Piccolo of Taranto, even if the sudden construction of the Iron and Steel Center of Taranto had by now decreed their death. Before others, in an article written in 2014 in collaboration with Tiziana Banini, for the Springer types, I denounced the senselessness of that location in a very flourishing agricultural area, as Antonio Cederna wrote in *Corriere della Sera*. Now there is a struggle between the need for production and the health of the inhabitants, between transfers of property and rebellion of workers and citizens. The abatement of pollution had to be done from the beginning. I remember with emotion my mother who collected the clothes from the clothesline, once colored black and another red. In Taranto we could understand well the direction of the winds from the direction of the dense fumes of the blast furnaces. It was necessary to clean up, but now it is late due to high costs. In the 19th and twentieth centuries, the cholera epidemics that erupted in Italy were six, in 1835–37, 1848–49, 1854–55, 1865–67, 1884–86, 1893, and two slight occurrences, in 1911 and 1973.

In the 30 s of the nineteenth century, when cholera began to spread in Europe, the health authorities and the governments of the Italian states, especially those who had commercial relations with other countries, began to take suitable measures. The Kingdom of Sardinia and the Kingdom of the Two Sicilies established maritime sanitary cordons and quarantine days for ships coming from infected or suspicious areas. The Grand Duchy of Tuscany sent doctors to the countries affected by the epidemic to study the disease and the measures taken there. In 1835, when cholera spread to their borders, some states in northern Italy established lazarettos as a precaution (Tognotti 200, p. 47). The sanitary cordons did not hold up due to the political fragmentation of the Italian states and were the cause of bankruptcies of seasonal family businesses, that had commercial relations with abroad. On July 27, the sanitary cordon was broken by some smugglers and the epidemic spread to Turin and Cuneo and on August 2 broke out in Genoa (Tognotti 2000, p. 53). In September, cholera reached the Lombardy-Veneto Kingdom, which had not established any health cordon, in October it activated in Venice and in November in Trieste, Padua, Verona, Vicenza (Tognotti 2000, p. 55) and Bergamo. In the spring and summer of 1836 it spread to Como, Brescia, Cremona, Pavia and Milan. In summer, anger reached the territory of current Tentino-Alto Adige, Parma, the Ligurian coast including Genoa, Livorno, the Papal Marches, Modena, Ancona, Trani and Bari (Tognotti 2000, p. 56–7). In 1837 cholera spread to Naples, Calabria, Malta, Sicily and the Ligurian coast. The duchy of Benevento and the papal state were infected again. The papal state in Sicily was surrounded by sanitary cordons. In late 1837 the first wave of cholera ended. The affected cities managed to break down the epidemic in a hundred days (Tognotti 2000, p. 58–9).

About 10 years after the first epidemic wave, a second wave raged in Europe. The carriers of the epidemic were the soldiers of the Austrian and Russian armies engaged in the revolutionary uprisings of 1848. In this year the infection spread throughout the Austrian empire including Vienna. In the summer of 1848 the infection came to Italy. The first areas affected were Austrian Lombardy, Veneto, Istria and the territories involved in the First World War. Cholera immediately spread to Treviso, Padua, Vicenza, Verona, Udine, Rovigo, Venice and Trieste (Tognotti 2000, p. 170).

In 1854 the third cholera epidemic broke out brought by a ship from India to London and from here to southern France and Italy. The infection spread from Genoa over the whole Ligurian and Tyrrhenian coast to Naples and Palermo (Tognotti 2000, p. 189). Sardinia was also infected. In Sassari, home to an ancient university, 5000 of its 23,000 inhabitants died. In 1855, cholera spread throughout Italy (Tognotti 2000, p. 189). In 1856 the third epidemic outbreak died after having infected 4468 Italian municipalities against 2998 of the first epidemic and 364 of the second. The deaths from this third epidemic were 284,514. 146,383 had been in the first infection and 13,359 in the second. Don Bosco with 44 young people went to the rescue of the sick, including his pupil Domenico Savio, who died of tuberculosis before turning 15 (Tognotti 2000, p. 195).

In the summer of 1865 France, Southern Italy, Genoa, Marseille and Toulon were invaded by the fourth cholera epidemic (Tognotti 2000, p. 221).

In the two-year period 1865–67, the epidemic revealed important characteristics, had spread to much less areas than previous epidemics. Especially in Italy it was limited to port areas and southern Italy (Tognotti 2000, p. 223–4). Some large cities, in fact, had implemented a public and private sanitation, that decreased the affected cases. However, there were no therapeutic aids, as more than 60% of the cases became lethal. The practice of bloodletting was banned, but the use of opium, zinc flowers, astringents, enemas, hot baths, alcoholic drinks such as rum and mulled wine especially in the algid stage remained (Tognotti 2000, p. 227).

In spite of scientific discoveries and awareness of the relationship between bad socio-economic conditions and disease, the European states were unable to defeat the two epidemic waves of 1884 and 1893. Above all Italy, intent on solving problems of realization of the railway network, the fight against literacy and administrative reorganization, did not devote much effort to treating the disease. The epidemic that broke out in 1884–86 plagued Naples above all (Borghi 212, p. 124). The epidemic that broke out in 1884–86 plagued Naples above all (Borghi 2012, p. 124). During that period 91% of the population of Naples was gathering in the center of the city, in the so-called "*bassi*" (slums), in very precarious hygiene conditions (Tognotti 2000, p. 251). During the 1884–87 four-year period, three Italian provinces were affected by the epidemic, Cuneo with 1655 deaths, Genoa with 1438 deaths and Naples with 7994 (Tognotti 2000, p. 253).

On January 15, 1885, the so-called "law for Naples" allocated many resources for sanitation standards (sewers and urban sanitation) not only in Naples, but also in many other cities, such as Genoa, La Spezia, Turin, Caltanissetta, Trapani, Milan, Catania and about 60 other municipalities (Tognotti 2000, p. 267).

In 1893 very few urban centers were affected by cholera, Genoa recorded 414 deaths, Rome, Turin and Milan had no serious consequences, and Naples and Palermo registered lower deaths than in previous epidemics (Tognotti 2000, p. 255).

At the end of June 1911, during the Giolitti government, a light cholera outbreak made its appeared in Casagiove in the province of Caserta. On the evening of June 29, the lawyer Giovanni Tescione proposed the establishment of a group of willing people, who would alleviate the misfortune. On June 30, 1911 the Public Assistance Committee was made up of young people from the best families of Casagiove. The members of the Promoting Committee organized a rescue team and found and reported the suspected cases, patrolled the affected houses, supervised the transportation of the victims, the disinfection of houses and streets, the preparation, the discipline and the provisions of the premises. This appearance of the epidemic however had short duration with few infected.

The 8th cholera pandemic, based on the strain called El Tor, occurred mainly from 1961 to 1975, spreading to Asia, Soviet Union, and northern Africa. The spread of the infection was due to the mass movements of the population, but with low mortality rates, thanks to the preventive curative measures adopted by the different countries. By the end of 1972, cholera was widespread in 59 countries around the world (Luzzi 2004, p. 94).

Table 1 N. of cases and deaths during the choler a epidemic in Southern Italy in 1973

Region	N. of cases	N. of deaths	Case fatality rate (%)
Campania	*130*		
Naples	119	15	–
Caserta	11	0	–
Apulia	*125*		
Bari	110	6	5.5
Brindisi	2	0	–
Foggia	4	1	25.0
Lecce	1	0	–
Taranto	8	0	–
Sardinia	*13*		
Cagliari	13	1	7.7
Other regions	9	1	11.1
Total	**277**	**24**	**8.7**

Source Italian Ministry of Health. Report to WHO on onset of Cholera cases in Italy, 1973

This epidemic of El Tor biotype, Ogawa serotype spread in southern Italy—regions of Campania, Puglia and Sardinia—from 20 August 1 to 12 October 1973 with 278 cases and 24 deaths. The blame was given to the consumption of raw mussels, which, as is known, filter the sewage of the ports and sewers (Barbuti et al. 2017) (Table 1).

I certainly must not ignore other serious infectious diseases such as plague, AIDS, sleeping sickness and Ebola. Unfortunately, these are not the only ones.

Plague

Plague is an infectious disease of bacterial origin, caused by the bacillus *Yersinia pestis*. It is a zoonosis, the area of development of which is linked to various species of rodents and whose only vector is the rat flea (*Xenopsylla cheopis*), which can also be transmitted from human to human. Hippocrates probably associated it with consumption, as it was also called in past centuries. The word used by Hippocrates in the Aphorisms is precisely φθίσις, or consumption. Many plague pandemics have spread around the world. Famous are the literary descriptions of the plague in Florence in 1348 which is the background to Giovanni Boccaccio's *Decameron* and the very detailed one in Milan in 1629–30, narrated by Alessandro Manzoni in chapter XXXI of the *Promessi Sposi*. The plague described by Manzoni was probably caused by the results of prolonged famines, invasions and wars. According to the WHO, from 2010 to 2015, 3248 cases occurred worldwide, with 584 deaths.

The plague is the pandemic that has a long tradition, which we can follow in many classic texts (Homer, *Iliad*, I. 61), Hesiod, *Works and Days*, 243, Aeschylus, *Oedipus Rex*, vv. 1-150, Herodotus, 7.171.2. Thucydides, *History of Peloponnesian War*, 2.47-54, Lucretius, *De Rerum Natura*, 6. vv. 1138–1286).[1]

The name of the disease in Greek is λοιμός and in Latin *pestilentia*, from which the English word pestilence originates through the ancient French, which means both an epidemic disease especially the bubonic plague and someone or something morally corrupting.

The most famous plague narrated by most of the classic authors mentioned above is that of Athens, which occurred during the second Peloponnesian war in 430 BC, killing Pericles.

Since in ancient times the term "plague" was used to indicate generically a "misfortune", a "ruin", there is a doubt as to whether to attribute many past epidemics to *Yersinia pestis* or other pathogens. Perhaps the famous Athens plague of 430 BC was possibly *Salmonella* or other infectious diseases (Shapiro et al. 2006, pp. 334–335 vs. Papagrigorakis et al. 2006, pp. 206–214).

More likely, however, the *Yersinia pestis* in the so-called Justinian plague,that broke out in 541 AD. in Constantinople, is the first documented plague pandemic. Told in great detail by the historian Procopius of Caesarea (Cunha Ujvari 2002, p. 45), it seems to be responsible for the death of about 40% of the population of the Byzantine capital. Then spread, in waves, throughout the area Mediterranean up to around 750, causing 50 to 100 million victims It is considered the first pandemic in history (Naphy and Spicer p. 13.). Even the Muslim world was not spared. At least five pestilences have been known since the Egira, the plague of Shirawayh (627–628), the plague of 'Amwas (638–639), the violent plague (688–689), the plague of the virgins (706) and the plague of notables (716–717) (Naphy and Spicer 2006, p. 16–17, *Conrad* (1981, pp. 51–93).

The most famous and devastating pandemic spread around the mid-fourth century, known as black plague and considered the second after that of Justinian. Imported from northern China through the Mongol Empire, it spread in successive stages to Asian and European Turkey and then reached Greece, Egypt and the Balkan peninsula. In 1347 it was transmitted to Sicily and from there to Genoa, in 1348 the black plague had infected Switzerland, except the canton of Grisons and the whole Italian peninsula, except Milan. The epidemic in Florence was particularly violent. It served as a frame of the Giovanni Boccaccio's *Decameron*. From Switzerland it expanded

---

[1] The plague narrated by Thucydides in book II of the Peloponnesian War was taken up by the Latin poet Lucretius in book VI of the poem *De rerum natura*. In the first book of the Iliad, the wrath of the god Apollo is described, which triggered an epidemic of plague in the camp of the Greeks for the refusal of Agamemnon to return his slave Chryseis to father Crise, priest of Apollo. And Sophocles in the tragedy Oedipus king deals with a plague epidemic in Thebes in Greece. In the Bible, in Book of Samuel is described an epidemic of plague sent by God against the Philistines for stealing the Ark of the Covenant: "The hand of the Lord was heavy against the people of Ashdod, and he terrified and afflicted them with tumors, both Ashdod and its territory" (1.5). The French painter Poussin in the seventeenth century depicted this biblical event in the painting *The Plague of Azoth*, now preserved in the Louvre Museum in Paris.

to France and Spain, in 1349 it reached England, Scotland and Ireland, in 1363, after infecting all of Europe, the outbreaks of the disease shrank until disappearing. According to some studies, it killed at least a third of the continent's population, probably from 45 million to 35–37.5 million (Alchon 2003, p. 21, Bergdolt 1997).

At the end of the great black plague pandemic of the European population, a period of continuous recurrence of the disease began with the consequent numerous victims, albeit to a lesser extent than in the first wave. It has been observed that, between 1347 and 1480, the plague struck the major European cities at intervals of about 6–12 years afflicting, in particular, young people and poorest sections of the population. Starting from 1480 the frequency began to decrease, reaching an epidemic every 15–20 years, but with effects on the population certainly not less (Naphy and Spicer 2006, p. 67). Other European cities, such as Paris, Reims, Amsterdam and London, sought remedies to counter the epidemic. Milan was one of the first cities to move in this direction, establishing a permanent health office in 1450 and building the San Gregorio hospital in 1488 (Naphy and Spicer 2006, p. 71). In 1486 Venice to established permanent control bodies, but in Florence it was necessary to wait for 1527. Despite the adoption of all these measures, the plague continued to recur and to reap victims.

In general, the plague waves of 14th and 15th centuries not only brought about radical changes in the appearance of the cities or in the patrimony of the survivors, but more importantly, the way of thinking of the men of the time changed. The taste for luxury and fun that spread immediately after the infection, was born from the fact that the experience of the plague had dramatically highlighted the uncertainty of tomorrow, so much so that it seemed to most senseless to worry about the future by investing their possessions in new productive activities or in the education of children. The patrimony was thus used essentially for the satisfaction of one's personal pleasure and the capital accumulated as a consequence of the plague was squandered in most cases. However, the effects of the new conception of life born during the scourge of the scourge were not entirely negative: if certainly the credit for the cultural renewal that characterized the following period cannot be attributed exclusively to the great plague, changing the mentality of the survivors, however, made a fundamental contribution to the rise of what will be among the most flourishing eras of our history, namely Humanism and the Renaissance. Furthermore, the age of great geographical discoveries begins at the end of the fifteenth century. An hope of a renewal of society and life is also aroused with the end of the COVID-19 pandemic.

Considerable epidemics occurred in the Milan area in the two-year period 1576–1577, in northern Italy in 1630. Alessandro Manzoni described it in the *Promessi Sposi*. But the plague spread in those years throughout northern Italy, in Turin, Venice, etc. The main testimonies that handed down the events of 1630 of the Duchy of Milan are the chronicles of the doctor Alessandro Tadino (1580–1661) and of the canon, historian and writer Giuseppe Ripamonti (1573–1643), source of Manzoni. Both were direct witnesses of the great pestilence of 1630, of which they left two fundamental works for understanding the events. In 1648 Tadino published the *Raguaglio* of the origin and the successes of the great contagious, poisonous, and malefic plague followed in the City of Milan, Ripamonti printed the chronicle in Latin *Iosephi*

Ripamontii canonical Scalensis chronistae urbis Mediolani De peste quae fuit year 1630 in 1640.

The epidemic of 1630 was particularly virulent. Between July and October of that tragic year 150,000 people died between the city and the rest of the Venetian *dogado* [2], equal to 40% of the population (Casoni 1830). In July and August 1630 the records of the Supreme Magistrate of Health reported that 48 people died in the city of Venice, touching the peak of 14,465 deaths in November of the same year (Casoni 1830). The dramatic painting by Giambattista Tiepolo preserved in the church of Santa Tecla in Este, Padua, where the saint is depicted outside the city walls, while praying to God among abandoned corpses and scenes of despair, is testimony of the plague in the hinterland. The plague was finally declared eradicated on November 21, 1631. As a vote for the end of the terrible plague, the government of Venice decreed that the basilica of Santa Maria della Salute be erected. The church was finished in 1632 (Frari 1830). Likewise, in Vicenza, at the end of the plague, the extension of the Basilica of Monte Berico was decided. The great plague that struck Venice was meticulously described by a future doctor, eyewitness of the plague, Cecilio Folio, who around 1680 described the events of 1630–1631 in one of his works cited by Giovanni Bianchi in 1833 (Bianchi 1630). In the absence of detailed data, it is estimated that in northern Italy between 1630 and 1631 1,100,000 people died from the plague out of a total population of about 4 million (Table 2).

As can be seen from the above table, the greatest contribution in absolute values and in percentage to mortality is due to Milan.

In 1894, on the occasion of the outbreak of the epidemic in Hong Kong, the French-Swiss bacteriologist Alexandre Yersin managed to discover and isolate the bacterium responsible for the disease, later called *Yersinia pestis*. In 1898 Paul-Louis Simond explained that this bacillus could be transmitted through the bite of fleas that had become infected by rodents (Simond et al. 1998, pp. 101–104). For the rest of the twentieth century, the outbreaks of the disease continued, but with mortality rates much lower than previous epidemics, thanks to the introduction of effective public health measures and, starting from the 1950s, antibiotics. However, the plague has remained as an enzootic rodent disease almost all over the world except Australia.

The plague epidemic of 1656 affected part of the Italian peninsula, in particular the Viceregno (Vice-kingdom) of Naples. In Naples the plague came from Sardinia, causing about 200,000 deaths out of a total of 450,000 inhabitants. Also in the rest of the kingdom the mortality rate ranged between 50 and 60% of the population (De Renzi 1867). On May 5, 1656 the plague struck two areas of the city of Rome, a house in the Ponte district and a farm on the slopes of Monte Mario. Nonetheless, the authorities did not begin to worry seriously until ten days later, following three more deaths between 15 and 26 May. Before Rome, the deadly disease had exterminated Sardinia, Genoa and finally the South of Italy causing almost 300,000 victims. In comparison,

---

[2]*Dogado* or Duchy of Venice was the homeland of the Republic of Venice, headed by the Doge. It comprised the city of Venice and the narrow coastal strip from Loreo to Grado, though these borders later extended from Goro to the south, Polesine and Padovanoto the west, Trevisano and Friuli to the north and the mouth of the Isonzo to the east.

Table 2 Mortality of plague data in some Italian cities in the period 1630–1631

City	Population	Dead	%
Verona	54.000	33.000	61
Padua	32.000	19.000	59
Modena	20.000	11.000	55
Parma	30.000	15.000	50
Milan	250.000	186.000	74
Cremona	37.000	17.000	46
Brescia	24.000	11.000	45
Piacenza	30.061	13.317	44
Como	12.000	5.000	42
Bergamo	25.000	10.000	40
Vicenza	32.000	12.000	38
Venice	140.000	46.000	33
Turin	25.000	8.000	32
Bologna	62.000	15.000	24
Florence	76.000	9.000	12

Source Cipolla 2005, p. 190–1

the nearly 15,000 dead Romans seem very little, although the contagion had raged for a long time devastating entire neighborhoods and destroying almost all the activities. However, the interest of travelers and diplomats then present in the capital focused on the efficiency, despite the delays, of the Roman organization. Alexander VII Chigi, the pontiff of that time did not hesitate to mobilize the Congregation of Health, 8 cardinals and 28 commissioners, two per ward. Each commissioner had a specific role, to supervise the disinfestations of houses and furnishings, to protect the non-infected and those in quarantine, to procure goods and care of families and the infected, to maintain public order and, of course, to eliminate corpses. Cardinal Girolamo Gastaldi sensed that it was essential to immediately concentrate infected individuals in a single hospital, identified in the broader and central structure of the Fatebenefratelli on the Tiber Island (Capparoni 1935, pp. 1–12).

Plague is often identified as a disease of the past, since over the centuries it has been responsible for large epidemics, which sometimes have reached the size of the pandemic, causing over 50 million deaths overall (*Plague*, in *World Health Organization*, October 2017). However, the bacterium has not yet been eradicated and remains a threat with which many populations of the world, and in particular the resident in Africa, have to deal, so much so that since the nineties of the twentieth century there has been an increase in cases by so that the plague was classified as a re-emerging disease. However, the bacterium has not yet been eradicated and remains a threat with which many populations of the world, and in particular those residing in Africa, have to deal, so much so that since the 90 s of the twentieth century an increase in cases was registered, by so that the plague was classified as a re-emerging disease (Schrag and Wiener 1995, pp. 319–324; Stenseth et al. 2008.

COVID-19 Pandemic

Our health system, while depleted for years of equipment and personnel, has been valiantly supported by doctors and nurses valiantly and relentlessly dedicated to doing everything possible to meet the needs of COVID-19 patients, while patients of other diseases they had to postpone treatment.

After the fear, with the so-called second phase or better resumption, the justified need for work has forced governments, sometimes in conflict with regions, often with a different degree of morbidity from the virus, to open some activities, not always with the right timing differentiation, and with strong inequalities. Also in the case of the school, the inequalities, which have always existed, have become accentuated, between students able and those unable to use electronic media and parental aids.

This pandemic has certainly produced a gap between the previous life and the next. A crisis, that derives from the Greek κρίσις, i.e. separation, choice, such as the one we are going through, is an important moment for future decisions. We must take advantage of it.

The present pandemic, forcing us to the so-called lockdown, as they say in erudite language, but simply and more prosaically confinement, accompanied by physical distancing, but some prefer social, has forced us to live very differently in a few months. The smart working possible for some workers has certainly decreased traffic volumes, which has certainly almost been canceled, because only some movements of proven necessity were allowed. Seeing Rome deserted and the Pope blessing in St. Peter's Square without faithful people was traumatic, even if from the balconies and windows many occasional singers improvised pieces of operas in front of applauding listeners.

Many elderly people died in nursing homes, without the comfort of relatives. Great impression was the long line of military trucks carrying the dead from Bergamo to nearby cemeteries, once the capacity of the local cemetery was exhausted.

Virologists, epidemiologists, immunologists dictated their rules and the citizens were resigned, more from fear than from intimate conviction. Unlike the time in which the action of the Betrothed takes place, today's scientists have the appropriate tools to attack evil, studying, looking for new drugs and hoping to find the vaccine (Capua 2020). Especially the younger ones, convinced that they were immune, due to their generally good immune defenses, wanted to return to normal.

At the date of end of August 2020 the data were as follows (Table 3):

The epidemic curve passed the peak and tended to drop in almost all regions, reaching the value of 1 infected which infects less than another.

"The epidemic curve continues to decrease in the number of both cases and symptoms, The Rt contagion index is below 1 in all Regions, a consequence of the measures adopted and the adhesion to these by citizens, But we are still in an epidemic phase".

Table 3 Cases of COVID-19 at end of August 2020

Region (Official name in Italian)	Confirmed cases	Healed	Active cases
Lombardia	98,545	16,857	5787
Piemonte	32,440	4143	1142
Emilia/Romagna	31,094	4458	2189
Veneto	22,190	2107	2119
Toscana	11,253	1139	1039
Liguria	10,683	1571	413
Lazio	10,236	875	2284
Trentino-Alto Adige/Südtirol	7925	697	224
Marche	7124	987	248
Campania	5976	442	1164
Puglia	5119	555	548
Sicilia	4091	286	947
Abruzzo	3662	472	342
Friuli-Venezia Giulia	3651	348	298
Sardegna	1859	134	463
Umbria	1679	80	195
Calabria	1408	97	170
Valle d'Aosta	1223	146	13
Molise	511	23	56
Basilicata	505	28	73
ITALY	**261,174**	**35,445**	**19,714**

[a]The nasopharyngeal swab is an examination that is used to search for the virus and therefore to diagnose the infection in progress, As can be seen, 4 regions of northern Italy (Lombardy, Piedmont, Emilia Romagna and Veneto) recorded 74.4% of confirmed cases and 77.8% of deceased, These are the most populous regions, with most of the economic activities and population movements
Source National Institute of Civil Protection

Thus the president of the Higher Institute of Health (Iss), Silvio Brusaferro, at the opening of one of the latest press conferences on the epidemiological analysis of Covid-19.

Unfortunately, after the summer, at the date of 16th October 2020, the confirmed cases are 391,611, the healed/discharged number is 247,872, with 36,427 deaths (Table 4).

As it is shown, the regions with the highest ratio between cases and population are Valle d'Aosta (12.94 per 1000), Lombardy (12.11), Liguria (10.91), Trentino-Alto Adige (10.61), Emilia Romagna (8.92), while Calabria (1.31), Basilicata (2.03), Sicily (2.21), Puglia (2.62) and Molise (2.76) register the lowest percentages.

The regions with the highest number of cases are Lombardy, Piedmont, Emilia Romagna, Veneto, Lazio, Campania and Tuscany, with a total of 310,086 cases

Table 4 Cases of COVID-19 at 16th October 2020 per Region, Population and per 1000 inhab

Region	Total cases	Population	Per 1000 inhabitants
Abruzzo	5825	1,331,574	4.37
Basilicata	1169	576,619	2.03
Calabria	2589	1,976,631	1.31
Campania	23,033	5,861,529	3.93
Emilia-Romagna	39,692	4,450,508	8.92
Friuli-Venezia Giulia	6093	1,227,122	4.97
Lazio	22,796	5,892,425	3.87
Liguria	17,281	1,583,263	10.91
Lombardia	121,130	10,002,615	12.11
Marche	9133	1,550,796	5.89
Molise	864	313,348	2.76
Piemonte	41,895	4,424,467	9.47
Puglia	10,734	4,090,105	2.62
Sardegna	5736	1,663,286	3.45
Sicilia	11,269	5,092,080	2.21
Toscana	21,017	3,752,654	5.60
Trentino-Alto Adige	11,208	1,055,934	10.61
Umbria	4210	894,762	4.71
Valle d'Aosta	1660	128,298	12.94
Veneto	34,277	4,927,596	6.96
ITALY	**391,611**	**60,795,612**	**0.64**

Sources National Institute of Civil Protection and National Institute of Statistics

in total. These are regions of northern Italy, but also central (Lazio) and southern (Campania), which make up 79% of cases.

The resurgence of the Sars-Cov-2 disease, generically called Covid-19, during the summer is due to the end of the lockdown throughout Italy, to the return of tourists from holidays to their homeland, in regions strongly dominated by the virus, by nightlife evening of young people and above all from the non-observance of the rules that require the use of a face mask, physical distancing, and gatherings. The biggest problem arises from asymptomatic people, who can infect the people they come into contact with. Of course, the data of the increase in infections depend on the simultaneous increase in the number of medical checks carried out on all those who frequent places and public transport vehicles. Furthermore, as we approach the winter season with low temperatures and high relative humidity, it is to be assumed that the infection increases and for this it is necessary that we quickly run to the prevention.

The national government and the regions plan to take increasingly stringent measures to avoid new lockdowns especially during the upcoming holidays at the end of the year.

Regions not yet much contaminated by the virus are now also affected by the disease. New rules have been enacted to intensify restrictions, the non-compliance of which will be punished with fines.

Unfortunately, the negative effects of the disease also affect economic conditions of many categories of workers, despite social protection, with the prolongation of the redundancy fund and the impossibility of dismissal at least for a certain period of time. The health emergency has not yet reached the levels of the early stages of the disease, although there is still a shortage of doctors and nurses in sufficient numbers. A recent Caritas study highlights the notable growth in poverty even among middle-class people.[3]

When It All Started

The first two cases of the 2020 COVID-19 pandemic in Italy were confirmed on January 30, 2020, when two tourists from China tested positive for the SARS-CoV-2 virus in Rome. An outbreak of COVID-19 infections was subsequently detected on 21 February 2020 starting from 16 confirmed cases in Lombardy, in Codogno, in the province of Lodi, which increased to 60 the following day with the first deaths reported on the same days.

Italy was the first European country to suspend all direct flights to and from China. Once the first internal outbreak was discovered, among the measures adopted there was the quarantine of 11 municipalities in Lombardy and Veneto. On 23 February the Council of Ministers with a decree sanctioned the total closure of municipalities with active outbreaks and the suspension of demonstrations and events in the same municipalities. From 25 February to 26 April the restrictive measures are extended to the whole national territory. With the decree of May 16, the start of phase two is announced, with the resumption of many commercial activities. The shift between regions was granted from 3 June with a further relaxation of restrictions.

Unfortunately, new small outbreaks opened in the second half of July and in August, mainly due to the return from holidays of young people, generally asymptomatic, from Italian and foreign tourist regions. At that time, many restrictions and health checks were decided at airports and ports for travelers from abroad and from regions with active outbreaks.

---

[3]On the occasion of the World Day to Fight Poverty, the Italian Caritas published a report, 'The antibodies of solidarity', which highlights a dramatic fact. In 2020, due to the pandemic, at least 450,000 more people must be supported by diocesan listening centers. These are important numbers, an increase of 45%. And almost one in two people do it for the first time. And it is, Caritas stresses, an increase that is "certainly underestimated" and different from the past, "when poverty was increasingly chronic, multidimensional, linked to complex experiences".

Today the biggest problems concern schools and pupil attendance, gatherings in theatrical and sporting events. The lack of people at the shows is a serious handicap to their success. The hope of young people is the search for normality, with the possibility of resuming entertainment in discos and clubs until late at night, without constraints of any kind.

This resulted in their asymptomatic infection spreading the virus among the older population less resistant to attacks of the disease for their possibly precarious health problems.

But What Does Normality Mean?

Is it normal to see our roads and highways clogged with traffic and unsafe?

Tragedies caused by road accidents? Crowding late into the night discos and streets? See the streets of our cities ruined and invaded by heaps of waste of various kinds? Visiting museums and places of artistic and cultural interest with crowds of tourists intent on taking selfies rather than enjoying the artistic beauties, not allowing others to stop to do so, because they are driven by a tumultuous crowd? Waiting a long time for public transport that is often overcrowded? Not breathing well due to polluted air? And seeing schools and hospitals run down and not adequately equipped to support too many students and patients respectively? The indiscriminate use of fossil fuels? And the invasion of plastic, even legitimized by recent measures, of polystyrene barriers to create the distance? (Palagiano 2020).

What the Virus Has Taught Us

It took only a few weeks of confinement because we managed not wanting to consider the environment with greater respect, much more than the unheard proclamations of climate change have done and are making.

I remember the conclusions of the 1992 Rio de Janeiro conference (about thirty years ago), regarding Agenda 21: (1) fight poverty; (2) change the logic of production and consumption for the conservation and management of natural resources which are the basis of life; (3) protect the atmosphere, oceans and biodiversity; (4) prevent deforestation; (5) promote sustainable agriculture, As can be seen, none of these indications have been observed.

So the first, minimum, desire is to see a near future in which greater attention is given to the preservation of the life of our cities, a great attention to safe work, for all and adequately paid, to the maintenance of our fragile territory, starting with from waterways and plant heritage, to avoid crowding in places of leisure, tourism and culture, to a rational management of means of transport and sanitation, to restructure industries and small factories in order to safeguard the safety and health of workers and inhabitants of the surrounding areas, to pay more attention to school buildings and education, not always adequate.

For those who think big and dream, we were able to represent livable cities and territories, well managed and supervised by rapid justice, without pollution and with prevention for the so-called "natural disasters", which are often caused by us. Even earthquakes, which are not predictable, can be softened in their destructive energy with adequate technical systems, in the places of their greatest and historically proven frequency.

All this is a dream, because long-term projects are needed, not taken into consideration for decades, by an inadequate political class, aimed more at the too frequent renewal of their offices and less at the well-being of citizens, And then what about corruption and tax evasion, always present in many business sectors?

Hoping for a Vaccine

But the safest remedy is the treatment with the vaccine, which is the subject of experimentation by several scientists linked together in international research and trials. The Coalition for Epidemic Preparedness and Innovations (Cepi), is coordinating the numerous projects for the preparation of vaccines against the virus, But the strategies adopted are very different from each other.

In particular, the researchers are working on three types of vaccines:

1. Rna vaccine: a sequence of Rna, synthesized in the laboratory, which should induce an immune response in the human body with antibodies active against the virus.
2. DNA vaccine: a fragment of DNA synthesized in the laboratory is introduced into the patient and is capable of stimulating the cells to activate the immune response through a protein.
3. Protein vaccine: through the RNA sequence of the virus in the laboratory, proteins or protein fragments of the viral capsid are synthesized, which are injected into the body with substances that enhance the patient's immune response.

Despite the strong pressure exerted by the pandemic, and the hope that each of us places in scientific research, the future use of a vaccine must necessarily be preceded by rigorous studies that require the time necessary to evaluate its efficacy and safety. Of course, the major pharmaceutical companies have an interest in producing their vaccine as quickly as possible, but they must also undergo all safety checks.

The World Health Organization has announced that there are currently around 80 potential vaccines against Covid-19 under evaluation, of which just over 10% are approved for human testing through clinical studies.

According to the Milken Institute, 202 vaccines are currently under development, of which 24 are in clinical trials. The process of developing a vaccine would take a decade, but in the case of COVID-19, times could be compressed due to the global urgency of the pandemic. However, vaccines are not used to cure, but to prevent disease, nor are they an alibi for disregarding the rules.

Addendum

The vaccine has arrived

Finally some pharmaceutical companies have prepared vaccines, of which the most used are: Pfizer BioNTech, Moderna, AstraZeneca, Johnson & Johnson. Vaccinations started from health personnel, the elderly and people with precarious health. However, deaths have not decreased, which are about 300-400 per day. Unfortunately, as the vaccinations are slow, some variations have occurred, including English, South African and Brazilian. Especially the first has a high rate of contagiousness. Therefore, the restrictions, lockdowns and curfews continue in a different and variable way between the regions, with different colors, depending on the spread of the virus and the hospitals' ability to respond. On February 9, 2021 the current positives were 413,967, the total healed 2,149,350, the dead 92,002. Total cases 2,655,319. The regions with the highest number of deaths are Lombardy, with 27,559, Veneto (9,364), Piedmont (9049), Emilia Romagna (9,914), Lazio (5,324), Campania (3,928), Sicily (3,728), Tuscany (4,360). Naturally, Lombardy and Piedmont are regions with a high population density.

References

Barbuti, S., Martinelli, D. & Prato, R. (2017). Bari in the seventh cholera pandemic. In *Hektoen international*, 24 March 2017.

Bergdolt, K. (1997). La peste nera e la fine del Medioevo: la storia della più spaventosa epidemia che mai abbia attraversato l'Europa, Casale Monferrato, E, Piemme.

Borghi, L. (2012). Umori, Roma, Società Editrice Universo.

Capparoni, P. (1935). *La difesa di Roma contro la peste del 1656–57 come risulta dall'opera del cardinale Gastaldi, in Tractatus de avertenda et profliganda peste, Atti e memorie dell'Accademia di storia dell'arte sanitaria*, XXXIV, *3*, 1–12.

Capua, I. (2020). *Il Dopo: Il virus che ci ha costretto a cambiare mappa mentale* (p. 135). Milano: Mondadori.

Casoni, G. (1830). La peste di Venezia nel MDCXXX, Origine della erezione del tempio a S, Maria della Salute, Venezia, *Dalla* Tipografia di Alvisopoli a spese di Giacomo Girardi.

Conrad, L. I. (1981). Arabic plague chronologies and treatises: Social and historical factors in the formation of a literary genre. *Studia Islamica, 54*(54), 51–93.

Cunha Ujvari, S. (2002). Storia delle epidemie, Città di Castello (Perugia), Odoya.

De Renzi, S. (1867). *Napoli nell'anno 1656, ovvero documenti della pestilenza che desolò Napoli nell'anno 1656, preceduti dalla storia di quella tremenda sventura*. Napoli: Tipografia di Domenico De Pascale.

Frari A. A. (1830). Cenni storici sopra la peste di Venezia del 1630–31: Per la quale si celebra in questi giorni la festa del secolo votiva, Con un compendio storico di tutte la altre pesti che afflissero la stessa città, *Venezia, Graziosi*.

Luzzi, S. (2004). Salute e sanità nell'Italia Repubblicana, Roma, Donzelli Editore, Biblioteca del Museo Provinciale Campano di Capua, *Comitato d'assistenza pubblica costituito a Casagiove durante l'epidemia colerica dell'estate 1911,* S, Maria Capua Vetere 1912.

Naphy, W. & Spicer, A. (2006). *La peste in Europa.* Bologna, *Il Mulino.*

Palagiano, C. (2020). Covid-19, Il parere di un geografo. in Bozzato, S. (ed). *Geografie del Covid-19,* Roma, Università di Roma Tor Vergata, pp. 845–849.

Papagrigorakis, M. J., Yapijakis, C., Synodinos, P. N., & Baziotopoulou-Valavani, E. (2006). DNA examination of ancient dental pulp incriminates typhoid fever as a probable cause of the Plague of Athens. *International Journal of Infectious Diseases, 10*(2006), 206–214.

Schrag, S. J., Wiener, P. (August 1995). Emerging infectious disease: what are the relative roles of ecology and evolution? In *Trends Ecol, Evol, (Amst,),* (vol, 10, n, 8,, pp. 319–24). 101016/s0169–5347(00)89118-1, PMID 21237055.

Shapiro, B., Rambaut, A., & Gilbert, M. T. P. (2006). No proof that typhoid caused the Plague of Athens (a reply to Papagrigorakis et a,). In *International, Journal of Infect, Diseases* (2006) 10, pp. 334–340.

Simond, M., Godley, M. L., & Mouriquand, P. D. E. (1998). Paul-Louis Simond and his discovery of plague transmission by rat fleas: A centenary. *Journal of the Royal Society of Medicine, 91,* 101–104.

Stenseth, N. C., Atshabar, B. B., Begon, M., Belmain, S. R., Bertherat, E., Carniel, E., Gage, K. L., Leirs, H., Rahalison, & L., Plague, past, present, and future. *PLoS Med,* 5(1), January 2008. 101371/journal,pmed,0050003, PMC 2194748, PMID 18198939.

Tognotti, E. (2000). *Il mostro asiatico, Storia del colera in Italia,* Bari, editori Laterza.

COVID-19 in Italy in the Context of the Pandemic Induced by SARS-Cov-2. Is There a Relationship Between COVID-19 and Atmospheric Air Pollution?

Gennaro D'Amato, Isabella Annesi-Maesano, Rosaria Valentino Maria, and Maria D'Amato

Abstract Following the outbreak of the disease in China, Italy was the first European country to be heavily involved in the pandemic defined as COVID-19 and induced by SARS-Cov-2 virus. During March 2020, many patients in the surrounding areas were diagnosed with COVID-19, often as severe cases. Another cluster was identified in the regions of Lumbardy, Veneto, and Piemonte, with an exponential increase in cases, mostly in northern Italy. Although the disease spreads throughout the whole country, it was the most prevalent in the north, reaching incidence and mortality rates among the highest in the world. Many factors explain differences from other countries, including Italy having a larger elderly population, different applications of detection tests, different prevention policies, and probably also the role of atmospheric air pollution. In Italy, the possibility of performing autopsies or post-mortem diagnostic studies on confirmed COVID-19 cases has been intensively debated; however, while post-mortem pathological analysis of COVID-19 patients in China has shown findings consistent with interstitial pneumonia and in some cases acute respiratory distress syndrome (ARDS), in Italian autopsies the involvement of endothelial, prevalently of microvascular system has been identified for the recruitment of multiple cytokine-activated inflammatory cell lineages. These anatomo-pathological and clinical observations, together with pharmacological and immunological observations, have improved the approach to the treatment of COVID-19. Of course, we hope to

G. D'Amato (✉)
Division of Respiratory and Allergic Diseases, Department of Chest Diseases, High Specialty A. Cardarelli Hospital, Naples, Italy
e-mail: gdamatomail@gmail.com

Medical School of Specialization in Respiratory Diseases, University on Naples Federico II, Naples, Italy

I. Annesi-Maesano
Directeur de Recherche (DR1) INSERM / INSERM Research Director, Co-Directrice Institut Desbrest d'Epidémiologie et Santé Publique (IDESP), INSERM et Université de Montpellier, Montpellier, France

R. Valentino Maria · M. D'Amato
Division of Pneumology, AO DeiColli and University Federico II, Naples, Italy

R. Akhtar (ed.), *Coronavirus (COVID-19) Outbreaks, Environment and Human Behaviour*, https://doi.org/10.1007/978-3-030-68120-3_14

239

have in future other treatments such as anti-COVID-19 vaccines and monoclonal antibodies.

Keywords Air pollution · Asthma · Rhinitis · COVID-19 · Interaction between air pollution and COVID-19

Introduction

COVID-19 has been declared a pandemic by the WHO (2020a). Following the outbreak of the disease in China, Italy was the first European country to be heavily involved in the pandemic (Livingston and Bucher 2020). Initially, three COVID-19 cases were reported in early February 2020, which were related to individuals who had traveled to China; then, on the 20th, a young man who had not traveled abroad presented with severe SARS-CoV-2-induced pneumonia in Lombardy, a region in the northern part of Italy (Onder e al 2020). Over the next two weeks, many patients in the surrounding areas were diagnosed with COVID-19, which was often severe, while another cluster was identified in the nearby regions of Veneto and Piemonte (WHO 2020b) with an exponential increase in cases., disease spread throughout the whole country, most commonly in the northern part (WHO 2020b) with the country's incidence and mortality rates among the highest in the world at the time (WHO 2020b).

Many factors could explain these differences from other countries, including different applications of detection tests, a larger elderly population, and different preventions policies, but the role of atmospheric air pollution (Dutheli 2020) is also probably important.

In Italy, the possibility of performing autopsies or post-mortem diagnostic studies on confirmed COVID-19 cases has been intensively debated. (Sapino et al. 2020) However, post-mortem pathological analysis of COVID-19 patients in China has shown findings consistent with acute respiratory distress syndrome (ARDS) (Pomara et al. 2020), while in Italian autopsies, the involvement of endothelial, prevalently of microvascular system, has been identified for the recruitment of multiple cytokine-activated inflammatory cell lineages (McAuley et al. 2020).

Other possibilities that deserve further experimental investigation include an exaggerated antibody-mediated response with complement activation and/or a hypothetical cytopathic effect of the virus (McAuley et al. 2020).

The latter could explain the microvascular damage leading to disseminated intravascular coagulation (manifested as thrombosis, thrombocytopenia, and gangrene of extremities), anti-phospholipid syndrome, and mimicry of vasculitis, which have been described in both Chinese (Huang et al. 2020) and Italian patients (Xu et al. 2020).

Epidemiology

Results

As of March 8, 2020, a total of 5830 positive cases were reported in all provinces of the Lombardy region. The first case (patient 1) was detected and laboratory confirmed at the hospital of Codogno in the Lodi Province on February 20. By the following day, 28 positive cases were identified in the same area.

On February 23, given the upsurge of positive cases in the area, a Prime Ministerial Decree (DPCM Law 23/02/20 n.6) introduced strict measures aimed at containing the spread of the disease. The implemented measures included the lockdown of a number of the municipalities which appeared at the center of the outbreak (Red Zone), the closure of schools of all grades and universities, and the suspension of public activities, sports events, and social gatherings across the region. The rapid intensification of regional surveillance that occurred in the following days, through contact racing and testing of both symptomatic and asymptomatic exposures to positive cases, provided critical information for the detection of possible epidemiological links, and uncovered ongoing transmission prior to the identification of patient 1. From various analyses, a highly different figure of the Lombardy outbreak emerged, with ongoing transmission in January appearing much less steep in terms of the number of cases as had been apparent in the temporal trend in the number of notified cases.

During the early stages of the COVID-19 epidemic in Lombardy, three major clusters were identified around the cities of Codogno, Bergamo, and Cremona. Later on, the epidemic began to become widespread in the entire region. However, from March 5 onward, the majority of cases (72%) were observed in the provinces of Bergamo, Lodi (where Codogno is located), and Cremona.

Were found a median age for positive cases of 69 years (range, 1 month to 101 years), with over half of the cases occurring in the over 65 (34% in the 75+) and 62% of the overall number of positives being males (Table). Almost half of the cases were hospitalized (47%), although this percentage increases when the data is stratified dataset considering to distinguish between those reporting a date of symptom onset and those who did not thought more likely to be less severe cases.

In the former group (those reporting a date of symptom onset), 63%, 61%, and 56% of the cases were hospitalized in, respectively, the three considered periods, whereas in the latter group, the percentage was 29%.

Among those hospitalized, 18% required intensive care. Overall, 346 deaths occurred in the region and the case fatality rates (CFRs) were highest in the older age groups (14% in the 75 + years). It was estimated that 95% of cases developed symptoms within 16.1 days of their infection. For the entire region of Lombardy, and in each of the three analyzed major clusters, we identified an initial phase lasting between one and two weeks and characterized by exponential growth.

Host Response and Possible Outcomes of SARS-CoV-2 Infection

Viral infection seems to occur mainly upon SARS-CoV2 engagement of angiotensin I converting enzyme 2 (ACE2), which acts as a functional receptor for the spike glycoprotein of the coronavirus. The HLA genetic system acts as a key player in determining the anti-viral immune response. In particular, the ability of HLA to trigger an adequate cytotoxic T-lymphocyte (CTL) response will result in viral clearance and host healing, along with the development of the IgM, IgA, and IgG humoral response. Conversely, an inadequate HLA asset will result in an inefficient CTL response and, consequently, incomplete viral clearance. In this context, various factors underlie increased COVID-19 severity, including an exaggerated antibody response, complement activation, leukocyte-mediated antibody-dependent cell-mediated cytotoxicity (ADCC), and T-cell-mediated inflammation, as discussed in the text.

Without a protective immune response, the virus is able to migrate, propagating into other ACE2-expressing tissues, while the damaged lung cells induce high inflammation, triggering the 'cytokinestorm'that represents the main cause of the acute respiratory distress syndrome (ARDS) and subsequent multi-organ failure. Incomplete viral clearance can also lead to the virus hiding in sanctuary sites and patient relapse with symptoms arising in new districts.

≈20% of patients develop interstitial pneumonia, and a subset of these (≈5%) develop ARDS that especially when so serious as to require invasive ventilation, is mostly fatal. The risk of ARDS rises with age, and almost all deaths occur in patients with pre-existing chronic conditions (Xu et al. 2020). Pre-admission hypertension, in particular, has been reported as a key mortality risk factor (Xu et al. 2020). The risk of death increases further in cases where there is a lack of ventilators or ventilation is refused, as described in Xu et al. (2020).

Moreover, an increasing number of clinical reports describe a biphasic behaviour: a first phase where COVID-19-infected patients are completely asymptomatic, which lasts on average seven days, and a second phase where the patients present mild to moderate flu-like symptoms, anosmia, ageusia, and blind conjunctivitis, which may last 10–15 days (Guan et al. 2020). A minority of patients who are unable to achieve complete virus coverage develop severe cardio-respiratory symptoms with radiological signs of pneumonitis, ARDS, and then multi-organ failure (Guan et al. 2020).

The last phase occurs, on average, 15–30 days after infection. In the latter case, patients may test negative for COVID-19 genome research standard molecular tests. Altogether, these clinical findings, as well as the available pathology studies, support the hypothesis of an inappropriate immune-related inflammatory response to COVID-19 epitopes and consequent auto-antigen release and T-cell cross-presentation in the damaged alveolar tissue. Consistently, recent results indicate that a systemic immune dysregulation that triggers auto-sustaining inflammatory lung damage, causing fatal respiratory failure and consequent multi-organ failure, is the main virus-related death cause in patients who develop SARS-CoV-2 (Guan et al. 2020).

COVID-19 and Air Pollution: Is There a Relationship?

In late 2019, a new infectious disease (COVID-19) was identified in Wuhan, China, which has now turned into a global pandemic (Zhu et al. 2020).

Countries around the world have all implemented some type of blockade to lessen their infection and mitigate it. The national lockdown to prevent the spread of the novel coronavirus resulted in less traffic and industrial activity (Dantas et al. 2020).

Monitoring of satellite pollution from the *National Aeronautics and Space Administration* (NASA) and the European Space Agency found that in cities in northern China such as Beijing, where much of the pollution comes from heating during the winter, there have been no reductions (Hopkins Univ 2020). However, in southern cities such as Shanghai and Wuhan, where there are heavy industrialization and intense air pollution, the decline in pollution was very substantial. However, recent data released by the National Aeronautics and Space Administration (NASA), European Space Agency (ESA), Copernicus Sentinel-5P Tropomi Instrument, and Center for Research on Energy and Clean Air (CREA) indicate that the pollution in some of the epicenters of COVID-19, such as Wuhan, Italy, Spain, USA, and Brazil, reduced by up to 30% (Liu et al. 2020). These levels of pollution were higher in the north than in the south of these countries for various reasons (cattle ranching, heating, and transport). These declines appear to demonstrate a clear correlation of COVID-19 infestation in China with levels of pollution and local climatic conditions consisting of a low temperature, moderate daytime temperature range, and low humidity that would favour the transmission of the virus (Iriti et al. 2020).

Declining levels of pollution have also been noted in northern Italy. A recent study identified correlations between COVID-19 mortality in northern Italy, including Lombardy, Veneto, and Emilia-Romagna, and the high levels of pollutants in those regions (Conticini et al. 2020). This decrease in environmental pollution was also seen in Barcelona, Spain, after two weeks of lockdown due to the novel coronavirus pandemic (Iriti et al. 2020). It is known that the transmission of SARS-CoV-2 by aerosol and fomites is plausible, and the virus can remain viable and infectious in aerosols for hours and on surfaces for days (Van Doremalen et al. 2020).

Setti et al. (2020) suggest that PM could act as a carrier for droplet nuclei, boosting the spread of the virus. The results were corroborated by Coccia (2020a, b) who demonstrated that the accelerated transmission dynamics of COVID-19 resulted mainly from transmission by contaminated air to humans, besides the transmission from person to person. The exhaled droplets would likely be stabilized in the air by the aerosol fusing with the PM in highly concentrated stable conditions. Under normal conditions with clean air and atmospheric agitation, a small droplet meets with the virus and evaporates or disperses rapidly in the atmosphere. However, in a stable atmosphere with high concentrations of PM, viruses have a high probability of agglomerating with the particles and reducing their diffusion coefficient, increasing their duration and permanence in the atmosphere, and becoming more contagious as a result. In England, Travaglio et al. (2020) provided further evidence of the relationship between air pollution and SARS-CoV-2 lethality by showing an association between pollutants released by fossil fuels and susceptibility to viral infection. This association suggests that individuals exposed to chronic high levels of air pollution may become more susceptible to SARS-CoV-2 infection. This would occur as a result of compromised immune defense responses due to pollution, as has already been demonstrated in patients affected by severe viral pneumonia during other pandemics (Cui et al. 2003; Min et al. 2016).

Recent studies suggest that the cause of death of many COVID-19 patients would be related to cytokine storm syndrome, a severe reaction of the immune system that leads to a chain of destructive processes in the body that can end in death (Air Quality News 2020).

Similar trends were observed in other regions of Europe, such as France, where COVID-19 distribution maps depict areas with a very large number of severely infected patients requiring hospitalization. This trend was seen in Paris, Lyon, and Toulouse and in regions like the Belgian-Swiss border near Lyon and the coast from Nice to Montpellier, where there are significant problems with air pollution and large agglomerations.

Following the epidemic of the disease in China, Italy was the first European country to be heavily struck with need of a heavy lockdown (Wu et al. 2020). The frequency of affected patients was much higher in northern Italy, in particular in Lombardy, than in southern Italy where COVID-19 has been less frequent. One plausible significant difference between these two areas could be the different air pollution (inhalable particulate matter (PM) and gaseous components like nitrogen dioxide (NO_2) and ozone (O_3)) which are higher in northern areas than in southern Italy (Kirby 2020).

It can therefore be credibly speculated that air pollution may contribute to the damage caused by the COVID-19 pandemic, by carrying virus particles and by rendering people more susceptible to SARS-CoV-2. According to The Lancet Commission on pollution and health, formed on the basis of data from the Global Burden of Disease (GBD) study1, air pollution (both indoor and outdoor) represents the single most important environmental factor presenting a risk to health (Report 21 2020). Environmental pollution has been shown to have alarming consequences for human health and be responsible for episodes of exacerbations of chronic respiratory diseases and contribute to the onset of allergic rhinitis and asthma (Nakada and Urban 2020).

The viability and long-range transport of respiratory viruses in open air, and thus their potential to infect individuals, have also been observed under real-life conditions (Myllyrta and Thieriot 2020).

Experimentally, it has also been observed that SARS-CoV-2 remains viable in aerosols for up to 3 h, with a reduction in infectious charge comparable to that found for SARS-CoV-1, confirming another possible route of infection transmission other than respiratory droplets and contact via contaminated surfaces (Implementation Mitigation strategies 2019).

Air pollution may be an important factor in viral disease transmission, and further investigations into the interaction of SARS-Cov-2 and air pollution are warranted.

These data suggest that air pollution—regardless of origin (traffic, heating, agricultural spreading, etc.)—should be considered as a contributing factor to the dynamics respiratory epidemics, and thus, there is an important need to advocate for cutting air pollution emissions in the interest of public health (Domingo 2020).

There are also reported observations that air pollution may contribute to the damage caused by the COVID-19 pandemic by rendering people more susceptible to SARS-CoV-2.

It is well established that gases such as NO_2 or O_3 and respirable PM can modify the permeability of airway mucosa. In particular, exposure to either fine PM or ultrafine carbon black particles can significantly enhance respiratory virus-induced inflammation, which may at least partially be caused by the decreased ability of macrophages to phagocytize the virus and mount an effective immune response against the infection. In addition, there is now evidence that SARS-CoV-2 is exhaled as a result of breathing regularly, becoming airborne and hence inhalable (Comunian et al. 2020).

Once in the air, it can either be carried by fine particles or mix with secondary ultrafine aerosols (Comunian et al. 2020).

It is important to consider also that long-term endemic exposure to air pollution is responsible for chronic systemic inflammation at the origin of the comorbidities that put COVID-19 patients at higher risk of severe events and death. SARS-Cov-2 virus and other respiratory viruses survive longer and become more aggressive in an immune system already aggravated by exposure to harmful substances, especially in patients affected by chronic respiratory diseases, such as chronic obstructive pulmonary disease (COPD) (Scafetta 2020).

In southern Italy, it appears likely that the action of UV radiation and winds coming from the sea has contributed to reduce the widespread of COVID-19 infection. At the general population level, infection rates for SARS-CoV-2 have sky-rocketed in countries, regions, and towns characterized by high levels of air pollution (Lombardy, Iran, China, South Korea, New York, Madrid, etc.) where COVID-19 hospitalizations and deaths have occurred in excess (Scafetta 2020).

One study positively related the proportion of daily excess for particulate matter of 10 μm of diameter (PM_{10}) with the number of COVID-19 admissions reported within the following 14 days (Comunian et al. 2020). The existing research had already shown the significant impact of air pollution on other respiratory viruses, such as SARS, influenza, syncytial virus, and measles, which are also transmissible via the respiratory route.

Urbanization with its high levels of vehicle emissions and a westernized lifestyle are linked to the rising frequency of respiratory allergic diseases and bronchial asthma observed over recent decades in most industrialized countries (D'Amato et al. 2015).

It is important to consider that these potential correlations can be reduced by reducing deforestation and by reducing old components that increase air pollution and climate change (D'Amato et al. 2015).

More than 297,000 people have died from COVID-19 infection, and more than 4.6 million have fallen ill (Wu et al. 2020). However, for those who have not been contaminated, the disease has caused significant changes to their lifestyle. COVID-19 also caused some unexpected consequences, such as the closure of industries, trade of non-essential items, transportation networks, and companies, accompanied by drastic declines in environmental pollution as governments introduced strict restrictions to combat the novel coronavirus pandemic.

Air pollution is known to cause damage to many organs and body systems, especially the respiratory and cardiovascular systems, being responsible for 4.2 million deaths (7.6% of total global deaths) in 2015. A particularly extensive and dangerous encounter with the virus must be feared in areas where air pollution is an important factor, since it appears to play an important role in determining the extent and lethality of COVID-19 in highly polluted areas around the world.

This is an important topic for the future of our countries and of the planet. The education of the population and the emergence of governmental decisions to prevent environmental pollution and climate change are urgent measures to be dealt with throughout the world. Although some adaptation and mitigation measures have been adopted over the years to limit air pollution induced by PM and gaseous components, however much remains to be done.

Conclusion

The relationship between COVID-19 and air pollution is supported by growing evidence, and therefore, further research efforts are urgently needed to clarify the exact mechanism of interaction (Tables 1 and 2).

Table 1 Number of patients affected by COVID-19 and death with COVID-19 infection, until the 30 of July

Germany	196.944	9.024
Italy	241.819	34.869
French	168.335	29.920
Spain	251.789	28.388
UK	285.768	44.236
USA	2938.625	130.306
Brasil	1623.284	65.487

Table 2 Research questions on air pollution as a contributing factor to COVID-19 infection

Research questions	Certainty versus uncertainty
Is SARS-Cov-2 exhaled?	Yes
Is SARS-Cov-2 airborne?	Yes
Is SARS-Cov-2 found in PM aerosol?	Potentially, as it can be carried by or mixed with secondary aerosols
Is SARS-Cov-2 viable once in suspension?	Yes but needs to be confirmed
Is air pollution a co-factor of respiratory viral infections?	Yes, in the case of SARS, influenza, syncytial virus measles
IS COVID-19 significantly related to air pollution?	Yes, but lacking sufficient data

References

Coccia, M. (2020b). *Labn. Working paper 48/2020, CNR—National Research Council of Italy.* Available in: https://ssrn.com/abstract=3567841. (Accessed April 2020).

Coccia, M. (2020a). *Diffusion of COVID-19 Outbreaks: The interaction between air pollution-to-human and human-to-human transmis-sion dynamics in hinterland regions with cold weather and lowaverage wind speed.*

Comunian, S., Dongo, D., Milani, C., & Palestini, P. (2020). Air pollution and COVID-19: The role of particulatematter in the spread and increase of COVID-19's morbidity and mortality. International Journal of Environmental Research and Public Health, *22*;17(12):4487. https://doi.org/10.3390/ijerph17124487.

Conticini, E., Frediani, B., & Caro, D. (2020). Can atmospheric pollution beconsidered a co-factor in extremely high level of SARS-CoV-2 lethality in Northern Italy? *Environmental Pollution, 4,* 114465. https://doi.org/10.1016/j.envpol.2020.114465.

World Health Organization, (2020a). Corona-virus disease (COVID-19) outbreak. Available in: https://www.who.int/westernpacific/emergencies/covid-19.

Cui, Y., Zhang, Z., Froines, J., Zhao, J., Wang, H., Yu, S., et al. (2003). Air pollution and case fatality of SARS in the people's Republic of China: an ecologic study. *Environ Health, 2,* 15. https://doi.org/10.1186/1476-069X-2.

Dantas, G., Siciliano, B., Boscaro França, B., da Silva C. M., & Arbilla, A. (2020). The impact of COVID-19 partial lockdown on the air quality of the city of Rio. *Science of the Total Environment. 2020*;729:139085. http://dx.doi.org/10.1016/j.scitotenv.2020.139085.

Domingo, J. L. (2020). Effects of air pollutants on the transmission and severity of respiratory viral infections. *Environmental Research, 187,* 109650 May 11. https://doi.org/10.1016/j.envres.2020.109650.

Dutheli, F. (2020). COVID-19 as a factor influencing air pollution? Environmental Pollution 2020;263 Pt. http://dx.doi.org/10.1016/j.envpol.2020.114466, 114466.

Guan, W. J., Ni, Z. Y., Hu, Y., Liang, W. H., Ou, C. Q., & He, J. X. et al. (2020). Clinical characteristics of coronavirus disease 2019 in China. *The New England Journal of Medicine, 30*(18), 1708–1720. http://dx.doi.org/10.1056/NEJMoa2002032.

Huang, C., Wang, Y., Li, X., Ren, L., Zhao, J., Hu, Y., et al. (2020). Clinical features of patients infected with 2019 novel coro-navirus in Wuhan China. *Lancet, 395*(10223), 497–506. https://doi.org/10.1016/S0140-6736(20)30183-5.

Implementation of mitigation strategies for communities with local COVID-19 transmission. Available in: https://www.cdc.gov/coronavirus/2019-ncov/downloads/community-mitigation-strategy.pdf.

Iriti, M., Piscitelli, P., Missoni, E., Miani, A. (2020). Air pollution and health the need for a medical reading of environmental monitoring Data. *International Journal of Environmental Research and Public Health,* 17(March7)): 2174. http://dx.doi.org/10.3390/ijerph17072174.

Johns Hopkins University and Medicine. (2020). *Coronavirus resource center. COVID-19 Dashboard by the center for systems science and engineering (CSSE) at Johns Hopkins (JHU).* Available in: https://coronavirus.jhu.edu/map.html.

Kirby, T. (2020). South America prepares for the impact of COVID-19. *Lancet Respiratory Medicine, 2020,* 29. https://doi.org/10.1016/S2213-2600(20)30218-6.S2213-2600(20)30218-6.

Liu, J., Zhou, J., Jinxi Ya, J., Zhang, X. , Li, L., & Xu X.. (2020). Impact of meteorological factors on the COVID-19 transmission: A multi- city study in China. *Science of the Total Environment,* *726,* 138513. http://dx.doi.org/10.1016/j.scitotenv.2020.138513.

Livingston, E., & Bucher, K. (2020). Coronavirus disease 2019 (COVID-19) in Italy. *JAMA. 323,* 1335–1335. 10.1001/jama.2020.4344 PubMed Abstract|CrossRef Full Text|Google Scholar.

McAuley, D. F., Brown, M., Sanchez, E., Tattersall, R. S., Man-son, J. J., et al. (2020). COVID-19: consider cytokine storm syn-dromes and immunosuppression. *Lancet, 395,* 1033–1034. https://doi.org/10.1016/S0140-6736(20)30628-0.

Min, C. K., Cheon, S., Ha, N. Y., Sohn, K. M., Kim, Y., & Aigerim A. et al. (2016). Comparative and kinetic analysis of viral shedding and immunological responses in MERS patients representing a broad spectrum of disease severity. *Scientific Reports, 6,* 25359. http://dx.doi.org/10.1038/sre p25359.

Myllyvirta, L., & Thieriot, H. (2020). 11.000 air pollution-related deaths avoided in Europe as coal, oil consumption plum-met. Available in: https://energyandcleanair.org/wp/wp-content/uploads/2020/04/CREA-Europe-COVID-impacts.pdf. (Accessed May 2020).

Nakada, L. Y. K, & Urban, R. C. (2020). COVID-19 pandemic: Impacts onthe air quality during the partial lockdown in São Paulo, Brazil. *Science of the Total Environment.* http://dx.doi.org/10.1016/j.scitotenv.2020.139087 (Accessed May 2020).

Onder, G., Rezza, G., & Brusaferro, S. (2020). Case-fatality rate and characteristics of patients dying in relation to COVID-19 in Italy. *JAMA, 323,* 1775–1776. 10.1001/jama.2020.4683PubMed Abstract|CrossRef FullText|GoogleScholar.

Pomara, C., Li, Volti, G., & Cappello, F. (2020). COVID-19 deaths: are we sure it is pneumonia? Please, autopsy, autopsy, autopsy!. *Journal of Clinical Medicine, 9,* E1259. 10.3390/jcm905125PubMed Abstract|CrossRefFullText|Google Scholar.

Report 21. (2020). Estimating COVID-19 cases and reproduction number in Brazil. Imperial College COVID-19 ResponseTeam 2020. Available in: https://www.imperial.ac.uk/mrc-global-infectious-disease-analysis/covid-19/report-21-brazil/. (Accessed May 2020).

Sapino, A., Facchetti, F., Bonoldi, E., Gianatti, A., & Barbareschi, M. (2020). The autopsy debate during the COVID-19 emergency: the Italian experience. *Virchows Arch.* https://doi.org/10.1007/s00428-020-02828-2. [Epub ahead of print]. PubMedAbstract|CrossRef FullText|GoogleScholar.

Scafetta. N. (2020). Distribution of the SARS-CoV-2 Pandemic and its monthly forecast based on seasonal climate patterns 2020 May 17, *International Journal of Environmental Research and Public Health,* 17(10), 3493. https://doi.org/10.3390/ijerph17103493.

Setti, L., Passarini, De Gennaro, G. F., Barbieri, P., Perrone, M. G., & Piazzalunga, A., et al. (2020). The potential role of particulate matter in the spreading of COVID-19 in Northern Italy: first evidence-based research hypotheses. 2020. Available in: https://www.medrxiv.org/content/10.1101/2020.04.11.20061713v1.

Tobias, A., Carnerero, C., Reche, C., Massagué, J., Via, M., Min-guill´on, M. C., et al. (2020). Changes in air quality during the lockdown in Barcelona (Spain) one month into the SARS-CoV-2 epidemic. *Science of the Total Environment. 726,* 138540. http://dx.doi.org/10.1016/j.scitotenv. 138540.

Travaglio, M., Yu, Y., Popovic, R., Santos, Leal, N., & Martins, L. M. (2020). Links between air pollution and COVID in England. medRxiv. http://dx.doi.org/10.1101/2020.04.16.20067405.

Urrutia-Pereira, M., Mello-da-Silva, C. A., & Solè, D. (2020). COVID-19 and air pollution: A dangerousassociation? AllergolImmunopathol (Madr) 2020 Jul 1;S0301-0546(20)30109-9. https://doi.org/10.1016/j.aller.2020.05.004. Online ahead of print.

Van Doremalen, N., Bushmaker, T., Morris, D. H., Hol-brook, M. G., Gamble, A., Williamson, B. N., et al. (2020). Aerosol and surface stability of SARS-CoV-2 as comparedwith SARS-CoV-1. *New England Journal of Medicine, 382*(16), 1564–1567. https://doi.org/10.1056/NEJMc2004973.

World Health Organization. (2020). Coronavirus Disease 2019 (COVID-19) Situation Report—57. Available online at: 2020 https://www.who.int/docs/default-source/coronaviruse/situation-reports/20200317-sitrep-57-covid-19.pdf?sfvrsn=a26922f2_2.

Wu, X., Nethery, R. C., Sabath, B., Braun, D., & Dominici, F. (2020). Exposure to airpollution and COVID-19 mortality in the United States. MedRxiv. 2020. http://dx.doi.org/10.1101/2020.04.05.20054502.

Wu, F., Zhao, S., Yu, B., Chen, Y. –M., Wang, W., & Song, Z. –G. et al. (2020). A new coronavirus associated with human respiratory disease in China. *Nature. 579*, 265–269. 10.1038/s41586-020-.2008.

Xu, Z., Shi, L., Wang, Y., Zhang, J., Huang, L., & Zhang, C., et al. (2020). Pathological findings of COVID-19 associated with acute respiratory distress syndrome. *Lancet Respiratory Medicine, 8*, 420–422. 10.1016/S2213-2600(20)30076.

Zhu, X. N., Zhang, D., Wang, W., Li, X., Yang, B., & Song, J., et al. (2020) A novel coronavirus from patients with pneumonia in China. *The New England Journal of Medicine, 382*(8), 727.

United Kingdom: Anatomy of a Public Policy Failure

David Humphreys and Colin Lorne

Abstract The UK has one of the highest COVID-19 fatality rates due to legacy problems in government and healthcare management, slowness in recognising the severity of the outbreak that allowed the disease to gain hold across the country (despite scientific warnings) and serious public policy failings, including a delayed lockdown, insufficient provision of personal protective equipment and errors in introducing a test and trace system. The disease has disproportionately affected vulnerable populations, especially care home staff and residents, key workers, people from poorer socioeconomic groups and black, Asian and minority ethnic (BAME) citizens. The pandemic has had a serious effect on the UK economy, including unemployment and a record level of public debt.

Keywords Exercise Cygnus · Herd immunity · Lockdown · Science–politics interface · PPE · Test and trace

Introduction

The failure of UK authorities to deal effectively with the COVID-19 outbreak of 2020 is due to a combination of legacy problems and failings across the highest levels of government. Drawing from policy reports and media analysis published during the early months of the pandemic, this chapter will outline the UK response to the pandemic focusing primarily on the government in Westminster but also considering the responses of the devolved national governments within the UK. The impact of the disease on different population groups will be examined. It will be argued that interference in the science–politics interface weakened the integrity of scientific input to policy makers and seriously eroded public trust in government.

D. Humphreys (✉) · C. Lorne
School of Social Sciences and Global Studies, The Open University, Milton Keynes, UK
e-mail: david.humphreys@open.ac.uk

Context and Background

In the years before the pandemic struck, the UK had an internationally respected reputation in pandemic preparedness. In 2008, the first UK National Risk Register of Civil Emergencies listed a pandemic as a high consequence risk. That same year the Labour government's Global Health Strategy emphasised the importance of global cooperation in dealing with health risks in a globally interconnected world. Central to the strategy was improving the health of people across the world (Department of Health 2008). There were four rationales for this strategy: to reduce the risks to the UK population of a global pandemic; to support a global humanitarian mission to improve the lives of vulnerable people in the Global South; to enhance UK political influence overseas; and to market British health expertise. British medical experience was sought by governments during the SARS outbreak of 2002–3, the MERS outbreak of 2012 and the Ebola outbreak in Sierra Leone of 2014.

Four reasons explain why the UK government forfeited this leading position to the extent that the country was unable to respond effectively to the COVID-19 pandemic in 2020. The first relates to the politics of austerity. While the Conservative governments that have ruled since 2010 (including five years in coalition with Liberal Democrats) have not cut health funding in absolute terms, spending per capita has fallen sharply due to an increasing population and rising demands on health care through increased life expectancy. Per capita funding on health fell for ten consecutive years, leading to the most prolonged funding squeeze in NHS history (King's Fund 2017).

Second, health systems have suffered from continuous reorganisation under the Conservatives. In England, the Health and Social Care Act 2012 further embedded the logic of market competition into healthcare provision. Politically contested reforms that have received widespread criticism include NHS reorganisation and fragmentation, the abolition of regional structures that strategically coordinated care at scale between the local and national levels and the transfer of many public health functions to local government at a time of austerity-driven cuts to local government (Timmins 2012; Checkland et al 2018; Lorne et al. 2019; Hammond et al. 2017). In August 2020, the government announced a further round of restructuring in the middle of the pandemic, with Public Health England to be abolished and a new agency to be created focusing on infectious diseases.

Third, the results of Exercise Cygnus, a 2016 training exercise to simulate government and health sector responses to a major pandemic, were ignored. Exercise Cygnus identified a shortage of surge capacity in the NHS, including insufficient ventilators, personal protective equipment (PPE) and hospital beds, and a deficit in the capacity of care homes to handle a pandemic. Related to this was Brexit, a problem on which the government had become fixated. Contingency planning for a pandemic was set aside to prepare for the possibility of a 'no-deal' Brexit (Wenham 2020). On 3 March, the prime minister, Boris Johnson, declared 'Our country remains extremely well prepared, as it has been since the outbreak began in Wuhan several months ago' (gov.uk 2020). This statement leads to one of two conclusions. Either Johnson was unaware of the findings of Cygnus, which would be a serious failure of the machinery of government, especially as the previous day Johnson had chaired a meeting of COBRA, the government committee that handles national emergencies. The alternative explanation, as noted by Richard Horton, is that Johnson was aware of Cygnus, knew that the country was not 'extremely well prepared' and deliberately lied to the public (Horton 2020: 49).

Fourth, there was a belief within the government that the risks of a pandemic were overstated. The SARS, MERS and Ebola outbreaks threatened to become pandemics, although the threat never materialised on a global scale. These 'near misses' seem to have induced a sense of complacency (Leslie 2020). In 2020, this translated into procrastination and delay, despite evidence that the threat needed to be taken seriously. In contrast, countries such as China, Taiwan and Singapore that had suffered from the previous outbreaks such as SARS showed a far better preparedness than the UK when COVID-19 broke. These countries recognised that the outbreak was a novel coronavirus, whereas the initial UK response was premised on the assumption that the outbreak was a form of influenza (Horton 2020: 102).

The World Health Organisation (WHO) declared COVID-19 a public health emergency of international concern on 30 January. The first case in England was confirmed the following day. It is now clear that no 'patient zero' introduced the virus to the UK. Research by the COVID-19 Genomics UK Consortium (Cog-UK) found that at least 1300 people introduced the virus to the UK. Significant less than 1% came from China, with most importations coming from Italy in late-February, Spain in early-March and France in late-March (Pybus et al. 2020). Most of these arrived in the UK by air, although significant numbers entered via the Channel Tunnel and cross-channel ferries. Infected people were able to enter the UK unimpeded with the government failing to introduce infection screening of passengers at airports and other points of entry.

The failure to introduce travel restriction between the UK and countries with high infection rates was responsible for increasing the UK infection rate at the start of the pandemic. Approximately 3000 Atlético Madrid football fans travelled to Liverpool for a Champions League fixture on 11 March shortly after the Spanish authorities had banned fans from attending football matches in Spain. The mass gathering in close proximity of over 50,000 fans from Spain and England significantly increased the risk of infection. There was a surge of deaths in Liverpool one month after the fixture (Thorp 2020). Other mass gatherings in early-to-mid-March included Six Nations Rugby matches in London, Cardiff and Edinburgh, the Manchester United-Manchester City football match, two Stereophonics concerts in Cardiff and the Cheltenham horse racing festival, the latter attended by a quarter of a million spectators over four days.

The next section will outline the government's reaction to the pandemic. Government policy failed to control the spread of the virus and should therefore be considered a policy failure as defined by McConnell. A policy fails 'even if it is successful in some minimal respects, if it does not fundamentally achieve the goals that proponents set out to achieve' (McConnell 2015: 221).

Initial Government Response

Although it was never an officially declared strategy, government scientific advisors spoke publicly of 'herd immunity'. Such a policy would have allowed some spread of the virus in order to build up increased immunity among the population at large, but would have challenged the capacity of the NHS and resulted in increased deaths. In a BBC interview, government scientific advisor Graham Medley talked of the need to 'manage this acquisition of herd immunity' (cited in Horton 2020: 51). Johnson appeared to lend support to a herd immunity policy by suggesting 'One of the theories is, perhaps you could sort of take it on the chin, take it all in one go. And allow the disease, as it were, to move through the population without taking as many draconian measures'(cited in Grayling 2020). The resulting backlash to this bizarre statement led the government health secretary to announce to the BBC, 'Herd immunity is not our policy. It's not our goal' (Staunton 2020).

There was a lengthy period of delay before freedom of movement restrictions was introduced. During this period, the number of infected cases was doubling every three to four days. On 17 March, the government issued advice to avoid non-essential overseas travel. On 23 March, lockdown measures were finally announced. People were ordered to stay at home and not to meet friends and family. Leaving home was permitted only to buy food and medicine, exercise once a day and seek medical attention. A social distance rule of two metres in public spaces was introduced. People were told to work from home if they could do so. The UK government's Chief Scientific Advisor, Patrick Vallance, announced that 20,000 deaths would be a 'good outcome'. That figure was passed on 25 April (Kelly 2020). The figure of 40,000 was passed on 5 June.

There is some ambiguity on the UK death toll. In August, the running total was reduced by 5377 deaths following concerns that the previous total for England had included people who had contracted the virus, recovered then died later of other causes (Duncan et al. 2020). However, once excess deaths are taken into account (namely, deaths in excess of the rolling three year average), the UK total has still risen significantly. From the end of February until mid-June, the UK had the highest level of excess deaths in Europe with an increase of 6.9% compared to the previous years, followed by Spain at 6.7% (BBC 2020b). (The UK figure is the average of the four nations. The level of excess deaths was worst for England at 7.5%, followed by Scotland 5.1%, Wales 2.8% and Northern Ireland 2.0%.) In June, a former government scientist, Neil Ferguson, told a House of Commons select committee that if Britain had locked down a week earlier the number of deaths would have been halved (BBC 2020a).

A lack of PPE in the early weeks of the pandemic placed medical staff and patients at heightened risk. The UK missed an opportunity to join an EU scheme for PPE procurement. At the time of the outbreak, the central PPE stockpile was suitable only for a relatively small-scale influenza outbreak and with insufficient quantities for all staff working in care homes. The leader of the opposition, Keir Starmer, has called for an inquiry into the purchase in April 2020 of 50 million face masks that were unsuitable for NHS use. The order was part of a £252 million medical supplies contract awarded to Ayanda Capital, an investment strategy firm with no expertise in medical supplies (Rawlinson 2020). Local efforts to coordinate the purchasing of PPE, such as in Greater Manchester, were undermined by demands to route all purchasing through the Department of Health and Social Care (Dunhill 2020). In May, it was reported that 48% of doctors had bought or otherwise provided their own PPE, while two-thirds did not feel fully protected (British Medical Association 2020). One of the dozens of medical staff to lose their life was Dr Abdul Mabud Chowdhury who died in April 2020. He had publicly appealed to the government for 'appropriate PPE' to protect medical staff, noting that healthcare workers had a 'human right like others to live in this world disease-free with our family and children' (cited in Weaver 2020).

A further problem was the absence of an effective test and trace system under which anyone who had been in contact with a newly infected case should be traced, tested and, if infected, isolated. The government specifically rejected the WHO recommendation of test and trace at the start of the pandemic before announcing that a NHS-branded contact tracing smartphone app would be ready by mid-May to complement the work of human tracers in identifying local outbreaks and tracing those who had been in contact with newly infected patients. Johnson took refuge in jingoistic statements that Britain would develop a 'world-beating' test and trace system, but efforts to boost testing capacity were erratic and slow, with several target dates for the implementation of an operational system missed. Only in May was a national level NHS test and trace system introduced. However, this received criticism that it was overly centralised and failed to tap into local knowledge by excluding local councils. Parts of the scheme were outsourced to Serco, a private sector provider with a poor record in delivering public services to contract (Observer 2020).

The system originally adopted thus reflected the Conservative government's ideological predisposition of contracting out public services to private providers while marginalising local government. The £108 m contract awarded to Serco was not put out to open tender (Murphy and Marsh 2020) and was awarded months after Serco has been fined £2.6 million for failings on an earlier government contract (Taylor 2020).The Serco contract represents one of many occasions when private contracts for pandemic-related work were handed to businesses without competition. Over £1 billion of such contracts have been handed to the private sector, both British and overseas, without open tender (Evans et al. 2020a, b).

Within weeks, two changes to government test and trace policy were made. First, trials of the app on the Isle of Wight, which began in early May, were abandoned in mid-June after it became clear the app did not work. The government announced that it would switch to an alternative app using Apple and Google technology, which at the time was months from being ready (Sabbagh and Hern 2020). Second, the government conceded that the national test and trace system was ill-equipped to deal with localised flare-up of the virus, which should be led by local people in the affected communities. This led to the national test and trace system being scaled down in August, with the number of national contact tracers reduced from 18,000 to

12,000. A new system of regional teams working closely with local authorities was introduced (Marsh and Halliday 2020).

In England, the first easing of the lockdown took place in May. By 4 July, pubs, restaurants, cinemas and hairdressers could reopen in England. The two-metre social distancing rule was dropped in England and replaced with a rather confusing 'one-metre plus' rule, namely people should stay two metres apart where possible, but if not at least one metre apart. The two-metre rule was retained in Scotland, Wales and Northern Ireland, placing England, which favoured more libertarian measures, on a different track to the rest of the UK.

If a pandemic is to be brought under control, daily new cases need to be suppressed. A strict lockdown can achieve that. However, a lengthy lockdown comes at significant economic cost as well as psychological and emotional strain for citizens, which can lead to political pressure to ease the lockdown.

The question then becomes what other interventions should be introduced to keep new infections low as lockdown is eased. The Westminster government gambled that some lockdown easing could take place before an effective test and trace system was in place, in effect hoping there would be no second wave before a system was fully established. There was unease within the NHS at the decision to ease the lockdown in England. The NHS Confederation argued that no further lockdown easing should take place before a comprehensive test and trace system was operational at national and local levels (Savage et al. 2020). Scotland, Wales and Northern Ireland were more cautious than England, wanting to suppress daily new infections further before easing the lockdown so that a second wave, and hence a second period of lockdown, would be unnecessary.

Localised lockdowns were imposed in England following local spikes in the number of new cases. Leicester was the first city to impose a local lockdown. Outbreaks were concentrated among workers from local garment factories that operated as normal during the lockdown, with workers forced to work in close proximity to each other. A report from the charity Labour Behind the Label found that many workers were paid less than the minimum wage, with workers expected to report for work even when sick (Labour Behind the Label 2020). Other towns and cities that imposed a second lockdown in the summer included Greater Manchester, Burnley, Bradford and Preston.

The Science–Politics Interface

Particularly contentious during the pandemic was the relationship between scientists and politics. The Scientific Advisory Group on Emergencies (SAGE) was established to provide advice to government in times of emergency. It has been activated on several occasions to deal with epidemics, flooding and the 2011 Fukushima nuclear accident. Scientists who serve on SAGE are not instructed by the government and are drawn from academia and practice. Membership is fluid and depends on the emergency being addressed. SAGE's role is to provide objective and neutral advice to government free of political interference.

Richard Horton, the editor of the *Lancet,* criticised SAGE for ignoring warnings published in January which concluded that the virus could pose a global public health risk (e.g. Wu et al. 2020). By failing to heed such warnings, the government wasted nearly two months. Horton criticised the relationship between scientists and politics during the early months of the pandemic as 'the greatest science policy failure for a generation' (cited in Anthony 2020) with scientific advisors becoming the 'public relations wing of a government that had failed its people' (Horton 2020: 59). However, ministers frequently claimed to be 'following the science' to legitimise their position and deflect media attention from flawed policies.

Evidence emerged of political undermining of the work of SAGE at an early stage of the pandemic. The prime minister's political advisor, Dominic Cummings, attended several SAGE meetings as a participant (as opposed to an observer). Some scientists felt his presence influenced the discussions and prevented dissenting views being expressed (Lawrence et al. 2020). The daily media briefings usually took the form of a cabinet minister flanked by two scientific advisors, a format that inhibited advisors from giving scientific advice that might run counter to statements from government ministers. So serious were concerns about the transparency of SAGE that a retired chief government scientific advisor, David King, set up an Independent Sage committee of 12 scientists to present the government with independent advice free of political interference.

This is not to criticise the scientists on SAGE who gave their time and expertise to mitigating and addressing the pandemic, but it is to say that senior political figures weakened the integrity of scientific advice by interfering in scientific meetings. A further problem, as noted by Nobel Laureate Paul Nurse, is that the scientific basis for many government decisions was unclear. He called for greater transparency in decision making without which public trust in government would be at risk (Sample 2020). Public trust in the government was further eroded by the conduct of senior members of the administration. Johnson undermined the advice of some of his advisors on social distancing by continuing to shake hands and being filmed openly doing so on TV, then joshing with the media about it. He also attended the England–Wales rugby international in London on 7 March at a time when many scientists favoured a ban on mass gatherings (so-called super-spreader events).

Johnson tested positive for the virus on 27 March after the lockdown was introduced. The health secretary, some senior scientific advisors and Johnson's political adviser Cummings also contracted the virus. Cummings broke the rules of the lockdown to travel and meet his parents. Despite media and public pressure, he refused to resign or even apologise, and Johnson declined to sack him. Johnson then prevented his two most senior scientific advisors, Patrick Vallance and Chris Whitty, from answering press questions that related to Cummings' actions at a daily media briefing on the basis that they could not comment on 'political issues'.

Published research has found that Cummings' actions had a lasting effect on public trust in the handling of the pandemic. In a paper published in the *Lancet,* researchers from University College London found that Cummings' actions 'under-mined confidence in the government to handle the pandemic specifically' (Fancourt et al. 2020). A report from the think tank British Future also found that public trust in the government fell, with citizens from across the political spectrum disapproving of Cummings' behaviour (British Future 2020). The result was a widespread feeling among the public that they had been treated in bad faith, with the enormous collec-tive effort of the public in observing the restrictions as part of a civic duty in which all made sacrifices for the common good being undermined by the irresponsibility and arrogance of a self-serving political elite in London. In contrast, there were higher levels of trust in the governments of the devolved nations, with a June poll reporting 71% confidence in Nicola Sturgeon, the Scottish leader, compared to 43% for Johnson (Marshall 2020).

The police did not prosecute or fine Cummings for breaching lockdown rules but subsequently reported increased difficulties in enforcing the lockdown, with rule-breakers citing Cummings' actions as a reason for visiting friends and family (Duncan 2020). A few weeks later, it emerged that police were more likely to fine black, Asian and minority ethnic (BAME) people for lockdown violations. Most police forces were two to three times more likely to fine BAME people compared to white people. In Cumbria, BAME people were 6.8 times more likely to be fined than a white person. This reflects a trend whereby citizens from BAME backgrounds, especially black people, are over-represented in the criminal justice system relative to white citizens. Only in Merseyside and Northumbria was there no difference (Dodd and Gidda 2020).

Vulnerable Population Groups

The failure of the government to suppress the virus affected some population groups more severely than others. Especially vulnerable to infection were elderly care home residents, BAME citizens, frontline workers and people living in poverty.

Three factors explain the crisis in care homes during the pandemic which led to thousands of people, mainly residents but also staff, losing their lives. First, there was no testing system in place for patients and staff. SAGE had anticipated the risks to people living in care homes in February and concluded that the only way to protect residents was to introduce extensive testing (BBC 2020a). To free NHS capacity, twenty-five thousand patients were discharged from hospitals to care homes without testing during the height of the pandemic (Syal 2020). Second, many care homes employ casual workers, some on zero-hour contracts who move around in search of work, sometimes moving between care homes in the same week, thus increasing the risks of transmission. Third, most care home staff lacked access to the necessary PPE during the early weeks of the pandemic. In 2018, the Association of Directors of Adult Social Care Services (ADASS) had recommended an increase in PPE supply for care homes, warning that demand could exceed supply (Peart 2020). According to the National Audit Office (2020, 12), there was a failure to stockpile gowns and visors, despite advice to do so from an independent committee advising the Department of Health and Social Care in 2019.

Elderly people were also vulnerable to infection outside care homes. A study from the New Policy Institute (Kenway and Holden 2020) found that a significant factor in the variation in COVID-19 cases across England was the proportion of people over 70 living in a multigenerational household with people of working age, especially in households where it was difficult or impossible to maintain social distancing. The research focused on two London boroughs (Ealing and Richmond) and four outside London (Luton, Medway, Halton and Rochdale).

In June 2020, a report from Public Health England concluded that BAME Britons were more likely to die from COVID-19 if infected by the virus, with black males at greatest risk. Many BAME individuals were less likely to seek health care when they needed it due to historic racism. Within the NHS BAME staff were 'less likely to speak up when they have concerns about Personal Protective Equipment (PPE) or risk' (Public Health England 2020, 5). There is evidence that the increased health risks faced by BAME people were not unique to COVID-19 but reflected and perpetuated pre-existing health inequalities between different ethnic groups. A report by the Deaton Review of the Institute of Fiscal Studies found that ethnic inequalities manifested themselves for COVID-19 in two distinct ways: exposure to infection and health risks; and risk of loss of income which often translated into enhanced health risks. The review found that men from ethnic groups are more likely to be affected economically from the lockdown, for example, from unemployment and income loss for the self-employed. It also found that per capita hospital deaths for COVID-19 are highest among the black Caribbean population at three times that of white Britons (Platt and Warwick 2020).

A report from the Runnymede Trust corroborated many of the conclusions of the Deaton Review, finding that compared to white people, many Britons from BAME backgrounds found it more difficult to shield from COVID-19 as they were more likely to take public transport, to live in larger, multigenerational households or overcrowded households, and are over-represented in key worker roles, including health and social care (Haque et al. 2020). Concern that the above average fatality rates for BAME people and medical staff were due to structural racism was expressed at the Black Lives Matter protests of June 2020 following the murder of a black citizen, George Floyd, by police in the USA.

Certain occupations are more at risk than others. The Trades Union Congress (TUC) found that certain types of worker were better able to insulate themselves from the risks of infection by working from home, namely those working in information/communications, scientific and technical professions and education. Males working as security guards, taxi drivers, chefs and care workers are on average more than three times as likely to die from the virus (Collinson 2020). Analysis for Transport for London (TfL) carried out by University College London's Institute of Health Equity found that the mortality rate for London bus drivers from March to May 2020 was 3.5 times higher than those of men of the same age (20–64) in all occupations in England and Wales. Drivers were at heightened risk both because of the frontline nature of their work and because the stressful nature of their job led to health issues such as high blood pressure. The report argued that an earlier lockdown would have saved the lives of many TfL drivers (Goldblatt and Morrison 2020).

It is well established that those from lower income groups are more likely to die of disease due to poor underlying health and, in some cases, problems in accessing health care. Analysis by the Office for National Statistics (2020a) found a correlation between socioeconomic deprivation and mortality rates. Mortality rates are normally higher in poorer areas, but in the three-month period from March to May 2020, COVID-19 infection drove mortality rates in the poorest areas even higher. In England, the COVID-19 mortality rate for the poorest 10% of the population was 128.3 per 100,000 people, compared with 58.8 per 100,000 for the richest 10%, indicating that the poorest people in England were twice as likely to die from the virus compared to the wealthiest (Devlin and Barr 2020). An infection hotspot in Greater Manchester was the Broughton area of Salford, an area with a high unemployment rate and low life expectancy (Proctor et al. 2020).

A full survey of those who were affected by the pandemic in the UK must include those whose health and safety was indirectly affected by the pandemic. Many patients died prematurely as a delayed consequence of missed screening and cancelled operations as hospital capacity was freed to create space for virus victims. Cancer deaths in England increased significantly in 2020 (MacKenzie 2020: 180). It is likely that the increased confinement in private addresses due to the lockdown will include increases in domestic violence, child abuse and mental health problems (Dodds et al. 2020). A study from University College London found that the lockdown had led to increased mental health problems within the LGBTQ+ community, especially among younger people facing homophobia or transphobia from relatives (Kneale and Becares 2020).

One impact of the economic slowdown has been increased poverty. The Food Standards Agency (2020) has reported that many of the increased numbers of people suffering income loss were experiencing food insecurity. As many as one in ten people were forced to rely on food banks, many of whom missed meals with a negative effect on nutritional health (Butler 2020).

Responses of the Nations and Regions

The Westminster government announced an easing of the lockdown for England on 11 May, changing the 3 point slogan from 'Stay Home-Protect the NHS-Save Lives' to the more ambiguous 'Stay Alert-Control the Virus-Save Lives'.

London had higher infection rates than most of the rest of the country at the start of the pandemic, notwithstanding the infection spike in parts of south Wales and north-west England (Devlin 2020). Thereafter there was a steep reduction in London infections, which contributed to the decision to start easing the lockdown in mid-May, despite the publication of a study from the University of Oxford that the UK was in one of the worst positions globally for meeting WHO criteria for lockdown easing (Hale et al. 2020). The failure to meet WHO criteria informed the decisions by the national governments of Scotland, Wales and Northern Ireland to continue with the original lockdown measures for a further period. These three nations were dismayed both at the change of policy of the Westminster government and the lack of consultation that preceded it. Lockdown easing in England marked the end of a unified consensus on how to tackle the virus across the four nations. Political leaders in Scotland and Wales announced that anyone travelling to their countries from England without a legitimate reason could be in breach of the law and at risk of arrest (Helm et al. 2020).

Lockdown easing was also resisted by English regions with high infection rates, such as the north-west, where rates of infections were falling far less steeply than London. The mayor of Greater Manchester, Andy Burnham, stated that the restrictions had been eased too soon with no fully operational test and trace system in place. There was also protest from the mayor of Liverpool that the lockdown was eased without sufficient consultation with local authorities. As well as the north-west councils in the north-east, especially Newcastle, Sunderland and Gateshead, were unsure about easing restrictions when the infection rate remained stubbornly high, with many ex-workers vulnerable with secondary illnesses related to work in ship-building and mining. The leader of Gateshead council said that had they possessed the same legislative powers as Wales, Scotland and Northern Ireland, they would have overridden the Westminster decision and retained the existing lockdown (Coman 2020, 30).Overall, there was a feeling among political leaders in both the nations and regions that the Westminster government had adopted an autocratic approach to handling the pandemic.

There was concern that a resurgence in cases would impose on local councils increased costs for adult social care provision coupled with a decline in income as council tax revenues fell as people became unemployed and registered for benefits. Overall, the decision by Westminster to ease the lockdown in England was a national decision that did not take into account regional particularities. It was criticised by one commentator for exposing a dysfunctional relationship between the regions and central government (Coman 2020, 31).

The Westminster government did increase funding to councils during the early months of the pandemic with £3.2bn of emergency funding. In mid-June, the mayors of London, Liverpool, Greater Manchester and Sheffield issued a joint call for increased funding without which councils, which according to the Local Government Association (LGA) faced increased costs from the pandemic of nearly £13bn, would be forced to cut services or risk bankruptcy (Proctor 2020). The Centre for Progressive Policy found that eight out of ten English councils providing adult social care services were at risk of bankruptcy and unable to meet the financial pressures caused by the pandemic without further cuts to services (Billingham 2020).

Following the easing of the first lockdown, the four nations of the UK adopted different approaches. A second lockdown, less severe than the first and lasting four weeks, came into effect in England on 5 November. Meanwhile, Scotland introduced a system of tiered restrictions, with Wales and Northern Ireland also adopting their own rules to limit the spread of the virus.

Economic Impacts

The government's handling of the pandemic also represents a profound failure of economic policy. The UK was late entering lockdown because the Johnson administration feared the economic consequences of early tough measures. However, and as predicted, the failure to act more resolutely had more serious economic consequences than would have been incurred from an earlier lockdown (Micklethwait and Wooldridge 2020). Once restrictions were imposed, they had to be held in place far longer than would have been necessary had they been introduced earlier, and longer than most other western European countries.

Financial markets fell and by the beginning of March, the government faced the prospect that it would be unable to finance further government expenditure, including a furlough scheme under which the government committed itself to paying the wages of some 9.4 million workers unable to work because of the pandemic. The Bank of England intervened and created £200 billion of additional funds through quantitative easing. By June, the total Bank of England stimulus package stood at £745 billion (Partington 2020). In July, the government debt exceeded £2 trillion for the first time, and the first time UK debt had exceeded 100% of GDP since 1961 (Office for National Statistics 2020c). In the three months to June 2020, GDP shrank by an unprecedented 20.4%, twice as much as the shrinkage of the US economy in the same period and worse than any EU or G7 country (Office for National Statistics 2020b).

A further reason for the high economic costs to the UK is that lockdown measures were imposed on an economy that was ailing due to political uncertainty over Brexit. No economic sector has been left unaffected by the pandemic. British Airways used the pandemic as an opportunity to push forward changes to staff contracts and pay, despite receiving £35 m to furlough staff and a £300 m Bank of England loan (Topham 2020). Other airlines that made staff redundant during the pandemic included EasyJet and Virgin Atlantic. High street retailers to have announced job cuts of 1000 or more include John Lewis, Boots, Marks & Spencer, WH Smith and Debenhams. By mid-August, over 183,000 job losses had been announced in the UK since the start of the pandemic (Partridge and Butler 2020).

A report commission by the Creative Industries Federation (CIF) found that Britain's creative businesses, such as music, theatre, film, television, fashion and museums, many of which closed in March, faced a projected £74bn drop in revenues throughout 2020, a drop of 30% compared to 2019, and the loss of one in five jobs, a total of 400,000 (Oxford Economics 2020). The chair of the CIF said the UK faced a 'cultural catastrophe' and called for a cultural renewal fund to help creative industries return to work (Brown 2020). In July, the government announced a financial stimulus package for the cultural sector, although this was criticised for concentrating on theatres, concert halls, arts centres, heritage sites and museums and neglecting smaller independent arts organisations and creative businesses (Higgins 2020).

Conclusions

The UK response to the pandemic represents a serious policy failure. This is not due solely to poor policy making in the moment. The policy failure needs to be understood as a combination of two things: long-term legacies from the previous Conservative administrations, in particular NHS reorganisation, austerity and the distraction of Brexit; and delay, poor decision making and ineptitude once the pandemic struck.

The faltering response of the UK government, for example, on herd immunity, was due in part to political interference in the science–politics interface, which slowed the provision of timely and clear scientific advice to government. The UK response also brings into focus the tensions and contradictions between different power centres, including the relationship between Westminster, the regions and the devolved nations of the UK, and the undermining of local authorities through interference from central government and the allocation of contracts to inexperienced or incompetent private sector contractors.

If the UK is to respond more effectively to the next pandemic, a number of things need to change. Scientific warnings must be respected and heeded, as should the results of training simulations such as Exercise Cygnus. Disregarding Cygnus led to the country being wholly unprepared. There needs to be serious stockpiling of PPE and medical equipment such as ventilators, and the industrial capacity to manufacture such equipment, if necessary at short notice, rather than rely on external sources. A viable nationwide test and trace system needs to be established, including comprehensive testing of passengers at points of entry such as airports. There needs to be improved research on treatments for emerging diseases, along with improved capacity to manufacture vaccines and drugs. Finally, the UK needs to be a more prominent international actor, promoting international cooperation within the auspices of the WHO on disease surveillance, information sharing and vaccine and drug research.

The government lost moral authority early in the crisis by locking down too slowly. Trust was further eroded by a cavalier disregard for the rules within the highest levels of government. Had travel restrictions, lockdown, social distancing and an effective test and trace system been introduced earlier, the numbers of infections and deaths would certainly have been significantly lower, and the costs to the economy would have been less severe. The slow government response and the serious mistakes in the management of the pandemic should be seen as a gross breach of the duty of care that would be considered criminally negligent in a business enterprise and lead to charges of corporate manslaughter for serious failings that resulted in multiple unnecessary deaths.

Addendum

In January 2021 there were four significant developments in the United Kingdom. First, the country became the first in Europe to pass 100,000 deaths. Second, a new COVID-19 variant was detected (the Kent variant) that is more transmissible than the original strain. Third, the government introduced quarantine measures for those arriving in the UK. These measures were strengthened in February, with all visitors required to quarantine for 10 days, many of whom were required to pay for accommodation in a quarantine hotel. Fourth, mass vaccination was introduced, first of the Pfizer vaccine then of the Oxford/AstraZeneca vaccine. By mid-February over 16 million people had received the first dose of one of these vaccines, with over half a million receiving a second dose. The government aims to offer a first vaccine dose to about 32 million people in nine priority groups by 15 April, 2021.

References

Anthony, A. (2020). The Lancet's editor: "The UK response to coronavirus is the greatest science policy failure for a generation". *Observer*, 14 June. Available at: https://www.theguardian.com/politics/2020/jun/14/the-lancets-editor-the-uk-response-to-coronavirus-is-the-greatest-science-policy-failure-for-a-generation (August 1, 2020).

BBC. (2020a). Coronavirus: "Earlier lockdown would have halved death toll", 10 June. Available at: https://www.bbc.co.uk/news/health-52995064 (June 14, 2020).

BBC. (2020b). Coronavirus: England highest level of excess deaths, 30 July. Available at: https://www.bbc.co.uk/news/health-53592881 (August 1, 2020).

Billingham, Z. (2020). Why the government needs to pay up before levelling up, 23 June. *Centre for Progressive Policy*. Available at: https://www.progressive-policy.net/publications/why-the-government-needs-to-pay-up-before-levelling-up (August 1, 2020).

British Future. (2020). *Remembering the kindness of strangers: Division, unity and social connection during and beyond COVID-19*. London: British Future. Available at: https://www.thelancet.com/journals/lancet/article/PIIS0140-6736(20)31690-1/fulltext (August 8, 2020).

British Medical Association. (2020). BMA survey reveals almost half of doctors have relied upon donated or self-bought PPE and two-thirds still don't feel fully protected, BMA media team, 3 May. Available at: https://www.bma.org.uk/news-and-opinion/bma-survey-reveals-almost-half-of-doctors-have-relied-upon-donated-or-self-bought-ppe-and-two-thirds-still-don-t-feel-fully-protected (June 15, 2020).

Brown, M. (2020). UK's creative industries face £74bn "cultural catastrophe". *Guardian*, 17 June, p. 8.

Butler, P. (2020). UK's poorest "skip meals and go hungry" during coronavirus crisis. *Guardian*, 13 August. Available at: https://www.theguardian.com/uk-news/2020/aug/12/coronavirus-lockdown-hits-nutritional-health-of-uks-poorest (August 14, 2020).

Checkland, K., Dam, R., Hammond, J. O. N., Coleman, A., Segar, J., Mays, N., et al. (2018). Being autonomous and having space in which to act: Commissioning in the "New NHS" in England. *Journal of Social Policy, 47*(2), 377–395.

Collinson, A. (2020). Coronavirus doesn't discriminate—But the UK labour market certainly does. *Trades Union Congress*, 18 May. Available at: https://www.tuc.org.uk/blogs/new-class-divide-how-covid-19-exposed-and-exacerbated-workplace-inequality-uk (August 1, 2020).

Coman, J. (2020). London doesn't always know best: How the north is choosing its pace out of lockdown. *Observer*, 31 May, pp. 30–32.

Department of Health. (2008). Health is global: A UK government strategy 2008–13. Available at: https://webarchive.nationalarchives.gov.uk/20130105191920/http://www.dh.gov.uk/en/Publicationsandstatistics/Publications/PublicationsPolicyAndGuidance/DH_088702 (June 9, 2020).

Devlin, H. (2020). Infection rate: Uncertainty over figures raises fears of second wave. *Guardian*, 16 May, pp. 4–5.

Devlin, H., & Barr, C. (2020). Poorest twice as likely to die of virus—ONS. *Guardian*, 13 June, p. 15.

Dodd, V., & Gidda, M. (2020). Police in England and Wales far more likely to fine BAME people in lockdown. *Guardian*, 17 June. Available at: https://www.theguardian.com/world/2020/jun/17/police-in-england-and-wales-six-times-more-likely-to-fine-bame-people-in-lockdown (June 17, 2020).

Dodds, K., Broto, V. C., Detterbeck, K., Jones, M., Mamadouh, V., Ramutsindela, M., et al. (2020). The COVID-19 pandemic: Territorial, political and governance dimensions of the crisis. *Territory, Politics, Governance, 8*(3), 289–298.

Duncan, C. (2020). Dominic cummings: Policing coronavirus lockdown "made much harder by No 10 aide's actions". *Independent*, 23 May. Available at: https://www.independent.co.uk/news/uk/politics/dominic-cummings-latest-news-lockdown-coronavirus-resign-boris-johnson-travel-durham-a9530081.html (June 16, 2020).

Duncan, P., Barr, C., & McIntyre, N. (2020). Covid toll in England is reduced by 5400. *Guardian*, 13 August, pp. 1, 11.

Dunhill, L. (2020). Top hospital chief wades in after government "strong-arming" on PPE. *HSJ*, 27 May. Available at: https://www.hsj.co.uk/manchester-university-nhs-foundation-trust/top-hospital-chief-wades-in-after-government-strong-arming-on-ppe/7027715.article (August 21, 2020).

Evans, R., Garside, J., & Smith, J. (2020). £1bn in state contracts given to companies without public tender. *Guardian*, 16 May, 4.

Evans, R., Geohegan, P., & Pegg, D. (2020). Revealed: £56m bill for consultants to help deal with virus. *Guardian*, 21 August, p. 2.

Fancourt, D., Steptoe, A., & Wright, L. (2020). The cummings effect: Politics, trust, and behaviours during the COVID-19 pandemic. *Lancet*, August 6. Available at: https://www.thelancet.com/journals/lancet/article/PIIS0140-6736(20)31690-1/fulltext (August 8, 2020).

Food Standards Agency. (2020). The COVID-19 consumer research. Available at: https://www.food.gov.uk/print/pdf/node/4406 (August 12, 2020).

Goldblatt, P., & Morrison, J. (2020). *Initial assessment of bus driver mortality from COVID-19*. London: Institute of Health Equity. Available at: http://content.tfl.gov.uk/initial-assessment-of-london-bus-driver-mortality-from-covid-19.pdf (August 9, 2020).

gov.uk. (2020). Prime Minister's statement on coronavirus (COVID-19): 3 March 2020. Available at: https://www.gov.uk/government/speeches/pm-statement-at-coronavirus-press-conference-3-march-2020 (August 1, 2020).

Grayling, A. C. (2020). Boris Johnson's masterclass in how not to respond to an outbreak: The British leader loves to imitate Churchill, but this will not be his finest hour. *Washington Post*, 10 March. Available at: https://www.washingtonpost.com/outlook/2020/03/10/boris-johnsons-masterclass-how-not-respond-an-outbreak/ (August 1, 2020).

Hale, T., Phillips, T., Petherick, A., Kira, B., Angrist, N., Aymar, K., & Webster, S. (2020). Lockdown rollback checklist: Do countries meet WHO recommendations for rolling back lockdown? *Research note, version 3.0*, 29 May. Available at: https://www.bsg.ox.ac.uk/sites/default/files/2020-06/Lockdown%20Rollback%20Checklist%20v3.0.pdf (June 14, 2020).

Hammond, J., Lorne, C., Coleman, A., Allen, P., Mays, N., Dam, R., Mason T., & Checkland, K. (2017). The spatial politics of place and health policy: Exploring Sustainability and Transformation Plans in the English NHS. *Social Science and Medicine* 190, 217–226.

Haque, Z., Laia, B., & Treloar, N. (2020). *Over-exposed and under-protected: The devasting impact of COVID-19 on black and minority ethnic communities in great Britain*. London: Runnymede Trust.

Helm, T., Wall, T., & McKenna, K. (2020). Spirit of Churchill in short supply as national consensus fractures. *Observer*, 17 May, pp. 6–7.

Horton, R. (2020). *The Covid-19 catastrophe: What's gone wrong and how to stop it happening again.* Cambridge: Polity.

Higgins, C. (2020). Rishi Sunak's arts bailout is more divisive than it looks. *Guardian,* 14 July. Available at: https://www.theguardian.com/commentisfree/2020/jul/14/cultural-rescue-package-artists-institutions-covid-19 (August 2, 2020).

Kelly, J. (2020). Coronavirus: The "good outcome" that never was. *BBC,* 25 April. Available at: https://www.bbc.co.uk/news/stories-52419218 (June 16, 2020).

Kenway, P., & Holden, J. (2020). Accounting for the Variation in the Confirmed Covid-19 Caseload across England: An Analysis of the role of multi-generation households. *London and time,* 11 April. Available at: https://www.npi.org.uk/files/2115/8661/6941/20-04-11_Accounting_for_the_variation_in_Covid_cases_across_England.pdf (August 1, 2020).

King's Fund. (2017). Does the NHS need more money? *King's Fund,* 20 November. Available at: https://www.kingsfund.org.uk/publications/articles/does-nhs-need-more-money (June 9, 2020).

Kneale, D., & Becares, L. (2020) The mental health and experiences of discrimination of LGBTQ+ people during the COVID-19 pandemic: Initial findings from the Queerantine Study. *MedRxiv(pre-print),* pp. 1–20. Available at: https://www.medrxiv.org/content/10.1101/2020.08.03.20167403v1.full.pdf+html (August 14, 2020).

Labour Behind the Label. (2020). *Boohoo & COVID-19: The people behind the profits.* Bristol: Labour Behind the Label. Available at: https://labourbehindthelabel.net/wp-content/uploads/2020/06/LBL-Boohoo-WEB.pdf (August 13, 2020).

Lawrence, F., Carrell, S., & Pegg, D. (2020). Attendees of Sage meetings worried by presence of Cummings. *Guardian,* 26 April. Available at: https://www.theguardian.com/world/2020/apr/26/attendees-of-sage-coronavirus-meetings-worried-by-presence-of-dominic-cummings (June 16, 2020).

Leslie, I. (2020). Sars, Ebola and Mers were near misses that led us to believe Covid-19 would pass us by too. *New Statesman,* 27 May. Available at: https://www.newstatesman.com/international/coronavirus/2020/05/sars-ebola-and-mers-were-near-misses-led-us-believe-covid-19-would (June 9, 2020).

Lorne, C., Allen, P., Checkland, K., Osipovic, D., Sanderson, M., Hammond, J., & Peckham, S. (2019). Integrated care systems: What can current reforms learn from past research on regional co-ordination of health and care in England? A literature review, pp. 1–84. Available at: https://prucomm.ac.uk/assets/uploads/PRUComm_-_Integrated_Care_Systems_-_Literature_Review.pdf (August 21, 2020).

MacKenzie, D. (2020). *COVID-19: The pandemic that never should have happened and how to stop the next one.* London: The Bridge Street Press.

Marsh, S., & Halliday, J. (2020). NHS test and trace to cut 6000 jobs in shift towards regionalised system. *Guardian,* 11 August, p. 11.

Marshall, M. (2020). Scotland could eliminate the coronavirus—If it weren't for England. *New Scientist,* 30 June. Available at: https://www.newscientist.com/article/2247462-scotland-could-eliminate-the-coronavirus-if-it-werent-for-england/ (July 2, 2020).

Micklethwait, J., & Wooldridge, A. (2020). *The wake up: Why the pandemic has exposed the weakness of the West—and how to fix it.* London: Short Books.

Murphy, S., & Marsh, S. (2020). Leaked email shows Serco "cherry picked" for £108 m contract, Labour MP claims. *Guardian,* 12 August, p. 10.

National Audit Office. (2020). Readying the NHS and adult social care in England for COVID-19. *HC 376,* 12 June. Available at: https://www.nao.org.uk/wp-content/uploads/2020/06/Readying-the-NHS-and-adult-social-care-in-England-for-COVID-19.pdf (June 16, 2020).

Observer. (2020). The government's response to Covid-19 has been dire. A public inquiry is needed now. *Observer,* 7 June, p. 38.

Office for National Statistics. (2020a). Deaths involving COVID-19 by local area and socio-economic deprivation; deaths occurring between 1 March and 31 May 2020. Available at: https://www.ons.gov.uk/peoplepopulationandcommunity/birthsdeathsandmarriages/deaths/bulletins/

deathsinvolvingcovid19bylocalareasanddeprivation/deathsoccurringbetween1marchand31may 2020 (June 16, 2020).

Office for National Statistics. (2020b). GDP first quarterly estimate, UK: April to June 2020. Available at: https://www.ons.gov.uk/economy/grossdomesticproductgdp/bulletins/gdpfirstquarterlye stimateuk/apriltojune2020 (August 14, 2020).

Office for National Statistics. (2020c). Public sector finance, UK: July 2020. Available at: https://www.ons.gov.uk/economy/governmentpublicsectorandtaxes/publicsectorfinance/ bulletins/publicsectorfinances/july2020 (August 21, 2020).

Oxford Economics. (2020). The projected impact of COVID-19 on the UK creative industries, 15 June. Available at: https://www.creativeindustriesfederation.com/sites/default/files/inline-images/The%20Projected%20Economic%20Impact%20of%20Covid-19%20on%20the%20C reative%20Industries%20Report%20-%20Creative%20Industries%20Federation%202020.pdf (June 16, 2020).

Partington, R. (2020). "Near meltdown": Britain was close to going bust as virus took hold, says Bank chief. *Guardian*, 23 June, p. 9.

Partridge, J., & Butler, S. (2020). Marks & Spencer: Retailer to cut 7000 jobs as high street gloom deepens. *Guardian*, 19 August, p. 8–9.

Peart, L. (2020). Council directors issued care home COVID-19 warning two years ago. *Care Home Professional*, 14 May. Available at: https://www.carehomeprofessional.com/council-directors-iss ued-care-home-covid-19-warning-two-years-ago/ (June 15, 2020).

Platt, L., & Warwick, R. (2020). *Are some ethnic groups more vulnerable to COVID-19 than others (Inequality: The IFS Deaton Review)*. London: Institute of Fiscal Studies. Available at: https:// www.ifs.org.uk/inequality/wp-content/uploads/2020/04/Are-some-ethnic-groups-more-vulner able-to-COVID-19-than-others-V2-IFS-Briefing-Note.pdf (August 11, 2020).

Proctor, K. (2020). Pandemic could drive mayors to bankruptcy, mayors warn. *Guardian*, 12 June, p. 14.

Proctor, K., Parveen, N., & Pidd, H. (2020). Regional differences widen as north-west overtakes London for numbers in hospital. *Guardian*, 4 May, p. 7.

Public Health England. (2020). *Beyond the data: Understanding the impact of COVID-19 on BAME groups*, London: Public Health England. Available at: https://assets.publishing.service. gov.uk/government/uploads/system/uploads/attachment_data/file/892376/COVID_stakeholder_ engagement_synthesis_beyond_the_data.pdf (June 16, 2020).

Pybus, O., Rambaut, A. with 11 others (2020). Preliminary analysis of SARS-COV-2 importation and establishment of UK transmission lineages. *COG-UK Consortium*, 8 June. Available at: https://virological.org/t/preliminary-analysis-of-sars-cov-2-importation-establishment-of-uk-transmission-lineages/507 (June 15, 2020).

Rawlinson, K. (2020). Labour calls for inquiry into purchase of 50 m unusable face masks. *Guardian*, 6 August. Available at: https://www.theguardian.com/world/2020/aug/06/fifty-mil lion-face-masks-bought-government-cannot-be-used-nhs (September 26, 2020).

Sabbagh, D., & Hern, A. (2020). Government abandons NHS contact tracing app. *Guardian*, 19 June, pp. 1, 13.

Sample, I. (2020). Top scientist attacks "shroud of secrecy" over UK virus decisions. *Guardian*, 3 August, pp. 1, 5.

Savage, M., Tapper, J., & McKie, R. (2020). PM told: Dump the rhetoric and plan for a new wave of Covid. *Observer*, 7 June, pp. 1, 5.

Staunton, D. (2020). "Herd immunity" is not our policy, says UK's health secretary. *Irish Times*, 15 March. Available at: https://www.irishtimes.com/news/world/uk/herd-immunity-is-not-our-policy-says-uk-s-health-secretary-1.4203637 (August 2, 2020).

Syal, R. (2020). Government failed to heed PPS warnings—watchdog. *Guardian*, 12 June, pp. 1, 15.

Taylor, D. (2020). Serco given £108 m contact-tracing job after failings on asylum seeker work. *Guardian*, 17 August, p. 10.

Topham, G. (2020). BA changes to staff terms "a national disgrace", say MPs. *Guardian*, 13 June, 33.

Thorp, L. (2020). Liverpool deaths soared after Atletico Madrid match. *Liverpool Echo*, 3 June. Available at: https://www.liverpoolecho.co.uk/news/liverpool-news/liverpool-coronavirus-deaths-soared-after-18354705 (June 15, 2020).

Timmins, N. (2012). *Never again?: The story of the health and social care act 2012: A study in coalition government and policy making*. London: Institute for Government and the King's Fund.

Weaver, M. (2020). Doctor who pleaded for more hospital PPE dies of coronavirus. *Guardian*, 9 April. Available at: https://www.theguardian.com/world/2020/apr/09/consultant-who-pleaded-for-more-nhs-hospital-ppe-dies-of-coronavirus (July 22, 2020).

Wenham, C. (2020). The UK was a global leader in preparing for pandemics. What went wrong with coronavirus? *Guardian*, 1 May. Available at: https://www.theguardian.com/commentisfree/2020/may/01/uk-global-leader-pandemics-coronavirus-covid-19-crisis-britain (June 9, 2020).

Wu, J. T., Leung, K., & Leung, G. M. (2020). Nowcasting and forecasting the potential domestic and international spread of the 2019-nCoV outbreak originating in Wuhan, China: A modelling study. *Lancet*, *395*(10225), 698–697 (29 February). Available at: https://www.thelancet.com/action/showPdf?pii=S0140-6736%2820%2930260-9 (June 16, 2020).

COVID-19 Spread in the Iberian Peninsula during the "First Wave": Spatiotemporal Analysis

Ricardo Almendra, Paula Santana, Ana Santurtún,
and Pablo Fdez-Arroyabe

Abstract COVID-19 was identified in December 2019 in Wuhan, China, and has spread worldwide at an unprecedented speed. On January 31, 2020, the first case was identified in Spain and on March 2 in Portugal, on the following months, transmission increased rapidly. This work aims to identify different spatiotemporal patterns in the evolution of COVID-19 incidence in the regions (NUT III) of the Iberian Peninsula (Portugal and Spain) from January until June, covering the so-called first wave of COVID-19 crisis. To identify areas with significant high COVID-19 incidence from January until June, a retrospectivespatialtemporal cluster analysis was conducted. Important spatial disparities across (and within) the Iberian Peninsula regions were established during the first months of 2020. In the period analyzed, the COVID-19 crude incidence rate was of 385.2 cases per 100,000 inhabitants in Portugal and 608.5 in Spain. The spread was significantly faster in Spain where by the end of March 63% of cases have been identified (23% in Portugal); a second increasing period was fond in Portugal in June, but not in Spain. The spatial–temporal analysis of COVID-19 incidence allowed the identification of six high-risk clusters which account for 50% of the total cases analyzed; the clusters identified in march include regions from Spain and north of Portugal (one cluster groups the North of Portugal and the north-western regions of Spain), in June one cluster was identified in Portugal, in the Lisbon Metropolitan Area. The spatiotemporal analysis of a specific geographical region that includes two countries can contribute to support the development of future interventions and the assessment of the impact of the policies applied.

R. Almendra (✉) · P. Santana
Centre of Studies on Geography and Spatial Planning (CEGOT), Coimbra, Portugal
e-mail: ricardoalmendra85@gmail.com; uc41942@uc.pt

Department of Geography and Tourism, University of Coimbra, Coimbra, Portugal

A. Santurtún
Unit of Legal Medicine from the University of Cantabria, Santander, Spain

P. Fdez-Arroyabe
University of Cantabria, Santander, Spain

Department of Geography and Planning, Faculty of Philosophy and Letters, Geobiomet Research Group, Santander, Spain

© The Author(s), under exclusive license to Springer Nature Switzerland AG 2021
R. Akhtar (ed.), *Coronavirus (COVID-19) Outbreaks, Environment and Human Behaviour*, https://doi.org/10.1007/978-3-030-68120-3_16

Keywords COVID-19 incidence · COVID-19 spread · Space–time clustering · Geographical inequalities · Portugal · Spain

Introduction

COVID-19 has struck the world with force, by October 2020, 3.5 million cases have been confirmed worldwide, and the death toll has exceeded 1 million victims (WHO 2020a). Since the outbreak of the virus, governments and local institutions have been struggling to limit its spread and ensure a measure of continuity in the lives of their citizens (Furman 2020).

While population mobility seems to be the major factor contributing to the rapid spread of COVID-19, namely through daily commuting to neighboring areas (Kang et al. 2020) and large-scale migrations (Adegboye et al. 2020), the progression of the disease and its lethality rates show important geographical disparities across and within countries (WHO 2020b).

In April 2020, the World Health Organization recommended the implementation of comprehensive measures, adapted to individuals' capacities and characteristics, to slow down transmission and reduce mortality associated with COVID-19. However, the spread of the disease evolved unevenly between and within countries, raising the question whether variables that were not previously considered may be affecting transmission rates and should have been taken into account in the analysis (WHO 2020b).

Several environmental (e.g. atmospheric conditions, indoor thermal comfort, seasonal climatic variability) and socio-demographic factors (e.g. population density, population age structure, population mobility) can strongly impact spatiotemporal patterns of infectious disease outbreaks (Couceiro et al. 2011; Dalziel et al. 2018; Ficetola and Rubolini 2020). Based on previous research on influenza, it was hypothesized that COVID-19 transmission might be influenced by atmospheric relative humidity and temperature (Lowen et al. 2007). Anomalous atmospheric circulation could favour the virus's propagation, by short-range droplet transmission, as well as by long-range aerosol transmission (Sanchez-lorenzo et al. 2020)). Air pollutants are risk factors for respiratory infection by carrying microorganisms, making pathogens more invasive to humans and affecting body's immunity to make people more susceptible to them, and the same pathway is presented by Zhu et al. (2020), indicating that SARS-CoV-2 could remain viable in aerosols for hours. The work developed by Setti et al. (2020) in Italy identified a significant association between the geographical distribution of daily PM10 exceedances and the spread of COVID-19 during the early stages of the pandemic threat; these results suggest that the virus transmission is likely affected by airborne PM10 (Wang et al. 2020a, b, c). Also, the impact of population mobility, including pendulum movements in these regions, may have contributed to the high concentration of cases; the transmission of infectious diseases has been reported to be enhanced by individual mobility during previous pandemic crises such as acute respiratory syndrome (SARS) in 2003 (Hufnagel et al. 2004) and influenza A (H1N1) in 2009 (Bajardi et al. 2011).

Biometeorological distress generated by atmospheric changes and indoor thermal conditions (Fdez-Arroyabe 2012) can also be associated with the development of infectious respiratory diseases; however, the results presented so far do not provide a clear answer (Briz-Redón and Serrano-Aroca 2020; Ficetola and Rubolini 2020).

Less controversial seems to be the influence of demographic factors, and places with higher population density are the principal locations for the transmission between humans and the main context where the drivers of transmission interact (Dalziel et al. 2018; Sharifi and Khavarian-Garmsir 2020). Also, factors related to the heath status of the population (e.g. diabetes, obesity) and its age structure seem to play an important role in the shape of the pandemic's progression (Dowd et al. 2020; Horton 2020; Onder et al. 2020).

Unequal geographical patterns may also arise as a consequence of deprivation inequalities (Almendra et al. 2017; Marmot et al. 2012; Santana et al. 2015). Individuals in more deprived circumstances may be more exposed to infection due to a greater prevalence of precarious employment and overcrowded conditions in the home, which leads Horton (2020) to highlight the syndemic nature of the threat. Air pollution and poor quality public spaces are also well-known factors contributing to co-morbidities and leading to a possible accentuation of infection rate disparities. Moreover, the impact of isolation measures may be unevenly distributed, as adverse environmental conditions will likely correspond to an increasing severity of their impact (Brooks et al. 2020; Wang et al. 2020a, b, c).

Understanding the spatiotemporal patterns of the COVID-19 epidemic at multiple scales could contribute to the gathering of valuable knowledge for controlling the outbreak. Specifically, the study of the Iberian Peninsula can be especially clarifying since the two countries that make it up are experiencing the pandemic in particularly contrasting ways. In Spain, the cumulative incidence rate since the beginning of the pandemic is 57% higher than that registered in Portugal, and there are marked differences between their neighboring regions (WHO 2020b). Taking into account that all regions of the Iberian Peninsula have similar sociocultural characteristics and health management models (National Health Service based on primary health care), carrying out a spatial study of the COVID-19 cases can provide insight into how SARS–CoV2 is spread, or what circumstances favour its expansion.

Thus, this work aims to identify different spatiotemporal patterns in the evolution of COVID-19 incidence in the regions (NUT III) of the Iberian Peninsula (Portugal and Spain) between January and June 2020, covering the so-called first wave.

Although the differences in the restrictions applied in each country are factors that undoubtedly affected the dissimilarities in spread of cases, the spatial distribution of cases, especially when lockdown characteristics were similar (and in some cases identical between regions of the same country), may be useful to hypothesize about the involvement of socioeconomic, climatic or biological factors in the spread of the virus. In addition, considering the incubation period of the virus, the first weeks of the pandemic, when the spread begins, are of special interest, since no measures had yet been taken to limit mobility.

Data and Methods

The Iberian Peninsula mainland, excluding the islands, (hereinafter referred to as the Iberian Peninsula) is constituted of 70 NUT III administrative areas from two different countries (Portugal and Spain) (Fig. 1a). In 2017, the Portuguese resident population was almost 10 million inhabitants, 70% of whom reside in urban spaces (Santana and Almendra 2018); according to national statistics (Ine 2018), at NUT III level, the population density (hab/km^2) varies between 13.6 and 949.6 (with an average population density of 110 hab/km^2). Lisbon and Porto are the two most populous cities with higher population densities (Lisbon Metropolitan Area with 1003 hab/km^2 and Porto metropolitan area with 847 hab/km^2); the two cities are the

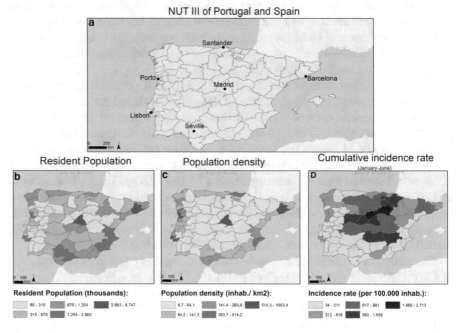

Fig. 1 **a** NUT III administrative areas of Portugal and Spain; **b** Resident population; **c** Population density; **d** COVID-19 cumulative incidence rate. *Source* based on data from the Portuguese Nation Statistics, Spanish National Statistics Institute, Portuguese Directorate-General of Health and the Spanish National Epidemiological Surveillance Network, National Center for Epidemiology at Carlos III Health Institute

capitals of the correspondent metropolitan areas, accounting together for nearly 4.5 million inhabitants (Fig. 1a, c).

Spain's population in 2019 was around 47 million inhabitants with a population density of approximately 92 hab/km^2. At a NUT III level, population density ranges from 9 hab/km^2 in Soria province, to 833 hab/km^2 in Madrid. Most of the population live in urban areas (82%), with rural spaces occupied by roughly 18%. According to the Spanish National Institute of Statistics, the most populated city in 2019 was Madrid, with 3.2 million citizens, followed by Barcelona, with 1.6 million. Both cities are at the core of the biggest metropolitan zones in Spain, with 6.1 and 5.1 million inhabitants, respectively. Less populated metropolitan areas are Valencia and Seville, with 1.5 and 1.3 million people, respectively. The metropolitan zones of Malaga and Bilbao each represent nearly 1 million inhabitants.

To identify areas with a significantly high risk of COVID-19 incidence toward time and space, secondary data, at NUT III level, between January and June, was analyzed: (a) resident population was collected from the Portuguese and Spanish Statistics Institutes; (b) weekly COVID-19 incidence data was provided by official Portuguese and Spanish institutions (Portuguese Directorate-General of Health and the Spanish National Epidemiological Surveillance Network, National Center for Epidemiology at Carlos III Health Institute, respectively). It is important to highlight that during the first weeks the testing structure was still not in place, and the ability to test was bellow of what was requested. Diagnostic tests were done in severe cases where the clinical manifestations of the disease were much more serious, contributing to the high rates recorded among elders (furthermore, this also explains why the lethality rate was initially so high). According to data made available from the European Centre for Disease Prevention and Control, the number of tests was very low in the early months of the pandemic: Portugal reached 500 tests per 100.000 inhabitants in week 14 (Spain in week 17) and 900 in week 17.

The retrospective spatial–temporal method of clusterization, developed by Kull-dorff, (1997) was applied to cluster the NUT's III regions with significantly higher or lower COVID-19 incidence rates, when compared to the expected value, considering the period under analysis. It assumes a Poisson probability model, estimating the relative risk (RR), with significance levels of 5%, applying the Monte Carlo method. The spatial structure of the model was defined considering: (i) the centroid of each NUT III; (ii) a circular spatial window; (iii) 13% of the population, as the maximum cluster size; (iv) 2 weeks as the minimum temporal cluster dimension; (v) the impossibility of clusters overlapping.

SaTScan v9.6.1 was used to develop the spatiotemporal analysis and ArcMap 10.6 to map the results.

Results

The first cases of COVID-19 were detected in Spain on January 31 and in Portugal on March 2. Between January and June, the COVID-19 crude incidence rate was 385.2

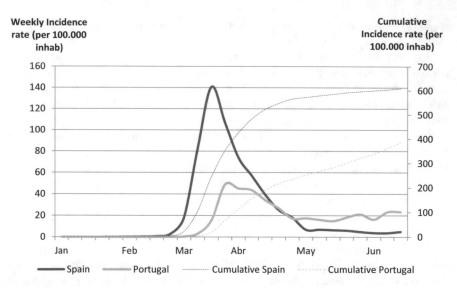

Fig. 2 Weekly and cumulative COVID-19 incidence rate, between January and June. *Source* based on data from the Portuguese Nation Statistics, Spanish National Statistics Institute, Portuguese Directorate-General of Health and the Spanish National Epidemiological Surveillance Network, National Center for Epidemiology at Carlos III Health Institute

cases per 100,000 inhabitants in Portugal and 608.5 in Spain. The highest incidence rates were recorded in the NUT III of the center of the Iberian Peninsula, with a particular highlight to the eastern border between Portugal and Spain (Fig. 1d).

Starting from the identification of the first cases, and continuing until the first week of May, the incidence rate was higher in Spain, when compared with Portugal; after that, and until the end of the study period, Portugal presented a higher incidence rate (Fig. 2). Two distinct patterns can be seen in the spread of COVID-19 in the two countries: In Spain there was an extremely fast increase, followed by a progressive diminution of reported cases, eventually falling to very low rates; in Portugal the peak was less intense, as was the decrease in cases, with incidence rates never reaching Spain's low levels.

The spatial–temporal analysis of COVID-19 incidence allowed the identification of six high-risk clusters which account for 50% of the total cases analyzed.

The spatial and temporal patterns of the spread of COVID-19 throughout the Iberian Peninsula can be observed in Fig. 3. The number of infected individuals increased rapidly in the first two weeks of March (weeks 10 and 11). Clusters A, B and C were identified in Spain, indicating significantly higher incidence rates in those NUT III. One week later, in week 12, two more clusters were established: Cluster D gathers 3 NUT III around Barcelona, and Cluster E covers the northern regions of Portugal and the majority of the north-western regions of Spain. From the third week of June (week 22), and during the following 4 weeks, Lisbon Metropolitan Area registered a significantly higher incidence rate. The three earliest clusters identified

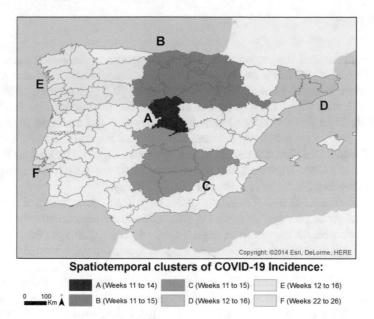

Fig. 3 Spatiotemporal clusters of COVID-19 incidence

(constituted by NUT III located in the center regions of Spain) present the higher relative risk (RR); particularly highlighted by Cluster A, with a RR of 9.02, and Cluster B with an average incidence rate of 137.9 cases per 100,000 inhabitants (Table 1).

Table 1 Relative risk of the spatiotemporal clusters of COVID-19 incidence

Clusters (p-value <0.05)	Temporal frame (weeks)	N.º of NUT III	Population	Population density (hab./km^2)	RR	Cumulative incidence rate (per 100,000 inhab.)
A	11–14	2	6,901,653	460.1	9.02	32.5
B	11–15	11	5,846,829	67.6	5.08	137.9
C	11–15	5	2,996,250	38.6	4.69	109.5
D	12–16	3	6,837,769	260.6	4.67	20.6
E	12–16	14	6,908,353	89.7	2.65	57.3
F	22–26	1	2,863,272	949.6	2.11	9.4

Discussion

This study analyzed the evolution of the spatial pattern of incidence rates due to COVID-19 during the first wave (between January and June), in the Iberian Peninsula. The methods adopted allowed the identification of NUT III administrative areas with significantly high COVID-19 incidence rates. The results show the fast spread of COVID-19 in Spain from the beginning of the crisis and highlight the cultural, economic and social proximity between the north of Portugal and north-western Spain. The trend verified in Portugal after May is also captured by Cluster E, in the Lisbon Metropolitan Area (in Spain the incidence rate was very low between May and June, while the number of cases in Portugal increased).

The Clusters A, D and F cover the most populated metropolitan areas of the Iberian Peninsula. This may be related to the social and environmental consequences of the high concentration of economic, cultural, educational and political activities. Clusters E and B also include metropolitan areas located in the northern part of the peninsula such as Porto and Bilbao, Nevertheless, some more populated metropolitan areas, such as Valencia and Seville, are not included in any cluster. Finally, Cluster C does not contain any metropolitan zones "strictosensu", being constituted by less populated provinces. Moreover, Cluster C includes four provinces of Castilla la Mancha, which is the Autonomous Community with the lowest population density in Spain (with 25.7 pop./Km^2; country average is 91 pop./Km^2); economically, the primary sector has a heavier representation in Cluster C than in the average of the country. Both factors undoubtedly affect the characteristics of social interaction and thus the potential spread of Sars Cov2.

From a biometeorological point of view, the temporal frame of five clusters coincides with the end of wintertime and the starting of the spring season in the northern hemisphere when, in general terms, weather is more variable. Some studies link the availability of sunlight with vitamin D production, (Moan et al. 2009), which prevents infections, and with the inactivation of some respiratory viruses such as influenza (Martineau et al. 2017). During the spring period, solar radiation increased progressively over the Iberian Peninsula and the cumulative incidence curve flattened in Spain but not in Portugal.

The intensity and type of measures applied by national and local governments, as well as the timing of their execution, influence the spatiotemporal evolution of COVID-19 incidence (Dowd et al. 2020).

Despite the relatively lower COVID-19 peak observed in Portugal, when compared with Spain, the incidence rate did not decrease below 15 cases per 100.000 inhabitants. On March 12, the first measures were implemented in Portugal, and 6 days after, the emergency state was decreed, imposing stronger containment measures, such as mandatory confinement. By the end of April, a steadily decreasing trend could be seen, and, by the beginning of May, the Portuguese Government presented the plan of measures for post-confinement.

In Spain, the first two cases were detected in tourists (the first case was detected on January 31 in the Canary Islands and the second on February 9 in Palma de Mallorca).

Subsequently, the number of cases increased, reaching 3869 infected and 90 dead by March 13 (140 cases per 100.000 inhabitants and the cumulative incidence rate was over 200), at which time the state of alarm was decreed. During the last week of March, the number of detected cases reached its peak, exceeding 9000 daily cases; this major decision had a clear impact on the incidence curve, which decreased rapidly in the following weeks, reducing the incidence rate in Spain to less than 10 cases per 100.000 inhabitants by the beginning of May. Due to the limitations in diagnostic testing (availability, sensitivity, specificity), it is believed that many people who had mild clinical symptoms were not diagnosed (and therefore not accounted for in this work). However, it should be noted that the Spanish Ministry of Health conducted a seroprevalence study after the state of alarm and concluded that the national prevalence of antibodies was around 5%, presenting geographical inequalities (Instituto de salud carlos III 2020). The study included 68,296 participants on three follow-up phases (27/04-11/05, 18/05-01/06 and 08/06-22/06). Although the results show a rate of antibody presence that is far from herd immunity, the Government of Spain is starting a second study at this time to evaluate the impact of the second wave. The results show that the highest values of prevalence of antibodies IgG clearly are found in regions where continental climate is predominant (Madrid et al.).

Cluster F is mostly related to cases of working-age individuals, living in the municipalities of the outskirts of the Lisbon Metropolitan Area (70% of the cases recorded during week 22 and 26 were in individuals aged between 24 and 65 years old); Portuguese national press discussed the role played by the usage of public transportation connecting the peripheral neighborhood to the inner city. The impacts of the COVID-19 pandemic are being felt the most by those living in deprivation, exposing the existing and persistent health inequalities in our societies (Eurohealthnet 2020; Wang and Tang 2020). People living in the more deprived neighborhoods of Lisbon, Amadora, Loures or Sintra (municipalities of the Lisbon Metropolitan Area) may be more exposed to infection, as it is more likely they live in overcrowded conditions, or in shared houses, and experience insecure working conditions, i.e., unable to take statutory sick or care leave (Eurohealthnet 2020; Horton 2020).

Moreover, the evolution of the incidence of COVID-19 in Portugal and Spain was (and still is) marked by outbreaks of cases inside retirement homes. The aging trend that has been reported in Europe is particularly intense in Portugal and Spain (two of the countries with the highest proportion of elders) where nearly 20% of the population is over 64 years old (Portugal: 22%; Spain: 20%) (United Nations 2020). COVID-19 tends to affect the elderly more severely, being less frequently observed in children (who tend to present milder symptoms) (World health organization 2020), with increasing age there is a progressive loss of psychological resilience and deterioration of health, adoption of less healthy lifestyles and a tendency toward loneliness and social isolation (Carter et al. 2016).

It is important to highlight that the results presented in this study maybe impacted by differences in the way health management is developed, between countries and even between administrative health regions (e.g., different testing strategies; availability of diagnosis tests).

Not only the infection rate (and the consequences of the disease) has important geographical inequalities, but the impacts of the current emergency situation and the isolation measures are also unevenly distributed throughout the Iberian Peninsula. It is undeniable that the COVID-19 crisis will have many multifaceted impacts on people and communities. Constant fear and worry as well as other stressors felt during the pandemic can have severe impacts on mental health (Brooks et al. 2020; Torales et al. 2020; Usher et al. 2020; Wang et al. 2020a, b, c). Related stressors may include the fear of being infected and infecting others, the possible deterioration of one's physical and mental health, uneasiness related to family well-being (e.g., divorce, domestic violence, alcohol abuse), deterioration of social networks, local dynamics and economies (Bradbury-Jones and Isham 2020; Ornell et al. 2020; Usher et al. 2020), the closure of essential public spaces and facilities and lack of knowledge and the appearance of misinformation (IFRC Reference Centre for Psychosocial Support 2020; Teixeira and Doetsch 2020).

At the time of this writing, seven months have passed since the end of the studied period (January–June of 2020), and the Iberian Peninsula has faced a so-called second wave and is currently struggling with the third. If during the first wave the COVID-19 impacts were milder in Portugal compared to Spain, its consequences during the second and third waves have been harsher to Portugal.

According to the data published by the COVID-19 Data Repository by the Center for Systems Science and Engineering at Johns Hopkins University, Spain has faced a second wave after summer, with a peak of cases at the beginning of November, and the third wave after Christmas, which peaked at the end of January 2021. While the temporal pattern observed in Portugal is similar to the one registered in Spain, the magnitude of the events strongly deviates: At the peak of the second wave, the incidence rate in Portugal was of 630.4 cases per million hab. (November 9, 2020) and of 451.9 cases per million hab. (November 4, 2020) in Spain; at the peak of the third wave, it was of 1264.2 cases per million hab. (January 28, 2021) in Portugal and of 791.6 in Spain (January 26, 2021). Moreover, the COVID-19 incidence rate registered in Portugal at the end of January 2021 was the highest in the world (Spain ranked fourth).

In both, Portugal and Spain, the high incidence values in the new waves have been associated with the reduction of restrictions (to facilitate economic activities) and increased social interaction during Christmas. Moreover, the introduction of new and more infective strains of the virus (British strain) has been also confirmed (INSA 2021).

In Spain, the second and third COVID-19 wave dwarfed the first in terms of cases, but gladly not in terms of mortality (according to the data published so far); in this sense, it is important to highlight the role played by the increase of tests made during the second and third waves. In Portugal, the second and third waves led the National Health Service to its limit by narrowing its response capacity. It brought a death toll (mainly during the third wave) that, at its peak, largely overcame mortality observed in Spain during the first wave. Biweekly data from the Johns Hopkins University show that in Portugal the mortality rate reached values around 374 deaths per million habitants during the sixth week of 2021 (in Spain, the highest value was recorded during week 15 of 2020 with 240 deaths per million habitants). Simultaneously, the test positivity rate reached higher values during the second and third waves than during the first, despite the large increase of daily tests since March 2020. Within this period, Portugal was the country with the highest mortality rate worldwide.

The events recorded in the first wave seem to have no influence on incidence rates observed at the end of January: Regions classified as high-risk cluster in this study are currently facing high incidence rates, and, in the same way, regions that have escaped the first wave are struggling with the amount of new cases in the second and third waves. The vaccination against COVID-19 just recently started, and its impacts on the evolution of the disease are still to be felt.

Limitations

It is important to highlight that the results express official and identified cases of COVID-19. Incidence data, based on the outcome of the tests, depends on the variability of the testing coverage (and its availability), on the reliability of the result (e.g., false positives) and therefore do not exactly reflect the spread of the disease. Due to these factors, there was a period of uncertainty on testing availability and validity of test used which affected data quality during first weeks of the spreading of the diseases.

It would have been important to complement the analysis of COVID-19 incidence with lethality and hospital morbidity data; due to lack of available information, the authors did not address these aspects in this text. Finally, in future studies, when the data is available, it would be important to analyze the differences by age group, since the population pyramid of the different regions should be taken into account in the analysis.

Conclusions

The spatiotemporal pattern of the COVID-19 incidence in the Iberian Peninsula is marked by important disparities. The most populated metropolitan areas of the Iberian Peninsula, where the population and economic density are higher, were classified as high-risk cluster, while regions with low density seem to have been less affected during the first wave.

The spatiotemporal analysis of two contiguous countries can contribute to the growing knowledge about the COVID-19 dissemination and the impact of the policies applied, therefore supporting the development of future interventions. The complexity of the current COVID-19 syndemic crisis demands a whole range of actions, covering multiple dimensions (e.g., social, economic, spatial, physical and environmental) to protect people. Despite shortage of vaccines, nearly 2.8 million citizens completed the process of vaccination, according to Spanish Health Ministry, with two needed doses in Spain by the end of March 2021. In Portugal, according to the Directorate General of Health, nearly 1.7 million vaccines have been administered. In both countries nearly 5 % of the entire population received the second dose.

Acknowledgements R. Almendra and P. Santana are members of CEGOT, which received support from the Centre of Studies in Geography and Spatial Planning (CEGOT), funded by national funds through the Foundation for Science and Technology (FCT) under the reference UIDB/04084/2020.

A. Santurtun and P. Fdez-Arroyabe are members of Geobiomet Research Group at University of Cantabria and thank the support received by the Spanish National Funds to develop the study "Health cities, biometeorological alerts and acute respiratory infections in Spain" under the grant CSO2016-75154-R (AEI- FEDER-UE).

The authors would like to acknowledge Linda Naugton for her native English revision.

References

Adegboye, O., Adekunle, A., Pak, A., Gayawan, E., Leung, D., Rojas, D., Elfaki, F., McBryde, E. & Eisen, D. (2020). *Change in outbreak epicenter and its impact on the importation risks of COVID-19 progression: a modelling study, MedRxiv.* 2020.03.17.20036681.

Almendra, R., Santana, P., & Vasconcelos, J. (2017). Evidence of social deprivation on the spatial patterns of excess winter mortality. *International Journal of Public Health, 62*(8), 849–856.

Bajardi, P., Poletto, C., Ramasco, J. J., Tizzoni, M., Colizza, V., & Vespignani, A. (2011). Human mobility networks, travel restrictions, and the global spread of 2009 H1N1 pandemic. *PLoS ONE, 6*(1), e16591.

Bradbury-Jones, C. & Isham, L. (2020). The pandemic paradox: the consequences of COVID-19 on domestic violence. *Journal of Clinical Nursing.*

Briz-Redón, Á., & Serrano-Aroca, Á. (2020). A spatio-temporal analysis for exploring the effect of temperature on COVID-19 early evolution in Spain. *Science of the Total Environment, 728,* 138811.

Brooks, S. K., Webster, R. K., Smith, L. E., Woodland, L., Wessely, S., Greenberg, N. & Rubin, G.J. (2020). The psychological impact of quarantine and how to reduce it: rapid review of the evidence. *Lancet (London, England), 395*(10227), 912–920.

Carter, T. R., Fronzek, S., Inkinen, A., Lahtinen, I., Lahtinen, M., Mela, H., O'Brien, K. L., Rosentrater, L. D., Ruuhela, R., Simonsson, L. & Terama, E. (2016). Characterising vulnerability of the elderly to climate change in the Nordic region. *Regional Environmental Change, 16*(1), 43–58.

Couceiro, Luisa, Santana, P., & Nunes, C. (2011). Pulmonary tuberculosis and risk factors in Portugal: A spatial analysis. *International Journal of Tuberculosis and Lung Disease, 15*(11), 1445–1454.

Dalziel, B. D., Kissler, S., Gog, J. R., Viboud, C., Bjørnstad, O. N., Metcalf, C. J. E. & Grenfell, B. T. (2018). Urbanization and humidity shape the intensity of influenza epidemics in U.S. cities. *Science 362*(6410), 75–79.

Dowd, J. B., Rotondi, V., Andriano, L., Brazel, D. M., Block, P., Xuejie D. et al. (2020). Demographic science aids in understanding the spread and fatality rates of COVID-19, MedRxiv. 2020.03.15.20036293.

EuroHEalthNet. (2020). What covid-19 is teaching us about inequality and the sustainability of our health systems. https://eurohealthnet.eu/COVID-19.

Fdez-Arroyabe, P. (2012). Influenza epidemics and Spanish climatic domains. *Health, 04*(10), 941–945.

Ficetola, G. F. & Rubolini, D. (2020). Climate affects global patterns of COVID-19 early outbreak dynamics. MedRxiv 2020.03.23.20040501.

Furman, J. (2020). Protecting people now, helping the economy rebound later. In Richard, B. & Beatrice i M. (Eds.), *Mitigating the COVID economic crisis: Act fast and do whatever it takes* (CEPR, pp. 191–196). London: VoxEU.org.

Horton, R. (2020). *Offline: COVID-19 is not a pandemic. The Lancet*. Lancet Publishing Group.

Hufnagel, L., Brockmann, D., & Geisel, T. (2004). Forecast and control of epidemics in a globalized world. *Proceedings of the National Academy of Sciences of the United States of America, 101*(42), 15124–15129.

IFRC Reference Centre for Psychosocial Support. (2020). *Mental health and psychosocial support for staff, volunteers and communities in an outbreak of novel coronavirus*. Hong Kong. Retrieved from https://pscentre.org/wp-content/uploads/2020/02/MHPSS-in-nCoV-2020_ENG-1.pdf.

INSA. (2021). Diversidade genética do novo coronavírus SARS-CoV-2 (COVID-19) em Portugal. https://insaflu.insa.pt/covid19/relatorios/INSA_SARS_CoV_2_DIVERSIDADE_G ENETICA_relatorio_situacao_2021-01-12.pdf.

Kang, D., Choi, H., Kim, J.-H. & Choi, J. (2020). Spatial epidemic dynamics of the COVID-19 outbreak in China. *International Journal of Infectious Diseases*.

Kulldorff, M. (1997). A spatial scan statistic. *Communications in Statistics—Theory and Methods, 26*(6), 1481–1496.

Lowen, A. C., Mubareka, S., Steel, J., & Palese, P. (2007). Influenza virus transmission is dependent on relative humidity and temperature. *PLoS Pathogens, 3*(10), e151.

Instituto de Salud Carlos III, (2020). *ESTUDIO ENE-COVID: INFORME FINAL ESTUDIO NACIONAL DE SERO-EPIDEMIOLOGÍA DE LA INFECCIÓN POR SARS-COV-2 EN ESPAÑA*. Madrid.

Marmot, M., Allen, J., Bell, R., Bloomer, E. & Goldblatt, P. (2012). WHO European review of social determinants of health and the health divide, The Lancet.

Martineau, A. R., Jolliffe, D. A., Hooper, R. L., Greenberg, L., Aloia, J. F., Bergman, P., Dubnov-Raz, G., et al. (2017) Vitamin D supplementation to prevent acute respiratory tract infections: Systematic review and meta-analysis of individual participant data, *BMJ (Online), 356*.

Moan, J. E., Dahlback, A., Ma, L., & Juzeniene, A. (2009). Influenza, solar radiation and vitamin D. *Dermato-Endocrinology, 1*(6), 308–310.

Onder, G., Rezza, G., Brusaferro, S. (2020). Case-fatality rate and characteristics of patients dying in relation to COVID-19 in Italy. *JAMA—Journal of the American Medical Association*. American Medical Association.

Ornell, F., Schuch, J. B., Sordi, A. O., & Kessler, F.H. P. (2020). "Pandemic fear" and COVID-19: mental health burden and strategies. *Brazilian Journal of Psychiatry*, (AHEAD).

Sanchez-Lorenzo, A., Vaquero-Martínez, J., Calbó, J., Wild, M., Santurtún, A., Lopez-Bustins, J.-A., Vaquero et al. (2020). Anomalous atmospheric circulation favored the spread of COVID-19 in Europe. *MedRxiv*, 2020.04.25.20079590.

Santana, P. & Almendra, R. (2018). The health of the Portuguese over the last four decades. *Méditerranée* (130).

Santana, P., Costa, C., Marí-Dell'Olmo, M., Gotsens, M., & Borrell, C. (2015). Mortality, material deprivation and urbanization: exploring the social patterns of a metropolitan area. *International Journal for Equity in Health, 14*(1), 55.

Setti, L., Passarini, F., De Gennaro, G., Barbieri, P., Licen, S., Perrone, M. G., Piazzalunga, A., Borelli, M., Palmisani, J., Di Gilio, A., Rizzo, E. & Miani, A. (2020). Potential role of particulate matter in the spreading of COVID-19 in Northern Italy: first observational study based on initial epidemic diffusion. *BMJ Open, 10*(9), e039338.

Sharifi, A. & Khavarian-Garmsir, A. R. (2020). The COVID-19 pandemic: Impacts on cities and major lessons for urban planning, design, and management. *Science of the Total Environment*. Elsevier B.V.

Teixeira, R., & Doetsch, J. (2020). The multifaceted role of mobile technologies as a strategy to combat COVID-19 pandemic. *Epidemiology and Infection, 148,* 1–11.

Torales, J., O'Higgins, M., Castaldelli-Maia, J. M. & Ventriglio, A. (2020). *The outbreak of COVID-19 coronavirus and its impact on global mental health.* International Journal of Social Psychiatry, 002076402091521.

United Nations. (2020). *World Population Ageing 2019* (Department), (United Nations, Ed.), *Economic and Social Affairs, Population Division.* New York.

Usher, K., Bhullar, N. & Jackson, D. (2020). Life in the pandemic: Social isolation and mental health. *Journal of Clinical Nursing.*

Wang, Z. & Tang, K. (2020). Combating COVID-19: health equity matters. *Nature Medicine,* 1–1.

Wang, B., Liu, J., Li, Y., Fu, S., Xu, X., Li, L. et al. (2020a). Airborne particulate matter, population mobility and COVID-19: a multi-city study in China. *BMC Public Health, 20*(1), 1585.

Wang, C., Pan, R., Wan, X., Tan, Y., Xu, L., Ho, C. S. & Ho, R. C. (2020b). Immediate Psychological Responses and Associated Factors during the Initial Stage of the 2019 Coronavirus Disease (COVID-19) Epidemic among the General Population in China. *International Journal of Environmental Research and Public Health, 17*(5).

Wang, G., Zhang, Y., Zhao, J., Zhang, J. & Jiang, F. (2020c). Mitigate the effects of home confinement on children during the COVID-19 outbreak. *The Lancet, 395*(10228), 945–947.

WHO. (2020a). *WHO coronavirus disease (COVID-19) dashboard.* https://covid19.who.int/.

WHO. (2020b). *COVID-19 strategy update—14 April 2020.* Geneva. Retrieved from https://www.who.int/publications/i/item/covid-19-strategy-update—14-april-2020.

World Health Organization. (2020). *Clinical management of severe acute respiratory infection (SARI) when COVID-19 disease is suspected. Who.* Retrieved from https://www.who.int/publications-detail/clinical-management-of-severe-acute-respiratory-infection-when-novel-coronavirus-(ncov)-infection-is-suspected.

Zhu, Y., Xie, J., Huang, F., & Cao, L. (2020). Association between short-term exposure to air pollution and COVID-19 infection: Evidence from China. *Science of the Total Environment, 727,* 138704.

COVID-19 in the Russian Federation: Regional Differences and Public Health Response

Svetlana M. Malkhazova, Fedor I. Korennoy, Natalia V. Shartova, and Tamara V. Vatlina

Abstract This study presents a medical-geographical analysis of the SARS CoV-2 pandemic development in the Russian Federation as by August 31, 2020. In general, the initial course of pandemic in Russia was characterized by a basic reproductive ratio (R_0) of 2.41 (2.22–2.60), which is relatively low as compared to most affected countries. A spatial regression analysis demonstrated that the onset of the epidemics by the regions of Russia was determined by their proximity to major international airports and connectivity of transportation network, while morbidity and mortality rates show a pronounced relationship with the population density, urban population proportion and proportion of the population over working age.

Keywords COVID-19 · Public health · Medical geography · Spatial regression · Transmission · ArcGIS

Introduction

Human coronaviruses have been known since the middle of the twentieth century, but in the twenty-first century, they have become one of the new threats to humanity. Among the family of coronaviruses, the genus Betacoronavirus, the 2002–2003 SARS-CoV outbreak deserves special mention, as it infected more than 8000 people, of whom 745 died. The pathogen was a new coronavirus, which had never previously been isolated from either animals or humans (Vabret et al. 2009). The outbreak began in Guangdong province in southern China and quickly spread to 26 countries around the world dubbed the "first pandemic of the twenty-first century." Although the Southeast Asia region was affected the most, the infection was reported in the

S. M. Malkhazova · N. V. Shartova
Lomonosov Moscow State University, Moscow, Russia

F. I. Korennoy (✉)
Federal Center for Animal Health (FGBI ARRIAH), Vladimir, Russia
e-mail: korennoy@arriah.ru

T. V. Vatlina
Smolensk State University, Smolensk, Russia

USA, Canada, many countries of Western Europe, as well as Russia (Banos and Lacasa 2007). This outbreak had significant consequences not so much on public health, but on the economy. The disproportionate scale and negative impact of such a relatively small epidemic raised concerns at that time that outbreaks of more serious diseases could have catastrophic consequences for the global economy (Smith 2006).

Ten years after the emergence of SARS-CoV, another coronavirus became the cause of a respiratory illness reported in the Middle East, i.e., the Middle East respiratory syndrome (MERS). In total, the virus has spread to 27 countries, causing large outbreaks in some countries. The total number of confirmed cases was about 2000 with a high mortality rate of about 35% (Reperant and Osterhaus 2017).

The SARS-CoV-2 coronavirus, originating from Wuhan, China in 2019 and causing the COVID-19 coronavirus pandemic across the world, has become a new threat. Despite being directly adjacent to China, Russia received the first infection cases relatively late. While two isolated cases[1] were detected in Russia on January 31, 2020, the country officially remained free from COVID-19 for more than one month longer.

Observing the development of the pandemic in other countries, the authorities in Russia believed in early spring that the chances of a wide spread of the virus in the country were low. This opinion was formed since the Russian-Chinese border was closed early on (from January 28, 2020, the organized tourist groups from China were banned; from February 20, a ban was introduced on the entry of any Chinese citizen), the population density in the country is low, and the public health system in Russia has many years of successful experience in combating infectious diseases (The Lancet 2020).

However, the closure of the Russian-Chinese border and the timely stopping of the first cases did not prevent the spread of infection in the country. The registration of infection among Russian citizens began in early March. At the same time, sanitary control was strengthened at all airports in the country that received flights from Italy, Iran, and South Korea. Since March 27, Russia has completely stopped charter and regular flights with all countries, and three days later, crossing the state border at other checkpoints (road, water, pedestrian) was restricted.

At the time of writing[2], about 33 million coronavirus tests had been done in the country and, according to the official COVID-19 screening data, about one million cases were detected. Out of these, about 15,000 people died (Table 1).

Despite the measures taken, Russia turned out to be one of the most affected countries (Fig. 1). The goal of the study was to analyze the dynamics of COVID-19 spread across the country, to identify regional differences in the epidemiological situation, and to determine its possible geographical drivers. The principal objectives were as follows: (1) to create a database of the main epidemic and geographical parameters in the regions of Russia; (2) to model the initial rate of epidemic evolution

---

[1] These were two Chinese citizens who stayed in Siberia separately, in Zabaykalsky Krai and Tyumen Oblast. They safely recovered and did not serve as s source of mass infection (StopCoronavirus 2020).

[2] See the Addendum for the most recent update.

Table 1 Main indicators of COVID-19 epidemics in Russia as of August 31, 2020

	Absolute number	Per 10,000 population
Registered cases	871,894	59.4
Recovered	676,357	46.1
Deceased	14,606	1.0
Active cases	180,931	12.3

Source StopCoronavirus (2020)

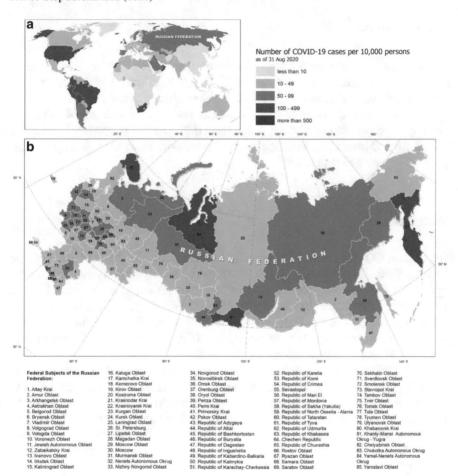

Fig. 1 Total number of registered COVID-19 cases per 10,000 population, across the world (**a**) and in the regions of Russia (**b**) as of August 31, 2020

in Russia in comparison with some other countries; (3) to identify the statistical correspondence between epidemic indicators and geographical factors in the regions of Russia, and (4) to visualize the results by a series of choropleth maps in order to identify the regional difference.

Materials and Methods

Study Region

The Russian Federation is a country in the northern part of Eurasia, covering an area of over 17 million km^2 and having a population of over 146 million people. Russia is territorially divided into 85 federal subjects (hereinafter regions)—oblasts, republics, krais, autonomous oblasts, autonomous okrugs, and federal cities (Fig. 1). The population density by regions varies from 0.07 to 171.5 people/km^2 excluding federal cities (Moscow, St. Petersburg, and Sevastopol), in which it is, respectively, 4852; 3846; and 492 people/km^2.

Data Sources

The data on COVID-19 cases[3] in Russian regions as of July 28, 2020, were taken from the Yandex[4] web-portal (Yandex 2020), which shows the daily dynamics of the epidemiological situation according to official information of the Russian Federal Service for Surveillance on Consumer Rights Protection and Human Wellbeing "Rospotrebnadzor" (2020). Data on the dynamics of the disease worldwide for the period from December 31, 2019, to August 31, 2020, were obtained from the website of the World Health Organization (WHO) (2020).

As potential factors influencing the spread of infection, the following geographic indicators were selected: the built environment factors (characterizing the possibility of movement of people), demographic factors (characterizing the vulnerability of the population), and an indicator of the quality of health care (as in the ability to resist infection). The data sources on geographic factors are the official website of the Federal State Statistics Service "Rosstat" (Federal State Statistics Service 2020) and the analytical report of the Higher School of Economics "rating of the availability and quality of medical care in the regions of the Russian Federation" (Shishkin et al. 2019).

---

[3] In this study, we use the term "case" to an officially confirmed infection of an individual.

[4] Yandex is an information technology company that has created an Internet dashboard to help track the coronavirus epidemic in the regions of the Russian Federation and related events around the world, including displaying epidemic data and infographics on COVID-19 based on official data from Rospotrebnadzor and WHO (Yandex Coronavirus statistics 2020).

Data on Russian airports and their passenger traffic were taken from the statistical reports of the Federal Air Transport Agency "Rosaviatsia" (Federal Air Transport Agency 2020). The locations of airports were georeferenced using "World Airports" dataset from Esri Data and Maps resources (World Airports 2020).

Modeling of the Pandemic Onset Russia and Some Other Countries Based on the Basic Reproduction Number R_0

One of the main quantitative characteristics in epidemiology is the effective reproduction number R_t, defined as a number of secondary infections that one primary infected individual can cause during its infectious period in a given population (Anderson and May 1991). At the stage of a fully susceptible population, the concept of a basic reproduction number R_0 is used, which is widely employed in quantitative epidemiology to model the possible size of epidemics at the stage of their initial growth before passing the peak value and introduction of intensive countermeasures. One of the basic methods or R_0 calculation for the current epidemic is an estimate based on the assumption of an exponential increase in the number of cases at the initial stage of the epidemic (the so-called exponential initial growth approach) (Dietz 1993; Iglesias et al. 2011; Zhang et al. 2013; Korennoy et al. 2016).

In this study, the value of R_0 was calculated at a country level for Russia, and, for comparison, for some other countries that demonstrated the highest rates of increase in morbidity and mortality at the time of introduction of restrictive measures in Russia. These countries are China, Italy, the USA; Sweden was also considered as a country with relatively low morbidity combined with less restrictive measures. For the purpose of modeling for each country, a time interval of one month was used starting from the date of the official first case registration.

The calculations were made by fitting an exponential model of the epidemic and then calculating R_0 based on the model curve. The quality of the model fit to real data is assessed by the coefficient of determination R^2, which demonstrates the proportion of data variation explained by the regression model, as well as the corresponding level of significance (p-value). Knowing the value of R_0 allows us to estimate the proportion of the population that must be immune to infection in order to stop its spread (the so-called herd immunity threshold—p_c). Immunity is provided by both the development of natural immunity in humans after illness and vaccination (Anderson and May 1991).

The exponential epidemic model was built in the R software environment (R Core Team 2017). The calculation of R_0 and p_c values was carried out using the Monte–Carlo modeling Microsoft Excel add-in @Risk (Palisade 2020).

Statistical Analysis of the Correlation Between Epidemic Indicators and Geographical Factors to Identify Regional Differences in the Regions of Russia

Spatial regression analysis using the Ordinary Least Squares method (OLS) was implemented to reveal the correlation between the COVID-19 epidemic indicators and geographic features in the regions of Russia. The following parameters were considered as epidemic indicators (dependent variable): (a) the registration date of the first COVID-19 case in the region; (b) the duration of the period from the detection of the first case to the peak of the incidence; (c) the number of cases at the peak of incidence; (d) the total number of detected cases from the beginning of the epidemic; (e) the total number of deaths from the beginning of the epidemic. The indicators in items (c)–(e) were log-transformed to approximate the normal distribution and provide a better model fit. Additionally, the relative rates of infection and mortality (per 10,000 population) were calculated for each region. The maps were compiled for all these characteristics (Figs. 2, 4, 5, 7 and 8).

Dates of registration of the first COVID-19 cases

6 March - 12 March	20 March - 26 March	3 April - 9 April	17 April
13 March - 19 March	27 March - 2 April	10 April - 16 April	

Federal Subjects of the Russian Federation:

1. Altay Krai
2. Amur Oblast
3. Arkhangelsk Oblast
4. Astrakhan Oblast
5. Belgorod Oblast
6. Bryansk Oblast
7. Vladimir Oblast
8. Volgograd Oblast
9. Vologda Oblast
10. Voronezh Oblast
11. Jewish Autonomous Oblast
12. Zabaikalsky Krai
13. Ivanovo Oblast
14. Irkutsk Oblast
15. Kaliningrad Oblast
16. Kaluga Oblast
17. Kamchatka Krai
18. Kemerovo Oblast
19. Kirov Oblast
20. Kostroma Oblast
21. Krasnodar Krai
22. Krasnoyarsk Krai
23. Kurgan Oblast
24. Kursk Oblast
25. Leningrad Oblast
26. St. Petersburg
27. Lipetsk Oblast
28. Magadan Oblast
29. Moscow Oblast
30. Moscow
31. Murmansk Oblast
32. Nenets Autonomous Okrug
33. Nizhny Novgorod Oblast
34. Novgorod Oblast
35. Novosibirsk Oblast
36. Omsk Oblast
37. Orenburg Oblast
38. Oryol Oblast
39. Penza Oblast
40. Perm Krai
41. Primorsky Krai
42. Pskov Oblast
43. Republic of Adygeya
44. Republic of Altai
45. Republic of Bashkortostan
46. Republic of Buryatia
47. Republic of Dagestan
48. Republic of Ingushetia
49. Republic of Kabardino-Balkaria
50. Republic of Kalmykia
51. Republic of Karachay-Cherkessia
52. Republic of Karelia
53. Republic of Komi
54. Republic of Crimea
55. Sevastopol
56. Republic of Mari El
57. Republic of Mordovia
58. Republic of Sakha (Yakutia)
59. Republic of North Ossetia - Alania
60. Republic of Tatarstan
61. Republic of Tyva
62. Republic of Udmurtia
63. Republic of Khakassia
64. Chechen Republic
65. Republic of Chuvashia
66. Rostov Oblast
67. Ryazan Oblast
68. Samara Oblast
69. Saratov Oblast
70. Sakhalin Oblast
71. Sverdlovsk Oblast
72. Smolensk Oblast
73. Stavropol Krai
74. Tambov Oblast
75. Tver Oblast
76. Tomsk Oblast
77. Tula Oblast
78. Tyumen Oblast
79. Ulyanovsk Oblast
80. Khabarovsk Krai
81. Khanty-Mansi Autonomous Okrug - Yugra
82. Chelyabinsk Oblast
83. Chukotka Autonomous Okrug
84. Yamal-Nenets Autonomous Okrug
85. Yaroslavl Oblast

Fig. 2 Dates of registration of the first COVID-19 cases in the regions of Russia

The main demographic and socio-economic indicators which could have had an impact on the dynamics of the epidemic situation at the initial stage were considered as potential geographic factors (explanatory variables): (1) the number and density of the population (as of 2019); (2) the proportion of the urban population (as of 2019); (3) the proportion of the population over working age (60 years for men and 55 years for women, as of 2018); (4) total length of roads (km); (5) density of highways (km per 1000 km^2 of the territory); (6) density of paved roads (km per 1000 km^2 of the territory); (7) availability of roads (per 10,000 people); (8) Engel's coefficient[5]; (9) index of quality of medical service for 2017; (10) distance between the regional center and the nearest major international airport (km). Seven Russian airports were considered as major international airports, providing a total of 2/3 passenger traffic: Sheremetyevo, Domodedovo, Vnukovo (Moscow), Pulkovo (St. Petersburg), Koltsovo (Yekaterinburg), Tolmachevo (Novosibirsk), and Sochi.

The pre-selection of dependent and explanatory variables was performed by iterative fitting of OLS models with various combinations of variables using the Exploratory Regression software tool (ArcGIS, Esri). The best variables' combinations were identified based on the lowest Akaike criterion (AIC), the highest adjusted coefficient of determination (Adjusted R-squared, R^2), and the lowest Variance Inflation Factor (VIF). The regression residuals were tested for randomness and for absence of spatial autocorrelation by Jarque–Bera test and Moran's I test, respectively (Mitchell 2005). For the passing combinations of variables, regression coefficients, their standard errors, and standardized coefficients were calculated, enabling judgment about the direction and strength of the correlation for each factor, which was the main goal of the analysis.

Geographic information systems ArcMap Desktop 10.8.1 and ArcGIS Pro 2.6.1 (Esri 2020) were used to carry out spatial statistical analysis as well as to visualize the results.

Results and Discussion

Analysis of the Initial Rate of the Pandemic Onset in Russia and Some Other Countries

Modeling of the theoretical epidemic curve based on epidemic data for Russia and some other countries showed a very good fit of the exponential model to the real data, which is evidenced by high R^2 and their statistical significance (Table 2). Out of the variety of analyzed countries, Russia demonstrates the lowest value of R_0, which suggests a slower development of the epidemic process throughout the country. The herd immunity threshold p_c for Russia based on the R_0 value was estimated at 0.58

---

[5]The Engel's coefficient is used to assess the level of availability of transport network: $K_e = \frac{L}{\sqrt{S \times H}}$; where L is the overall length of roads, S is the area, H is the population (Plotnikov et al. 2019).

Table 2 Basic reproduction number (R_0) values for Russia and some other countries

	R_0	95% confidence interval (CI)	R^2	p-value
Russia	2.41	2.22–2.60	0.88	<0.001
Sweden	2.64	2.51–2.78	0.94	<0.001
USA	3.10	3.02–3.19	0.98	<0.001
Italy	3.38	3.15–3.61	0.92	<0.001
China	4.10	3.78–4.41	0.93	<0.001

(95% confidence interval 0.55–0.62), meaning that at least 58% of population needs to become immune in order to cease the disease transmission.

The R_0 value is a characteristic of the initial stage of the epidemic process and cannot serve as an exceptional indicator that predetermines its further course. This value depends both on the selected baseline parameters (in particular, the duration of the infectious period D) and on the method of calculation. Based on data from publications, we valued D as nine days (To et al. 2020; McIntosh 2020). As the epidemic situation develops, the number of susceptible individuals decreases, which leads to a decrease in the reproduction rate R compared to the baseline value of R_0. Additionally, this value is directly influenced by quarantine measures aimed at reducing the frequency of contacts between people. Therefore, the parameter R_t defined as the effective reproduction number is used by authorities as a basis for decision making on introducing or removing restrictions (Methodical Recommendations … 2020).

Dynamics and Regional Differences of the COVID-19 Spread in Russia

In the first decade of March, cases were registered in the European territory of Russia: Moscow and Moscow Oblast, St. Petersburg, as well as Nizhny Novgorod, Lipetsk, and Belgorod oblasts (Fig. 2). Starting from the second half of March, the infection began to spread across the regions rapidly, and by the first days of April, COVID-19 cases were already registered in 73 out of 85 federal subjects of Russia, including the North Caucasus, Siberian, and Far Eastern regions. The latest cases of the disease were registered in Tyva Republic (April 10), Chukotka Autonomous Okrug (April 15), and the Altai Republic (April 17). Thus, it took less than two months for the coronavirus to spread across the largest country.

A sharp increase in the number of daily cases was observed throughout April and until mid-May (Fig. 3). A sudden jump took place in late April–early May when the number of cases increased from 6000 to 11,000. The decrease in the number of new cases which has been observed since mid-May continues to this day. As by the end of August, 2020, the number of daily cases was slightly less than 5000 (StopCoronavirus 2020).

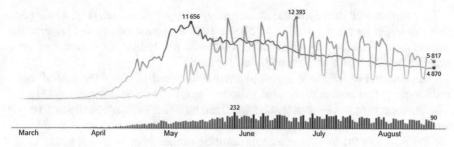

Fig. 3 Number of daily cases (red), recoveries (green), and deaths (black histograms) in Russia from March to late August 2020. *Source* Yandex Coronavirus statistics (2020)

It can be seen that a fairly inconsistent regional picture is emerging in terms of the relative number of officially registered cases. As of the end of July, no obvious spatial pattern of the most affected regions (over 100 infections per 10,000 population) was observed (Fig. 4). A relatively low infection rate (less than 20 per 10,000 population) was registered mainly in the European territory of Russia.

Fig. 4 Number of registered COVID-19 cases per 10,000 population in the regions of Russia as of July 28, 2020, and its density diagram

The distribution density of the total number of registered cases (Fig. 4) suggests that this indicator ranges from 25 to 50 per 10,000 population in most regions; the second most frequently recorded range is 60 to 80 per 10,000; an indicator of more than 100 per 10,000 is only found in few regions.

The duration of the initial period of a pandemic (starting from the date of registration of the first case until the peak) also varies greatly across the country (Fig. 5). The European regions are mainly characterized by a shorter duration (up to 70 days) while the maximum duration (over 120 days) is typical for the regions of Siberia and the Far East. For the most regions, this duration ranges from 50 to 100 days.

The dynamics of daily cases during the pandemic varied significantly across the country (Fig. 6). For example, in Moscow Oblast, geographically surrounding Moscow, a strong increase in new cases was extended over time as compared to a sharper increase in Moscow. In St. Petersburg, the second wave of growth was recorded after the first peak. The graph for another region of the European Russia, Nizhny Novgorod Oblast, which is one of the most affected regions, has a similar appearance in general. Thus, although in said regions the first cases were registered almost simultaneously, the further dynamics of the development of the epidemic

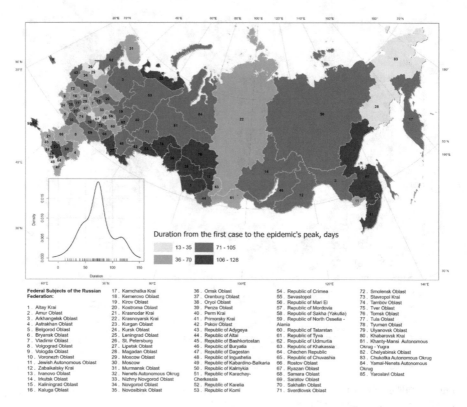

Duration from the first case to the epidemic's peak, days

13 - 35	71 - 105
36 - 70	106 - 128

Federal Subjects of the Russian Federation:

1. Altay Krai
2. Amur Oblast
3. Arkhangelsk Oblast
4. Astrakhan Oblast
5. Belgorod Oblast
6. Bryansk Oblast
7. Vladimir Oblast
8. Volgograd Oblast
9. Vologda Oblast
10. Voronezh Oblast
11. Jewish Autonomous Oblast
12. Zabaikalsky Krai
13. Ivanovo Oblast
14. Irkutsk Oblast
15. Kaliningrad Oblast
16. Kaluga Oblast

17. Kamchatka Krai
18. Kemerovo Oblast
19. Kirov Oblast
20. Kostroma Oblast
21. Krasnodar Krai
22. Krasnoyarsk Krai
23. Kurgan Oblast
24. Kursk Oblast
25. Leningrad Oblast
26. St. Petersburg
27. Lipetsk Oblast
28. Magadan Oblast
29. Moscow Oblast
30. Moscow
31. Murmansk Oblast
32. Nenets Autonomous Okrug
33. Nizhny Novgorod Oblast
34. Novgorod Oblast
35. Novosibirsk Oblast

36. Omsk Oblast
37. Orenburg Oblast
38. Oryol Oblast
39. Penza Oblast
40. Perm Krai
41. Primorsky Krai
42. Pskov Oblast
43. Republic of Adygeya
44. Republic of Altai
45. Republic of Bashkortostan
46. Republic of Buryatia
47. Republic of Dagestan
48. Republic of Ingushetia
49. Republic of Kabardino-Balkaria
50. Republic of Kalmykia
51. Republic of Karachay-Cherkessia
52. Republic of Karelia
53. Republic of Komi

54. Republic of Crimea
55. Sevastopol
56. Republic of Mari El
57. Republic of Mordovia
58. Republic of Sakha (Yakutia)
59. Republic of North Ossetia - Alania
60. Republic of Tatarstan
61. Republic of Tyva
62. Republic of Udmurtia
63. Republic of Khakassia
64. Chechen Republic
65. Republic of Chuvashia
66. Rostov Oblast
67. Ryazan Oblast
68. Samara Oblast
69. Saratov Oblast
70. Sakhalin Oblast
71. Sverdlovsk Oblast

72. Smolensk Oblast
73. Stavropol Krai
74. Tambov Oblast
75. Tver Oblast
76. Tomsk Oblast
77. Tula Oblast
78. Tyumen Oblast
79. Ulyanovsk Oblast
80. Khabarovsk Krai
81. Khanty-Mansi Autonomous Okrug - Yugra
82. Chelyabinsk Oblast
83. Chukotka Autonomous Okrug
84. Yamal-Nenets Autonomous Okrug
85. Yaroslavl Oblast

Fig. 5 Duration from the first case to the peak of the COVID-19 pandemic in the regions of Russia, and its density diagram

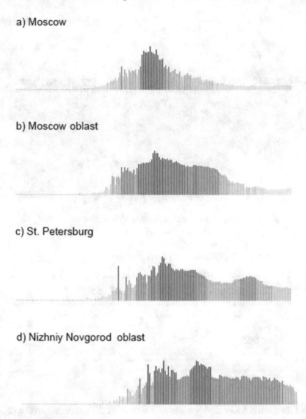

Fig. 6 Daily number of COVID-19 cases from the beginning of the spread of the infection until the end of August using the example of different regions: **a** Moscow; **b** Moscow Oblast; **c** St. Petersburg; **d** Nizhniy Novgorod Oblast. The number of cases per day is shown vertically, and days horizontally. The intensity of the color reflects the intensity of the increase per day. *Source* Yandex Coronavirus statistics (2020)

situation was different. This may be due not only to the peculiarity of the registration of infection in different regions and the scale of testing but also to the different social behaviour of the population (maintenance of social distancing, refraining from attending various events involving big crowds, etc.) in the regions during the period of restrictive measures. The restrictive measures in the regions of Russia were taken separately and independently of each other that also contributed to the observed epidemic patterns.

The daily increase in the number of cases at the peak of the pandemic widely ranged from 6703 cases (Moscow) down to 9 (Chukotka Autonomous Okrug). In the majority of the regions, the maximum increase was up to 150–250 cases per day. Moscow and the neighboring Moscow Oblast (1133 cases) differ significantly from other regions in the number of registered cases (Fig. 7). However, the data present on the maximum increase cannot be considered final, since it is possible that at the time of data inclusion in the model, not all regions have passed the peak of the pandemic.

Fig. 7 Number of new daily COVID-19 cases at the peak of the epidemic in the regions of Russia, and its density diagram

The highest relative mortality rates from COVID-19, according to official data by the end of July, were registered in the two largest cities of the country—St. Petersburg and Moscow (3.6 and 3.5 per 10,000 population, respectively) (Fig. 8). The mortality rate of more than 1 per 10,000 population is registered in 12 regions, mainly situated in the European part of Russia, but also outside of it (in Yamal-Nenets Autonomous Okrug, Kamchatka and Krasnoyarsk krais). The lowest mortality rate (less than 0.06 per 100,000) was registered in Karelia Republic, Kemerovo Oblast, Tatarstan Republic, Bashkortostan Republic, and Altai Republic. No deaths were recorded in Nenets Autonomous Okrug (the Arctic region) and Sakhalin Oblast (the Far East region).

The analysis showed that the mortality rate is no less dynamic in space and time than the infection rate. As of the end of August, Moscow and St. Petersburg still take the leading positions, but Tyva was surpassed by the regions of European part of Russia.

Regional differences are also evident when analyzing the graph of the daily increase of mortality, in particular in Moscow and St. Petersburg (Fig. 9). In St.

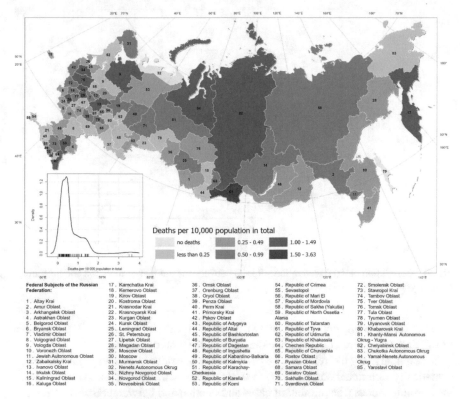

Fig. 8 Number of deaths from COVID-19 per 10,000 population in the regions of Russia as of July 28, 2020, and its density diagram

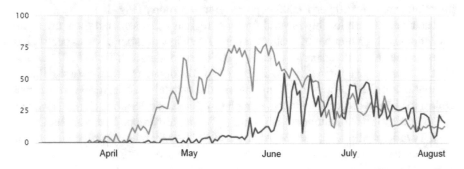

Fig. 9 Daily increase of deaths from COVID-19 in Moscow and St. Petersburg. Moscow is shown in blue, St. Petersburg is shown in red. *Source* Yandex Coronavirus statistics (2020)

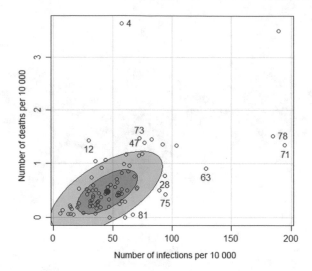

Fig. 10 Correlation plot (concentration ellipses) between the number of infection and the number of deaths per 10,000 population

Petersburg, there is a noticeable delay (in comparison with Moscow) in the number of officially registered deaths since the beginning of an outbreak.

When analyzing the relationship between the considered epidemic parameters, a statistically significant correlation ($r = 0.61$; p-value < 0.01) was found both between the absolute values (the number of cases and deaths) and between respective relative (per 10,000) indicators ($r = 0.45$; p-value < 0.01). Most regions are concentrated in the range of values up to 55 per 10,000 population for the number of cases of infection and up to 1.0 per 10,000 population for mortality (Fig. 10).

The revealed relationship is valid for most regions of Russia, but there are some exceptions. A high number of deaths combined with low number of cases per 10,000 is registered in St. Petersburg (4 on Fig. 10) and Arkhangelsk Oblast (47), in the North Caucasus (Dagestan (12), Ingushetia (73)). The opposite situation (a significant number of infected people following low mortality rate) is observed in Murmansk Oblast (63), in the North Caucasus (Karachay-Cherkessia Republic (75)), as well as in the Khanty-Mansi Autonomous Okrug (28), Yamal-Nenets Autonomous Okrug (71), Tyva Republic (78), and the Altai Republic (81). A more detailed analysis of the local data is required in order to discover the reasoning for this situation. No association was found between other epidemic indicators.

Correlation Between COVID-19 Epidemic Indicators and Geographic Factors

Analysis of the connections between the indicators of the spatial dynamics of the epidemic process in Russia and a number of geographical factors showed the following most pronounced statistically significant[6] correlations:

a. the registration date of the first cases in the regions is statistically related to the distance to the nearest airport, the proportion of the urban population, the overall length of roads, and the road density (adjusted $R^2 = 0.40$). The regression coefficients are presented in Table 3. The most significant factor is the proportion of the urban population, and the revealed correlation is negative; in more urbanized regions, the pandemic onset was earlier. The connection with the total length of roads is also negative, which means that a more developed road network also contributes to faster delivery of infection.

b. the maximum daily increase in cases at the peak of the pandemic showed a statistically significant correlation (adjusted $R^2 = 0.60$) with the population number, distance from the airport, the proportion of the urban population, and the proportion of the population over working age. The regression coefficients are presented in Table 4. The most significant factor, in this case, was the population number, demonstrating an obvious relationship with the largest increase in the number of infections.

Analysis of the obtained regression correlations of epidemic indicators with some selected geographical factors, in general, enables asserting that:

(1) the rate of spread of the pandemic in the regions of Russia depends on the remoteness of the region from the main airports with international passenger traffic; the intensity of the regional transport network and the proportion of

Table 3 Regression metrics for the "registration date of the first cases" indicator

Factor	Regression coefficient	Standard error	Standardized coefficient	p-value
Intercept	32.55	4.86	0	0
Proportion of urban population	−0.24	0.06	−0.37	<0.001
Road length	-2.02×10^{-4}	6.50×10^{-5}	−0.28	0.002
Availability of roads	0.03	0.01	0.27	0.005
Airport distance	2.00×10^{-6}	0.06	0.24	0.011

---

[6] If relax the requirements for the statistical significance of the residuals' normality and spatial distribution, the dependency may also be indicated between the mortality rate and population density, proportion of urban population, proportion of the population over working age, and Engel's coefficient (adjusted $R^2 = 0.59$). In this case, clustering pattern of the regression residuals was revealed that requires further investigation.

Table 4 Regression metrics for the "maximum daily increase in cases" indicator

Factor	Regression coefficient	Standard error	Standardized coefficient	p-value
Intercept	4.09	0.41	0	0
Population number	1.00×10^{-6}	4.86	0.63	<0.001
Airport distance	1.00×10^{-6}	0.06	−0.27	0.002
Proportion of urban population	0.02	0.01	0.24	0.004
Proportion of population over working age	−0.03	0.01	−0.16	0.043

the urban population in the region, which may be associated with the financial wellness of residents and their ability to travel internationally;

(2) the daily increase in morbidity and mortality rates shows a pronounced correlation with the population of the region and its density, as well as the proportion of the urban population and the proportion of the population over the working age.

The obtained correlations can be interpreted as (a) a confirmation of the fact that the virus was carried by air through the main airports and was subsequently spread across the country when people move through ground transport; (b) an evidence of a more intensive spread of the disease in regions with a higher population density (i.e., an increased likelihood of contacts) and a higher proportion of the urban population (which, in turn, is highly concentrated).

The important role of airports in this pandemic identified by our analysis is also confirmed by the results of the genomic analysis of the virus (Komissarov et al. 2020). It has been shown that the virus has been imported into Russia many times, with the overwhelming majority of imports from the European countries. The European traffic, apparently due to the absence of travel bans between the European countries at the beginning of the pandemic, turned out to be more vulnerable, and further strengthening of sanitary controls at the borders did not help limit the spread of infection. Thus, according to Park et al. (2020), entry screening at airports can detect only 34% of infected passengers at best. At the same time, apparently due to the early restriction of passenger traffic with China, it was possible to avoid the import of sources of infection from this area.

It should be noted that the virus was brought to Russia in late February to early March,[7] and after the import almost immediately (from March 11), it began to circulate throughout Russia. At the same time, no traces of the export of Russian forms abroad (in contrast to, for example, the United Kingdom) were found (Komissarov et al. 2020).

---

[7] As mentioned before, two single cases in January were successfully isolated and did not result in the further spread of the virus.

The intensity of international traffic is a significant factor affecting the emergence of the virus in the country and the speed of its spread in the initial period. In the case of small island states, it was shown that continuous human entry was an important factor in the spread of infection within the country (Filho et al. 2020). In Thailand, the number of infected was also statistically significantly associated with the number of tourists (Tantrakarnapa et al. 2020).

In the future, the deceleration of the spread of the disease may be associated with the efficiency of the functioning of the health care services, compliance with quarantine measures, ethnic, and social characteristics of the population, and coordinated actions of the central government and local communities. For example, in China, strict quarantine, urban isolation, and local public health measures introduced in late January have significantly reduced the rate of transmission of the virus. By mid-February, the spread of the virus was stopped (Qiu et al. 2020).

The results of the regression analysis, which showed that socio-economic factors can have a significant impact on the formation of an inconsistent picture of the spread of infection within the state, are in accordance with the results of other studies (Zemtsov and Baburin 2020), including those obtained for other countries. A positive correlation between morbidity and mortality from COVID-19 and socio-economic factors, including population density, the proportion of elderly residents, and low income, is characteristic of the USA (Zhang and Schwartz 2020). Socio-economic factors also play an important role at a higher level, a level of individual regions or cities, which was illustrated by the example of China (You et al. 2020; Xiong et al. 2020). Urban density and socio-economic variables were important factors in the incidence rate in Germany. The strongest predictors were associated with geographic location, interregional links, transport infrastructure, and labour market characteristics (Scarpone et al. 2020).

Social inequality can play a significant role in the spread and impact of COVID-19 in highly segregated countries. In the USA, the risk of infection and mortality is higher among groups affected by racial inequality, religious or other differences (ethnicity, language, living conditions, etc.) (Kim and Bostwick 2020; Kopel et al. 2020; Rozenfeld et al. 2020).To a certain extent, this factor may be important for Russia as well. Even though social stratification of the population is less typical for Russia due to the historical legacy of the Soviet past, the multinational composition and associated cultural and religious characteristics of the population can form areas with an increased risk of infection.

Analysis of the Restrictive Measures During the Pandemic in the Regions of Russia

During the 2020 pandemic, the international community has applied various mechanisms of restrictive measures, the effectiveness of which has yet to be evaluated. However, it is clear that measures to promote social distancing and reduce unnecessary travel can be important means of reducing infection (Scarpone et al. 2020).

In Russia, official non-working days were established with the preservation of wages for employees from April 4 to April 30, 2020, by the decree of the President of the Russian Federation (The President signed ... 2020). During this period, restrictive measures were in force, including the closure of enterprises, except for those continuously operating, the establishment of restrictions on the movement of citizens based on the access system, etc. After the end of the non-working period, a set of measures was determined, providing for a three-stage removal of restrictions associated with the improvement of the epidemiological situation that is based on the current R_t value, infectious hospital bed occupancy, and PCR testing coverage of the population[8] (Methodical Recommendations ... 2020). At all stages of restrictions removal, self-isolation is maintained for people with an increased risk of infection (over 65 years of age, people with underlying chronic conditions, primarily cardiovascular, respiratory diseases, as well as diabetes), remote work is preferable, and the mask requirement and social distancing are still in effect.

By the end of August, most regions of Russia moved over to the second stage of restrictions removal (Fig. 11). The first stage remains in 16 regions, and the third one is in effect in 23 regions.

In order to monitor the epidemic situation and maintain public awareness, an integrated indicator of the level of self-isolation was introduced, based on data of various Yandex applications and services (Yandex self-isolation index 2020). Based on a comparison of the urban activity on a specific day during the pandemic with the same day before the pandemic, a score ranging from 0 to 5 is given, where 0 represents a low level of self-isolation, 5 is a high level. Figure 12 presents a sharp increase of the self-isolation level after introducing the restrictive measures in the beginning of April with a gradual decline until June. The self-isolation index of Moscow was significantly higher in comparison with other cities, which indicates a different social behaviour of Moscovites. The dynamics of the level of self-isolation was indirectly reflected in the development of the epidemic process.

Yandex also calculates the level of activity based on data on the movement of people and car traffic. 0% is the lowest score registered in the city since the virus began to spread, and 100% is the activity level on the busiest weekday in February–March. If we compare the level of activity in Moscow and St. Petersburg, it is obvious that after the introduction of non-working days, the level of activity in early April in St. Petersburg was lower than in Moscow. The indicator for Moscow was 17–19%, while in St. Petersburg 13–15% (Fig. 13). However, subsequently, the level of activity

---

[8]The main condition for the transition to the first stage of restrictions removal is a steady downward trend in morbidity, determined by $R_t < 1.0$; no more than 50% of the infectious hospital bed occupancy and polymerase chain reaction (PCR) testing coverage of at least 70 per 100,000 population in a day. At the same time, the work of enterprises in the service sector and trade in non-food products is resumed with a restriction on the number of people simultaneously present in the office. Street walks and sports are possible with the condition of no more than two people walking together. To proceed to the second stage of restrictions removal, the R_t must be less than 0.8 and less than 0.5 to the third stage.

Federal Subjects of the Russian Federation:

1. Altay Krai
2. Amur Oblast
3. Arkhangelsk Oblast
4. Astrakhan Oblast
5. Belgorod Oblast
6. Bryansk Oblast
7. Vladimir Oblast
8. Volgograd Oblast
9. Vologda Oblast
10. Voronezh Oblast
11. Jewish Autonomous Oblast
12. Zabaikalsky Krai
13. Ivanovo Oblast
14. Irkutsk Oblast
15. Kaliningrad Oblast

16. Kaluga Oblast
17. Kamchatka Krai
18. Kemerovo Oblast
19. Kirov Oblast
20. Kostroma Oblast
21. Krasnodar Krai
22. Krasnoyarsk Krai
23. Kurgan Oblast
24. Kursk Oblast
25. Leningrad Oblast
26. St. Petersburg
27. Lipetsk Oblast
28. Magadan Oblast
29. Moscow Oblast
30. Moscow
31. Murmansk Oblast
32. Nenets Autonomous Okrug
33. Nizhny Novgorod Oblast
34. Novgorod Oblast

35. Novosibirsk Oblast
36. Omsk Oblast
37. Orenburg Oblast
38. Oryol Oblast
39. Penza Oblast
40. Perm Krai
41. Primorsky Krai
42. Pskov Oblast
43. Republic of Adygeya
44. Republic of Altai
45. Republic of Bashkortostan
46. Republic of Buryatia
47. Republic of Dagestan
48. Republic of Ingushetia
49. Republic of Kabardino-Balkaria
50. Republic of Kalmykia
51. Republic of Karachay-Cherkessia
52. Republic of Karelia
53. Republic of Komi

54. Republic of Crimea
55. Sevastopol
56. Republic of Mari El
57. Republic of Mordovia
58. Republic of Sakha (Yakutia)
59. Republic of North Ossetia - Alania
60. Republic of Tatarstan
61. Republic of Tyva
62. Republic of Udmurtia
63. Republic of Khakassia
64. Chechen Republic
65. Republic of Chuvashia
66. Rostov Oblast
67. Ryazan Oblast
68. Samara Oblast
69. Saratov Oblast
70. Sakhalin Oblast
71. Sverdlovsk Oblast
72. Smolensk Oblast

73. Stavropol Krai
74. Tambov Oblast
75. Tver Oblast
76. Tomsk Oblast
77. Tula Oblast
78. Tyumen Oblast
79. Ulyanovsk Oblast
80. Khabarovsk Krai
81. Khanty-Mansi Autonomous Okrug - Yugra
82. Chelyabinsk Oblast
83. Chukotka Autonomous Okrug
84. Yamal-Nenets Autonomous Okrug
85. Yaroslavl Oblast

Fig. 11 Stages of the restriction removal in the regions of Russia as of August 31, 2020. *Source* StopCoronavirus (2020)

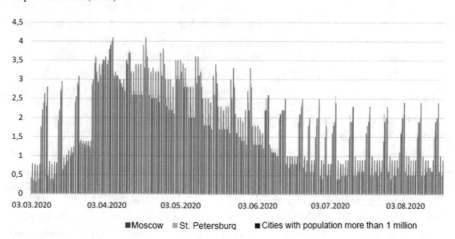

Fig. 12 Dynamics of the self-isolation index in the cities of Russia (from March to late August 2020)

Fig. 13 Dynamics of the level of activity in Moscow (**a**) and St. Petersburg (**b**) from February to late August 2020. Level of activity: green—high; blue—low. *Source* Yandex Coronavirus statistics (2020)

during the period of restrictions in Moscow was never higher than 29%, and in St. Petersburg, it reached 38–41%.

A distinctive attribute of the period of restrictive measures was that the regions independently made decisions about specific restrictions and their duration, while the role of the federal government was minimal (Åslund 2020). Probably, compliance with restrictive measures has become one of the most important tools for reducing the risks of developing an epidemic situation. In Moscow, known for some of the most severe restrictive measures, there were obvious positive trends in the course of the development of the pandemic, compared to other regions.

Limitations

The study has several limitations. The first limitation is related to the completeness of statistical data, which depends on the scale of testing, the accuracy of the diagnosis, and the cause of death. Also, it is impossible to record asymptomatic cases. For example, the underestimation of the extent of the pandemic due to the insufficient number of tests in Republics of Dagestan and Karachay-Cherkessia has been revealed and led the emergence of publications that cast doubt on the reliability of official statistical information in Russia (Coronavirus in Russia … 2020; Embury-Dennis 2020; Sevrinovsky 2020).

Another limitation is related to the set of factors used in the model. The present study includes several variables that can be attributed to the key demographic parameters affecting the initial stage of a pandemic. In the future, it is necessary to consider other factors, including environmental and climatic ones, that may be related to the further development of COVID-19 epidemics and will make it possible to identify with a greater degree of certainty the existing regional differences for the development of differentiated preventive measures. Of those factors, a degree of public confidence in the authorities can be particularly mentioned that may have an ambiguous effect on the epidemic situation and should be modeled using appropriate quantitative measures (Wong and Jensen 2020).

Conclusion

It was revealed that the infection was first brought to the European part of Russia, from where it spread to other regions of the country. The rate of development of the COVID-19 pandemic in Russia was slower than in some other countries, for example, Sweden, Italy, the USA, and China. Geographical factors such as the remoteness from major international airports of the region, the transport network of the region, population density, and the proportion of the urban population in the region played a leading role in the spread of the infection across Russia. Further study of the role of various factors including those related to the environment, climate, cultural and social characteristics of the population, as well as profile of the health status of local populations, will help better understand the possibilities of controlling infectious diseases. In addition, it could help identify the most vulnerable regions and population groups for targeted health improvement measures. A separate task for further research is to analyze the effectiveness of restrictive measures, in turn, depending on many factors: the level of education of the population, ethnic and religious specifics, the degree of wellbeing, etc. The use of other methodological solutions seems to be promising, including the use of spatial autoregressive models that allow considering dependencies introduced by interactions between neighboring regions. It should be noted that more complete and reliable results, especially concerning population mortality, can be obtained after the end of the COVID-19 pandemic.

Acknowledgements The spatial analysis was funded by the Russian Science Foundation (Grant 17-77-20070 "Assessment and Forecast of the Bioclimatic Comfort of Russian Cities under Climate Change in the 21st Century"), the GIS mapping was supported by Russian Foundation for Basic Research (Grant 18-05-60037).

Conflict of Interests The authors declare no conflicts of interests with regard to the present study.

Addendum 07.02.21

After the decline of the COVID-19 epidemics in Russia with decrease in the number of new cases below 5,000 a day by the end of August 2020, a new increase in the incidence was recorded, which became most intense since the second half of September 2020. The second wave of the disease reached its peak at the end of December 2020, when the number of daily detected cases of COVID-19 amounted to almost 30,000, more than 2.5 times higher than the corresponding figure during the first wave (Yandex Coronavirus Statistics 2020). The incidence increased to

267 cases and the mortality rate rose to 5 deaths per 10,000 population. Despite significantly higher rates of incidence and mortality, the country did not introduce a general lockdown. In the whole country, the general mask regime and the ban on holding mass events were preserved. The decision to apply additional restrictive measures was taken by the regional authorities.

According to the Russian Federal Service for Surveillance on Consumer Rights Protection and Human Wellbeing (Rospotrebnadzor), at the end of December 2020, a single case of the disease caused by a new variant of the SARS-CoV-2 VUI 202012/01 virus was registered in the country (Interfax, 2021; WHO SARS-CoV-2 Variant, 2020). The disease was diagnosed in a passenger arriving from the UK. The patient subsequently recovered safely.

In August 2020, the Gam-COVID-Vac (Sputnik V) vaccine was registered in the Russian Federation, representing a combined vector vaccine (Logunov et al. 2020, 2021). The drug has also been registered in a number of other countries. In December 2020, mass vaccination with this vaccine was launched in the Russian Federation and other countries. Vaccination is carried out on a voluntary basis within the framework of compulsory health insurance. As of 7 February 2021, one million people have been vaccinated, which is 69 persons per 10,000 population. In October 2020, the registration of the peptide antigenic vaccine EpiVacCorona was announced in Russia, which is in phase III clinical trials as of February 2021 (ClinicalTrials 2021; Dai & Gao 2021). One more vaccine is being clinically tested, namely COVI-VAC, a whole-virion inactivated vaccine developed by the Chumakov Federal Scientific Center for Research and Development of Immune and Biological Products (ClinicalTrials 2021a).

References

ClinicalTrails. Safety and Immunogenicity of COVI-VAC, a Live Attenuated Vaccine Against COVID-19 (2021a). URL: https://clinicaltrials.gov/ct2/show/NCT04619628. Accessed 07.02.2021.

ClinicalTrails. Study of the Safety, Reactogenicity and Immunogenicity of "EpiVacCorona" Vaccine for the Prevention of COVID-19 (EpiVacCorona) (2021b). URL: https://clinicaltrials.gov/ct2/show/NCT04527575. Accessed 07.02.2021.

Dai, L., Gao, G.F. Viral targets for vaccines against COVID-19 (2021). Nat Rev Immunol 21, 73–82. DOI: https://doi.org/10.1038/s41577-020-00480-0.

Interfax. The only British COVID variant carrier in Russia has already recovered (2021). URL: https://www.interfax.ru/russia/744984 (in Russian). Accessed 07.02.2021.

Logunov, DY., Dolzhikova, IV., Shchebluakov, DV., et al. Safety and efficacy of an rAd26 and rAd5 vector-based heterologous prime-boost COVID-19 vaccine: an interim analysis of a randomised controlled phase 3 trial in Russia (2021). The Lancet, ISSN: 0140-6736, Published online. DOI: 10.1016/S0140-6736(21)00234-8.

Logunov, DY., Dolzhikova, IV., Zubkova, OV., et al. Safety and immunogenicity of an rAd26 and rAd5 vector-based heterologous prime-boost COVID-19 vaccine in two formulations: two open, non-randomised phase 1/2 studies from Russia

(2020). The Lancet, ISSN: 0140-6736, Vol: 396, Issue: 10255, Page: 887–897. https://doi.org/10.1016/S0140-6736(20)31866-3.

WHO. SARS-CoV-2 Variant – United Kingdom of Great Britain and Northern Ireland (2020). URL: https://www.who.int/csr/don/21-december-2020-sars-cov2-variant-united-kingdom/en/. Accessed 07.02.2021.

Yandex Coronavirus statistics (2020). URL: https://yandex.ru/covid19/stat (in Russian). Accessed 31.12.2020.

References

Anderson, R. M., & May, R. M. (1991). *Infectious diseases of humans: Dynamics and control* (757 pp.). Oxford: Oxford University Press. https://doi.org/10.1002/hep.1840150131.

Åslund, A. (2020). Responses to the COVID-19 crisis in Russia, Ukraine, and Belarus. *Eurasian Geography and Economics*. https://doi.org/10.1080/15387216.2020.1778499.

Banos, A., & Lacasa, J. (2007). Spatio-temporal exploration of SARS epidemic. *CyberGeo*. https://doi.org/10.4000/cybergeo.12803.

Coronavirus in Russia: The Latest News. (2020, May 22). *Moscow Times*. Retrieved September 24, 2020, from https://www.themoscowtimes.com/2020/05/22/coronavirus-in-russia-the-latest-news-may-22-a69117.

Dietz, K. (1993). The estimation of the basic reproduction number for infectious diseases. *Statistical Methods in Medical Research*. https://doi.org/10.1177/096228029300200103.

Embury-Dennis, T. (2020, March 21). Coronavirus: Reported spike in pneumonia cases in Moscow as Russia accuses critics of fake news. *The Independent*. Retrieved September 14, 2020, from https://www.independent.co.uk/news/world/europe/coronavirus-russia-cases-pneumonia-deathsputin-kremlin-fake-news-a9415771.html.

Esri. (2020). *ArcGIS Pro: Release 2.6.1*. Redlands, CA: Environmental Systems Research Institute. URL: https://www.esri.com/en-us/arcgis/products/arcgis-pro/overview.

Federal Air Transport Agency "Rosaviatsia". (2020). Retrieved August 31, 2020, from https://favt.ru/dejatelnost-ajeroporty-i-ajerodromy-osnovnie-proizvodstvennie-pokazateli-aeroportov-obyom-perevoz/ (in Russian).

Federal State Statistics Service "Rosstat". (2020). Retrieved August 31, 2020, from https://eng.gks.ru/.

Filho, W. L., Lütz, J. M., Sattler, D. N., & Nunn, P. D. (2020). Coronavirus: COVID-19 transmission in Pacific Small Island developing states. *International Journal of Environmental Research and Public Health*. https://doi.org/10.3390/IJERPH17155409.

Iglesias, I., Perez, A. M., Sánchez-Vizcaíno, J. M., et al. (2011). Reproductive ratio for the local spread of highly pathogenic avian influenza in wild bird populations of Europe, 2005–2008. *Epidemiology and Infection*. https://doi.org/10.1017/S0950268810001330.

Kim, S. J., & Bostwick, W. (2020). Social vulnerability and racial inequality in COVID-19 deaths in Chicago. *Health Education & Behavior*. https://doi.org/10.1177/1090198120929677.

Komissarov, A. B., Safina, K. R., Garushyants, S. K., et al. (2020). Genomic epidemiology of the early stages of SARS-CoV-2 outbreak in Russia. *medRxiv*. https://doi.org/10.1101/2020.07.14.20150979.

Kopel, J., Perisetti, A., Roghani, A., et al. (2020). Racial and gender-based differences in COVID-19. *Frontiers in Public Health* (preprint).

Korennoy, F. I., Gulenkin, V. M., Gogin, A. E., et al. (2016). Estimating the basic reproductive number for African swine fever using the Ukrainian historical epidemic of 1977. *Transboundary and Emerging Diseases*. https://doi.org/10.1111/tbed.12583.

McIntosh, K. (2020). Coronavirus disease 2019 (COVID-19): Epidemiology, virology, and prevention. Retrieved August 31, 2020, from https://www.uptodate.com/contents/coronavirus-disease-2019-covid-19-epidemiology-virology-and-prevention.

Methodical Recommendations (MR) 3.1.0178-20. (2020). Determination of a set of measures, as well as indicators that are the basis for the gradual removal of restrictive measures in the context of the epidemic spread of COVID-19. Retrieved September 10, 2020, from https://47.rospotrebnadzor.ru/sites/default/files/ctools/mr_poetapnoe_snyatie_ogranich._08.05.2020.pdf (in Russian).

Mitchell, A. (2005). *The ESRI guide to GIS analysis* (Vol. 2). ESRI Press.

Palisade. (2020). @Risk add-in for Microsoft Excel. Retrieved August 31, 2020, from https://www.palisade.com/risk/.

Park, M., Cook, A. R., Lim, J. T., et al. (2020). A systematic review of COVID-19 epidemiology based on current evidence. *Journal of Clinical Medicine.* https://doi.org/10.3390/jcm9040967.

Plotnikov, V., Makarov, I., et al. (2019). Transport development as a factor in the economic security of regions and cities. *E3S Web of Conferences, 91.* https://doi.org/10.1051/e3sconf/20199105032.

Qiu, Y., Chen, X., & Shi, W. (2020). Impacts of social and economic factors on the transmission of coronavirus disease 2019 (COVID-19) in China. *Journal of Population Economics.* https://doi.org/10.1007/s00148-020-00778-2.

R Core Team. (2017). *R: A language and environment for statistical computing.* Vienna: R Foundation for Statistical Computing. https://www.R-project.org/.

Reperant, L. A., & Osterhaus, A. D. M. E. (2017). AIDS, Avian flu, SARS, MERS, Ebola, Zika… what next? *Vaccine.* https://doi.org/10.1016/j.vaccine.2017.04.082.

Rospotrebnadzor. (2020). Federal service for surveillance on consumer rights protection and human wellbeing. Retrieved August 31, 2020, from https://www.rospotrebnadzor.ru/en/.

Rozenfeld, Y., Beam, J., Maier, H., et al. (2020). A model of disparities: Risk factors associated with COVID-19 infection. *International Journal for Equity in Health.* https://doi.org/10.1186/s12939-020-01242-z.

Scarpone, C., Brinkmann, S. T., Große, T., et al. (2020). A multimethod approach for county-scale geospatial analysis of emerging infectious diseases: A cross-sectional case study of COVID-19 incidence in Germany. *International Journal of Health Geographics.* https://doi.org/10.1186/s12942-020-00225-1.

Sevrinovsky, V. (2020). After the peak How Dagestani communities hit hard by a COVID-19 cover-up are mourning, recovering, and bracing for the rest of the fight. Retrieved August 31, 2020, from https://meduza.io/en/feature/2020/06/11/after-the-peak.

Shishkin, S. V., Ponkratova, O. F., Potapchik, E. G., & Sazhina, S. V. (2019). Rating of the availability and quality of medical care in the regions of the Russian Federation. In *Series WP8 "State and Municipal Administration"* (96 p.). Preprint WP8/2019/01/National Research University Higher School of Economics. Moscow: Ed. House of the Higher School of Economics.

Smith, R. D. (2006). Responding to global infectious disease outbreaks: Lessons from SARS on the role of risk perception, communication and management. *Social Science and Medicine.* https://doi.org/10.1016/j.socscimed.2006.08.004.

StopCoronavirus official governmental information board of the Russian Federation. (2020). Retrieved August 31, 2020, from https://xn--80aesfpebagmfblc0a.xn--p1ai/ (in Russian).

Tantrakarnapa, K., Bhopdhornangkul, B., & Nakhaapakorn, K. (2020). Influencing factors of COVID-19 spreading: A case study of Thailand. *Journal of Public Health.* https://doi.org/10.1007/s10389-020-01329-5.

The Lancet. (2020). Salient lessons from Russia's COVID-19 outbreak. *Lancet.* https://doi.org/10.1016/S0140-6736(20)31280-0.

The President signed executive order on declaring non-work days in the Russian Federation of March 25, 2020 No. 206. Retrieved September 24, 2020, from https://en.kremlin.ru/events/president/news/63065.

To, K. K. W., Tsang, O. T. Y., Leung, W. S., et al. (2020). Temporal profiles of viral load in posterior oropharyngeal saliva samples and serum antibody responses during infection by SARS-CoV-2:

An observational cohort study. *The Lancet Infectious Diseases*. https://doi.org/10.1016/S1473-3099(20)30196-1.

Vabret, A., Dina, J., Brison, E., et al. (2009). Coronavirus humains (HCoV). *Pathologie Biologie*. https://doi.org/10.1016/j.patbio.2008.02.018.

WHO Coronavirus Disease (COVID-19) Dashboard. (2020). Retrieved August 31, 2020, from https://covid19.who.int/.

Wong, C. M. L., & Jensen, O. (2020). The paradox of trust: Perceived risk and public compliance during the COVID-19 pandemic in Singapore. *Journal of Risk Research*. https://doi.org/10.1080/13669877.2020.1756386.

World Airports. (2020). Esri data and maps. Retrieved August 31, 2020, from https://arcg.is/0zPDve.

Xiong, Y., Wang, Y., Chen, F., & Zhu, M. (2020). Spatial statistics and influencing factors of the COVID-19 epidemic at both prefecture and county levels in Hubei Province, China. *International Journal of Environmental Research and Public Health*. https://doi.org/10.3390/ijerph17113903.

Yandex Coronavirus statistics. (2020). Retrieved August 31, 2020, from https://yandex.ru/covid19/stat (in Russian).

Yandex COVID spread interactive map in Russia and in the world. (2020). Retrieved August 31, 2020, from https://yandex.ru/maps/covid19 (in Russian).

Yandex self-isolation index. (2020). Retrieved August 31, 2020, from https://yandex.ru/company/researches/2020/podomam (in Russian).

You, H., Wu, X., & Guo, X. (2020). Distribution of COVID-19 morbidity rate in association with social and economic factors in Wuhan, China: Implications for urban development. *International Journal of Environmental Research and Public Health*. https://doi.org/10.3390/ijerph17103417.

Zemtsov, S., & Baburin, V. (2020). COVID-19: Spatial dynamics and factors of regional spread in Russia. *Izvestiya Rossiiskoi Akademii Nauk Seriya Geograficheskaya, 84*(4), 485–505. https://doi.org/10.31857/S2587556620040159 (in Russian).

Zhang, C. H., & Schwartz, G. G. (2020). Spatial disparities in coronavirus incidence and mortality in the United States: An ecological analysis as of May 2020. *The Journal of Rural Health*. https://doi.org/10.1111/jrh.12476.

Zhang, Z., Chen, D., Ward, M. P., & Jiang, Q. (2013). Transmissibility of the highly pathogenic avian influenza virus, subtype H5N1 in domestic poultry: A spatio-temporal estimation at the global scale. *Geospatial Health*. https://doi.org/10.4081/gh.2012.112.

COVID-19 in South Africa

Louis Reynolds

Abstract COVID-19 exacerbates a long-standing health crisis in South Africa. Before the pandemic, health outcomes were worse than in almost all other upper-middle-income countries. The vast majority of South Africans have inadequate access to the social determinants of health, resulting in an enormous burden of ill health. The health services are unequally fragmented into a complex, predatory private sector and a largely dysfunctional, under-resourced, and demoralised public sector. The government's policy to improve the health system by introducing a National Health Insurance is in limbo. Massive socio-economic inequality and a growing culture of corruption erode trust in the government and undermine the state's ability to deliver services. Lockdown exacerbated inequality and joblessness and further undermined routine health care. "Disaster capitalism" and corruption linked to the lockdown aggravated distrust among citizens and the state. However, the pandemic galvanised civil society and stimulated creative citizen action from below.

Keywords Health · COVID-19 · South Africa · Social determinants · Disease burden · Health systems · Inequality · Pandemic · Social solidarity · Civil society · Mobilisation · Public health · Corruption

Introduction

The SARS-CoV 2 pandemic exposed deep fault lines in many countries. South Africa presents a striking example of this. Like many countries, South Africa was unprepared for the pandemic. Beyond mere lack of preparation, COVID-19 compounded an underlying and multilayered health crisis, rooted in the country's history (Coovadia et al. 2009).

L. Reynolds (✉)
People's Health Movement, South Africa, Cape Town, South Africa
e-mail: louisgeorger@gmail.com

Advocacy Committee, Department of Paediatrics and Child Health, University of Cape Town, Cape Town, South Africa

309

Even before the coronavirus arrived, health outcomes in South Africa were poor for an upper-middle-income country. Figure 1 shows trends in life expectancy, a key indicator of population health, around the world between 1960 and 2016, with countries grouped by income. While people in richer countries live longer than those in poor countries and inequality persists, countries of all categories, rich and poor, showed steady improvement throughout the period, with global life expectancy improving from 52 years in 1960 to 70 years in 2016. The South African trend lags behind the global average and even that of poorer countries.

As Fig. 1 shows, South Africa's life expectancy in 1960 was equal to the world average at 52 years and above the average of upper-middle-income countries (UMICs). Over the next three decades, it fell behind other countries in its rate of improvement, dropping below UMICs in the mid-1960s and below MICs in 1970. This period occurred during the oppressive white supremacist apartheid era. Apartheid perpetuated and deepened colonial subjugation for the vast majority of the population. It imposed racial and gender discrimination, land dispossession, an exploitative migrant labour system, the destruction of family life, vast income inequalities, and extreme violence on the majority of the population. These factors profoundly affected the health and access to health services of the majority of the population, and this became manifest in the county's failure to improve health outcomes.

The next phase saw a dramatic downturn in life expectancy, starting in the early 1990s and coinciding with the ending of apartheid. Tragically, the advent of

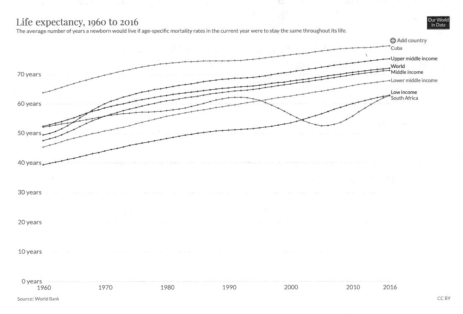

Fig. 1 Trends in life expectancy for South Africa, Cuba, and other countries grouped by wealth. *Source* UN Population Division, World Bank. From: Roser et al. (2019)

democracy in South Africa coincided with the arrival of the devastating HIV/AIDS epidemic. The impact of HIV/AIDS was catastrophic. Life expectancy dipped below that of LMICs in 1995, below LICs in 2002, and reached its nadir of 52.6 years in 2006. Unfortunately, the first democratically elected government became trapped in AIDS denial and failed to respond appropriately to the epidemic, resulting in more than 300 thousand unnecessary deaths (Chigwedere et al. 2008).

The third phase saw a remarkable improvement in health outcomes with life expectancy increasing from 52.6 years in 2006 to 62.77 years in 2016. This improvement can be ascribed to a reversal of the government's AIDS denial and the success of what became the world's biggest antiretroviral programme. The programme started in 2004 and led to a dramatic decrease in mother-to-child transmission of the AIDS virus and an improvement in child survival. Young child mortality has a strong influence on life expectancy.

Despite this impressive gain, life expectancy in South Africa remains lower than almost all MICs and many LICs. In 2016, life expectancy in South Africa (with a per capita gross domestic product (GDP) of 12,054 $US) was 62.65 years; the average for LMICs is 70.37 years and 62.94 years for LICs. Comparative examples of life expectancy alongside GDP include the Democratic Republic of the Congo (63.1 years; GDP 4526 $US); India (68.69 years; GDP 5639 $US); and Cuba (78.56 years; GDP 7889 $US) (Roser et al. 2019). Furthermore, the improvement has slowed down.

The national health crisis was a legacy of apartheid. Three key failures of the state underpin its persistence in the democratic era (Reynolds 2020). Firstly, it failed to address inequalities in access to the social determinants of health (SDH) it inherited from apartheid. Secondly, it failed to unite the fragmented apartheid health system into a single, equitable national health service around the common goal of achieving health for all and building social solidarity. Finally, it failed to address inequality in income and wealth.

Social Determinants of Health

The social determinants of health (SDH) are the goods and services we need to be healthy. They include education; adequate housing; safe environments; personal security; adequate food, water and sanitation; and good nutrition. Inequitable access to the SDH in South Africa originated in colonialism and apartheid. Apartheid came to an end in 1994 when the first democratic election brought the African National Congress (ANC) to power. The new South African Constitution, adopted two years later in 1996, includes a Bill of Rights that specifies a comprehensive range of socio-economic rights that broadly cover the SDH. The constitution obliges the state to "take reasonable legislative and other measures, within its available resources, to achieve the progressive realisation" of these rights. Unfortunately, progress in South Africa has been too slow, and the COVID-19 crisis has thrown this starkly into the light.

For example, Section 27.1.b of the Constitution gives everyone the "right to sufficient food and water". However, 24 years after its adoption, there remains a widespread lack of food security at the household level (Statistics South Africa 2019). More than a quarter of South African children under five years old are stunted. In 2017, almost 20% of South African households had inadequate or severely inadequate access to food, and more than half a million households with children aged five years or younger experienced hunger. Inadequate access to the SDH has profound impacts on health. Taking food as an example, in South Africa, wealthier households enjoy healthier and more diverse diets than poorer households. For healthy foods such as vegetables, fruits, meat, and dairy, the socio-economic gradient is particularly steep (Jonah and May 2019).

Addressing access to the SDH lies beyond the ambit of the health sector. It requires collaborative action by all state sectors to deliver the goods and services essential for health. The health sector could take a leadership role. In the context of COVID-19, the Coronavirus Command Council provides the leadership. Full community participation through organised civil society, from local to national levels, is essential. However, as generally happens, civil society is largely marginalised in the governance processes around the management of the COVID-19 pandemic—these processes tend to go on behind closed doors. However, as described later, a whole range of civil society organisations is involved in campaigns around the SDH, though some might not see the links between health and their struggles around education, housing, transport, water and sanitation, access to information and data, among others.

A Broken Health Service

South Africa's health system is highly inefficient and inequitable. Its resources remain fragmented between the private and public sectors. Total health expenditure (THE) in the country amounts to 8.9% of gross domestic product (GDP). In 2015, government (public) health expenditure provided 48.4% of THE, while pooled private financing provided 51.6% (Health Policy Project 2016). With almost equal expenditure between the two sectors, the large private sector caters only for a small, relatively healthy section of the population—the 16% of the population that can afford voluntary private health insurance. Thus, the private sector plays a minor role in dealing with the burden of disease, effectively isolating it from the long-standing national health crisis. In contrast, the public sector provides health care to the 84% of the population that bears a disproportionate share of the disease burden with the same amount of money. This inequity in the allocation of a nation's healthcare resources is staggering in its extent.

Both the private and public sectors are in crises of their own. The private sector is fragmented between multiple byzantine and unaffordable medical schemes, hospitals, and competing corporate stakeholders. It is inefficient, and even the quality of care is substandard and often unethical. A recent Health Market Inquiry by the Competition Commission found that the private sector is unaccountable, lacks proper regulation, and is increasingly unaffordable (Competition Commission 2019). In its current form, it is unsustainable.

The crisis in the public sector is profound. The South African Office of Health Standards Compliance's Annual Inspection Report for 2016/17 reported that of the 696 public-sector facilities it inspected, only five were fully compliant with National Core Standards. At the same time, 412 were either unconditionally or critically non-compliant (Office of Health Standards Compliance 2018). After more than two decades of austerity budgets under neoliberalism, the public sector is under-equipped, understaffed, and demoralised. Services are often unreliable and do not provide adequate specialist services (WHO 2012).

Nevertheless, when it comes to population health and dealing with the disease burden, the public sector is capable of significant achievements. Over the past decade, it has implemented the largest antiretroviral (ARV) programme in the world, with 4,788,139 people receiving treatment by 2018. The recent improvement in life expectancy seen in Fig. 1 can be attributed mainly to the programme's implementation of the reduction of mother-to-child transmission of HIV.

The Most Unequal Country in the World

Ironically, in the same year that it adopted the progressive new rights-based constitution, the former liberation movement—now in control of the state—adopted a neoliberal macroeconomic policy that would impede its ability to deliver the goods and services necessary to address the SDH. The policy, known as Growth, Employment and Redistribution (GEAR), was a home-grown version of the economic structural adjustment programmes (ESAPs) devised by the World Bank and the International Monetary Fund during the post-colonial period (Riddell 1992). ESAPS proved to be disastrous throughout sub-Saharan Africa, impacting particularly harmfully on women's and children's health (Thomson et al. 2017).

In South Africa, GEAR abandoned the commitment to social justice and equality that had inspired both the liberation struggle and the drafting of the constitution. Instead, relying on free market ideology and "trickle-down" economic theory, it led to the weakening of the state, privatisation of public goods and services, rampant growth of the private sector, and growing inequality. On top of this came massive corruption involving all levels of the state and the private sector, further weakening the state's ability to fulfil the constitutional rights of the citizens (Corruption Watch 2020).

State capture—harnessing the state apparatus to serve private interests—became the central feature of this corruption (February 2019). A central feature of state capture is the hollowing out of key state institutions, including the treasury, the national prosecuting authority, the criminal justice system, and the public protector. Meanwhile, crippling factionalism and infighting within the ruling political party, the African National Congress (Van Heerden 2019); ineffective parliamentary opposition; a powerful, mercenary private sector; and entrenched corruption with little evidence of accountability became prevalent.

The result is that 25 years after "liberation", South Africa, with a Gini coefficient of 0.63, is more unequal than it was in the aftermath of apartheid in 1996 when the Gini coefficient was 0.61 (Duffin 2020). The nation is broken as a result of gross inequality, widespread poverty, a profound lack of social cohesion, and lack of trust in the state.

Large numbers of people lack adequate access to crucial SDH, including adequate housing, clean water, personal safety and security, safe environments, education, and nutrition. These issues underpin the burden of disease: communicable diseases, injuries, and trauma, including violence against women and children, maternal and perinatal mortality, non-communicable disease, and mental health problems.

The Arrival of the SARS-CoV-2 Virus

When the World Health Organisation (WHO) declared COVID-19 a Public Health Emergency of International Concern on 30 January 2020, the South African government activated an emergency operation centre and set up a clinician hotline. On 5 March, a 38-year-old man from the province of KwaZulu-Natal became the first person in South Africa to test positive for the COVID-19 virus. He had returned from a trip to northern Italy along with nine others who had visited Europe. On the same day, the Minister of Health, Dr. Zweli Mkhize, notified the country that the SARS-CoV 2 virus had arrived. Six of the nine other travellers tested positive in the days that followed. These first seven coronavirus-infected people had probably acquired the virus in Europe.

Lockdown, Militarisation, Unintended Consequences

On 15 March, the president expressed concern that "internal transmission of the virus had started" and declared a national state of disaster for three months. Next, the government established a "National Coronavirus Command Council" (NCCC) to "lead the nation's plan to contain the spread and mitigate the negative impact of the coronavirus". Parliament suspended all activities from 18 March. The two largest political parties, the African National Congress (ANC) and Democratic Alliance (DA), postponed their elective conferences.

Schools were closed on 18 March. Several universities suspended classes and cancelled or postponed their graduation ceremonies. After a wave of panic buying, retailers limited the number of specified items customers could buy. On 19 March, a government gazette imposed price controls on essential commodities and punitive measures for price gougers with a fine equivalent to 10% of their turnover or

12 months in prison. The president announced a national lockdown on 23 March, which commenced at midnight on 26 March and lasted for three weeks. On 27 March, the day the lockdown took effect, the health minister reported the first death from COVID-19 in South Africa.

President Ramaphosa received local and international praise for "calm and measured leadership" in enforcing the lockdown early and for basing decisions on science and the guidance of public health experts, and for "displaying the decisive leadership so many hoped for when they cast their ballot for him in May 2019" (Calland 2020; Vandome 2020). On the eve of the lockdown, on 26 March, the president, dressed in military regalia, deployed the military to help the police enforce the lockdown, and between March and April the state deployed more than 70,000 troops (Staff Writer 2020). The army, however, was unsuited for this task (Bailie 2020). The militarisation of the lockdown led to unacceptable violations of human rights by troops and the police. In the early days of the lockdown, more people died from police and military heavy-handedness than from the coronavirus itself (Gumede 2020). Civil society organisations condemned the violations and called publicly for a people-led, rather than an army-enforced, lockdown (Cairncross and Gillespie 2020).

Apart from the abuses of the police and army, the lockdown had indirect consequences that are less visible, but impact either the SDH or the function of the health system. These are likely to aggravate the COVID-19 health crisis. For example, UNICEF reports from the early months of the COVID-19 pandemic suggest a 30% reduction in the coverage of essential nutrition services in LMICs and declines of 75–100% under lockdown (UNICEF 2020). Impacts on the SDH include loss of the main daily meal for more than 9 million schoolchildren who depended on the National School Nutrition Programme (NSNP). There was no clear plan to replace the NSNP. Feeding schemes operating through early childhood development centres also stopped operating.

The National Income Dynamics Study—Coronavirus Rapid Mobile Survey is a monthly, nationally representative panel survey of 7000 South Africans. It reported that many households experience enormous hardship as a direct consequence of lockdown, despite a range of social protection plans to protect livelihoods. Almost half of adults interviewed reported that they ran out of money to buy food in April. Between May and June, two out of every five reported that their households had lost their primary source of income, with devastating consequences for household food security. Two out of ten reported that someone in their household had gone hungry in the previous seven days, and 15% reported that children went hungry (Wills et al. 2020).

The lockdown has impacted on the health system and its capacity to deliver routine services. The childhood immunisation rate in South Africa declined by over 20% in April during the lockdown, increasing the possibility of outbreaks of vaccine-preventable diseases in children. Since vaccine-preventable diseases like measles and meningitis are more lethal in young children than COVID-19, outbreaks of these diseases could overwhelm the already overburdened South African healthcare system and increase child mortality.

Cashing in on COVID-19: Pandemic Corruption

It was not long before unscrupulous people realised that the pandemic brought enormous opportunities for self-enrichment. Some of these people are in the private sector, some in government, and some are merely politically connected. Brazen perpetrators of corruption have unashamedly appropriated emergency measures put into place to deal with COVID-19. These include a R500-billion (US$ 30 billion) relief package to provide food parcels for the needy, a temporary social grant increase for over 16 million beneficiaries and the Temporary Employer/Employee Relief Scheme (TERS) for those whose salaries were affected. The desperate need for personal protective equipment provided another opportunity for cashing in on the pandemic (Transparency International 2020).

When media and social media reports revealed that in some parts of the country government officials—including councillors—have been accused of stealing food parcels and then distributing them in exchange for personal or political gain, the president promised that the state "will not hesitate to ensure that those involved in such activities face the full might of the law" (Mahlangu 2020). But this will not be easy. Corruption—harnessing the state apparatus to serve private interests—has been endemic in South African politics since the dawn of colonial rule in 1652. The end of apartheid did not bring it to an end. The ANC-run state refined it into a highly sophisticated form known as state capture (February 2019).

A central feature of state capture is the hollowing out of key state institutions, including the treasury, the national prosecuting authority, the criminal justice system, and the public protector. Meanwhile, crippling factionalism and infighting within the ruling political party, the African National Congress (ANC); ineffective parliamentary opposition; a powerful, mercenary private sector; and a culture of impunity have created a situation where it will be very difficult to eradicate corruption despite growing public anger (Van Heerden 2019).

Civil Society Response to the Crisis

South Africa's long tradition of civil society activism has continued into the post-apartheid era (Mottiar and Bond 2011). The COVID-19 crisis galvanised a wave of activism from the local community level to new multi-sectoral coalitions and issue-based initiatives.

Community Action Networks

In March, a group of community organisers, activists, public health professionals, and artists in Cape Town met and initiated a community-led response to the pandemic that became known as Cape Town Together (CTT) (CAN activists 2020). It spawned a growing network of neighbourhood-level Community Action Networks (CANs) that spread across the city. CANs are locally organised to support the specific needs relevant to community members and vulnerable neighbours through "collective thinking, action, and change" (van Ryneveld et al. 2020). Soon, CANs emerged in other parts of the country (Gauteng Together 2020).

CANs bring together people who live in the same area but come from widely different backgrounds and experiences. Members include seasoned community organisers; privileged people with no background in community organising; women who provide care in the neighbourhood; men in community security structures; the technologically disadvantaged elderly, and many more—in one virtual space. Through these networks, social solidarity spreads faster than the virus.

Other social solidarity networks emerged in response to the COVID-19 hunger crisis. Such networks are a vital part of the societal response because of their innovation, speed, and local relevance. Social networks have a critical complementary role in partnership with governments because most governments lack sufficient capacity, flexibility, and innovation for dealing adequately with the COVID-19 crisis. In South Africa, however, bureaucracy obstructs the development of such partnerships with civil society (Surmeier et al. 2020).

The C-19 People's Coalition

The C19 People's Coalition (The C19PC) emerged early in the course of the pandemic with the coalescence of a broad range of activist organisations around the COVID-19 crisis. The coalition includes community structures, trade unions, informal workers' organisations, civics, social movements, rural groups, national and provincial NGOs across all social sectors, frontline responders such as community health workers and shelters, migrants' and refugees' organisations, public interest law firms, and faith-based organisations. The coalition seeks to ensure that South Africa's response to the COVID-19 crisis is rooted in principles of social justice and democracy, prioritises the most vulnerable, the hungry, those with weakened immune systems, and those with poor access to housing, health care, and social security. By September 2020, more than 350 organisations had endorsed the C19PC's programme of action, "A Programme of Action in the time of COVID-19" (C19 Admin 2020).

To carry out its programme of action, the C19PC established a broad range of working groups. These fall into three established categories: geographic (one for each province); issue-based (e.g. health, education); and cross-cutting (gender, economics, media). *The C19PC Health Working Group* (HWG) comprises 25 organisations and individuals connected directly or indirectly with the health system. It meets online weekly, coordinated by the South African chapter of the People's Health Movement. Its aims include support of healthcare delivery; monitoring the health and safety of healthcare workers; advocacy for better conditions of service, remuneration, and formal recognition of community health workers (CHWs). It also supports advocacy for continued access to other health services (e.g. immunisation, cancer treatment, HIV and TB treatment, hypertension, and diabetes) during the COVID-19 pandemic.

Members develop critical analyses of the health system and the COVID-19 health response and publish critical commentary in the mainstream and social media. Recently, some members have been appointed the Ministerial Advisory Committee on COVID (MAC) which advises the Minister of Health during this pandemic. A longer-term aim is to use lessons from the pandemic in the process of developing a vision for building a unitary and equitable national health system for the country. The group will also draw from the COVID-19 experience to analyse the complex relationship and linkages between economics, health, and human development. These areas of focus may change as the pandemic unfolds.

The Health Justice Initiative

The newly established Health Justice Initiative (HJI) draws on the expertise of local and international researchers in law, economics, and public health to address inequity in access to health care for COVID-19, with race, class, and gender discrimination as key issues. It aims to initiate broad research-based local and global policy reform campaigns to protect classes of people and movements. Focus areas include the social determinants of ill health, advancing the right to access health care, and intellectual property and patent barriers.

Other Civil Society Actions

In June 2020, two NGOs, SECTION 27 and the Equal Education Law Centre (EELC) filed urgent papers in the High Court in an attempt to address the nutrition situation in poverty-stricken homes or schools where the COVID-19 lockdown removed the only daily meal for 9 million children. They hope to force the Department of Basic Education (DBE) to feed all learners irrespective of whether or not their classes resume (Pikoli 2020).

The Course of the Epidemic

In the early phases, the numbers of confirmed cases grew rapidly and soon reached a doubling time of around 14 days. Figure 2 shows the cumulative count by the date that the laboratory received the specimen (National Institute for Communicable Diseases 2020). At the time of writing (22 September 2020), a total of 661,936 people had tested positive. Daily counts have come down from between 13 and 14 thousand in mid-July to less than 2 thousand with only 725 in the previous 24 h.

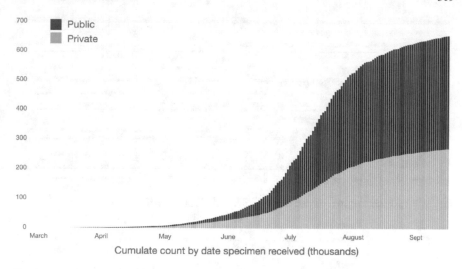

Cumulate count by date specimen received (thousands)

Fig. 2 Cumulative count between the first confirmed infected case in March and September 2020

The number of COVID-19 deaths stood at 15,992, with 39 occurring in the previous 24 h (Worldometer: https://www.worldometers.info/coronavirus/country/south-afr ica/). The true infection rate is likely to be far higher than the numbers above because, throughout the period, asymptomatic people, including contacts, were generally not tested.

Excess Deaths During the COVID-19 Epidemic

For some years, the South African Medical Research Council has tracked mortality from all causes as recorded in the National Population Register (NPR). During the COVID-19 pandemic, they have been issuing a weekly report on the data. This report shows a large gap between the weekly mortality as recorded by the NPR and deaths from confirmed cases of COVID-19. Figure 3 shows the total weekly deaths from natural causes in South Africa for people aged one year and over between January and September 2020 (thick black line) and deaths of people with confirmed COVID-19. The solid orange lines represent predicted numbers based on previous years, while the dotted orange lines show the upper and lower-bound prediction values. The striped area between the two lines represents 44,481 excess natural deaths that occurred during the COVID-19 pandemic (Bradshaw et al. 2020).

Figure 4 shows how the total number of natural deaths compares with deaths reported as being due to COVID-19. The baseline (zero) corresponds to the thin black line in Fig. 2. It reflects the expected number of natural deaths in South Africa between January and late September 2020, based on historical NPR data for 2018 and 2019. The blue line shows the number of weekly excess natural deaths during

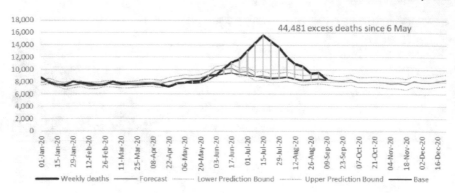

Fig. 3 Weekly deaths from natural causes in South Africa for people over 1 year old

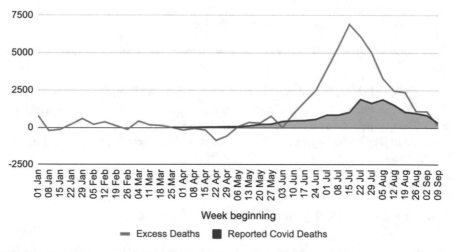

Fig. 4 Normalised excess mortality during COVID-19, shown as deviations from previous years as reflected in data from the National Population Register

the COVID-19 pandemic as recorded in the NPR. The area below the red line shows the number of COVID-19 deaths reported by the Minister of Health. The two series would be identical if all the excess deaths were due to COVID-19, and all COVID-19 deaths were accurately identified and reported. The number of estimated excess deaths started decreasing in mid-July, in keeping with the trend in the number of confirmed COVID-19 deaths.

While good data is lacking, it is highly likely that a significant share of the excess mortality is attributable to unconfirmed COVID-19 deaths. Other possible excess deaths include people with chronic diseases who are unable to access their usual healthcare services as the health system deals with COVID-19. In April, there was a drop in the natural death rate, probably due to the social distancing imposed by the lockdown (Medical Research Council 2020). Non-natural deaths fell by almost 50%

during the hard lockdown; this can be attributed to fewer traffic-related deaths and a ban on alcohol sales. On 22 September, the government relaxed the lockdown to level one, almost returning the country to normal. International travel resumed, though some countries were excluded on account of their COVID-19 rate and transmission.

The Way Forward: Adapting to the New Normal

With the trend towards fewer new confirmed daily cases in spite of increased testing, the COVID-19 pandemic in South African seems to be over the hump and on the decline at least at the time of writing, mid September, 2020. The overall mortality rate from COVID-19 was around 2.4%. If the downward trend continues, its spread appears to be less than the early predictions suggested. A preliminary recent survey in Cape Town suggests that around 37% of pregnant women and 42% of people living with HIV are seropositive for the SARS-CoV-2 virus, with higher prevalence in crowded periurban slums (Baleta 2020). Elsewhere, there is accumulating evidence of cross-immunity to SARS-CoV-2 generated by previous infection with coronaviruses known to cause the common cold (Stephens and McElrath 2020). Such coronaviruses are more likely to spread or be endemic in crowded areas. It is also possible that the high prevalence of HIV and tuberculosis in South Africa did not increase the severity of COVID-19 as had been expected. Furthermore, fake news may have had less of an impact here than in, for example, Brazil and the USA because the government's messaging was consistent and in keeping with that of the World Health Organisation.

It is too soon, however, to draw conclusions about the future of the pandemic in South Africa. The critical question now is whether the South African response to the crisis will open up the way to address the chronic health crisis, promote the human rights enshrined in the constitution, and lead to a future based on equality and social justice. The unfolding global ecological crisis makes further pandemics inevitable. Climate change is already adding to extreme weather events, growing food insecurity, climate migrations and refugees, and all this will add to the pressure on society and the state. There might be no return to the old pre-COVID-19 normal.

Three issues are immediate and critical. First, reducing the burden of disease and improving the nation's health require elimination of inequity in access to the SDH. Second, it is essential to harness available healthcare resources to develop an equitable health system based on the principles of primary health care (PHC) and capable of making rapid progress towards the delivery of universal health care (UHC). Third, social and economic inequality must be reduced by adopting a new economic order. It seems clear that so far neither the state, nor the private sector, can make these things happen. The government lacks both the capacity and the will to do so, given the scale of corruption, the lack of accountability, and the factional battles within the governing ANC. The profit-orientated private sector is mired in free market fundamentalism, and parts of it are also tainted by corruption.

This situation places an enormous responsibility on civil society. The many current grassroots struggles and campaigns in South Africa all have relevance to a broader struggle for health and its determinants, irrespective of whether they focus primarily on housing, education, food security, land, energy and the environment, public transport, crime and personal security, or health care. Many might not see the links between their struggles and health clearly, but the COVID-19 crisis and lockdown show how important it is for a broad range of civil society actors to work together towards a common goal. This calls for painstaking bottom-up organisation, education, and mobilisation.

Addendum

Major developments in South Africa regarding COVID-19 since September 2020 include the emergence of a larger and more deadly second wave of the epidemic; the detection of a more infectious new SARS-CoV 2 variant known as 501Y.V2 in samples collected in October 2020; and clinical trials suggesting that the AstraZeneca vaccine is relatively ineffective against the new strain.

The second wave and the emergence of the 501Y.V2 SARS-CoV 2 In October 2020 a second wave of COVID-19 began in the Nelson Mandela Metropolitan Area, on the Southern coast of the Eastern Cape Province. It soon spread to two other coastal provinces, the Western Cape and KwaZulu-Natal. Large numbers of inland people spend their traditional end-of-year holidays at resorts in the three provinces, while crowds of school leavers gather at "rave events" to celebrate the end of their school careers. Some of the raves became super-spreader events (Government News Agency 2020). Unsurprisingly, when inland holiday-makers and ravers returned home in early January, the second wave took off in other provinces. The new 501Y.V2 variant spread throughout the country, and soon accounted for over 90% of new cases. The daily number of newly confirmed cases peaked at 21,980 on 8 January 2021, while reported daily deaths peaked at 839 on 19 January (Worldometer 2021). During January 2021, average weekly hospital admissions peaked at 16,566 (National Institute of Communicable Diseases 2020).

By mid-January the 501Y.V2 variant first discovered in South Africa had been detected in at least 20 countries (Mahase 2021)

The vaccine saga On 1 February, when one million doses of the Oxford University AstraZeneca vaccine, produced by the Serum Institute of India, arrived in South Africa, President Ramaphosa announced the beginning of "a new chapter in our struggle against the coronavirus" (Ramaphosa 2021). The government had faced growing criticism over it's slow-moving COVID-19 vaccine acquisition plans and for agreeing to pay more than twice the price the European Union had earlier negotiated directly with AstraZeneca (Kahn 2021, January 21).

Six days later, on 7 February, a media briefing by the health minister and leading scientists revealed that a research study indicated that the AstraZeneca (AZ) vaccine offered only limited protection against mild-to-moderate COVID-19 due to the 501Y.V2 variant (Low 2021, February 10). The (as yet unpublished) study of 2000 people aged between 24 and 40 years old had begun in August 2020, before the new variant was identified. Results looked promising until late October with an efficacy rate of above 70%, but as the 501Y.V2 variant became dominant, efficacy dropped to around 10% — a clear indication that the vaccine was ineffective against it (Malan 2020). The principal investigator, Professor Shabir Madhi, stressed that it had only evaluated the vaccine's efficacy against mild to moderate disease. He said that it would "probably protect against severe COVID-19, hospitalisation and death" (Ellis 2021, February 9). There is no evidence yet to support this statement.

In response, the government suspended its plan to roll out the AZ vaccine despite the limitations of a small trial. On 10 February the health minister announced in an on-line public webinar that an implementation study with the Johnson & Johnson (J&J) vaccine would start in the following week. However, the South African Health Products Regulatory Authority has only approved the J&J vaccine for use in research settings. Clearly there are many uncertainties regarding the vaccine programme, while public scepticism grows about the state's ability to govern the large amounts of money it will need to implement its policies and plans in the face of persistent pandemic corruption (Corruption Watch 2021, February 8).

Other developments The country remains under a relatively mild lockdown that restricts large public gatherings amid growing lockdown fatigue. The economic impact of the pandemic has led to growing inequality, unemployment, hardship and hunger. Death rates in poor areas have doubled under COVID-19 (Fokazi et al. 2021, January 21). We are in a race between viral mutations on the one hand, and equitable access to effective vaccination on the other.

Civil society, in particular the C-19 People's Coalition and allied organisations, are organising a broad-based "People's Vaccination Campaign". Delegates are drafting a *People's Vaccine Manifesto* as a discussion and mobilising document. Among its demands, the draft Manifesto calls for equity in the distribution of effective vaccination and for openness, full transparency, and participation in all aspects of the government's vaccine from planning to delivery and follow-through (People's Vaccination Campaign 2021).

References

Bailie, C. (2020). South Africa's military is not suited for the fight against COVID-19. Here's why. *The Conversation*. Retrieved September 11, 2020, from https://theconversation.com/south-africas-military-is-not-suited-for-the-fight-against-covid-19-heres-why-138560.

Baleta, A. (2020, September 5). Covid-19: High antibody prevalence found in Cape Town study. *Citypress*. Retrieved September 23, 2020, from https://www.news24.com/citypress/news/covid-19-high-antibody-prevalence-found-in-cape-town-study-20200905.

Bhekisisa Team (2021, February 10). SA's first J&J jabs could arrive next week. Here's what our roll-out plan looks like. *Bhekisisa*. https://bhekisisa.org/health-news-south-africa/2021-02-10-what-does-sas-new-vaccine-roll-out-plan-look-like-heres-what-we-do-dont-know/.

Bhekisisa Team (2021, January 15). Keen to join the people's vaccine campaign? Here's what it's about. *Bhekisisa*. https://bhekisisa.org/resources/2021-01-15-keen-to-join-the-peoples-vaccine-movement-heres-what-its-all-about/.

Bradshaw, D., Laubscher, R., Dorrington, R., Groenewald, P., & Moultrie, T. (2020). Report on weekly deaths in South Africa. https://www.samrc.ac.za/sites/default/files/files/2020-09-22/weekly15September2020.pdf.

C19 Admin. (2020, March 24). A programme of action in the time of COVID-19. *C19 People's Coalition.* https://c19peoplescoalition.org.za/poa/.

Calland, R. (2020). All world leaders face mega COVID-19 crises: How Ramaphosa is stacking up. *The Conversation.* Retrieved July 22, 2020, from https://theconversation.com/all-world-leaders-face-mega-covid-19-crises-how-ramaphosa-is-stacking-up-134682.

Cairncross, L., & Gillespie, K. (2020, April 1). COVID-19: We need a people-led not army-enforced lockdown • Spotlight. https://www.spotlightnsp.co.za/2020/04/01/covid-19-we-need-a-people-led-not-army-enforced-lockdown/.

CAN Activists. (2020, June 30). Cape Town together: Organising in a city of islands. *Daily Maverick.* https://www.dailymaverick.co.za/article/2020-06-30-cape-town-together-organising-in-a-city-of-islands/.

Chigwedere, P., Seage, G. R. I., Gruskin, S., Lee, T.-H., & Essex, M. (2008). Estimating the lost benefits of antiretroviral drug use in South Africa. *JAIDS Journal of Acquired Immune Deficiency Syndromes, 49*(4), 410–415. https://doi.org/10.1097/QAI.0b013e31818a6cd5.

Competition Commission, South Africa. (2019). Healthcare market inquiry—The Competition Commission. Retrieved September 12, 2020, from https://www.compcom.co.za/healthcare-inquiry/.

Coovadia, H., Jewkes, R., Barron, P., Sanders, D., & McIntyre, D. (2009). The health and health system of South Africa: Historical roots of current public health challenges. *The Lancet, 374*(9692), 817–834. https://doi.org/10.1016/S0140-6736(09)60951-X.

Corruption Watch. (2020, March 31). CW annual report: Public fights corruption in key sectors. *Corruption Watch.* https://www.corruptionwatch.org.za/cw-annual-report-public-fights-corruption-in-key-sectors/.

Corruption Watch. (2021, February 8). SIU on the trail of govt officials abusing Covid-19 procurement. *Corruption News.* https://www.corruptionwatch.org.za/siu-on-the-trail-of-govt-officials-abusing-covid-19-procurement/.

Duffin, E. (2020). Gini Index—Inequality of income distribution—Country ranking 2017. *Statista.* Retrieved September 22, 2020, from https://www.statista.com/statistics/264627/ranking-of-the-20-countries-with-the-biggest-inequality-in-income-distribution/.

Ellis, E. (2021, February 9). How South Africa navigates the AstraZeneca Covid vaccine will have 'global repercussions'. *Daily Maverick.* https://www.dailymaverick.co.za/article/2021-02-10-how-south-africa-navigates-the-astrazeneca-covid-vaccine-will-have-global-repercussions/.

February, J. (2019). State capture: An entirely new type of corruption. Retrieved July 5, 2020, from https://issafrica.org/research/southern-africa-report/state-capture-an-entirely-new-type-of-corruption.

Fokazi, S. et al (2021, January 21). Death rates double in poor areas. *TimesLIVE.* Retrieved 13 February 2021, from https://www.timeslive.co.za/sunday-times/news/2021-01-17-death-rates-double-in-poor-areas/.

Friedman, S. (2020, August 28). How corruption in South Africa is deeply rooted in the country's past and why that matters. *The Conversation.* https://theconversation.com/how-corruption-in-south-africa-is-deeply-rooted-in-the-countrys-past-and-why-that-matters-144973.

Gauteng Together: No one is safe unless we are all safe. (2020, April 16). *Corruption Watch.* https://www.corruptionwatch.org.za/gauteng-together-no-one-is-safe-unless-we-are-all-safe/.

Government News Agency. (2020, December 7). Matric rage attendees urged to test for COVID-19. *SAnews.* https://www.sanews.gov.za/south-africa/matric-rage-attendees-urged-test-covid-19.

Gumede, W. (2020). Human rights focus needed in police and army's Covid-19 lockdown enforcement. *Democracy Works Foundation.* (n.d.). Retrieved September 11, 2020, from https://democracyworks.org.za/human-rights-focus-needed-in-police-and-armys-covid-19-lockdown-enforcement/.

Health Policy Project. (2016). Health financing profile. Available at: https://www.healthpolicypro ject.com/pubs/7887/SouthAfrica_HFP.pdf.

Jonah, C. M. P., & May, J. D. (2019). Evidence of the existence of socioeconomic-related inequality South African diets: A quantitative analysis of the 2017 general household survey. *World Nutrition, 10*(4), 27–42. https://doi.org/10.26596/wn.201910427-42.

Kahn, T. (2021, January 21). SA paying huge premium for Covid-19 shots from serum institute of India. *Business Day*. https://www.businesslive.co.za/bd/national/health/2021-01-21-exclusive-sa-paying-huge-premium-for-covid-19-shots-from-serum-institute-of-india/.

Low, M. (2021, February 10). Opinion: Pivot to J&J vaccine makes sense in light of new findings. *Spotlight*. https://www.spotlightnsp.co.za/2021/02/10/opinion-pivot-to-jj-vaccine-makes-sense-in-light-of-new-findings/.

Mahlangu, T. (2020, April 24). Government to tackle food parcel corruption. *Corruption Watch*. https://www.corruptionwatch.org.za/government-to-tackle-food-parcel-corruption/.

Mottiar, S., & Bond, P. (2011). *Social protest in South Africa*. Centre for Civil Society, University of KwaZulu-Natal. [Online]. Retrieved April 3, 2014, from, https://ccs.ukzn.ac.za/Files/Social% 20Protest%20in%20South%20Africa%20sept2011.pdf.

Malan, M. (2020, June 24). Q&A: Nine things to know about Africa's first COVID-19 vaccine trial. *Bhekisisa*. https://bhekisisa.org/article/2020-06-24-qa-nine-things-to-know-about-africas-first-covid-19-vaccine-trial/.

Mahase, E. (2021). Covid-19: What new variants are emerging and how are they being investigated? *BMJ*, 372, n158. https://doi.org/10.1136/bmj.n158.

National Institute for Communicable Diseases. (2020, October 29). Daily hospital surveillance report. *NICD*. Retrieved 11 February 2021, from https://www.nicd.ac.za/diseases-a-z-index/covid-19/surveillance-reports/.

Office of Health Standards Compliance. (2018, June 11). *OHSC Annual Inspection Report for the public-sector health establishments inspected during the 2016/17 financial year*. Office of Health Standards Compliance. https://ohsc.org.za/ohsc-annual-inspection-report-for-the-public-sector-health-establishments-inspected-during-the-2016-17-financial-year/.

People's Vaccination Campaign (2021, February). People's vaccine manifesto (unpublished draft), C-19 People's Coalition.

Pikoli, Z. (2020, June 12). Courts asked to compel government to feed learners. *Daily Maverick*. https://www.dailymaverick.co.za/article/2020-06-12-courts-asked-to-force-gov ernment-to-feed-learners/.

Ramaphosa, C. (2021, February 1). Presidential address: First consignment of a million doses of Covid-19 vaccines lands in SA; Level 3 restrictions eased. *Daily Maverick*. https://www.dailym averick.co.za/article/2021-02-01-first-consignment-of-a-million-doses-of-covid-19-vaccines-lands-in-sa-level-3-restrictions-eased/.

Reynolds, L. (2020). The coronavirus crisis and the struggle for health. *Amandla, 69*, 20–22. Retrieved July 22, 2020, from https://aidc.org.za/the-coronavirus-crisis-and-the-struggle-for-health.

Riddell, J. (1992). Things fall apart again: Structural adjustment programmes in sub-Saharan Africa. *The Journal of Modern African Studies, 30*(1), 53–68. Retrieved September 12, 2020, https://www.jstor.org/stable/161046.

Roser, M., Ortiz-Ospina, E., & Ritchie, H. (2019). Life expectancy. Published online at OurWorldIn-Data.org. Retrieved July 3, 2020, Available at: https://ourworldindata.org/life-expectancy [Online Resource].

SA Corona Virus Online Portal. (2021, February 9). Webinar: Public briefing: Current pertinent issues in relation to the department of health vaccination plan. In *SA Corona Virus Online Portal*. Retrieved 13 February 2021, from https://sacoronavirus.co.za/2021/02/09/webinar-pub lic-briefing-current-pertinent-issues-in-relation-to-the-department-of-health-vaccination-plan/.

South African Medical Research Council. (2020, September 22). Report on weekly deaths in South Africa. *South African Medical Research Council*. https://www.samrc.ac.za/reports/report-wee kly-deaths-south-africa.

Staff Writer. (2020, April 22). South Africa to deploy 73,000 more troops to enforce COVID-19 lockdown. *The Defense Post*. https://www.thedefensepost.com/2020/04/22/south-africa-deploy-73000-troops-covid-19-lockdown/.

Statistics South Africa. (2019). The extent of food security in South Africa. *Statistics South Africa*. Retrieved September 21, 2020, from https://www.statssa.gov.za/?p=12135.

Stephens, D. S., & McElrath, M. J. (2020). COVID-19 and the path to immunity. *JAMA*. https://doi.org/10.1001/jama.2020.16656.

Surmeier, A., Delichte, J., Hamann, R., & Drimie, S. (2020, May 4). Local networks can help people in distress: South Africa's COVID-19 response needs them. *The Conversation*. https://theconversation.com/local-networks-can-help-people-in-distress-south-africas-covid-19-response-needs-them-138219.

Thomson, M., Kentikelenis, A., & Stubbs, T. (2017). Structural adjustment programmes adversely affect vulnerable populations: A systematic-narrative review of their effect on child and maternal health. *Public Health Reviews, 38*(1), 13. https://doi.org/10.1186/s40985-017-0059-2.

Transparency International. (2020, September 20). In South Africa, COVID-19 has exposed greed and spurred long-needed…. *Transparency.Org*. https://www.transparency.org/en/blog/in-south-africa-covid-19-has-exposed-greed-and-spurred-long-needed-action-against-corruption.

UNICEF. (2020, April 14). Situation tracking for COVID-19 socioeconomic impacts. *UNICEF DATA*. https://data.unicef.org/resources/rapid-situation-tracking-covid-19-socioeconomic-impacts-data-viz/.

Vandome, C. (2020). *COVID-19 in South Africa: Leadership, resilience and inequality*. Chatham House. Retrieved July 22, 2020, from https://www.chathamhouse.org/expert/comment/covid-19-south-africa-leadership-resilience-and-inequality.

Van Heerden, O. (2019, April 23). With factionalism rife in the ANC and DA and the EFF confused and immature, whither the 2019 election? *Daily Maverick*. https://www.dailymaverick.co.za/opinionista/2019-04-24-with-factionalism-rife-in-the-anc-da-and-the-eff-confused-and-immature-whither-the-2019-election/.

van Ryneveld, M., Whyle, E., Brady, L., Radebe, K., Notywala, A., van Rensburg, R., et al. (2020, June 5). Cape Town together: Organizing in a city of islands. *ROAR Magazine*. https://roarmag.org/essays/cape-town-together-organizing-in-a-city-of-islands/.

Wills, G., Patel, L., van der Berg, S., & Mpeta, B. (2020). Household resource flows and food poverty during South Africa's lockdown: Short-term policy implications for three channels of social protection. In *National Income Dynamics Study (NIDS)—Coronavirus Rapid Mobile Survey (CRAM)*. https://cramsurvey.org/wp-content/uploads/2020/07/Wills-household-resource-flows-and-food-poverty-during-South-Africa%E2%80%99s-lockdown-2.pdf.

World Health Organization. Regional Office for Africa. (2012). *Health systems in Africa: Community perceptions and perspectives: The report of a multi-country study*. World Health Organization. Regional Office for Africa. https://apps.who.int/iris/handle/10665/79711.

Worldometer. (2020). Available at: https://www.worldometers.info/coronavirus/country/south-africa/.

Worldometer. (2021). South Africa Coronavirus: 1,482,412 Cases and 47,145 Deaths. Retrieved 11 February 2021, from https://www.worldometers.info/coronavirus/country/south-africa/.

Geographical Dynamics of COVID-19 in Nigeria

Stanley I. Okafor and Tolulope Osayomi

Abstract This paper examines the entry and spread of COVID-19 in Nigeria. It also discusses some policy responses of government as well as some of the consequences of the disease, and the perceptions and attitudes of the public. Government moved to establish or expand testing laboratories, and isolation and treatment centers. Available data indicate that the spread is on an upward trajectory, testing capacity is inadequate and the number of cases is not a true reflection of the pandemic situation. Ignorance, skepticism, mistrust of government and economic hardship are bedeviling efforts to combat the disease.

Keywords COVID-19 · Pandemic · Index case · Lockdown · Diffusion · Carriers

Introduction

Globalization and improvements in transportation technology have brought about increased interconnectedness of countries and set the stage for increased volume and intensity of movements of people, goods and services as well as the spread of diseases. The story of COVID-19 in Africa is a case in point. COVID-19 was first detected in Wuhan, China, on December 31, 2019. On February 14, 2020, it got to Egypt, the first country in Africa to be infected, and on February 27, 2020, it got to Nigeria. Nigeria is the third country to be infected on the continent, and the first in sub-Saharan Africa. All African countries have now been infected. The index case in Nigeria was an Italian who flew into Lagos from Italy and went on to Ogun State, just outside Lagos. He was then returned to Lagos for management and treatment. The second case was detected in Lagos, Nigeria's commercial/economic capital, followed by Ekiti State, and then the Federal Capital Territory (FCT-Abuja), Nigeria's administrative capital. Thereafter, COVID-19 started spreading to other

S. I. Okafor (✉) · T. Osayomi
Department of Geography, University of Ibadan, Ibadan, Nigeria
e-mail: stanikaf@yahoo.com

T. Osayomi
e-mail: osayomi@yahoo.com

© The Author(s), under exclusive license to Springer Nature Switzerland AG 2021   327
R. Akhtar (ed.), *Coronavirus (COVID-19) Outbreaks, Environment and Human Behaviour*, https://doi.org/10.1007/978-3-030-68120-3_19

cities and lower order settlements. Thus, there seems to be a discernible spatial pattern in the spread of the disease in the country. Nigeria is a federation of thirty six sub-national states and the FCT.

The Nigerian government has responded to the pandemic by providing or upgrading infrastructure alongside other policy measures in a bid to address the COVID-19 problem. The advent of the disease in Nigeria has had significant social and economic effects, partly consequent upon some policy responses to the pandemic. Total or partial lockdowns and prohibition or restriction of movement in some states severely disrupted livelihoods and social life. As in many developing countries, Nigeria has a very large informal sector which accounts for a large share of employment in the country and in which many of the workers are daily wagers. The lockdowns brought about loss or cessation of income for a vast majority of Nigerians, although restrictions on intra-state and inter-state movements have been relaxed, and domestic flights have resumed. This chapter therefore examines the spatial dynamics of COVID-19 in Nigeria, the policy responses, and some of the economic and social consequences of the disease on Nigerians, including their perceptions and behaviour. It employs aspects of the spatial diffusion theory as a framework for examining the spatial dynamics of the disease in the country and a brief description of the theory is the concern of the next section.

Spatial Diffusion and the Spread of COVID-19

Spatial diffusion refers to the movement of people, goods, services and innovations across space and through time (Sabel et al. 2010). As Bjelland et al. (2013, p. 14) observe, it '... is a process of dispersion of an idea or an item from a centre of origin to ... points with which it is directly or indirectly connected ...' Simply put, spatial diffusion is "the spread of a phenomenon over space and through time" (Johnston et al. 2003, p. 175). The phenomenon of interest in this chapter is COVID-19. In medical geography, spatial diffusion seeks to "... know not only how disease spreads in a population, but how that spread occurs over space; not only how many cases might occur at a future point in time, but where those are likely to occur" (Meade and Emch 2010, p. 35). It then means that, among other things, the pattern of spatial diffusion and the location of future occurrence are of utmost concern to medical geography. Over the years, the spatial diffusion theory has proven to be a useful framework for the study of infectious disease diffusion. It describes how, where and when infectious diseases spread (Sabel et al. 2010).

There are different typologies of the diffusion process, one of which recognizes three types, namely expansion, hierarchical and relocation diffusion. Another typology labels the three processes as contagious, hierarchical and relocation diffusion and regards the first two as subtypes of expansion diffusion because both of them lead to an expansion (i.e., an increase) in the number of persons infected and in the territory covered. Contagious diffusion occurs as a result of direct contact between individuals, that is to say, between infected individuals (carriers) and others. As a result, the disease expands outward from an original point, affecting more people and more places. Because contagious diffusion depends on physical contact or proximity between individuals for a disease agent to spread, the process is strongly influenced by friction of distance. Therefore, the probability of being infected is a function of geographical proximity to an infected individual or territory.

In geography, hierarchical diffusion is one in which diseases diffuse down the urban/settlement hierarchy. The largest city gets it first, followed by the next in rank, and so on. The reason is that large cities are usually linked by strong streams of movement of people, goods and services. In essence, the disease leapfrogs intervening people and places and physical distance/proximity is not an important influence on the diffusion process. Relocation diffusion is one in which the disease spreads from one place to another via the migration process, thereby relocating it. In the process, migrants who are carriers introduce the disease to the destination, vacating it in their place of origin because the migrants move with the disease. In a strict sense, relocation diffusion does not on its own bring about an increase in the number of infected persons. However, once at the destination, the contagious diffusion process takes over in the spread of the disease, leading to an increase in numbers.

It must be stressed that the typologies of the diffusion process are used mostly for pedagogical purposes. In reality, the different processes often operate in conjunction. Once a disease is introduced via the hierarchical or relocation process, the contagious process takes over. It is also important to note that without networks (roads, etc.) and carriers (infected persons), disease diffusion cannot occur. Also, diffusion does not happen unimpeded. It encounters barriers, some of which stop the spread (absorbing barriers), slow down the process (permeable barriers) or reflect it, thereby intensifying it in the local area (reflecting barriers). These barriers take different forms. They could be physical barriers (mountains, rivers, deserts, etc.), or cultural barriers (religion, language, international and sub-national boundaries, etc.). Barriers impede the movement and spatial/social interaction of people and can therefore affect the diffusion process and the spread of diseases because contact between people is impeded. Finally, if graphed, cumulative diffusion data approximate an S-shaped curve. This curve indicates that initially the diffusion process is slow, after which there is an exponential increase in diffusion until the process tapers off in the end, indicating that the diffusion process either slows down or ceases. In practical terms, this implies that the spread of diseases usually starts slowly, increases rapidly and then slows down or stops in the end. There is no doubt that some knowledge of spatial diffusion process informed some of the policy responses in Nigeria.

Spatial Dynamics and Patterns of COVID-19

Data from the Nigerian Center for Disease Control (NCDC) show that confirmed cases of COVID-19 rose from 1 in February to 139 in March, to 1932 in April and to 10,162 in May, 2020. In terms of geographical coverage, one state was affected in February; subsequently it spread to eleven states plus the FCT (Abuja) by the end of

March, to thirty-four states plus the FCT (Abuja) in April 2020 so that by the end of May, only Cross River State had no confirmed case (Fig. 1). Lagos had the largest concentration of cases in three of the four months. For instance, the state had no cases of COVID-19 in February but by the end of March, the number of cases rose to 82 and increased to 976 in April and to 4943 in May. However, the state's relative share of the disease burden presents a somewhat different picture of the pandemic than the rest of the country.

The relative contribution of Lagos rose from zero percent in February to 58.99% of all cases in March and steadily declined to 50.51% in April and to 48.64% in May. The emergence of new epicenters in Kano, and FCT (Abuja) as well as other centers of infection explain the diminished dominance of Lagos. All through March to May, 2020, Lagos was the leading epicenter. In order of importance, FCT (Abuja) with 28 cases (20.14%) in March was next. In April, Kano State surpassed FCT (Abuja) with 219 cases representing 11.33% and maintained that position but with a relatively smaller share of total confirmed cases: 954 or 9.38%. The data reveal that as the total number of confirmed cases and the number of infected states increased, the relative shares of states decreased (or increased).

The most recent map in Fig. 1 reveals that there were three epicenters of COVID-19 cases in Nigeria, namely Lagos, FCT (Abuja) and Kano. Lagos was by far the prime center of the infection, followed by Kano and the FCT. The map shows that to some extent, there is a clustering of states with relatively high numbers around Lagos and the FCT—Kano axis. Concerning the former, Ogun State and Oyo State

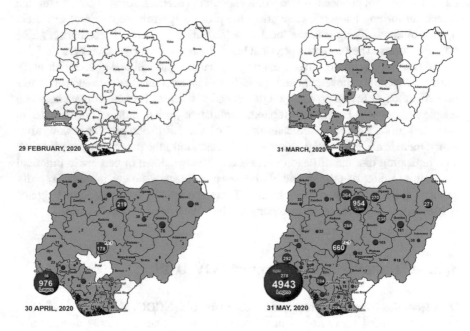

Fig. 1 Geographical pattern of COVID-19 in Nigeria from February to May 2020

are cases in point, while for the latter Kaduna, Bauchi, Katsina and Jigawa States are cases in point. This is probably a result of geographical proximity coupled with the intensity of movement of people between the epicenters and the surrounding states. This seems to suggest some element of contagious diffusion in the spread of the disease, but not to the exclusion of hierarchical diffusion. Ibadan, one of Nigeria's largest cities, is in Oyo State which is not geographically contiguous with Lagos State; yet Oyo State has a larger number of cases than Ogun State which is right next to Lagos. Lagos, Kano and the FCT (Abuja) are at the top of the urban hierarchy in Nigeria. Apart from having the busiest international airports in the country, they also have substantial domestic air passenger movements and road transport flows between them, and between them and other states in Nigeria.

Outside the two clusters above, there is a third one in the south-south geopolitical zone. This consists of Edo State, Delta State and Rivers State. Edo, Delta and Rivers are proximate and they are important oil producing states with comparatively healthy economies. Therefore, there are significant road passenger movements between them, as well as between them and Lagos and FCT (Abuja) by both land and air.

Table 1 and Fig. 2 show the diffusion of the virus across the country since it arrived on February 27, 2020. As earlier indicated, the first case of COVID-19 was detected

Table 1 Dates of index cases

S/N	Nigerian states	Date
1	Ogun	27-Feb-20
2	Lagos	16-Mar-20
3	Ekiti	18-Mar-20
4	FCT	21-Mar-20
5	Oyo	22-Mar-20
6	Edo	23-Mar-20
7	Bauchi	24-Mar-20
8	Osun	25-Mar-20
9	Rivers	25-Mar-20
10	Enugu	27-Mar-20
11	Benue	28-Mar-20
12	Kaduna	28-Mar-20
13	Akwa-Ibom	1-Apr-20
14	Ondo	3-Apr-20
15	Kwara	6-Apr-20
16	Delta	7-Apr-20
17	Katsina	7-Apr-20
18	Anambra	10-Apr-20
19	Niger	10-Apr-20

(continued)

Table 1 (continued)

S/N	Nigerian states	Date
20	Kano	11-Apr-20
21	Borno	19-Apr-20
22	Jigawa	19-Apr-20
23	Gombe	20-Apr-20
24	Abia	20-Apr-20
25	Sokoto	20-Apr-20
26	Adamawa	22-Apr-20
27	Plateau	23-Apr-20
28	Zamfara	24-Apr-20
29	Imo	25-Apr-20
30	Bayelsa	26-Apr-20
31	Ebonyi	26-Apr-20
32	Kebbi	26-Apr-20
33	Taraba	26-Apr-20
34	Nassarawa	28-Apr-20
35	Yobe	29-Apr-20
36	Kogi	27-May-20
37	Cross River	

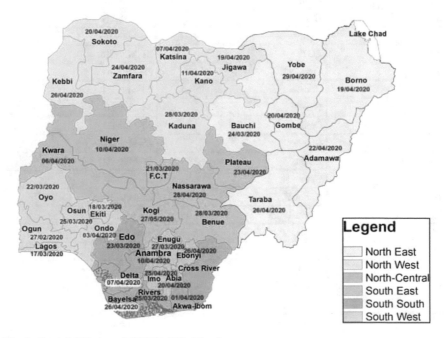

Fig. 2 Spatial diffusion of COVID-19 in Nigeria

in Ogun State on February 27. Sixteen or so days later, the second confirmed case was in Lagos State and Ekiti recorded its first case two days thereafter. In quick succession, FCT, Oyo, Edo and Bauchi had confirmed cases on March 21, 22, 23 and 24, 2020, respectively. On March 25, Bauchi Rivers and Osun States reported their first COVID-19 cases. From March 27 to April 7, COVID-19 had diffused to additional eight states, namely Enugu, Benue, Kaduna, Akwa Ibom, Ondo, Kwara, Delta and Katsina. Within a space of eleven days, it moved to Anambra, Niger, Kano, Borno, Jigawa, Gombe, Abia and Sokoto States. By May 27, the whole country, except Cross River State, had the infection.

By the end of May 2020, Nigeria had recorded a total of 10,162 confirmed cases of COVID-19, 3007 discharged cases, 287 confirmed fatalities and 6868 active cases (Table 2). Active cases are persons who are still in hospital receiving care. The May 2020 data from the NCDC show that males (6882; 68%) are more affected than females (3280; 32%). The most affected age group is the 31–40 year cohort (24%). Most of these confirmed cases (74%) had no epidemiologic link; the source of the infection could not be identified. Two percent of the cases had foreign travel history, while 24% were attributed to community transmission (NCDC 2020a, b). In order of magnitude, Lagos had 4943 cases, followed by Kano with 954 and FCT (Abuja) had 660 cases. The lowest figures are in the southeastern geopolitical zone of the country, particularly Anambra State (11 cases), Abia State (10) and the north central states—Benue State (7 cases) and Kogi State (2 cases) (Table 2).

Lagos, Kano and the FCT (Abuja) collectively account for nearly 64.5% of the total confirmed cases. They are therefore at the heart of the pandemic in Nigeria. Several reasons might explain the high prevalence of COVID-19 they recorded. As earlier noted, Lagos was virtually the first point of entry of the virus. In the early phase of the outbreak, most of the reported cases were persons who came into Nigeria from foreign countries. Subsequently, infections began to occur among persons who had contact with the imported cases and later on still, community transmission commenced. A combination of factors is at play here. Lagos arguably has the busiest seaports and domestic and international airports in the country. It is therefore the busiest transportation hub in Nigeria. The city generates more intra-state, inter-state and international traffic flows of people, goods and services than any other state in the country.

As far as intra-state traffic flows are concerned, Lagos State has 80 cars per 1000 people, with a high car density of 264 vehicles per kilometer of roadway (Oshodi et al. 2018). In addition, the state has a road network of 2600 km, often congested with a daily count of over 2 million vehicles (International Association of Public Transport and African Association of Public Transport cited in Oshodi et al. 2018). The volume of movement within Lagos metropolis is therefore quite high. Another reason why Lagos has a high risk of infection is because it is the second most populous state in the country (after Kano) and it is the most densely populated state, according to the 2006 national population census (which is the most recent) and subsequent projections (National Population Commission 2006). The state has a total population of 12,550,598 (2016 projection) and an area of 3577.28 km², giving

Table 2 Distribution of confirmed, discharged, active cases and deaths

States	Confirmed cases	Discharged cases	Deaths	Total active cases
Lagos	4943	825	54	4064
Kano	954	240	45	669
FCT	660	182	19	459
Katsina	364	68	14	282
Oyo	292	97	6	189
Edo	284	69	13	202
Ogun	278	149	9	120
Borno	271	167	26	78
Jigawa	270	135	5	130
Kaduna	258	157	8	93
Bauchi	238	220	8	10
Rivers	206	59	14	133
Gombe	161	122	6	33
Sokoto	116	96	14	6
Plateau	105	53	2	50
Kwara	88	37	1	50
Delta	83	17	8	58
Zamfara	76	71	5	0
Nasarawa	62	18	2	42
Yobe	52	24	7	21
Akwa-Ibom	45	14	2	29
Osun	45	35	4	6
Ebonyi	40	8	0	32
Adamawa	38	20	4	14
Imo	36	14	0	22
Kebbi	33	29	4	0
Niger	32	9	1	22
Ondo	25	20	2	3
Bayelsa	21	7	1	13
Ekiti	20	16	2	2
Enugu	18	12	0	6
Taraba	18	10	0	8
Anambra	11	3	1	7
Abia	10	3	0	7
Benue	7	1	0	6
Kogi	2	0	0	2

(continued)

Table 2 (continued)

States	Confirmed cases	Discharged cases	Deaths	Total active cases
Total	10,162	3007	287	6868

Source NCDC (2020a, b)

it a population density of 3508.42 persons per square kilometer. Available evidence shows that infection rates increase with increasing city size (Stier et al. 2020).

The population argument also applies to Kano. The 2006 population census shows that the state had the largest population in the country, and therefore, it is a major population center in Nigeria. The state has a total population of 13,076,892 (2016 projection) and a land area of 20,131 km. Its population density is 649.59 persons per square kilometer, which is much smaller than that of Lagos State, but higher than that of other states in the north. Besides, it is the commercial node of northern Nigeria and "… also a major transit point between (sic) other major urban centers in northern Nigeria. For instance, a federal highway connects Kaduna to Kano via Zaria. The Katsina-Maiduguri road passes through Kano. A rail line which originates from Lagos … goes through Kaduna, Zaria to Kano and ends in Nguru (Yobe state)" (Osayomi and Areola 2015, p. 96). It is therefore an important destination and transit point in that part of the country.

As indicated earlier, Lagos State, Kano State and the FCT (Abuja) have the largest number of active cases in the country. Together, these three account for 75.6% of active cases. Kebbi and Zamfara had no active cases due to the high recovery rates reported there. The distribution of discharged cases is slightly different from the former two. Although Lagos and Kano top the list of discharged cases, Bauchi recorded the third highest number of discharged cases in Nigeria. The numbers of discharged cases are comparatively large in the north, particularly Bauchi, Borno, Jigawa, Kaduna, Gombe and the FCT, while fewer numbers are reported in the south. Like discharged cases, the pattern of deaths associated with COVID-19 is somewhat different from those of confirmed and active cases. Again, Lagos and Kano have the largest number of fatalities in the country followed by Borno, the hotbed of the Boko Haram insurgency. Imo, Enugu, Taraba, Benue and Kogi States had no confirmed deaths arising from COVID-19 as at May 31, 2020.

As expected, there is some association between confirmed, active, discharged cases and deaths in the country. A correlation analysis was used to establish the association. The coefficients of correlation among these COVID-19 indicators reveal a positive and very strong association among them (Table 3). The strongest association is between confirmed and active cases ($r = 0.997$) while the weakest, in relative terms, is between deaths and active cases ($r = 0.775$). The comparatively weak association may be due to the fact that the preexisting health conditions of patients affect mortality from COVID-19. Therefore, the number of deaths is not necessarily proportional to the number of active cases. If a small population of active cases contains many patients with preexisting conditions, the number of deaths will be higher than in

Table 3 Correlation coefficients

	Active	Confirmed	Deaths	Discharged
Active cases		0.997**	0.775**	0.927**
Confirmed cases	0.997**		0.801**	0.951**
Deaths	0.775**	0.801**		0.849**
Discharged cases	0.927**	0.951**	0.849**	

a large population of active cases in which very few patients have preexisting conditions. The positive associations are not counterintuitive. For instance, all things being equal, it is expected that as the number of confirmed cases increases, the number of active cases will also increase. The same logic applies to the association between discharged and confirmed cases, and so on.

Another issue of interest is the growth in the aggregate number of confirmed cases over the months. From February 27, 2020 when COVID-19 was first discovered, the number of confirmed cases increased to 81 on March 27, 981 on April 27, and thereafter, it increased rapidly to 6175 on May 18, and 10,162 on May 31, 2020. Figure 3 indicates that COVID-19 has passed the early stage of the diffusion process in Nigeria and is now on an upward trajectory. It approximates an S-shaped curve that has not flattened out, indicating that the spread of the disease is not yet slowing down in the country. The exponential increase in the number of confirmed cases is largely due to the continued expansion of the network of testing laboratories and increased testing capacity across the country. So far, 63,882 samples had been tested as at May 31, 2020 (NCDC 2020a, b), up from 15,759 on April 30. Even the 63,882

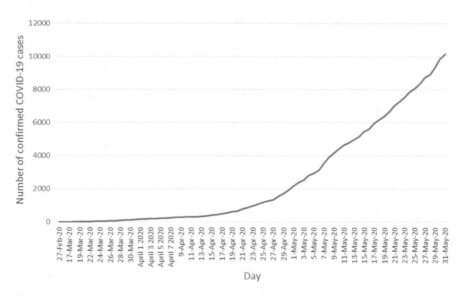

Fig. 3 Growth trend of COVID-19 from February 27 to May 31, 2020

tests are very minuscule in a country with an estimated population of about two hundred million persons. The potential for an explosive growth in the number of confirmed cases is a worrying prospect.

Policy Response

Following the emergence of the country's first case, the Federal Government of Nigeria set up a 12 member Presidential Task Force on COVID-19 (PTF-COVID19) on the March 9, 2020. The PTF-COVD19 is "… the central coordinating body on the COVID-19 response" charged with the responsibility of "developing a workable National Response Strategy …" (Buhari 2020). On March 18, the Federal Government of Nigeria imposed travel bans on thirteen countries that were considered to be highly infected with the virus: USA, the United Kingdom, South Korea, Switzerland, Germany, France, Italy, China, Spain, Netherlands, Norway, Japan and Iran (Punch Newspaper 2020). Two more countries, Sweden and Austria, joined the travel ban list on March 20, 2020.

In addition to the travel ban, there was a temporary suspension of visa-on-arrival policy for the countries that enjoyed the privilege. In order to slow down COVID-19 infections, the country ordered the closure of all public and private educational institutions from primary (elementary) to tertiary levels (Punch Newspaper 2020). Nigerian airspace was closed to both international and domestic commercial flights while intra- and inter-state road transport was prohibited or restricted. In addition, workplaces (except essential services) were shut for a while, and a ban was placed on religious gatherings, weddings and funeral ceremonies. These and some other measures were put in place in a bid to restrict the movement or congregation of people, which are important means of disease diffusion. As well as measures designed to restrict movement, social distancing and use of face masks were encouraged or mandated in different states.

The private sector also joined in the effort to contain the pandemic. Operatives in the sector formed the Coalition Against COVID-19 (CACOVID-19) at the onset of the pandemic . The cardinal objective of CACOVID-19 is to "mobilize private sector leadership and resources to support health facilities to respond to the crisis but also to use their reach to increase awareness about the pandemic." By April 6, the CACOVID-19 had raised over $55.7 m (African Business Magazine 2020).

One area where both the national and sub-national governments have been active is in the provision and upgrading of infrastructure, especially isolation centers and testing laboratories. For instance, at the onset of pandemic, the number of testing laboratories in Nigeria was just four. The NCDC and some state governments have rapidly expanded the network of molecular testing laboratories. Presently (July 1, 2020), there are forty laboratories across the country and another one is under construction (see Fig. 4). Southern Nigeria has a slightly greater share of the laboratories (21), while Lagos and Kano States have the largest number (5 each). The testing laboratories are still inadequate both in terms of numbers and their geographical distribution.

Fig. 4 Distribution of NCDC molecular testing laboratories in Nigeria as at July 1, 2020. *Source* NCDC

Many states in the middle belt of Nigeria as well as some states in the far north have none. This creates problems of geographical access as there is no income barrier to access. The tests are free of charge. The accessibility problem is partly responsible for the number of confirmed cases in Nigeria which is considered low against the backdrop of the country's population size. As at July 1, 2020, 141,525 samples had been tested in the country (NCDC 2020a, b). For a country with a population of about 200 million, this number is still very small.

Attitudes, Perceptions and Behaviour

The perception of a disease in a population influences its health seeking behaviour. Understanding public perceptions of COVID-19 and the lived experiences of people in the pandemic situation certainly has implications for the containment of the disease. This informed a collaborative study of risk perceptions of COVID-19 in Nigeria. The United Nations Development Programme (UNDP) Nigeria Office, in conjunction with NOI Polls carried out a telephone based survey of 5554 Nigerians across the six geopolitical zones of the country (UNDP 2020). The results of the poll revealed that 99% were aware of the disease; 73% were of the view that anyone is likely to be infected; 61% said the Federal Ministry of Health and the NCDC were the main sources of information on COVID-19 and 91% showed familiarity with the government and UN recommendations concerning preventive measures.

Twenty-five percent of the respondents believe that they had immunity from the disease. Most of those who held onto this belief within the different zones were in the southeast (40%) and southwest (31%); further down the scale are northwest (24%), south-south (23%), north-central (20%) and northeast (18%). This position on immunity is very puzzling and could be a reflection of religious and cultural influences (UNDP 2020), but it can lead to cavalier attitudes that may facilitate the spread of the pandemic. With respect to prevention methods, 78% washed hands with soap and water; 57% stayed at home, and 50% practiced social distancing. More of the respondents claimed to observe social distancing in the southwest (59%), northeast (58%) and north central (52%) than in the south-south (48%), northwest (48%) and southeast (34%). The challenges encountered by the respondents in implementing government/NCDC recommended protective measures include the unavailability of face masks and lack of financial resources.

The study revealed that the impacts of the pandemic identified by the respondents include slow business activity, less money, high food prices, hunger and lack of food as well as restriction of movement, restriction of religious activity, becoming closer to God and more family time. The impacts are both social and economic in nature. Concerning what government should do to stem the spread of COVID-19, the responses were largely in favour of the enforcement of containment measures (e.g., lockdowns) as well as increased testing. However, the responses also include prayers and provision of welfare support for citizens. In an economy with a large informal sector and in which many people are daily wagers, it is very difficult to observe lockdowns and other restrictions because it implies loss of income.

A recent BBC News Africa video documentary (BBC 2020) on COVID-19 shot in Lagos revealed the skepticism surrounding the disease among some Nigerians. The opinions of some of the people interviewed are reflections of ignorance and the trust deficit that bedevils both national and sub-national governments in Nigeria. One of the respondents stated that "*Coronavirus is abroad and not here.*" Others interviewed were of the view that there was no concrete evidence of its existence; if it was present in Nigeria, people would have been dying everywhere. One viewpoint is that COVID-19 is being exaggerated, and another, that it was a means of defrauding the citizenry of taxpayers' money and so on. As rightly observed by the BBC anchor, these opinions and perceptions mirrored the "historical mistrust between some Nigerians and the government."

The perceptions and attitudes of sections of the Nigerian population to the pandemic also extend to widespread myths concerning preventive/curative measures. The NCDC and some other government agencies continually undertake public education on several platforms in a bid to debunk the myths and misconceptions about COVID-19 (The Punch Newspaper 2020). Some of the myths include the following: drinking hot water, gargling with salt solution, drinking the fluid obtained by boiling pineapple peels, lime and ginger in water, drinking alcohol, drinking palm oil, consuming garlic and ginger, use of chloroquine and antibiotics, and spreading onions around homes. These myths influence behaviour, and as a result the use of face masks, social distancing, hand washing (or use of alcohol-based sanitizers) may not be adhered to religiously by some people.

Conclusion

COVID-19 has spread across Nigeria and no state is free of the pandemic presently as Cross River State now has active cases too. The infection started in Ogun/Lagos States, and leapfrogged to some other states, especially those with large cities that are at the top of Nigeria's urban hierarchy. The July 26, 2020, data from NCDC indicate that the total number of confirmed cases in the country were 40,532, discharged cases were 17,374 and deaths were 858. With 14,456 cases, Lagos remains the prime epicenter of the pandemic in Nigeria, followed at considerable distance by FCT (Abuja), with 3486 cases, Oyo State, with 2570 cases, and Edo State, with 2167 cases. Kano dropped to the sixth position in the league table of confirmed cases. It had 1520 cases on July 26. It is believed that these numbers are not a true reflection of the magnitude of the problem because of the limited testing capacity in the country. On July 26, there were only 59 laboratories and 12 GeneXpert laboratories in the country spread across thirty states and the FCT (Abuja). Six states had no testing facilities. At the same date, the total number of samples tested in Nigeria was only 266,323 and many rural communities are not captured in the data. The burden of COVID-19 is much higher in southern Nigeria than in the north. The July 26 data indicate that the southern states/geopolitical zones account for 70.19% of the confirmed cases. This is probably due to differences in levels of urbanization which are much higher in the south.

Apart from the early phase of the pandemic in the country, there are hardly any more imported cases because of the closure of Nigeria's air space and land borders. Since the closures, community transmission took over, and therefore, contagious and other forms of expansion diffusion are now prevalent. With the easing of the lockdown and of restrictions on intra- and inter-state movements, it is feared that the number of cases might increase because of increased migration and intra-city movements. The policy of closing internal and external borders was no doubt informed by knowledge of the effect of barriers in the diffusion process. Unfortunately, the diffusion curve is not yet tapering off, indicating that the diffusion of COVID-19 has not started slowing down. This situation suggests two things. First is the need to expand the number and network of laboratories so as to improve access and ramp up testing. Secondly, given the perceptions, myths and misconceptions of people concerning the disease, the need for more aggressive public education is urgent. This will lead to change of attitude and make more people embrace appropriate preventive measures.

References

African Business Magazine. (2020, April 14). Nigerian private sector donates more than most other African countries in fight against COVID-19. https://africanbusinessmagazine.com/sec tors/health-sectors/covid-19/nigerian-private-sector-donates-more-than-most-other-african-cou ntries-in-fight-against-covid-19/. Accessed online July 5, 2020.

BBC. (2020). Coronavirus in Nigeria: Inside a Lagos Corona virus ward. The child beggars at the heart of the outbreak. https://www.bbc.com/news/av/world-africa-53333056.

Bjelland, M. D., Montello, D. R., Fellmann, J. D., Getis, A., & Getis, J. (2013). *Human geography: Landscapes of human activities* (12th ed.). New York: McGraw Hill.

Buhari, M. (2020). Address by his excellency Muhammadu Buhari, President of the Federal Republic of Nigeria on the COVID-19 pandemic, Sunday, March 29, 2020.

Johnston, R. J., Gregory, D., Pratt, G., & Watts, M. (Eds.). (2003). *The dictionary of human geography* (4th ed.). Oxford: Blackwell Publishing.

Meade, M., & Emch, M. (2010). *Medical geography* (3rd ed.). New York: Guildford Press.

National Population Commission. (2006). *2006 population and housing census of the Federal Republic of Nigeria* (Vol. II).

NCDC. (2020a, May 31). *COVID-19 situation report no. 93*. Abuja: Nigeria Centre for Disease Control.

NCDC. (2020b, July 1). *COVID-19 situation report no. 124*. Abuja: Nigeria Centre for Disease Control.

Osayomi, T., & Areola, A. A. (2015). Geospatial analysis of road traffic accidents, injuries and deaths in Nigeria. *Indonesian Journal of Geography, 47*(1), 88–99.

Oshodi, L., Salau, T., Olatoye, M. S., & Unuigboye, R. E. (2018). Urban mobility and transportation. In F. Hoelzel (Ed.), *Urban planning processes in Lagos: Policies, laws, planning instruments, strategies and actors of urban projects, urban development and urban services in Africa's largest city*. Zurich: Heinrich Boll Stiftung.

Punch Newspaper. (2020, May 13). Video: NCDC debunks popular myths about COVID-19. https://healthwise.punchng.com/video-ncdc-debunks-popular-myths-about-covid-19/.

Sabel, C. E., Pringle, D., & Schaerstrom, A. (2010). Infectious disease diffusion. In T. Brown, S. McLafferty, & G. Moon (Eds.), *A companion to health and medical geography* (1st ed.). West Sussex: Wiley-Blackwell.

Stier, A., Berman, M., & Bettencourt, L. (2020). *COVID-19 attack rate increases with city size* (Mansueto Institute for Urban Innovation Research Paper). Available at DDRN: https://ssrn.com/abstract=3564464.

UNDP. (2020, May 17). *The COVID-19 pandemic in Nigeria: Citizen perceptions and the secondary impacts of COVID-19* (Brief 4).

Americas

Political Miscalculation: The Size and Trend of the COVID-19 Pandemic in Mexico

María Eugenia Ibarrarán, Tamara Pérez-García, and Romeo A. Saldaña-Vázquez

Abstract The COVID-19 pandemics have had a significant effect in Mexico, in terms of cases, deaths, and economic impact by various standards. Within Latin America, Mexico is the country with highest lethality. We analyze the differences in the trends of the pandemics at the national and state levels and relate them to variables that have been posed as driving forces of the COVID-19 contagion velocity, namely population density, water security, and the adoption of public policies intended to curb transmission. We run statistical analysis to establish relationships among these variables. Results show that COVID-19 effects on Mexico are strong because only a few of the policies that promote the reduction of COVID-19 contagions were adopted at the national and state levels and with little enforceability. Thus, the policy variables have had no effect on the COVID-19 cases reported. Therefore, the pandemic is far from being controlled. Policies will only be useful if most of the citizens abide by them and if the government enforces their adoption. People will only follow policies either if they are mandatory or if their basic needs are met, i.e., with a strong government commitment to protect their jobs and income. Finally, people follow leaders, so a consistent position between the federal authorities and local governments is necessary to make civil society commit to attend government requirements.

Keywords Mexico · COVID-19 · Lethality · Economic impacts · Subnational

Introduction

The SARS-CoV2 virus and the COVID-19 disease, which at the beginning of 2020 seemed so far from the Mexican reality, was identified for the first time in the country on February 27, in Mexico City (Vital Signs 2020). The pandemic contagion velocity was slow for some months, but in the summer of 2020, it skyrocketed and there is little evidence that numbers are truly falling, as opposed to what the federal and local governments have stated. Mexico is now the third with most accumulated deaths

M. E. Ibarrarán (✉) · T. Pérez-García · R. A. Saldaña-Vázquez
Xabier Gorostiaga SJ Environmental Research Institute, Universidad Iberoamericana Puebla, San Andrés Cholula, Mexico
e-mail: mariaeugenia.ibarraran@iberopuebla.mx

worldwide as of June 17th, and the second in Latin America, only behind Brazil (COVID-19 Observatory 2020). Using other standards, Mexico is 8th with most cases following the USA, India, Brazil, Russia, and Colombia. When comparing the number of cases per million inhabitants, Mexico ranks 64, and 9th in the case of deaths per million (Roser et al. 2020).

Within Mexico, different states have adopted policies at various stages, hoping to lower contagion rates. In this chapter, we analyze the differences in the trends of new cases, active cases, and deaths at the national level and state levels and relate them to variables that have been posed as driving forces of the COVID-19 contagion velocity, namely population density, water security, and the adoption of public policies intended to curb transmission (Hsiang et al. 2020). We run multivariate analysis tests and establish relationships among these variables that lead to several public policy recommendations.

This chapter is divided into four parts. Section "An Overview of the Pandemics" describes the evolution of the pandemics at the national level, the specific conditions of the country, and the federal policies adopted and their timing. Section "Trends at the Subnational Level" describes what has happened at the state level and analyzes possible driving forces that may have created regional differences in the behaviour of contagion and deaths of COVID-19. Section "Testing the Effect of Social and Policies Variables on COVID-19 Outcomes" shows results for our analysis among contagion, active cases, and death rates, and the suspected driving forces. We finally present a brief conclusion and policy implications for further handling this pandemic in Mexico.

An Overview of the Pandemics

The curve of the pandemics in Mexico has been increasing steadily since March 2020. For several weeks, the curve was rather flat, and both contagion and deaths attributed to COVID-19 were low. This led to a false complacency of the federal government that the epidemic would not be so harsh on the country. Time and again, the government has not paid attention to scientific evidence and this has reflected in inappropriate and late implementation of policies that have in turn led to Mexico being among the first places in contagions and deaths as of the summer of 2020, only behind the USA and Brazil (Fig. 1).[1]

For planning purposes, the federal government, through the Ministry of Health, recognized three epidemiological phases. The first was the importation of cases that lasted 27 days from the appearance of the first case. During Phase I, isolated cases appeared in several states, and infected people allegedly contracted the virus outside the country (Government of Mexico 2020b). Phase II, declared on March 24, was of community transmission, this meant that the infections began to occur locally,

---

[1]In late August 2020, India became third and Mexico dropped to the fourth place, but India's population is more than 10 times that of Mexico.

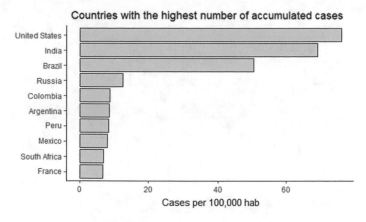

Countries with the highest number of accumulated cases

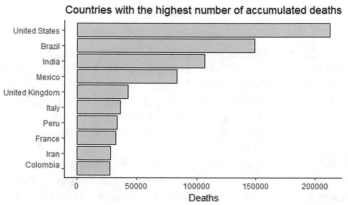

Countries with the highest number of accumulated deaths

Fig. 1 Trends of the pandemics. *Source* Owned based on Roser et al. (2020)

and during this period, the first sanitary protocols were applied (Ministry of Health 2020a). Schools stopped at all levels and non-essential economic activities were halted. Mobility also dropped and people were asked to stay home. Phase III began on April 21, when there were active sprouts in communities and more than 1000 cases within the country. At this stage, the number of cases kept increasing, which led to the strengthening of protection measures and protocols (Ministry of Health 2020b). The main actions undertaken at that time were social distancing and a stay-at-home policy (Government of Mexico 2020a). Figure 2 presents a timeline describing the adoption of policies.

In March, public events were canceled as well as school activities, and information campaigns with a one-hour televised nightly and daily conference on the pandemic, describing its trends, ways to control it and other related topics. The stay-at-home recommendation started on March 23 and since then has not been lifted, but it is weakly enforceable due to the economic situation and to the right to mobility. Schools and non-essential businesses were closed in March, and businesses started opening

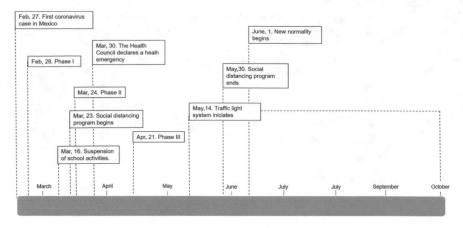

Fig. 2 Timeline of policies. *Source* Own elaboration

again in mid-May; as of August, many businesses operate at least partially, restaurants are take-out only, and bars and clubs open in some states. Schools are set to remain closed until probably October, depending on a regional "traffic light" that will be explained below. Outside activities were significantly reduced on April 6, by the mandatory closing of hotels and resorts. This was at the time of the Holy Week holiday, so it had a significant effect reducing travel. Specific actions designed under each phase are shown in Table 1.

The COVID-19 Observatory for Latin America, based in the University of Miami, reports a Public Policy Adoption Index at the federal and state level for Mexico, together with other Latin American countries. They include closing of

Table 1 Actions under the contingency phases

Phase I	Phase II	Phase III
– No constraint on social contact (greetings, massive events in open and closed spaces) – Preventive messages in schools and workplace – Basic hygiene measures	– Hiring of additional health personnel – Preparing areas for COVID patients – Purchase of medical material, medicines, and equipment – Social distancing campaign – Strengthening of hygiene measures – Suspension of school attendance – Suspension of public/private events with more than 100 participants	– Extension of the social distancing campaign – Temporal suspension of all non-essential activity in private and social sectors – Suspension of activities in public spaces

Source Owned, based on Ministry of Health (2020a, b, c)

schools and non-essential businesses; suspension of open-air activities; cancelation of public events; suspension of public transport; deployment of informative campaigns; constraints to mobility within the state; constraints to international traveling; stay-at-home orders; constraints to the size of gatherings; and mandatory use of facemasks. The Observatory compares the overall Public Policy Adoption Index across countries as well as the mobility reduction, use of facemasks, tests per 100,000 inhabitants, and the positive confirmed cases over the total tests. This latter indicator may summarize the effect of the adoption of the other policies (COVID-19 Observatory 2020).

Even though Mexico has adopted policies, the pace and enforcement has not been adequate. The countries with more policies and more strictly enforced are Honduras, Guatemala, Bolivia, El Salvador, Argentina, Colombia, Dominican Republic, Ecuador, Panama, Paraguay, and Peru. Mexico (together with Brazil and Nicaragua) are the countries with the weakest enforcement (COVID-19 Observatory 2020). This can be seen in Fig. 3.[2]

Thus, from this data, it can be inferred that Mexico had time to plan before hand, by observing what was happening in other places, namely China, Europe, and the USA. It was not until late March that it started buying equipment and converting hospital. As of September 2020, of all the infected by COVID-19, 17% were health providers, and so were 1320 of the deceased, thus surpassing countries such as Brazil, the USA and the United Kingdom (El Universal 2020). Mexico is, therefore, the country where health practitioners are most at risk.

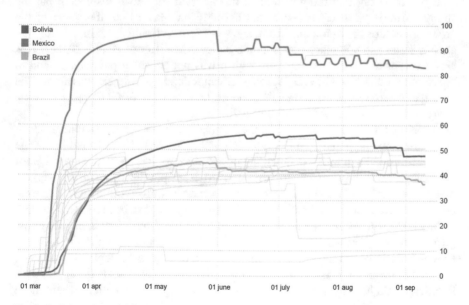

Fig. 3 Policies adopted: Mexico and other Latin American countries. *Source* COVID-19 Observatory (2020)

---

[2]For further detail visit https://observcovid.miami.edu/.

Additionally, the outset of the COVID-19 pandemics happened at the same time there was a profound conversion of the medical public system, that became federalized under the National Health Institute for Wellbeing (INSABI), in the first quarter of 2020. Over 50% of the population depend on public health services in Mexico (INEGI 2015a) and were most likely affected by such change.

Lethality and Informality

Even though Mexico ranks high in the cases of COVID-19 and deaths, there are strong arguments supporting that there is a severe under-reporting. This is the case for many countries, though. However, when comparing accumulated cases with accumulated deaths, with both sets of data from the same official source, Mexico reports that the lethality rate, i.e., the number of deaths as compared with contagion, is 11% (González 2020). This rate is higher than that of the USA and Brazil.

High lethality may be explained by several reasons. One is the very large informal sector where anywhere from 40 to 60% of the population participates. They have no unemployment insurance and no way to earn a living other than working. Social distancing and particularly a stay-at-home policy are impossible for them to survive money-wise.

It is fair to say that the government has advised for social distancing but the stronger campaign to enforce it ended abruptly on May 31, in the midst of the growing cases of contagion and death, and the "New Normality" set in, relaxing the already weak and voluntary stay-at-home policy[3] (Ministry of Health 2020c). This was done for economic reasons, although it was not accepted publicly. To add to the crisis associated to COVID-19, the Mexican federal government was among the ones that offered the least economic support, less than 1% of GDP, regardless the many voices requesting it (Vital Signs 2020).

Additional to the demographics of the population, the fact that a significant share of the working age population is in the informal sector has also reflected in the age structure of deaths in Mexico. Often the informal sector has been a scape valve for unemployment in the formal sector in times of crisis, employing those formal workers that lose their job. This time, the informal sector has also contracted; therefore, unemployment is soaring. About 1.1 million formal jobs have been lost from March to July 2020 (Flores 2020), and an estimated 5.4 million have lost their jobs in the informal sector (Tourliere 2020). Thus, the fact that government support to maintain economic activity and income is missing, the lack of unemployment insurance, and the need of people to make a living, has thrown people to the streets to support their families.[4] This has led to a different pattern of vulnerable population, where those

---

[3]Federal authorities have not enforced a mandatory constraint to stay home or wear facemasks because of their allegedly commitment to human rights.

[4]This can also be explained by the age pyramid of the population in Europe.

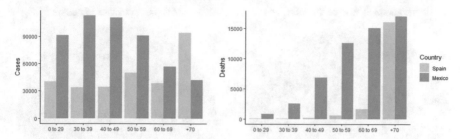

Fig. 4 Vulnerability of age groups. *Source* Own based on RENAVE (2020a, b) and Government of Mexico (2020a)

in their 30s and 40s are more exposed and face higher death rates that the elder in European countries like Spain, as Fig. 4 shows.

One added factor is that it was only in July that the federal government shyly recommended the use of facemasks as a cautionary measure. On this end, there has been great ambiguity by the government. On the one hand, there is information on the official website on the use of the mask, and how to wear it, and authorities often make it a mandatory policy to use it for public spaces and public transport, but at the same time in the daily conference, they say that it is not proved that it is an effective protection. Even the Mexican Nobel Laureate Dr. Mario Molina has appeared in social media and news channels recommending the use of facemasks to put pressure on the federal government.

However, the President of Mexico, together with those of the USA and Brazil, is among the ones that do not use them. Andrés Manuel López Obrador is quite popular among the voters and has over 50% approval. His actions are closely followed by a vast majority of the people in the country. This has turned into a significant controversy, since many of his followers, who need to get out to work daily, do not use facemasks either. This is at odds with some local governments that have high contagion rates in their states and are eager to enforce the use of facemasks as a protective measure.

Co-morbidities

Another important factor affecting Mexico is co-morbidities. The prevalence of these even from a young age maps out vulnerability in a very different way. Thus, this again alters the expected outcome of the elder being the highest risk group and the lethality rate.

The Mexican population is to a great extent plagued with obesity, type II diabetes, high blood pressure, and other cardiovascular and respiratory illnesses. Smoking is also a risk factor. These co-morbidities have led to higher probability of developing severe symptoms due to COVID-19. Since they are not constrained to the elder, but

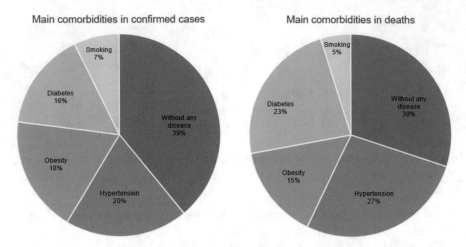

Fig. 5 Incidence of COVID-19 and other co-morbidities. *Source* Own, based on Government of Mexico (2020a)

across the population, the prevalence of cases and the incidence of death are also spread more evenly across the population.

Overall, this leads that of the total number of deaths confirmed by COVID-19, 73% are associated with at least one comorbidity. Of the latter, 67% have to do with either hypertension, diabetes, obesity, or cardiovascular disease and 6% with other causes (Ministry of Health 2020d). This is shown in Fig. 5.

This section has reviewed and synthesized the main trends and their causes at the national level of the COVID-19 pandemics in Mexico. In the next section, we will analyze what is happening at the state level and identify variables that can explain those regional trends.

Trends at the Subnational Level

Even though the Mexico is a federal republic with a strong central government, states dictate some of their own policies. Some state governments, primarily from opposition parties to the President, have stood their ground and managed the pandemics by differentiating their policies. Some have adopted policies earlier or have enforced them to a larger extent. In some cases, this has played a role in reducing the rate of contagion and deaths. Additionally, this may have had differentiated economic effects throughout the country. We now turn to analyzing the adoption of policies at the subnational level.

The Policies

States have adopted roughly the same policies, namely closure of schools and non-essential activities, stay-at-home requests, calls for reductions in mobility, the use of face masks, and testing for SARS-COV2, among others. The timing and enforcement of these policies has also been measured by the COVID-19 Observatory at the state level for several Latin American countries, Mexico being one of them. In this section, we discuss the COVID-19 Public Policy Adoption Index for all the 32 states of Mexico and then see their relation to cases and deaths. We also look at individual policies adopted that may help explain such differences.

According to COVID19 Observatory, the aggregate index of adoption of public policies is highest in Durango, Nayarit, Jalisco, Nuevo Leon, and Tamaulipas. The states with the lowest adoption are Campeche, Coahuila, CDMX, Querétaro y Quintana Roo (COVID-19 Observatory 2020). This index is shown in Fig. 6.[5]

Further disaggregation of the index is possible into the specific policies that each state has adopted. We will use this information to explain the links between policies and cases of contagion and death at the state level in Section "Testing the Effect of Social and Policies Variables on COVID-19 Outcomes".

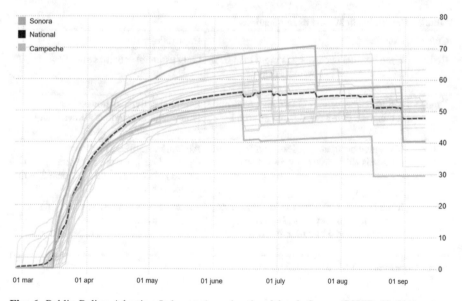

Fig. 6 Public Policy Adoption Index at the subnational level. *Source* COVID-19 Observatory (2020)

---

[5]For further detail visit https://observcovid.miami.edu/mexico/.

A Four-Colored "Traffic Light"

An interesting tool that the Mexican government has implemented is a "traffic light" using red, orange, yellow, and green. These colors are related to the occupancy rate of beds with ventilators and the number of cases registered the previous week. The color of the "traffic light" is determined on a weekly basis by federal and state authorities, and depending on the color, additional activities are allowed. This is shown in Table 2. Schools are not allowed to open until there is a green light and occupancy of schools and businesses at that stage can only be at 30%.[6] States and counties are given a color, and as there are achievements, the color changes from red, to orange, yellow, and green. Setbacks can make the entity fall into a previous color with the implications this may have on economic activity and stronger stay-at-home recommendations, that in turn affect the economic and physical survival of millions.

Based on this information, the country started in May as mostly red, with Zacatecas being orange. By June, all states were red. Some states have turned orange in August and yet some yellow in early September. No state is green, as of mid-September.[7]

We now look at the policies adopted at the state level and see if they make a difference in terms of deaths and contagion.

Table 2 The "traffic light" and its implications

Color	Conditions	Implications
Red	If the hospital occupancy is more than 65% and if there are two weeks of stable increase	Only essential activities are allowed
Orange	If hospital occupancy is less than 65% and if there is a two-week downward trend	Essential activities allowed Non-essential activities allowed with capacity of 30%
Yellow	If hospital occupancy is less than 50% and if there is a two-week downward trend	All work activities are allowed Open public space opens regularly, and closed public spaces can be opened with reduced capacity
Green	If the hospital occupancy is less than 50%, at least one month with stable low occupancy, and a one-month downward trend	All activities are allowed, including school activities

Source Government of Mexico (2020c)

---

[6]In August, a new "traffic light" system was discussed between federal and state authorities, but it has not been agreed upon; thus, it is not further discussed.

[7]Due to disagreements between the states and the federation, the indicators leading to the color of the traffic light are being revised.

Testing the Effect of Social and Policies Variables on COVID-19 Outcomes

We divide our analysis into two parts. In the first one, we evaluate the effect of population density and water security on the accumulated cases of contagion and deaths in Mexico due to COVID-19. We expect that states with lower population density and more water security have lower accumulated contagions and deaths per 100,000 habitants. Population density can be a proxy for exposure, while water security can contribute to lower contagion as has been indicated by the recommendation to escalate hygiene as a way of prevention.

In the second analysis, we evaluate the effect of several policies, i.e., use of masks, number of tests for SARS-COV2, and mobility reduction in each state. We expect that their adoption reduces the number of active cases of COVID-19 in Mexico. These policies are related with the reduction of COVID-19 contagion velocity in other countries (Hsiang et al. 2020).

Policy variables were obtained from the Observatory for the Containment of COVID-19 in the Americas at the state level, while the number of contagions, deaths, and active cases come from the Mexican official site for COVID19-related health statistics (https://coronavirus.gob.mx/). Both sets of variables were accessed at similar times. The social variables are population density (INEGI 2015b) and the water security index (IMTA 2017). In both exercises, social and policies variables were analyzed using generalized linear models (GLMs). This statistical procedure allows to test for correlations among variables of different kinds, count, and continuous.

In the first analysis, the predictive variables were population density and the water security index, and the response variables were the number of contagions cases and deaths by 100,000 inhabitants, respectively. Results shown that social variables have no effect over the number of contagions or on deaths. Other studies that evaluated the effect of population density of cities in different countries have shown that this variable is not related with confirmed cases (Bai et al. 2020), so our results confirm that population density is not a strong variable to predicts COVID-19 contagions and deaths in Mexico at the state level. No relationship was found either between water security and contagions and deaths.[8] The water security index has a mean of 0.70 and standard deviation of 0.11, showing low variation among the states in terms of water security.[9]

For the second analysis, we tested the effect of policies, namely the percentage of people that use masks, the percentage of mobility reduction, the number of tests by 100,000 inhabitants, on the number of active cases of COVID-19 per 100,000 habitants. Results show that only the number of tests was positively related with the number of active cases ($\beta = 0.042$, P-value $= 0.02$).[10] The null effect of the use of

---

[8]Results are not included but may be requested to the authors.

[9]Water security is quite debatable and official numbers often confront local perceptions.

[10]Results may be requested from the authors.

masks and mobility reduction on the number of active cases could be explained by the low adoption level of such policies. For example, on average, Mexican citizens had a reduction of 31% in their mobility with a standard deviation of 7%, showing a low reduction. Additionally, even though 69% of Mexican citizens allegedly adopted the use of face masks during this period, it is difficult to know how strongly this was enforced, if they used them during the entire time they should and correctly, or how compliant the face masks were in terms of quality standards. The fact is that these policies have not helped reduce the number of COVID-19 active cases. Finally, the positive relationship between the SARS-COV2 tests and active cases shown that tests are a good predictor of active cases, and therefore, if they are expanded, they may help identify contacts of those infected and isolate them. This in turn may help lower the number of active cases. Unfortunately, Mexico only applies 8.4 tests per 1000 inhabitants, the lowest in Latin America, of all OECD countries and in the lowest third of the world (Lozano 2020).

Even though the results we find are that policies seem to be meaningless, this certainly has to do with their timing and enforceability. Predictions by Washington University show that if sanitary measures are relaxed any further, Mexico could be facing 177 thousand deaths by early December that is 47 thousand more than under the current situation expected at that time, or over 56 thousand more as compared to the scenario where facemasks are generalized and properly worn (Reforma, August 23, 2020). Other studies suggest that if in March 2020, the Mexican government had promoted that 70% of Mexicans adopted the policies that led to the reduction of COVID-19 contagions, the peak of the pandemic would have arrived in June 2020 (Scudellari 2020), and its effects would have been much less devastating.

Conclusion and Policy Recommendations

This chapter shows that the COVID-19 pandemic is strongly affecting Mexican citizens, probably because only a few of the policies that promote the reduction of COVID-19 contagions were adopted at the national and state levels and with little enforceability. The policy variables, thus, have had no effect on the COVID-19 active cases reported. Therefore, the pandemic is far from being controlled.

Results presented here shed light on possible roads to be better equipped from what is to come. Policies will only be useful if most of the citizens abide by them and if the government enforces their adoption with sanctions to people that do not comply. People, on the other hand, will only follow policies either if they are mandatory or if their basic needs are met, i.e., with a strong government commitment to protect their jobs and income.

Finally, people follow leaders, particularly popular ones, so a consistent position between the federal authorities, particularly the President, and local governments is necessary, so civil society feel the commitment to attend government requirements.

In brief, Mexico was never over the first wave. Cases and deaths keep increasing due to the lack of congruent policies within the country. As of February 15, 2021, there is an accumulated 2 million cases of COVID-19 and over 175 thousand accumulated deaths officially reported (Government of Mexico 2020a). Due also to official statistics, the excess mortality for 2020 is of 326,609 deaths (Excelsior 2021).

New variants are not yet a significant factor in the country but will sure become one since government authorities have no intention of closing borders or requesting any tests or isolation of travelers from abroad (Fernández Menéndez 2021).

The vaccination campaign was announced at the beginning of December 2020. It is divided into five stages to immunize at least 70% of the population. During the first stage (December 2020- February 2021) health providers will be vaccinated. The second stage (February-April 2021) contemplates people aged 60+ years. The third stage (April-May 2021) will cover people from 50 to 59 years old, the fourth one (May-June 2021) people from 40 to 49 years old and from June 2021 to March 2022 the rest of the population will be vaccinated (Aristegui Noticias 2020).

Pre-purchase agreements were made with AstraZeneca for 77.4 million doses, with Pfizer for approximately 34.4 million doses, with CanSino Biologics for 35 million doses and according to the latest government announcements there will be 24 million doses of the Russian Sputnik V vaccine (Tamayo 2021). In addition, through the international COVAX mechanism, Mexico will have access to 51.5 million doses. However, after receiving four shipments from Pfizer in December, the government has had many complications in obtaining the vaccines due to the huge global demand. In Mexico, the shortage is so great that health personnel have not been able to receive the second dose necessary to be fully immunized (El País 2021). In addition to this, the rate of application of vaccines is slower than previously thought. For the first days of February, only 675,202 doses have been applied, with a rate of 4,728 daily applications (Tamayo 2021). The health providers have not been fully vaccinated and applications to older adults will begin 15 days after announced (El País 2021), which will lead to an obvious delay in the other stages.

References

Aristegui Noticias (2020). Así será en México el plan de vacunación contra el Covid-19. *Aristegui Noticias*. Retrieved from: https://aristeguinoticias.com/0812/mexico/asi-sera-en-mexico-el-plan-de-vacunacion-contra-el-covid-19/.

Bai, X., Nagendra, H., Shi, P., & Liu, H. (2020). Cities: Build networks and share plans to emerge stronger from COVID-19. *Nature, 584*(7822), 517–520.

COVID-19 Observatory. (2020). Observatorio para la Contención del COVID-19 en América Latina. Retrieved July 17, 2020, from https://observcovid.miami.edu/?lang=es.

El Universal. (2020, September 2). Mexico, country with the most deaths of medical personnel by COVID. Retrieved from https://www.eluniversal.com.mx/mundo/mexico-pais-con-mas-muertes-de-personal-medico-por-covid-amnistia-internacional.

El País (2021). México volverá a recibir vacunas contra la covid a partir del 15 de febrero. Retrieved from: https://elpais.com/mexico/2021-02-09/mexico-volvera-a-recibir-vacunas-contra-la-covid-a-partir-del-15-de-febrero.html.

Excelsior (2021). 2020, el año más mortal para México; hubo más de un millón de decesos. Retrieved from: https://www.excelsior.com.mx/nacional/2020-el-ano-mas-mortal-para-mexico-hubo-mas-de-un-millon-de-decesos/1432963.

Flores, Z. (2020, August 12). Mexico loses 1.1 million formal jobs so far in the pandemic, according to IMSS data. *El Financiero*. Retrieved from https://www.elfinanciero.com.mx/economia/mex ico-pierde-1-1-millones-de-empleos-formales-en-lo-que-va-de-la-pandemia-segun-datos-del-imss.

Fernández Menendez, J. (2021). Enfermos que no vuelan, brindis y entusiasmo de López-Gatell. *Excelsior*. Retrieved from https://www.excelsior.com.mx/opinion/jorge-fernandez-men endez/enfermos-que-no-vuelan-brindis-y-entusiasmo-de-lopez-gatell/1430815.

González, R. (2020, September 4). Mexico, first place in the world in fatality rate due to coronavirus. *Milenio*. Retrieved from https://www.milenio.com/ciencia-y-salud/mexico-primer-lugar-mundial-en-tasa-de-letalidad-por-coronavirus.

Government of Mexico. (2020a). COVID-19 Mexico. Retrieved from https://coronavirus.gob.mx/.

Government of Mexico. (2020b). Phases or contingency scenarios and level of spread of COVID-19 Mexico: COVID phases. Retrieved from https://educacionensalud.imss.gob.mx/es/system/files/Fases-COVID19.pdf.

Government of Mexico. (2020c). Traffic light system COVID-19. Retrieved from https://corona virus.gob.mx/semaforo/.

Hsiang, S., Allen, D., Annan-Phan, S., Bell, K., Bolliger, I., Chong, T., et al. (2020). The effect of large-scale anti-contagion policies on the COVID-19 pandemic. *Nature, 584*(7820), 262–267.

IMTA. (2017). Security water index. Retrieved from https://repositorio.imta.mx/handle/20.500.12013/1831.

INEGI. (2015a). Intercensal surveys 2015. Retrieved from https://www.inegi.org.mx/temas/derech ohabiencia/.

INEGI. (2015b). Population density by state. Retrieved from https://www.inegi.org.mx/app/tabula dos/interactivos/?px=Poblacion_07&bd=Poblacion.

Lozano, R. (2020). Millonario de pruebas y de infectados. *El Economista*. Retrieved from https://www.eleconomista.com.mx/opinion/Millonario-de-pruebas-y-de-infectados-20200810-0048.html.

Milenio. (2020, May 13). Coronavirus traffic light system in Mexico: This will be the return to the 'new normal'. Retrieved from https://www.milenio.com/politica/coronavirus-semaforo-reinicio-de-actividades-mexico.

Ministry of Health. (2020a). *Phase 2 begins by coronavirus COVID-19*. Mexico: Press. Retrieved from https://www.gob.mx/salud/prensa/095-inicia-fase-2-por-coronavirus-covid-19.

Ministry of Health. (2020b). *Phase 3 begins on COVID-19*. Mexico: Press. Retrieved from https://www.gob.mx/salud/prensa/110-inicia-la-fase-3-por-covid-19.

Ministry of Health. (2020c). Agreement establishing the specific technical guidelines for the reopening of economic activities. Retrieved from https://nuevanormalidad.gob.mx/files/Acu erdo_Salud_290520_VES-1.pdf.

Ministry of Health. (2020d). Informe Técnico 23 de julio. Retrieved from https://coronavirus.gob.mx/2020/07/23/conferencia-23-de-julio/.

RENAVE. (2020a). Informe nº 38. Situación de COVID-19 en España a 6 de agosto de 2020. Retrieved from https://www.isciii.es/QueHacemos/Servicios/VigilanciaSaludPublicaRENAVE/EnfermedadesTransmisibles/Documents/INFORMES/Informes%20COVID-19/Informe%20C OVID-19.%20N%c2%ba%2038_6agosto2020_ISCIII.pdf.

RENAVE. (2020b). Informe nº 30. Situación de COVID-19 en España a 11 de mayo de 2020. Retrieved from https://www.isciii.es/QueHacemos/Servicios/VigilanciaSaludPublicaRENAVE/EnfermedadesTransmisibles/Documents/INFORMES/Informes%20COVID-19/Informe%20n%c2%ba%2030.%20Situaci%c3%b3n%20de%20COVID-19%20en%20Espa%c3%b1a%20a%2011%20de%20mayo%20de%202020.pdf.

Roser, M., Ritchie, H., Ortiz-Ospina, E., & Hasell, J. (2020). Coronavirus pandemic (COVID-19). Published online at OurWorldInData.org. Retrieved from https://ourworldindata.org/coronavirus.

Scudellari, M. (2020). How the pandemic might play out in 2021 and beyond. *Nature, 584*(7819), 22–25.

Tamayo, K. (2021). La estrategia de vacunación contra COVID-19 en México: a tiempo de cambiar el rumbo. *Animal Político*. Retrieved from: https://www.animalpolitico.com/inteligencia-publica/la-estrategia-de-vacunacion-contra-covid-19-en-mexico-a-tiempo-de-cambiar-el-rumbo/.

Tourliere, M. (2020). Por Covid-19 6.5 millones de personas perdieron su empleo en México. *Proceso*. Retrieved from https://www.proceso.com.mx/634408/por-covid-19-6-5-millones-de-personas-perdieron-su-empleo-en-mexico-encovid-19.

Vital Signs. (2020). The pandemic in Mexico. Dimension of the tragedy. Retrieved from https://signosvitalesmexico.org.mx/reportes/.

Vitela, N. (2020). Liderea México en América Latina letalidad por COVID. *Reforma*. Retrieved from https://www.reforma.com/aplicacioneslibre/preacceso/articulo/default.aspx?__rval=1&urlredirect=https://www.reforma.com/lidera-mexico-en-al-letalidad-por-covid/ar1951152?referer=--7d616165662f3a3a6262623b727a7a7279703b767a783a--.

Evolution of the Coronavirus Disease (Covid-19) in Argentine Territory. Implications for Mental Health and Social and Economic Impacts

Daniel Oscar Lipp

Abstract The confirmation of a new disease called COVID-19 in Argentina was an outstanding novelty in the country, with accents of surprise, both for clinical and epidemiological reasons. The present situation of the disease according to the Ministry of Health of the Nation awakens the urgent need to respond with local inquiries to the knowledge of the dynamics of this pathology that at the moment is far from over. This document aims to make an analytical summary of what has happened so far in the country with COVID-19 and to take into account the results of monitoring the pandemic with the data obtained from official reports from the Ministry of Health. Aspects of its evolution, its spatial distribution and the temporal diffusion of COVID-19 will be investigated. In addition, relevant scientific evidence is added in relation to issues of mental health and social isolation that are emerging from this pandemic.

Keywords COVID-19 · Coronavirus · Epidemiology · Pandemics · Social isolation · Post-traumatic stress disorders

Introduction

Throughout the history of mankind, the human species has been invaded by numerous diseases that were often fatal. And although these diseases have not succeeded until today, even in their worst manifestations, to put an end to human beings as a living species, they can seriously damage a civilization and change the course of history. Careful hygiene measures and vaccination techniques have allowed those diseases to be controlled, thus preventing their spread. We might then believe that as long as our civilization survives and our medical technology does not falter, there is no longer the danger that a disease will produce a catastrophe like those of the past. However, I do not encourage myself to think in that sense. Today humanity is faced with the danger

D. O. Lipp (✉)
Universidad de Buenos Aires, Buenos Aires, Argentina
e-mail: daniellipp@arnet.com.ar

Universidad Católica de Salta, Salta, Argentina

© The Author(s), under exclusive license to Springer Nature Switzerland AG 2021
R. Akhtar (ed.), *Coronavirus (COVID-19) Outbreaks, Environment and Human Behaviour*, https://doi.org/10.1007/978-3-030-68120-3_21

of a new disease, COVID-19, highly contagious caused by an SARS-CoV-2, without yet finding a vaccine to eliminate it. From the beginning of this disease to the date of this report (07/29/2020), more than 16,000,000 people are infected worldwide and almost 179,000 in Argentina (CABA 2020).

The natural reservoir of the virus and the way in which it could be transmitted is unknown until today. Everything points to direct contact with infected animals or their secretions. The closest virus is the Bat CoV RATG13 isolated a couple of years ago from a horseshoe bat in Yunnan, southeast China. Bats harbor a great diversity of coronaviruses. For this reason, what is believed today or the most accepted hypothesis about the ancestral origin of SARS-CoV-2 is that a bat virus could have evolved into SARS-CoV-2, through intermediate hosts. The discovery of coronavirus in pangolins seized by the police in the Chinese provinces of Guangxi and Guandong has corroborated the idea that these animals could be said intermediate host (Ministerio de Sanidad 2020). However, more studies are needed in this regard.

The cases of COVID-19 occurred for the first time in Wuhan, China, were related to a locally known as wet market which probably suggests to us that the virus was initially transmitted from animals to humans. Transmission of the virus occurs through contact with infected secretions and/or through contact with surfaces contaminated by the virus. Researchers are still studying the spread of the virus among humans. The numbers indicate that SARS-CoV-2 is extraordinarily effective in human-to-human transmission, probably due to its incubation time that would reach 14 days after exposure to the virus, which provides it with great pre-symptomatic transmissibility. However, SARS-CoV-2 presents a fatality rate much lower than that of SARS-CoV and MERS-CoV, which is estimated at 2–4%, in addition to a low mutation according to the data accumulated in the already more than 850 sequenced genomes, which is undoubtedly very good news (Ramirez 2020).

Coronaviruses are a family of viruses that cause illness in man, affecting him with clinical pictures that range from the common cold to more serious ones. Only seven coronaviruses are known to cause disease in humans. Four of them, of the seven coronaviruses, cause symptoms of a common cold: types 229E and OC43 and serotypes NL63 and HUK1. The remaining three are the most serious and even fatal: SARS-Cov, MERS-Cov and SARS-Cov2 (Ministerio de Sanidad 2020).

SARS-CoV (severe acute respiratory syndrome-SARS) was initially detected in Guangdong province, China, in November 2002, from where it spread to more than 30 countries. More than 8,000 cases were reported worldwide, with 774 deaths mostly over the age of 65. But this outbreak subsided and no more cases of the disease were reported since 2004. The MERS-CoV (Middle East respiratory syndrome-MERS), on the other hand, was initially detected in Jordan and Saudi Arabia, which gave rise to its current name, in 2012 and later spread to neighboring countries. Cases have occurred in countries outside the Arabian Peninsula, such as France, Germany, Italy, Tunisia and the UK, in people who had been traveling or working in the Middle East. The virus does not appear to be easily transmitted from person to person unless there is close contact, for example when caring for a patient without proper protection. Already in 2018, there were 2220 confirmed cases of MERS and 790 deaths, mostly located in Saudi Arabia, where new cases of coronavirus continue to appear (Sachs 2020).

Materials and Methods

The methodology that has been developed in this work aims to determine the data provided by the COVID-19 disease in the country and in the City of Buenos Aires through qualitative–quantitative procedures and techniques. Line and column graphs were used as they are the most common and effective procedures for assessing disease progression and some other related data. In the case of the spatial distribution of the COVID-19 disease, a map of the Argentine territory has been used, discriminating it by province. Regarding the primary sources that have been used to carry out this work, there have been the Daily Report (morning and evening) of the COVID-19 disease of the Ministry of Health of the Argentine Nation[1] (Ministerio de Salud de la Republica Argentina 2020) and the Weekly Epidemiological Bulletins (BES) of the Ministry of Health of the City of Buenos Aires.[2] Regarding the axes investigated in the area of mental health associated with social isolation, we have resorted to the Applied Psychology Observatory of the Faculty of Psychology of the University of Buenos Aires (OPSA) Observatorio de Psicología Social Aplicada 2020)[3] whose main purpose is to collect information, in different sectors of the Argentine society (people and institutions), on a wide range of relevant psychosocial issues and problems, providing knowledge and understanding about them and enabling the development of diagnoses and approach strategies. It is noteworthy, on the other hand, that this study ranges from the first case registered in the country of COVID-19 (March 2020) until July 29, 2020. This allows us to compare data and compare them, improving their analysis.

Results

The source of information that has been used for this has been the Daily Report of the new coronavirus COVID-19 of the Ministry of Health of the Nation.[4]

On March 11, 2020, the World Health Organization declared the outbreak of the new coronavirus (COVID-19) a pandemic, and, as a result, our country expanded the Health Emergency and gave the Ministry of Health the power to take all necessary measures to minimize contagion and strengthen the response capacity of the health system. The approach was characterized by a graduated quarantine in flexible, early and strict phases and strict protocols of social distancing. Likewise, common provisions are established regarding groups at risk, limits to circulation, interurban and interjurisdictional public transport, controls carried out by the Ministry of Security, coordinated inspection to guarantee compliance with the established norms, infractions and verification of epidemiological and sanitary parameters.

---

[1] file:///C:/Documents%20and%20Settings/Master/Mis%20documentos/Downloads/15-07-20-reporte-matutino-covid-19.pdf.

[2] https://www.buenosaires.gob.ar/salud/boletines-periodicos/boletines-epidemiologicos-semanales-2018-2019.

[3] https://www.psi.uba.ar/opsa/.

[4] https://www.argentina.gob.ar/Coronavirus/informe-diario.

Temporary Evolution of Covid 19 in the Argentine Territory

The first confirmed case of the COVID-19 pandemic in the country was announced on March 3, 2020. In a few weeks, the SARS-Cov-2 virus has spread throughout almost the entire Argentine territory with the exception from some urban areas that were free of infection but only for a few days.

Figure 1 shows a graph of the evolution of the accumulated confirmed cases and deaths per day of COVID-19 during the period 3/3/20–7/29/20 in the Argentine territory. The number of confirmed cases rose to 178,966 in that period and in the case of the deceased to 3311. This period was accompanied by measures that the National State implemented as the disease progressed. As can be seen in the figure, on March 16, the teaching of classes in all primary, middle and university schools was suspended throughout the country, licenses and the possibility of remote work were also granted for the public and private sectors. These decisions were made with the purpose of limiting the spread of the virus and reinforcing care measures. Another important decision was to close all the country's borders. Already on March 20, as a consequence of the greater number of confirmed cases, social, preventive and mandatory isolation was decreed by the national government. This isolation was extended several times due to the advance of the disease in the national territory.

In an attempt to recognize stages during this period, it can be observed that from 3/3/20 to 4/26/20 the curve shows a very slight ascent, and it is enough to appreciate that the evolution line is very close to the abscissa of the graph with very little upward movement. Already from the day 4/26/20 and until the day 6/7/20, the increase in cases begins to be more evident, so we can assume the appearance of a second phase, since the curve detaches from the base at a greater angle. Finally, we recognize a third phase, from 6/7/20 to 7/29/20 where the evolution curve is steeper, outlining almost a 45° angle with the base of the graph. Another fact to take into account is the number of deaths. In this period, the average fatality rate reached 0.46%, which, according to Jeffrey Sachs (2020), is below that registered in Wuhan (China), the epicenter of the pandemic, where it reached 1.4% and also lower than that analyzed in the UK, where researchers from Imperial College London estimate that the fatality rate will be around 0.9% (Sachs 2020).

Spatial Distribution of Confirmed Cases

Figure 2 shows us the spatial distribution of the confirmed cases of COVID-19 in the Argentine territory as of July 29, 2020. As we can see, the conglomerates with the largest population in the country stand out due to the number of cases: the City of Buenos Aires, capital of the country, with 57,729 cases and province of Buenos Aires, with 106,247 cases. They are followed by Chaco (3428) and Córdoba (2057), two provinces where COVID-19 cases have been recognized already at the beginning of its introduction of the disease in the country. Chaco is highly affected by the virus

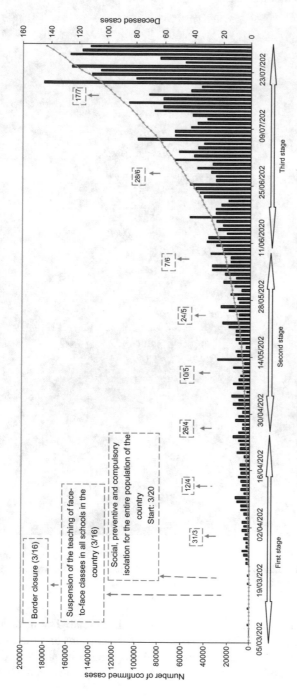

Fig. 1 Evolution of the accumulated confirmed cases and deaths per day of COVID-19 in Argentina. Period: 3/3/2020–7/29/2020. Observation: The mandatory quarantine began to take effect in Argentina at 00 h on Friday, March 20. Originally, confinement was in effect unit March 31, then extended until April 12, then until April 26, later until May 10, then until May 24, until June 7, until June 28, and until July 17. *Source* Daily Report, Ministry of Health, Argentina. https://www.argentina.gob.ar/coronovirus/informes-diarios/reportes/julio2020

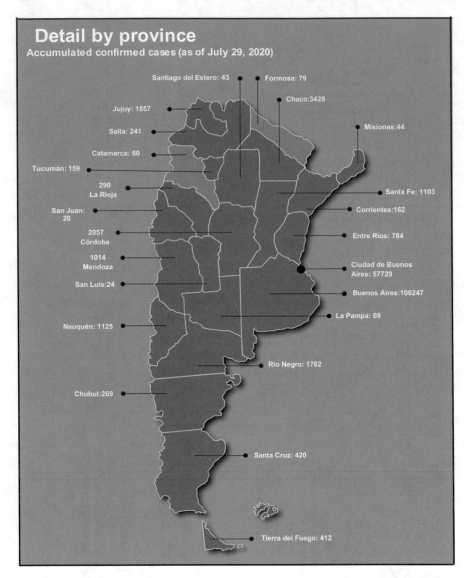

Fig. 2 Spatial distribution of the accumulated confirmed cases of COVID-19 in the Argentine Republic. *Source* Own elaboration based on the Daily Report, Ministry of Health, Argentina. https://www.argentina.gob.ar/coronovirus/informes-diarios/reportes/julio2020

since a combination of factors is combined to favour its spread: The most important outbreaks occurred early in health personnel and institutions and the isolation measures for positive cases were not followed. This, together with an inappropriate use of biosafety elements by health personnel, facilitated the circulation of the virus.

In the province of Formosa (79 cases), Catamarca (60 cases) and Misiones (44), no cases of coronavirus had been detected until very recently.

Case Study: City of Buenos Aires

From the date of notification of the first confirmed case of COVID-19 in Argentina (March 3, 2020) until Wednesday, July 29, 2020 inclusive, 155,446 suspected cases of COVID-19 have been reported in residents of the City of Buenos Aires. Of the 155,446 reported suspects, a total of 57,729 COVID-19 cases were confirmed in residents. 1,186 of those confirmed died. Figure 3 shows the total number of reported cases, noting an upward trend in the number of people infected by COVID 19 (confirmed cases), suspected of contracting it (suspected cases) and cases with negative results of the disease (cases discarded). Some observations should be made. In the City of Buenos Aires, there are 20 shantytowns (shanty towns) or precarious settlements, characterized by a dense proliferation of precarious homes, without gas, drinking water and electricity.[5] Its poverty conditions are extreme, and its inhabitants are unable to meet their basic needs. These villas, characterized essentially by their precariousness, are identified by their name or their number. In this vein, the Ministry of Health of the City of Buenos Aires implemented a special operation called "Operation Detect" for the villages 20, 15, 31 and 11–14 aimed at detecting cases of COVID-19 but only for those people who have been in close contact with confirmed cases.

On the other hand, Fig. 4 shows the curves of cumulative cases reported (suspected and confirmed) between 2/3 and 29/7 2020. Deaths from the same period are represented according to the day. As can be seen, the accumulated trend in the notification presents a steeper curve than the confirmed cases. Likewise, in recent days, the number of deaths has decreased, also taking into account the consolidation of information.

Figure 5 shows the distribution of confirmed cases and rates according to age groups. As can be seen, the highest absolute number of cases is found in the 20 to 49-year-old groups, representing 56% of the total confirmed cases. The most affected population in terms of rates corresponds to the groups between 20 and 39 years of age and the population over 80.

Figure 6 shows the registered symptoms of the disease in the 28,760 confirmed cases of COVID-19 in residents of the City of Buenos Aires. The most frequent symptoms were cough, odynophagia, fever (greater than or equal to 38°) and headache. In 8833 confirmed patients, no symptoms were reported in the surveillance system.

In Europe, with 14,011 confirmed cases reported to the European Surveillance System by 13 countries (97% from Germany), the most frequent symptoms were

---

[5]These slums are similar to the Brazilian favelas, the shacks of Spain, the Uruguayan cantegriles, the Chilean callampas populations, the Venezuelan ranches, the Costa Rican and Colombian slums, the Peruvian youth towns, the Paraguayan chacaritas and the Ecuadorian suburbs or guasmos.

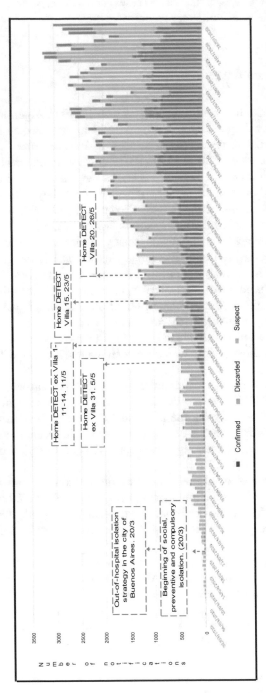

Fig. 3 Total reported cases (includes suspects, discarded and confirmed) of COVID-19 in residents of the city of Buenos aires according to notification date. $N = 155,446$ 2/3 to 7/29 of 2020. *Source* Weekly Epidemiological Bulletin, operational management of epidemiology, Ministry of Health of Autonomous City of Buenos Aires (CABA). No. 206 YEARV. https://www.buenosaires.gob.ar/salud/boletines-periodicos/boletines=epidemiologics-semanales-2018-2019. **Observations.** *Suspicious case:* (1) Any person with fever and one or more respiratory symptoms (cough, respiratory distress, odynophagia) without another etiology that fully explains the clinical presentation. (2) That has been in contact with confirmed or probable cases of COVID-19 or (3) have a history of travel outside the country or (4) have a history of travel or residence in local transmission areas (either community or conglomerate) of COVID-19 in Argentina. (5) All patients with a clinical and radiological diagnosis of pneumonia without another etiology that explains the clinical picture (6) all health personnel with fever and one or more respiratory symptoms (cough, odynophagia, respiratory distress). *Confirmed case:* Any suspected or probable case that presents positive results by rtPCR for SARS CoV-2. *Detect Operation:* it is a device created by the National Ministry of Health for the early detection of coronavirus cases. It is carried out in vulnerable neighborhoods (emergency villas) of the City of Buenos Aries (CABA)

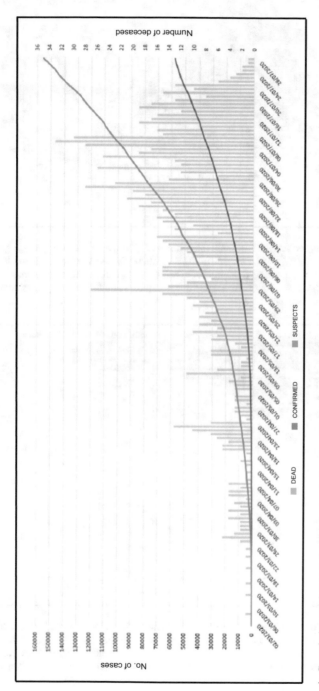

Fig. 4 Cumulative total reported cases (suspected and confirmed) and deaths per day of COVID-19 in residents of the city of Buenos aires according to notification date. Suspects ($N = 155,446$), confirmed ($N = 57,729$) and deceased ($N = 1186$) from 2/3 to 29/7/2020. *Source* Weekly Epidemiological Bulletin, Operational Management of Epidemiology, Ministry of Health of the Autonomous City of Buenos Aires. No. 206 Year V, https://www.buenosaires.gob.ar/salud/boletines-periodico/boletines-epidemiologicos-semanales-2018-2019

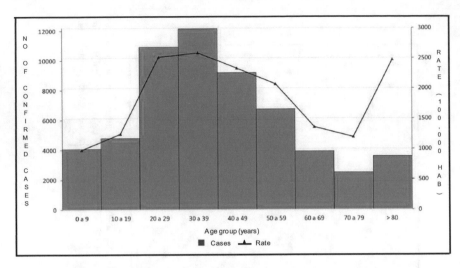

Fig. 5 Confirmed cases and rates per 100,000 inhabitants of COVID-19 according to age groups 9 years). Buenos aires city. $N = 57,729$. 2/3 to 29/7 of 2020. *Source* Weekly Epidemiological Bulletin, Operational Management of Epidemiology, Ministry of Health of the Autonomous City of Buenos Aires. No. 206 Year V, https://www.buenosaires.gob.ar/salud/boletines-periodico/boleti nes-epidemiologicos-semanales-2018-2019

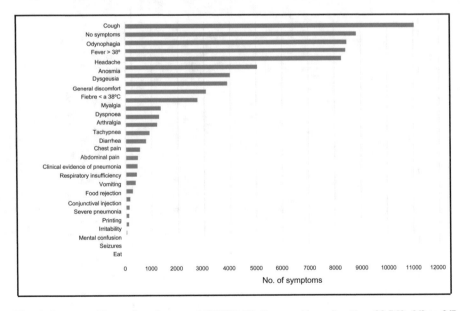

Fig. 6 Symptoms in confirmed cases of COVID-19. Buenos Aires city. $N = 28,760$. 3/2 to 2/7 of 2020. *Note* Each case can present more than 1 symptom. *Source* Weekly Epidemiological Bulletin, Operational Management of Epidemiology, Ministry of Health of the CABA No. 202 YEAR V, 2020. https://www.buenosaires.gob.ar/salud/boletines-periodico/boletines-epidemiologi cos-semanales-2018-2019

fever (47%), dry or productive cough (25%), sore throat (16%), asthenia (6%) and pain (5%) (146). In Spain, with 18,609 reported cases, the most frequent symptoms were fever or recent history of fever (68.7%), cough (68.1%), sore throat (24.1%), dyspnea (31%), chills (27%), vomiting (6%), diarrhea (14%) and other respiratory symptoms (4.5%). Other symptoms related to different organs and systems have also been described (HE et al. 2020).

Figure 7 shows the comorbidities data of confirmed and deceased COVID-19 patients. The most frequent comorbidities in the deceased cases were high blood pressure, diabetes and smoking. Of the total confirmed cases, 10,356 did not present any comorbidity and 12,174 cases did not have comorbidity data recorded in the surveillance system. Likewise, 36 of the deceased cases did not have comorbidities and 137 did not present comorbidity data recorded. At this time, it is unknown if COVID-19 will leave sequels in the survivors. By analogy with SARS, in a follow-up of people recovered from the disease, they showed signs of pulmonary fibrosis (Zhang et al. 2020).

Table 1 shows the detail of confirmed cases and rates per 100,000 inhabitants of each of the neighborhoods of the City of Buenos Aires. It is observed that the neighborhoods of Retiro, Flores, Villa Soldati, Barracas and Villa Lugano concentrate 47% of the confirmed cases of COVID-19 in the city. In addition, there is an important difference, in some neighborhoods, between their absolute cases and when they are linked to their population (rate). This is especially true in Palermo, where its 946 cases rank seventh, but the rate per 100,000 inhabitants appears in 32nd place. In addition, a relative increase is observed, with respect to the previous week, of the

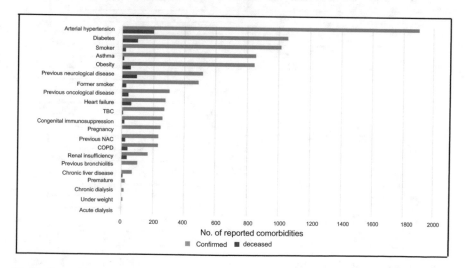

Fig. 7 Comorbidities of confirmed and deceased cases of COVID-19 with recorded data. Buenos Aires city. $N = 541$ (deceased); $N = 28,760$ (confirmed). 3/2–2/7 of 2020. *Source* Weekly Epidemiological Bulletin, Operational Management of Epidemiology, Ministry of Health of the CABA No. 202 YEAR V, 2020. https://www.buenosaires.gob.ar/salud/boletines-periodico/boletines-epidemiologicos-semanales-2018-2019

Table 1 Cases and rates per 100,000 inhabitants of confirmed cases of COVID 19 according to neighbourhood residence. Buenos Aires city. $N = 56{,}474$. 3/3–7/29 of 2020

n°	NEIGHBORHOOD	TOTAL CASES	RATES X 100,000 HAB	% CASES	PREVIOUS WEEK	PREVIOUS WEEK DIFFERENCE
1	VILLA SOLDATI	2477	5302,38	4%	2309	168
2	RETIRO	3449	5271,77	6%	3364	85
3	BARRACAS	3969	4433,35	7%	3691	278
4	NUEVA POMPEYA	1727	4089,24	3%	1616	111
5	VILLA LUGANO	4598	3634,48	8%	4296	302
6	FLORES	5876	3579,12	10%	5543	333
7	PATERNAL	632	3204,68	1%	534	98
8	LA BOCA	1380	3073,97	2%	1244	136
9	CHACARITA	787	2854,34	1%	714	73
10	CONSTITUCION	1238	2815,70	2%	1083	155
11	BALVANERA	3760	2702,18	7%	3396	364
12	VILLA RIACHUELO	371	2639,91	1%	331	40
13	PARQUE PATRICIOS	1074	2622,04	2%	981	93
14	MONSERRAT	989	2449,98	2%	881	108
15	PUERTO MADERO	160	2374,74	0%	148	12
16	PARQUE AVELLANEDA	1197	2249,72	2%	1099	98
17	SAN CRISTOBAL	1067	2193,76	2%	987	80
18	SAN NICOLAS	543	1840,09	1%	476	67
19	SAN TELMO	372	1839,11	1%	341	31
20	FLORESTA	656	1752,48	1%	587	69
21	ALMAGRO	2151	1629,31	4%	1942	209
22	MATADEROS	1039	1617,05	2%	927	112
23	BOEDO	718	1515,87	1%	644	74
24	PARQUE CHACABUCO	829	1473,58	1%	734	95
25	VILLA GRAL. MITRE	457	1308,28	1%	419	38
26	LINIERS	570	1289,62	1%	503	67
27	VELEZ SARSFIELD	452	1287,04	1%	406	46
28	VILLA LURO	403	1237,71	1%	363	40
29	VILLA SANTA RITA	407	1230,33	1%	376	31
30	COGHLAN	225	1201,70	0%	197	28
31	VILLA CRESPO	947	1157,63	2%	825	122
32	PALERMO	2613	1157,02	5%	2293	320
33	VILLA REAL	155	1151,50	0%	127	28
34	RECOLETA	1761	1109,98	3%	1523	238
35	CABALLITO	1903	1077,52	3%	1636	267
36	COLEGIALES	525	1003,63	1%	478	47
37	VILLA DEVOTO	641	963,16	1%	593	48
38	PARQUE CHAS	169	961,25	0%	145	24
39	BELGRANO	1198	944,58	2%	1063	135
40	AGRONOMIA	130	935,62	0%	108	22
41	VILLA URQUIZA	834	903,41	1%	721	113
42	MONTE CASTRO	295	874,18	1%	260	35
43	VILLA DEL PARQUE	445	807,82	1%	396	49
44	SAAVEDRA	393	784,01	1%	342	51
45	VILLA ORTUZAR	161	747,35	0%	144	17
46	VILLA PUEYRREDON	289	728,71	1%	249	40
47	VERSALLES	100	720,46	0%	86	14
48	NUÑEZ	342	651,45	1%	301	41
	Total	**56474**	**1952,82**	**100%**	**51422**	**5052**

Source Weekly Epidemiological Bulletin, Operational Management of Epidemiology, Ministry of Health of the Autonomous City of Buenos Aires No. 206 YEAR V. https://www.buenosaires.gob.ar/salud/boletines-periodico/boletines-epidemiologicos-semanales-2018-2019

total confirmed cases in the city of 19.5% (22,557–28,023). But more than half of the Buenos Aires neighborhoods exceed that percentage of increase. The neighborhoods of Flores, Barracas, Villa Soldati, Nueva Pompeya, Balvanera and Villa Lugano are those with the highest number of new cases in absolute terms. The percentage increase in Retiro is lower than the average for the City of Buenos Aires, since the increase from 2,927 to 2,996 cases represents 2.3% more.

Spatial Distribution in the City of Buenos Aires of the Confirmed Cases

Map 1 shows the rates of confirmed COVID-19 cases per 100,000 inhabitants according to the neighborhood of residence. As can be seen, the Retiro and Flores neighborhoods have the highest rates in the city. The increase in the rate in the Retiro neighborhood is due to a cluster of COVID-19 cases in the "Padre Mugica" neighborhood (or "Villa 31"), a vulnerable neighborhood in the town. The settlement arose in 1932 under the name of "Villa Desocupación" during the infamous decade and from then on various eradication attempts were made by the authorities, although they never managed to eliminate it completely.

Consequences of the Covid-19 Pandemic on Mental Health Associated with Social Isolation

The repercussions on mental health related to the context of a pandemic are evident in most studies at the international level. From the investigations carried out on previous epidemics and pandemics, unpleasant experiences of cases that are associated with worse mental health are reported, the most serious symptoms being post-traumatic stress (PTSD) , avoidance behaviours, anxiety and depression. The most highly vulnerable groups of people were those hospitalized for the disease, people with previous mental illness or with difficult situations caused by isolation and the economic crisis. In people hospitalized for COVID-19, a study evaluated the psychological comorbidities of 41 patients: 43.9% reported some psychiatric symptom, 12.2% had post-traumatic stress disorder (PTSD) and 26.8% had anxiety or depression (Ministerio De Sanidad 2020). These were more frequent in patients with coping difficulties and who reported being stigmatized.

Of course, direct attention to the physical repercussions and deaths that COVID-19 has caused in the population has masked public interest in the mental and emotional illnesses that pandemics also bring. Taking as reference the epidemics of the severe acute respiratory syndrome (SARS-CoV) in 2003 and the Middle East respiratory syndrome (MERS-CoV) in 2012, whose number of affected by the infection is not comparable to the current pandemic, was able to show that 35% of the survivors

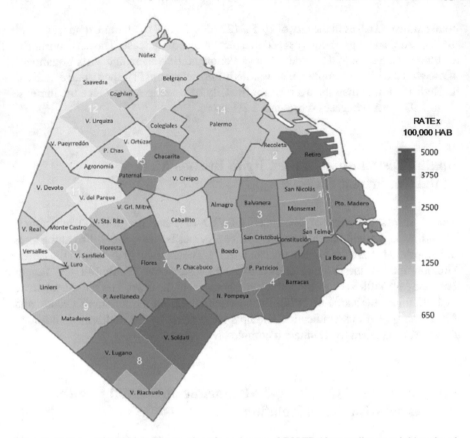

Map. 1 Rate per 100,000 inhabitants of confirmed cases of COVID 19 according to neighbourhood of residence. Buenos Aires city. $N = 56,474$. 3/3–29/7 of 2020. *Source* Weekly Epidemiological Bulletin, Operational Management of Epidemiology, Ministry of Health of the CABA No. 202 YEAR V, 2020. https://www.buenosaires.gob.ar/salud/boletines-periodico/boletines-epidemiologicos-semanales-2018-2019

of SARS-CoV reported psychiatric symptoms during the early recovery phase, and in the case of MERS-CoV, around 40% of the affected people required psychiatric intervention (Zhang et al. 2020).

Although many are affected by a pandemic, the degree of vulnerability to which they are exposed is extremely uneven. Health workers, especially nurses and doctors, who work directly with patients and quarantine, are the most likely to get sick, especially if they are not well cared for and with the proper instrumental equipment. This can precipitate risks that generate flare-ups of psychiatric symptoms ranging from a mental disorder such as insomnia, anxiety, depression and post-traumatic stress disorder (PTSD) . Patients and their families are those who are seen with depressive disorders or develop reactions with increased levels of anxiety. This is especially due to separation from loved ones, direct exposure to the disease and fear of contracting the infection. People who are in social isolation, on the other hand,

and not very close to contact with the disease, develop, among other complications, depressive disorders, being in environments overloaded with stress, panic and levels of symptoms attributable to PTSD. It is noteworthy that these people exposed and subjected to different degrees of vulnerability maximize their psychiatric pathologies if there are preexisting illnesses at the beginning of the pandemic.

Studies in the Country

Many countries, including Argentina, have had to take the necessary precautions to prevent the spread of the virus and combat the disease. To achieve the best results, measures, if not effective, sanitary measures have been proposed both in Argentina and in other latitudes, such as the one that refers to social, preventive and compulsory isolation. Indeed, since Friday, March 20, 2020, everyone in the country is obliged to stay at home in order to avoid the rapid spread of the virus and the collapse of the health system. Without a doubt, this measure becomes even more important in countries like ours, where the health system has serious structural deficiencies. However, it must be admitted that measures of this scope have an impact on the state of mental health of the population with a range of very critical illnesses. In this scenario, a group of Argentine researchers from the Observatory of Applied Social Psychology of the National University of Buenos Aires conducted a study with the aim of analyzing the psychological effects of mandatory quarantine. 3181 people over 18, residents in the country, of different genders and socioeconomic levels participated in it.[6] The work was carried out by means of a representative sample of the main urban centers of the country (Autonomous City of Buenos Aires, Greater Buenos Aires, Interior of the Province of Buenos Aires, Córdoba, Rosario, Mendoza and Tucumán). The technical sheet is indicated below:

Data sheet	
Kind of investigation	Quantitative
Modality	Online surveys geolocated
Contact ability	The invitation to complete the survey is made through social networks, according to geolocation parameters
Universe	General population, over 18 years of age
Sample size	3181 cases. Sampling error ±1.7; 95% confidence level
Sample design	Simple random probabilistic sample
Field date	May 7–11, 2020

---

[6]https://www.psi.uba.ar/opsa/informes/Crisis%20Coronavirus%206%2010-5-2020-FINAL.pdf.

Results

Figure 8 shows the percentage of participants who have experienced during their social isolation a significant degree of intensity of each of the emotional states indicated in the figure. After 50 days of compulsory social isolation in the country, the greatest impact is observed in the emotional-cognitive tripod made up of uncertainty, worry and anxiety. People show the same levels of concern, uncertainty and anxiety as indicated by a previous study, thirty days after compulsory social isolation, carried out by the same Applied Social Psychology Observatory (OPSA).

Figure 9 shows, on the other hand, the temporal evolution of each of the emotions/cognitions surveyed in Fig. 8. That is, 10 days after isolation (April 2), 30 days (April 21) and 50 days of compulsory social isolation (May 11). It is observed that the three curves corresponding to each survey are very similar in shape. There was an increase in intensity for all emotions/cognitions surveyed from the first to the second measurement. In this third measurement, after 50 days of quarantine, the intensity of the set of emotions and cognitions surveyed remains stable with respect to the levels obtained in the second survey. The highest levels of intensity continue to correspond to: worry–uncertainty–anxiety.

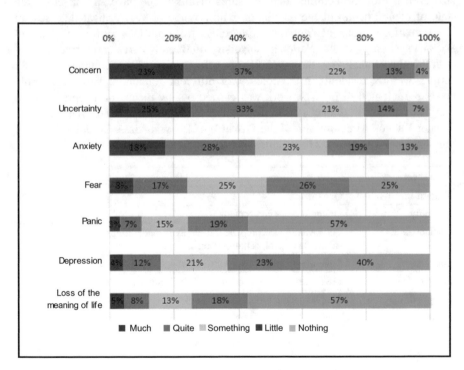

Fig. 8 Participants who have experienced feelings, emotions and thoughts during compulsory social isolation (quarantine). *Source* Observatory of applied social psychology, faculty of psychology, University of Buenos Aires. 2020. https://www.psi.uba.ar/opsa/#informes

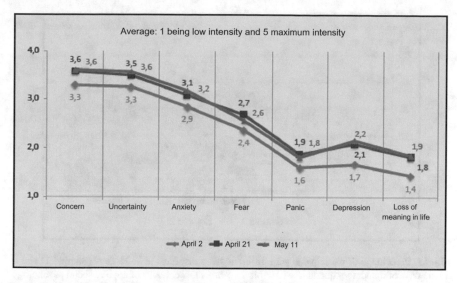

Fig. 9 Participants who have experienced feelings, emotions and thoughts during compulsory social isolation (quarantine) (Comparison April 2–April 21–May 11). *Source* Observatory of applied social psychology, faculty of psychology, University of Buenos Aires. 2020. https://www.psi.uba.ar/opsa/# informes

Figure 10 shows the percentage of people with and without sleep disturbances during the pandemic. An increase in the number of participants with sleep distur-bances is observed from 73.7% in sample 1 (7–11 days of mandatory quarantine) to 76.06% of participants in sample 2 (50–55 days of mandatory quarantine). These alterations are one of the most frequent symptoms during prolonged quarantines. The

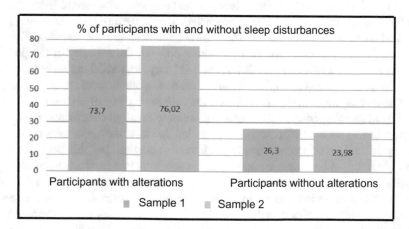

Fig. 10 Percentage of participants with sleep disturbances at 7–11 days (sample 1) and at 50–55 days of mandatory quarantine (Sample 2). *Source* Observatory of applied social psychology, faculty of psychology, University of Buenos Aires. 2020. https://www.psi.uba.ar/opsa/#informes

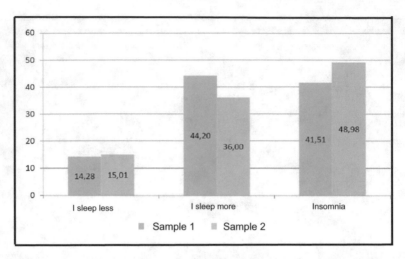

Fig. 11 Percentage of participants with different sleep disorders at 7–11 days (Sample 1) and at 50–55 days of mandatory quarantine (Sample 2). *Source* Observatory of applied social psychology, faculty of psychology, University of Buenos Aires. 2020. https://www.psi.uba.ar/opsa/#informes

decrease in physical activity and less exposure to sunlight in large cities alter sleep cycles. Among the most frequent sleep disturbances of the participants in sample 1 (7–11 days of mandatory quarantine) and sample 2 (50–55 days of mandatory quarantine), an increase in insomnia was found to the detriment of the option "I sleep more." The percentages can be seen in Fig. 11.

Sleeping too much is the most frequent sleep disturbance recorded at 7–11 days of quarantine, while insomnia is more frequent at 50–55 days of quarantine. Sleeping more than usual is considered an atypical depressive symptom that is usually accompanied by a lack of interest in the external world and low self-esteem. Insomnia, on the other hand, is the typical sleep disturbance in depressive disorders and is also associated with concerns about the course of events, financial impact, schedule disorder and diet. Regarding sexual life in a pandemic context, the majority of the population has reported a worsening of it both at 7–11 days and at 50–55 days (Fig. 12). Sexuality is considered one of the healthy behaviours along with sports and social life. The worsening of the same is associated with the indices of discomfort and the widespread social restriction. Mandatory quarantine for single or divorced/separated people prevents sexual encounters. It is expected that after the quarantine, these relationship difficulties will persist due to the fear of contagion. But this item should be taken with caution because the majority of respondents preferred not to answer.

With respect to the consumption of tobacco, alcohol and illegal drugs, a decrease in tobacco consumption and an increase in alcohol consumption were registered when comparing samples 1 and 2. No significant differences were observed with respect to the consumption of illegal drugs. The percentages can be seen in Fig. 13.

It is indicated that the consumption of all these substances is one of the problem behaviours that are implemented to manage psychological discomfort. Although they

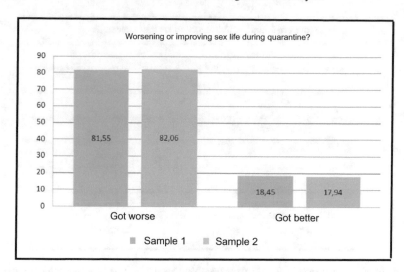

Fig. 12 Changes in sexual life during the 7–11 days (sample 1) and the 50–55 days of mandatory quarantine (sample 2). *Source* Observatory of applied social psychology, faculty of psychology, University of Buenos Aires. 2020. https://www.psi.uba.ar/opsa/#informes

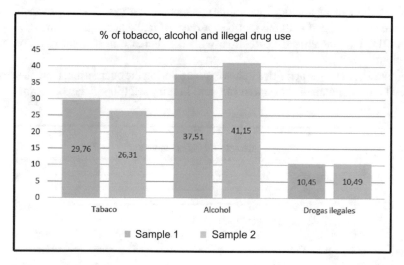

Fig. 13 Use of tobacco, alcohol and drugs during the 7–11 days (sample 1) and the 50–55 days of mandatory quarantine (sample 2). *Source* Observatory of applied social psychology, faculty of psychology, University of Buenos Aires. 2020. https://www.psi.uba.ar/opsa/#informes

provide relief because they impact neurotransmission producing pleasure or sedation, after their effect they deteriorate global health. Figure 14 shows the percentage of participants with suicidal ideas. When asked if they had thought about taking their own lives, 4.22 of sample 1 (7–11 days of mandatory quarantine) and 6.53 of sample

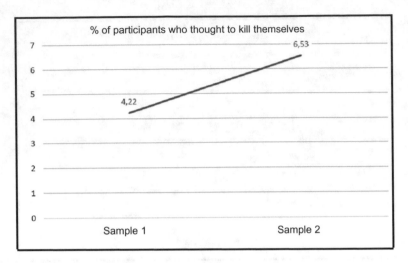

Fig. 14 Ideas of death/suicide at 7–11 days (sample 1) and 50–55 days of mandatory quarantine (sample 2). *Source* Observatory of applied social psychology, faculty of psychology, University of Buenos Aires. 2020. https://www.psi.uba.ar/opsa/#informes

2 (50–55 days of mandatory quarantine) answered that yes. A significant increase is observed associated with the number of days of mandatory quarantine ($z = 3.48$, $\rho < 0.01$). This increase is correlative with the increase in clinical psychological symptoms.

Figure 15 shows the activities that take place during the quarantine. Both in sample 1 (7–11 days of compulsory quarantine) and in sample 2 (50–55 days of compulsory

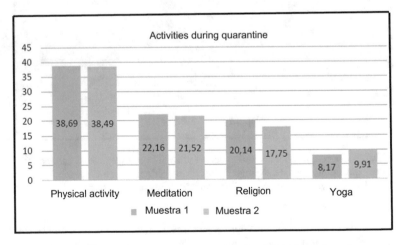

Fig. 15 Participants' activities at 7–11 days (sample 1) and at 50–55 days of mandatory quarantine (sample 2). *Source* Observatory of applied social psychology, faculty of psychology, University of Buenos Aires. 2020. https://www.psi.uba.ar/opsa/#informes

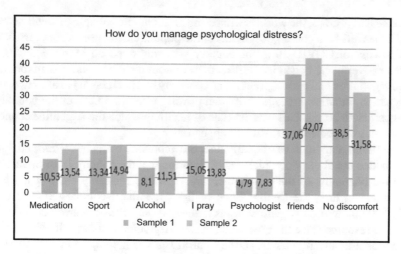

Fig. 16 Behaviours against psychological distress at 7–11 days (sample 1) and at 50–55 days of mandatory quarantine (sample 2). *Source* Observatory of applied social psychology, faculty of psychology, University of Buenos Aires. 2020. https://www.psi.uba.ar/opsa/#informes

quarantine), it is observed that less than half of the participants perform any of the following activities: physical activity, meditation, religious practice or yoga. An increase in the practice of yoga from 8.17 to 9.91% and a decrease in religious practice from 20.14 to 17.75% of the participants in both samples can be highlighted. In addition, as the quarantine time increases, there is a tendency to decrease healthy activities. Keep in mind that mandatory quarantine only allows most people to leave their homes to provide themselves with food and medicine. In some districts, children have been allowed to leave once a week accompanied by their parents on walks up to 500 m from their homes.

Figure 16 shows results to the question: How do you proceed during quarantine when you experience psychological distress or have emotional problems? The participants responded as follows: the use of non-prescription medication (self-medication) increased from 10.53 to 13.54%, alcohol consumption from 8.1 to 11.51%, psychological consultation with the 4.79 to 7.83% and conversations with friendly confidants from 37.06 to 42.07%. The use of sport increased from 13.34 to 14.94%. However, the group that does not experience discomfort was reduced from 38.5 to 31.58%. In another vein, almost 70% of the participants experience psychological discomfort 50–55 days after the mandatory quarantine.

Economic and Social Impact in the Country

The COVID-19 pandemic is a true tragedy whose consequences are still unknown, especially in terms of human lives. Simultaneously with the loss of life and the profound health crisis, the world is witnessing an economic collapse that will severely impact the well-being of large segments of the population in the coming years. When the first case of COVID-19 in the country was confirmed on March 3, 2020, the national government began its fourth month of administration and was in the midst of a renegotiation of the foreign debt. The economy had not rebounded after two years of recession, inflation exceeded 50% and poverty affected 35.5% of the population (Guerrieri et al. 2020). This was the state of health in which the Argentine economy was when, unexpectedly, COVID-19 appeared.

After more than 80 days of quarantine, the economic crisis has worsened and to this is added the already complex social situation that the country is going through. Since the beginning of the quarantine, the Argentine economy has lost US $ 8.4 billion; and in turn, the Central Bank of the Argentine Republic (BCRA) transferred 500,000 million Argentine pesos to the National Treasury, becoming the first financing resource since the quarantine began. Argentina is today without access to international financial markets, so the Central Bank is helping to finance the fiscal deficit, to which is added the pressure on inflation and exchange rates. The baseline scenario on which the Argentine Government is working today is a drop of 6.5% in GDP (internal gross product) by 2020 and a primary fiscal deficit of 3.1% that, the government specified, will be fully financed by contributions from the Central Bank (Naciones Unidas 2020). The real effect of the pandemic on the fall in GDP will depend on how long the social isolation measures will be applied, on their amplitude in sectoral and jurisdictional terms, and this is subject to the growth trajectory of the pandemic (Alzúa et al. 2020). In such a way, that the crisis caused in Argentina by the COVID-19 pandemic will have a strong impact on the labour market, which was already in a weak situation before the health emergency. According to the projections of the International Labour Organization (ILO) based on different scenarios of falling GDP, between 750,500 and 852,500 jobs would be lost in Argentina in 2020. However, compared to other countries in the region, the government has less room for maneuver given its lack of access to international financial markets and its difficult fiscal situation. In the post-pandemic world, Argentina must strengthen its productive apparatus and continue to eliminate the social inequities that are aggravated by COVID-19.

As for the social crisis, the pandemic will have a severe impact. The majority of workers and employers are expected to lose income as a result of the COVID-19 outbreak. Those who work in non-essential activities will suffer the greatest losses. However, those engaged in essential activities will also feel the impact of declining demand, as those who lose their jobs are likely to reduce consume spending, while those who still have income will be more cautious. About half of the workers have been unemployed for more than six months, and 37% have been unemployed for more than a year. Informal workers are the most vulnerable to a deep economic recession. Women, the young and the elderly are particularly affected by informality. Their income is low, so there is a strong link between informality and poverty. Therefore, the current scenario poses important challenges, especially in the Greater Buenos Aires (GBA) area—which includes a group of 33 districts that surround the City of Buenos Aires, with a population of about 10 million inhabitants—where unemployment, poverty and informality are high, and living conditions are poor, with overcrowding, difficulties in accessing healthcare and restricted access to drinking water. Greater Buenos Aires represents half of the total poor in Argentina (with about 5 million people below the poverty line in the second half of 2019) and 61.8% of the extreme poor (with 1.4 millions of extreme poor). Living conditions in Greater Buenos Aires are inadequate for most of its population. According to INDEC figures for the second half of 2019, 40.5% of the population of Greater Buenos Aires was poor, while 11.3% was extremely poor. Unemployment reached 10.8% in the last quarter of 2019, while underemployment was 13%, showing that almost a quarter of the workforce had labour problems. Therefore, one of the main challenges facing the government is to avoid a social and health crisis, especially in Greater Buenos Aires, which represents half of the country's poor population, in addition to being the epicenter of the COVID outbreak-19.

The New Variants of the Virus in Argentina

The emergence of viral variants is a natural process in the evolution of viruses. During the last fortnight of December 2020, three viral variants of SARS-CoV-2 aroused the attention of the scientific community and national governments: the VOC 202012/01 variants (lineage B.1.1.7), the most recent sample of which was detected in the UK on 09/20/2020; variant 501Y.V2 (lineage B.1.351), detected in South Africa since 10/08/2020, and the variant of Rio de Janeiro (derived from lineage B.1.1.28), detected in Rio de Janeiro, Brazil, since October 2020 (Argentina. gov.ar. 2021). The fact that new variants have emerged is not surprising: All viruses mutate as they generate copies of themselves to spread the infection. Viruses change their genetic sequence, and this is what causes the virus at a given moment to go from being a totally innocuous agent for the human species to being highly infectious.

The British variant of the coronavirus was detected in Argentina in a passenger who arrived at the Ezeiza airport without symptoms, and when the test was performed, it was positive. According to the information provided by the specialists, the passenger arrived "asymptomatic" at the Argentine airport at the end of December 2020 from Frankfurt: When the test was performed, he was positive for SARS-CoV-2 antigens. So far, in addition to Argentina, only a few countries reported the presence of this variant: Denmark, the Netherlands, Australia, Belgium, Italy, and Iceland, although it is not ruled out that it is present in others; in Argentina until October, in the sequences obtained from the detected provinces, this variant was not recorded.

Preliminary studies suggest that one of the 17 most important mutations in the UK virus (in the spike protein) allows the virus to better bind to a protein on the surface of human cells, thus facilitating infection. This mutation (known as N501Y) appears to be an important adaptation. Experts believe that the UK variant appeared in September, and that it may be up to 70% more transmissible or infectious, although the most recent research published by Public Health England estimates that this value ranges between 30% and 50%.

The Malbrán Institute of Argentina also found a strain with the Rio de Janeiro variant. Of the last genomes that were sequenced between November and December 2020, one was found in which the six mutations corresponding to the Rio de Janeiro variant could be identified. There are still no conclusive studies that allow us to affirm that this variant has any impact on the transmissibility, the severity of the infection or the efficacy of the vaccine. Of the six mutations it has, there is one that is in the spike protein (spicule or spike); in previous studies, it had been found that this variant decreased the neutralizing effects of monoclonal antibodies and of some convalescent plasmas, but there are no specific studies of the variant yet.

To determine the possible circulation of virus variants in our country, active surveillance is required, either through the sequencing of the entire genome, or through the analysis of partial sequences that include genetic markers of interest. While monitoring is ongoing, since active surveillance for the SARS-CoV-2 variants of interest began, none of the UK variant protein S marker changes had yet been detected.

What impact do these variants have on the effectiveness of vaccines?
Vaccines to prevent coronavirus disease 2019 (COVID-19) may represent the best hope to end the pandemic. For this reason, to date the Argentine National State has entered into agreements with two laboratories that develop vaccines against COVID-19. The Gam-COVID-Vac vaccine, known worldwide as Sputnik-V, developed by the Gamaleya Institute of the Ministry of Health of the Russian Federation and the ChAdOx1-S nCov-19 vaccine developed by the University of Oxford and the pharmaceutical company AztraZeneca. Additionally, an agreement has been established with the multilateral mechanism of COVAX, a coalition made up of International Organizations, such as the World Health Organization (Ibid, 2021).

Conclusion

The unexpected outbreak of the coronavirus disease or COVID-19 awakened in the country the urgent need to use control measures to limit the spread of the outbreak. Thus, it was, that by decision of the national authorities social, preventive and compulsory isolation was decreed in addition to other measures such as the suspension of major sporting events and cultural activities. Although these measures have allowed the contagion curve to remain flat for several weeks in its average and well controlled in certain jurisdictions of the country, it is increasing more rapidly in the month of June in the Autonomous City of Buenos and the Province of Buenos Aires. The next few days will be decisive in assessing the course of the pandemic in the country. What is important and urgent now is an escalation and intensification of certain key public health measures to put pressure on the virus and interrupt the chains of transmission. These measures are the testing and isolation of positive cases of COVID-19, and the exhaustive tracing of contacts and their quarantine for 14 days, while maintaining the community commitment to do their part in hand hygiene, respiratory etiquette and physical distancing.

However, the pandemic caused by COVID-19 will have a multidimensional impact in Argentina. The crisis unleashed by the pandemic is expected to have a direct impact on the economy and the most relevant social dimensions in the country. Although COVID-19 in the country had a very low death rate, compared to Spain, Italy and the USA, among other countries, and has allowed the contagion curve to remain flat for several weeks, this will exacerbate the risk of a economic crisis due to the absence of commercial, tourist and industrial activity never before equaled in Argentina. Specifically, a greater economic contraction is expected, with loss of employment, the corresponding increase in poverty and new challenges for sustainable development and social peace. In Argentina, the outbreak of the pandemic finds a country already affected by great socioeconomic challenges, which recognizes a food, social health and production crisis. At the same time, these challenges are marked by the fiscal constriction to meet the basic demands of the population and the complex negotiation of foreign debt. In addition, the effects of the pandemic on people's mental health— which can be affected for different reasons: stress, social isolation, family losses, economic losses or fear of being infected by them and/or their families—make this situation a real challenge for the health sector and for society as a whole.

References

Alzúa, M. L., Gosis, P. (2020). *Impacto Social y Económico de la COVID-19 y Opciones de Políticas en Argentina.* PNUD América Latina y el Caribe. Centro de Estudios Distributivos, Laborales y Sociales de la Universidad Nacional de la Plata y Partnership for Economic Policy (PEP).

Argentina. gov.ar. (2021). *Argentina vigila la presencia de las diferentes variantes de COVID-19 en el país.* Coronavirus Covid-19. Información, recomendaciones y medidas de prevención. https://www.argentina.gob.ar/noticias/argentina-vigila-la-presencia-de-las-difere ntes-variantes-de-covid-19-en-el-pais.

Guerrieri V., G. Lorenzoni, L. Straub., and I. Werning (2020), *Macroeconomic Implications of COVID-19: Can Negative Supply Shocks Cause Demand Shortages?* NBER Working Paper 26918, Cambridge, MA, April 2020.

He, X., Lau, E. H. Y., Wu, P., Deng, X., Wang, J., Hao, X., et al. (2020) *Temporal dynamics in viral shedding and transmissibility of COVID-19.* Nat. Med. 15 de abril de 2020.

Ministerio de Salud de la República Argentina. (2020). *Reporte Diario COVID-19.* https://www. argentina.gob.ar/Coronavirus/informe-diario.

Ministerio de Sanidad. (2020). *Centro de Coordinación de Alertas y Emergencias Sanitarias. Información científica-técnica (España).* Enfermedad por coronavirus, COVID-19. Actualización, 3 de julio 2020. https://www.mscbs.gob.es/profesionales/saludPublica/ccayes/alertasActual/nCov-China/documentos/ITCoronavirus.pdf.

Naciones Unidas. (2020). "COVID-19 EN ARGENTINA". IMPACTO SOCIOECONÓMICO Y AMBIENTAL. Actualizado al 19/06/2020. https://www.onu.org.ar/stuff/Informe-COVID-19-Argentina.pdf.

Observatorio de Psicología Social Aplicada (OPSA). (2020). Facultad de Psicología, Universidad de Buenos Aires. https://www.psi.uba.ar/opsa/#informes, https://www.psi.uba.ar/opsa/.

Ramirez, L. (2020). *Evolución, distribución y difusión del COVID-19 en Argentina: Primer mes (03/03/2020 - 02/04/2020).* Buenos Aires: Instituto de Investigaciones en Desarrollo Territorial y Hábitat Humano (IIDTHH-CONICET).

Sachs, J. (2020). *Nuestra mejor esperanza para combatir el Coronavirus.* Recuperado de. https://cnn espanol.cnn.com/2020/03/25/opinion-nuestra-mejor-esperanza-para-combatir-el-Coronavirus/.

Zhang, P., Li, J., Liu, H., Han, N., Ju, J., Kou, Y., et al. (2020). *Long-term bone and lung consequences associated with hospital-acquired severe acute respiratory syndrome: a 15 year follow-up from a prospective cohort study.* https://covid19.elsevierpure.com/es/publications/long-term-bone-and-lung-consequences-associated-with-hospital-acq.

Early Stages of the Coronavirus Disease (COVID-19) Pandemic in Brazil: National and Regional Contexts

Rhavena Barbosa dos Santos, Júlia Alves Menezes, and Ulisses Confalonieri

Abstract The first case of COVID-19 in Brazil occurred in February 2020, affecting mainly the country's large urban centers. The initial strategy to suppress the transmission of the virus was based on social isolation and the suspension of economic activities to flatten the incidence curve and allow health services to prepare for the expected increase in hospital admissions. However, there was no coordinated strategy between the federal government, states and municipalities to adopt effective control measures and to carry out epidemiological surveillance of the disease and/or contact tracing, favouring the rapid increase in the number of new COVID-19 cases. The spread of the disease occurred in a non-homogeneous manner throughout the territory, reaching small and medium-sized cities in the country and overloading local health services, which in many places reached their maximum capacity. There is still no clear trend toward stabilization of the epidemic curve, since the data available to estimate the behaviour of infection in the country are lagged by the low testing rate.

Keywords COVID-19 · Brazil · Health surveillance · Pandemic · Regional hubs · São paulo · Rio de janeiro · Amazon

Introduction

In December 2019, cases of pneumonia of unknown etiology began to be reported in Wuhan, China, which was later characterized as a new coronavirus (SARS-Cov-2 or COVID-19). It is estimated that in this region alone, around 75,000 people were infected with COVID-19 by the end of January, including patients who did not travel to Wuhan, which indicated the possibility of person-to-person transmission (Wu et al. 2020). By mid-February, about 30 countries had already reported cases of the new coronavirus, with varying severity of infection and symptoms, as well as deaths, forcing the World Health Organization to declare the COVID-19 epidemic

R. B. dos Santos (✉) · J. A. Menezes · U. Confalonieri
René Rachou Institute—Oswaldo Cruz Foundation, Transdisciplinary Study Group On Health and Environment, Avenida Augusto de Lima, 30.190-002, Barro Preto, Belo Horizonte, Minas Gerais 1715, Brazil
e-mail: rhavena.santos@gmail.com

R. Akhtar (ed.), *Coronavirus (COVID-19) Outbreaks, Environment and Human Behaviour*, https://doi.org/10.1007/978-3-030-68120-3_22

as an International Public Health Emergency (Malta et al. 2020; WHO n.d.). The new coronavirus quickly became a worldwide pandemic resulting in more than 10 million cases by July 2, 2020, with 516,000 deaths. Europe and the Americas, in addition to China, were considered the epicenters of the disease. In the Americas, Brazil and the USA have undoubtedly represented the places where the epidemic has had an exponential course since the first notification (WHO n.d.). In this chapter, the situation of the COVID-19 epidemic in Brazil in the first 6 months after the first confirmed case will be presented, explaining the burden of cases and deaths regionally, as well as the response capacity and the main impacts of the ongoing epidemic in the country.

Contextualizing the Social and Health Scenario of Brazil

Although Brazil is considered one of the greatest exponents of the world's economy in Latin America, inequalities related to health status, access to services and income are evident throughout the country. The South and Southeast regions concentrate the greatest economic and social developments, with the North and Northeast regions lacking investments in urban and health infrastructure. In 2018, approximately 13.5 million people had an income of less than US $ 1.90 daily per capita in purchasing power parity, a population contingent similar to the entire population of Bolivia for example, with the majority residing in the Northeast region. Access to sanitation is also substantially uneven; in 2018 only 7.7% of the population in the Southeast region lacked piped water, while in the Northeast and North regions, this percentage amounted to 20.7% and 41.8%, respectively (IBGE 2019a) (Fig. 1).

In the public health field, the country has one of the largest health systems in the world, the Unified Health System (SUS), which ranges from primary care to the high complexity of procedures (e.g., diagnostic; therapeutic; emergency services; epidemiological, health and environmental surveillances and pharmaceutical assistance) (Brasil 2020a). In parallel, there is a private healthcare network (health plans, hospitals, clinics, laboratories and private practices) which operates in a complementary manner to the SUS for the establishment of contracts/agreements between the public and the private sphere (Brasil 2016). However, there are growing challenges for a proper public health management, such as the overlapping of major emergencies related to vector-borne diseases, such as Dengue, Yellow Fever, Chikungunya and Zika, with consequences that endure during the COVID-19 epidemic (Cimerman et al. 2020). At the same time, the country suffers from a historical shortage of doctors in primary care. The "Mais Médicos" Program[1] and an underfinanced SUS have had a budget reduction in 2019 when compared to 2018 (Giovanella et al. 2019).

---

[1] To increase equitable and universal health coverage, the Mais Médicos (More Doctors) Program was launched in July 2013 with the aim of reducing the shortage of doctors in the interior cities and on the outskirts of major cities in Brazil. It was replaced in 2019 by "Médicos pelo Brasil" program.

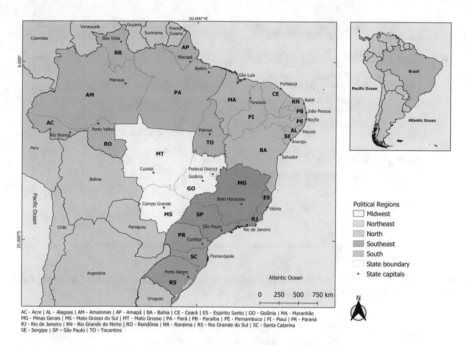

AC - Acre | AL - Alagoas | AM - Amazonas | AP - Amapá | BA - Bahia | CE - Ceará | ES - Espírito Santo | GO - Goiânia | MA - Maranhão
MG - Minas Gerais | MS - Mato Grosso do Sul | MT - Mato Grosso | PA - Pará | PB - Paraíba | PE - Pernambuco | PI - Piauí | PR - Paraná
RJ - Rio de Janeiro | RN - Rio Grande do Norte | RO - Rondônia | RR - Roraima | RS - Rio Grande do Sul | SC - Santa Catarina
SE - Sergipe | SP - São Paulo | TO - Tocantins

Fig. 1 Brazil according to regions, states and their respective capitals. *Source* Authors

The SARS-COV-2 Pandemic in Brazil

Regional Scenarios

According to the monitoring panel of the Ministry of Health, by August 4, 2020, the disease had already infected 2,750,318 people and caused the death of 94,665 people in the country (Brasil 2020b). In absolute terms, the number of cases was only smaller than the USA, which exceeded 4 million cases. Regarding mortality per 100 million inhabitants, the country presents values similar to its neighbors Peru and Chile, as shown in the graph below (Fig. 2).

The first confirmed case occurred on February 25, 2020, in the state of São Paulo. What was observed at the beginning of the epidemic in Brazil was the spread in large cities strongly connected by domestic and international airlines (e.g., São Paulo, Rio de Janeiro, Manaus). The subsequent growth occurred in a non-homogeneous way across the Brazilian territory, followed by the trend of an increase of cases in the interior of the country in mid-April to May, reaching even indigenous and riverine populations. Even with the implementation of a quarantine, about 44% of medium-sized cities (between 20,000 and 50,000 inhabitants) had new cases of COVID-19, a situation of concern due to the reduced health resources available in these places (Fiocruz 2020a).

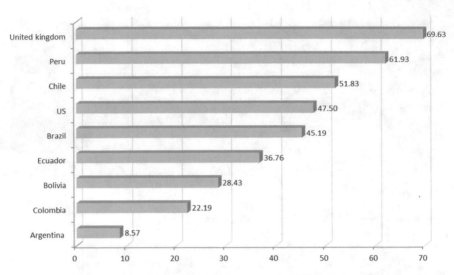

Fig. 2 COVID-19 deaths per 100,000 population—August 4, 2020. *Source* Adapted from John Hopkins University and Medicine, 2020

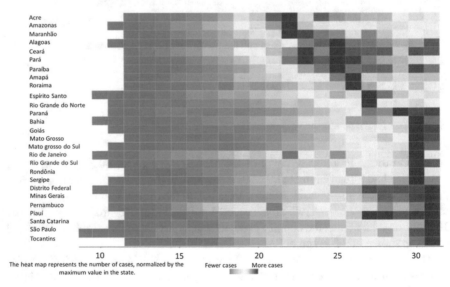

Fig. 3 Timeline for COVID-19 cases in the Brazilian states considering the period of the 9th and 30th epidemiological week (February 23–July 25, 2020). *Source* Fiocruz (2020b)

Figure 3 shows the incidence of cases by epidemiological week,[2] in the country: The red frames represent the epidemiological week with the highest concentration of cases for the period between the 9th and 30th week.

---

[2]By international convention, epidemiological weeks are counted from Sunday to Saturday. The first epidemiological week of each year is the one that contains the largest number of days.

Table 1 Data on cases and deaths by COVID-19 according to the Brazilian administrative regions—August 4, 2020

Region (population)	New cases	Accumulated cases	Accumulated cases/ 100 mi	New deaths	Accumulated deaths	Accumulated deaths/100 mi
Midwest (16,276,284)	4143	263,652	1620	80	5590	34
South (29,879,160)	2606	241,233	807	108	5262	18
North (18,430,980)	1744	416,236	2258	34	11,990	65
Northeast (57,022,185)	3857	881,957	1547	233	29,359	51
Southeast (88,362,248)	4291	947,240	1072	106	42,464	48
Brazil (209,970,857)	16,641	2,750,318	1310	561	94,665	45

Source Brasil (2020b)

In general, initially the states of the North and Northeast regions reached the highest cases records. In June 2020, these regions had the highest number of deaths and cases accumulated by 100 million inhabitants in the country (Table 1). In these places, poor human development is also observed—the Municipal Human Development Index (MHDI) of the North region is 0.667 and the Northeast region of 0.663—a condition that seems to be related to the greater burden of the COVID-19 epidemic(PNUD, IPEA, FJP 2016; Sirkeci and Yucesahin 2020).

In fact, the National Pandemic Confrontation Plan, prepared by national health entities, reinforces the excess of deaths that has been observed among some vulnerable groups, that is women, the poor, blacks and indigenous people, many of whom are quite representative in the north and northeast regions, when compared to groups of greater socioeconomic power in the south and southeastern regions of the country (Frente pela vida 2020).

Especially in the Amazon, indigenous peoples, who are historically more susceptible to exogenous diseases, have been extremely affected by the virus and see the threat of contamination potentialized by the processes of invasion of their lands, such as illegal occupation, mining and deforestation. A study produced by the Institute for Environmental Research in the Amazon (IPAM) and the Coordination of Indigenous Organizations in the Brazilian Amazon (COIAB), published on June 2020, showed that the spread rate of COVID-19 among indigenous people was higher than in the country by 84%, with a lethality of 8%, while the national rate was of 5% (IPAM & COIAB 2020).

Considering the national data on Severe Acute Respiratory Syndrome (SARS), which serves as an alert for COVID-19, what is currently perceived is a growth trend, with remarkably high weekly cases. States that previously had a decrease in the number of new weekly records, such as Rio de Janeiro, Maranhão and Amapá, are now observing an increase in that number, indicating a possible second wave of the disease (Fiocruz 2020c). At the same time, states such as Amazonas, Roraima and

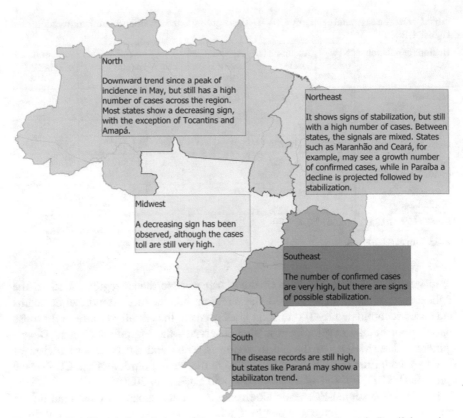

North

Downward trend since a peak of incidence in May, but still has a high number of cases across the region. Most states show a decreasing sign, with the exception of Tocantins and Amapá.

Northeast

It shows signs of stabilization, but still with a high number of cases. Between states, the signals are mixed. States such as Maranhão and Ceará, for example, may see a growth number of confirmed cases, while in Paraíba a decline is projected followed by stabilization.

Midwest

A decreasing sign has been observed, although the cases toll are still very high.

Southeast

The number of confirmed cases are very high, but there are signs of possible stabilization.

South

The disease records are still high, but states like Paraná may show a stabilizaton trend.

Fig. 4 Profile of hospital admissions of Severe Acute Respiratory Syndrome in Brazil, by region, epidemiological week n° 30 (July 19–25, 2020). *Source* Fiocruz (2020c)

Pará, after a downward trend, show signs of possible stabilization in the number of cases (Fiocruz 2020c). Figure 4 summarizes the SARS's hospital admissions profile of Brazilian regions.

Regarding the COVID-19 incidence and mortality trends, considering the epidemiological week 30, there is a fluctuation in the number of cases with peaks in some states. This may be the result, for example, of different measures and approaches to loosen social isolation in each location, the increase in the number of tests available in small and medium-sized municipalities, or even the interiorization of the pandemic (Fiocruz 2020d). Figure 5 shows these variations of COVID-19 incidence and mortality trends for the Brazilian states considering a threshold of 5%. In general, what is perceived is the stability and maintenance of the alert, in all regions of the country (Fiocruz 2020d). Overall, in the country as a whole, the epidemiological parameter Rt (effective reproduction number) in the past weeks is showing a trend toward the stabilization of the number of daily new infections. From July 7 up to August 7, it has varied between 1.14 and 0.82 (Institute of Global Health & Swiss Data Science Center 2020). In terms of bed occupancy, in eight federative units, the

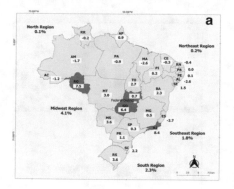
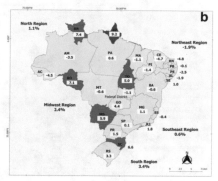

Fig. 5 Trends in incidence (**a**) and mortality (**b**) of COVID-19 by Brazilian federative unit. The value above 5% indicates a situation of high alert; variation between −5 and +5% indicates stability and maintenance of the alert and less than −5% indicates reduction, even if temporary, of transmission. The occupancy rate of COVID-19 beds for adults (**c**) by Brazilian federative unit. The ICU bed occupancy rates in Minas Gerais and Santa Catarina include the set of SUS ICU beds. The rate from Paraná state includes public and private beds. For Rio de Janeiro state, only the rate referring to the capital was identified. All the data are regarding epidemiological weeks 29 and 30 (July 12–25, 2020). *Source* Adapted from Fiocruz (2020d)

rate of adults intensive care unit (ICU) COVID occupancy was equal to or less than 60%, in other 15 locations, this value was between 61 and 79.9%, and in four this occupancy rate was over 80% (Fiocruz 2020d).

Regarding health inputs, in accordance with what occurred in other localities, the country adopted measures relating to the strengthening of infrastructure and human resources, in order to increase the responsiveness of the health sector before the disease reached its peak.

"O Brasil Conta Comigo" (Brazil counts on me), a strategy aimed at training health professionals and regulating the use of telemedicine, exceptionally in the context of the pandemic, is the example of strategies related to human resources(Cimini et al. 2020). At the same time, the amount of personal protective equipment increased: By June, the country had distributed 36.8 million gloves, 2.9 million aprons and 3.4 million N95 masks (Brasil 2020d).

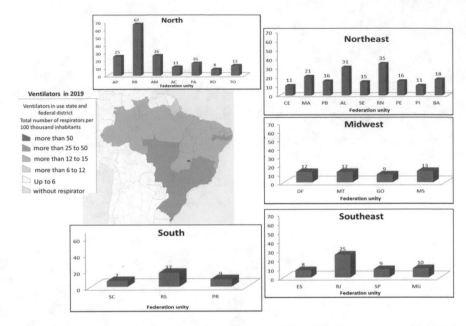

Fig. 6 Percentage increase in ventilators from January to May 2020. *Source* Adapted from BCG (2020) and IBGE (2019b)

As for ventilators, after the cancelation of the purchase of 15,000 items from China between April 8 and 13, the country closed an agreement to buy 15,300 respirators from national companies (Brasil 2020d). The distribution of this equipment has occurred according to the industry's production capacity. Between January and May 2020, the country had an average increase of 17% in the number of ventilators—the largest increases occurred in the North and Northeast regions, which in 2019 had the lowest number of ventilators per 100,000 inhabitants. (BCG 2020, IBGE 2019b) (Fig. 6).

The number of ICU beds has also increased in all regions of the country. However, for the North and Northeast, the number is still lower than the other regions, hampering access to health services in these locations. This situation is aggravated by the need to move around in search of highly complex health care, especially regarding residents from small cities, an average distance which reaches 276 km in the North, while in the South and Southeast it is around 100 km, close to the national average of 155 km (IBGE 2020b; Brasil 2020c). It is believed that the tendency of interiorization of the cases associated with shortage in the capacity of health services in these places can lead to an increase in the demand for medical care in regional urban centers, as is the case with state capitals such as Manaus, causing pressure on public services in major cities in a second wave or in a recrudescence situation of the epidemic (Fiocruz 2020a) (Table 2).

In fact, the measures implemented did not prevent the collapse of health systems in some locations. This is the case, for example, of Amazonas state, which in April

Table 2 Adult and pediatric ICU beds in December 2019 and June 2020, by region of the country.

Region	December 2019		June 2020	
	Total adult and pediatric ICU beds[a]	Adult and pediatric ICU beds/100,000 inhabitants[a]	Total adult and pediatric ICU beds[a]	Adult and pediatric ICU beds/100 mil inhabitants[a]
North	1855	10	2966	16
Northeast	6853	12	11,639	21
Southeast	18,727	21	27,693	31
Midwest	3306	21	4770	30
South	5080	17	7098	24

[a]Including SUS and non-SUS beds
Source Brasil (2020c)

presented 96% occupancy of ICU beds and overcrowding in hospitals in the capital Manaus; the state of Ceará, with 100% occupancy of public ICU beds; the city of Rio de Janeiro, also with 100% occupancy of public ICU beds and where a waiting list for critical care patients was organized; and the city of São Paulo, which reached 89% occupancy and adopted measures such as hiring private beds and sending patients to hospitals in cities in the interior of the state (Portal G1 2020; AMB 2020; Susam 2020; Fiocruz 2020e; Vieira 2020).

Control Measures and Observed Impacts

Quarantine

The implementation of quarantine in Brazil took place from March 20, in all states and in most municipalities, although there was no compulsory social isolation, quarantine or lockdown at the national level. Basically, the strategy was based on the suppression of contagion, given that in the absence of specific treatment or vaccine for COVID-19, the most appropriate thing was to avoid the circulation of the virus in the population with some measures such as social isolation, environmental disinfection, use of personal protective equipment, among others. Thus, the expectation was that the infection in the population as a whole would occur more slowly and that the incidence curve would be "flattened," giving more time for health services to prepare themselves to receive a possible growing number of patients, especially severe ones.

Considering a date after March 20, as the effective start of the "lockdown"—due to a delay in the perception of the population and the ability of local and federal governments to effectively enforce the quarantine—the following case growth data were observed: 91% increase between April 7 and 15, 40% increase between April 15 and 20, 112% increase between April 20 and 30 and 59% increase between April

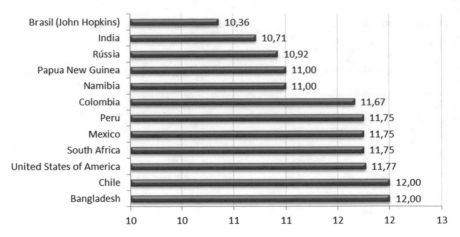

Fig. 7 Average days for doubling COVID-19 cases on 2020-08-04. *Source* Adapted from Fiocruz (2020b)

30 and May 5. Currently, the average number of days to double the number of new cases is among the lowest in the world, demonstrating that isolation does not seem to have been implemented effectively—Fig. 7.

In this sense, the COVID-19 Observatory, in order to estimate the rigidity of social isolation and lockdown measures in Brazil, analyzed: (i) the reduction in average automotive pollution in the country and (ii) the reduction in the Google's mobility index. When comparing both parameters between Brazil and several countries with a similar socioeconomic profile, the country showed lower average reductions in pollution and mobility, indicating that fewer people actually stayed at home at the beginning of the epidemic (Observatório COVID-19 2020).

Testing

Some facts about testing for COVID-19 in the country help in a more realistic interpretation of the evolution of the epidemic at the national level. What can be assessed so far in the Brazilian scenario is a growing daily increase in the cases (without considering possible effects of the scope of testing) , where there is not yet a clear trend of stabilization, since the available data to estimate the behaviour of the epidemic curve in Brazil are more related to the dynamics of notification than to the dynamics of the epidemic. In fact, at the beginning of the epidemic, there was a strangulation of the country's testing capacity (mostly RT-PCR), with slow distribution of tests and its effective performance and delivery of results in a timely manner by the National Network of Public Health Laboratories, especially outside the major regional centers. Thus, the information officially released in the country represented—and still represents—a retrospect of days or even weeks before the date of disclosure, making it difficult to adequately confirm and report the positive cases (Observatório COVID-19 2020).

It should be noted that the country has not adopted mass testing in the early stages of the epidemic, with most tests being restricted to health professionals, hospitalized patients or those with severe symptoms. In a first attempt to expand the testing for COVID-19 in the country, in May Brazil established the program "Diagnosticar para Cuidar" (Diagnose for Care), which foresaw the performance of 46 million rapid and molecular tests until October 2020 (Brasil 2020c). Divided into two strands, the action "Confirma COVID-19" (Confirm COVID-19) would use the RT-PCR molecular test to diagnose people with moderate or severe symptoms of the disease, the second phase, "Testa Brasil" (Testing Brazil), intended to leverage the use of rapid tests (serology) in the country to understand the progression of the virus nationally (Brasil 2020c). However, the actual testing carried out has been extremely low in the country. Information from Our World in Data showed that Brazil, until June 30, 2020, had a testing rate of 6.96/1000 inhabitants, while Denmark, the country that was testing the most, produced 183.76 tests/1000 inhabitants, and the USA, one of the epicenters of the disease at the time, performed 97.3/1000 on the same date (Our world in Data 2020). By the end of June, a testing extension was proposed and implemented for mild cases, as well as health and public safety professionals, but at that moment the confirmed cases already exceeded 1.5 million—considering underreporting, it is estimated that the actual number is about 11 times higher (Brasil 2020b; Prado et al. 2020).

Another important point refers to the percentage of positive tests, which has been higher than the 5% recommended by WHO—the positivity of tests in Brazil has been around 36.6%. The data indicate that only the most serious cases have actually been tested for the disease, hindering one of the key measures for monitoring COVID-19, which is the mass testing for the purpose of early detection of patients and isolation to suppress the spread of the disease (Tenente 2020). It is still necessary to take into account the fact that the country has not established a health policy that also focused on contact tracking between positive cases and potentially exposed people, which, together with the low rate of testing, makes it extremely difficult to draw a more realistic picture about the total number of active cases and the spread of the virus across the country.

The recent strategy of the Ministry of Health of counting serological tests in the statistics of new cases (along with RT-PCR) contributes to this difficulty. Although this strategy contributes to an increase in the capacity of exams by public and private systems, it compromises the clarity in the communication of active cases, since in some states, such as Paraná, even rapid diagnostic testing available in pharmacy has become part of the official numbers related to COVID-19. The concern is that this decision may blur the distinction between active cases and the prevalence of the disease, interfering in crucial epidemiological statistics such as fatality rate, as well as in public policies oriented to reopen the economy, threatening public health. It is possible that with the loosening of social isolation measures established from the second half of June onward by many states and municipalities, coupled with this testing scenario, there may be a worsening of the epidemic in the country. In fact, by August 2020, there was no sustained reduction in the number of new cases and deaths to allow a safely reopening for labour activities and commercial matters.

Economic Impacts

Social isolation, despite being the only strategy found so far to curb the spread of the disease, has a strong impact on economic activity, especially for those sectors and services that are not considered essential (FGV 2020). In general, estimates for the Brazilian scenario indicate a reduction of gross domestic product, in 2020, ranging from −0.68 to −4.5% (Júnior and Santa Rita 2020).

The guarantee of wages and employment was one of the most promoted issues worldwide to reduce the social impacts of COVID-19. The data show that this was a necessary measure, since 81% of the world workforce in April lived in countries that adopted some measure of social isolation, with 38% being employed in the sectors most severely affected by the pandemic: retail trade, food services and hospitality and manufacturing industry (Costa and Reis 2020).

The wage losses and unemployment resulting from the pandemic should compromise the family income of workers, especially the most vulnerable in the informal sector, which has led Brazil to adopt measures aimed at guaranteeing income for formal and informal workers. For the first group, the Emergency Employment and Income Maintenance Program, established by Provisional Measure (MP) N° 936, of April 1, 2020, was implemented (Costa and Reis 2020).[3] The program included a series of measures with the objective of preserving formal labour ties and guaranteeing workers' income in the midst of the economic and health crisis through the payment of an emergency benefit to the worker, based on the salary information for the last three months (Costa and Reis 2020). A study by the Institute for Applied Economic Research (IPEA) demonstrated the effectiveness of the program, which provided the maintenance of employment and income for a large part of the affected population, especially for workers who received up to 2 minimum wages—an estimated situation of about 70% of formal workers. However, for employees with a formal contract who received more than 3 minimum wages,[4] the maintenance of the job may have had a counterpart from small salary reductions—as in the case of the 25% decrease—to much more intense reductions resulting from the suspension of the contract of work (Costa and Reis 2020).

For the informal workers, individual micro-entrepreneurs, self-employed and unemployed, who comprise population groups that often cannot benefit from the protective measures of social distance and still have their jobs insured, an emergency

---

[3]The Emergency Employment and Income Maintenance Program authorizes, during the state of public calamity established until December 31, 2020, that employer and worker jointly plan, individually or collectively, the proportional reduction of working hours and wages or contractual suspension. The reduction in the employee's workday and salary could be 25, 50 or 70%, with a maximum term of 90 days. The suspension of employment contracts would have a maximum term of 60 days.

[4]The Brazilian minimum wage salary in 2020 was R$ 1,045.

aid was implemented. Established by law N° 13,982/2020, published on April 2, 2020, the aid provides for the payment of a benefit in the amount of R$ 600 for three months, for up to two people from the same family, or R$ 1200 for women heads of household for the same period.

Social Impacts

Both the pandemic itself and the measures adopted to contain it gave rise to new realities capable of raising or aggravating health problems, either directly or indirectly. Some examples would be mental disorders (increasing the risk of symptoms of stress, anxiety, depression and suicide, either in the general population or in health professionals), increased domestic violence and suicide attempts, pathological grief, post-traumatic stress disorder, excessive consumption of alcohol and other drugs, among others (Fiocruz 2020f, g; Wang et al. 2020; Zhang et al. 2020).

In the health sphere, there are still no national parameters to estimate the impact of the pandemic on the mental and behavioural health of the population (Ornell et al. 2020). However, it is estimated that the impacts will fall disproportionately on vulnerable population groups, such as residents of shantytowns, ethnic minorities, indigenous people, migrants and refugees, people deprived of their liberty, people with disabilities, LGBTQ, homeless people, among others (Fiocruz 2020g). These groups end up exposing a facet of the country evidenced by great socioeconomic disparities and low educational levels, which combined with the absence of a well-established culture of humanitarian cooperation makes psychological assistance to populations that have been victims of disasters and emergencies difficult (Fiocruz 2020g).

As for domestic violence, for example, in Brazil, although the official records of violence decreased in 2020 when compared to 2019, data show that the complaints and the action taken by the police for this type of case increased. Feminicide registrations grew by 22.2% in 2020 and The Call Center for Women in Situations of Violence (Ligue 180)[5] saw complaints of domestic violence grow by 27–37.6% of these only in April, when all Brazilian states have already adopted isolation measures (FórumBrasileiro de SegurançaPública 2020). Calls to the Police regarding domestic violence between March and April 2020 grew 3.5% in Rio de Janeiro, 22.3% in Acre and 44.9% in São Paulo (FórumBrasileiro de SegurançaPública 2020).

---

[5]The Call Center for Women in Situations of Violence (Ligue 180) is a free and confidential public service offered since 2005. Its purpose is to receive reports of violence and to guide women about their rights and current legislation. The Central operates 24 h a day, every day of the week, including weekends and holidays and can be activated from anywhere in Brazil and 16 other countries. Since March 2014, Ligue 180 has acted as a hotline, with the capacity to send complaints to public security with a copy to the District Attorney Office of each state.

Steps to Overcome the Current Situation

Considering that there are no scientifically proven preventive or curative technologies yet, epidemiological control measures remain the key to curbing the spread of COVID-19. However, there are important points to be considered by political and health authorities as well as managers of the SUS and civil society by conducing the epidemic in the country (Frente pela vida 2020).

Right now, it is important that strategies are drawn for the recovery of health and reconstruction of the population's living conditions, which can enhance the preparedness for future risks. Many measures presented so far, such as emergency aid, field hospitals and the purchase of health inputs, have been showing flaws and fraud in their implementation at the state and municipal levels; at the same time other important emergency measures have not yet managed to get off the ground (e.g., assistance for small companies, cultural sectors, among others). In addition, investments in input production capacity, according to the country's needs, and in technologies for testing and vaccines are extremely important (Frente pela vida 2020). In this sense, advances have been made, such as the agreement between the Oswaldo Cruz Foundation and the biopharmaceutical AstraZeneca for the purchase of lots and technology transfer of the vaccine developed by the University of Oxford (about 100 million expected doses), as well as the partnership between Sinovac and Butantan Institute, which signed a technology transfer agreement for the production of another vaccine with free supply to SUS of about 120 million doses. However, shortages of basic supplies such as drugs used to sedate and intubate critically ill COVID-19 patients, for example, need to be tackled, as their availability has been scarce and highly costly in most of the health units. Drug protocols should also be proposed based on scientific evidence, since speculations regarding new COVID-19 treatment and prevention protocols have spread across the country, fostering misinformation.

It is essential that the health system implements lines of care for patients at different stages of the disease, ranging from primary care (for patients with mild or asymptomatic conditions) to the ICU beds, as the rehabilitation of those affected by the disease (Frente pela vida 2020). The urgent need to increase the number of tests performed on the population must also be stressed, since, even with the low testing rate and the lack of accurate information available to understand the true epidemiological picture, the country is among the most affected in the world. (Frente pela vida 2020).

Finally, the need to implement a strategic intervention plan at national level, elaborated with the participation of scientific communities and SUS social control bodies, is reiterated (Frente pela vida 2020). Currently, a fragmented scenario is presented, in which decision makers from states and municipalities act at their own discretion, often not based on scientific evidence and without counting on a national public health guidance to cope with the epidemic. This is the result of a decision by the Supreme Court which ruled that the state and municipal governments should decide on issues such as quarantine, social distancing measures, use of face masks and the closure of schools (Supremo Tribunal Federal 2020).

Considerations

As in other countries, the pandemic policy in Brazil has not shown reasonable results. In the USA and UK, for example, considered by a study published by the Global Health Security Index as countries with greater "preparedness" to face COVID-19, there was a lack of personal protective equipment and ICU beds, the spread of a large volume of fake news and confusion about the measures that should be taken. On the other hand, in Asian countries such as China (which according to the same study ranked 51st in terms of preparedness) and South Korea, more effective policies have been observed to contain the pandemic nationally (Ortega and Behague 2020).

The fact is that the measures of social isolation in Brazilca used an enormous economic loss whose recovery will last for years, a crisis that affects and will affect mainly the poorest people who should be more protected by government measures. The epidemic evolved considerably during the writing this chapter. In January 2021, after the holidays, new peaks of cases and deaths began to be registered in cities the Southeast and North regions of the country. The most emblematic situation occurred the capital of the state of Amazonas, Manaus, in which the health system collapsed, with a lack of beds for hospitalization and oxygen for patients. In February 2021, almost a year after the first confirmed case of Covid-19 in Brazil, the country had already accumulated over 9.8 million new cases and 238.5 thousand deaths (surpassing, in many days, the mark of 1,000 daily deaths). This new reality highlights the need to rethink Brazil's public policies, both economic and social and related to the strengthening of the SUS, which will keep its demand increased after the end of the pandemic, given the accumulation of treatments in the background at this time.

That said, another important challenge in the country has been the achievement of collaborative and articulated governance between the federal, state and municipal levels. Considering these different conditions are essential for strategies to cope with the pandemic and its local repercussions.

Considering the presented scenarios, the Brazilian challenges facing the pandemic become even more complex because they coexist in a context marked by social vulnerability, economic and geographical inequalities, political instability and public health weaknesses. Clearly, the measures adopted so far have not had the desired effect: New cases have increased rapidly, and health services in many cities have reached their maximum capacity.

References

Associação Médica brasileira – AMB. (2020). Saúde em Manuas entra em colapso. Available from: https://amb.org.br/noticias/amb/saude-de-manaus-entra-em-colapso/. Accessed on July 05, 2020.

Boston Consulting Group – BCG. (2020). BCG Brazil Covid Task Force. Henderson Institute.

Brasil. (2016). Portaria GM/MS n° 2.567 de 25 de novembro de 2016 - Dispõe sobre a participação complementar da iniciativa privada na execução de ações e serviços de saúde e o credenciamento de prestadores de serviços de saúde no Sistema Único de Saúde (SUS). Ministério da Saúde. 2016. Availablefrom: https://observatoriohospitalar.fiocruz.br/sites/default/files/biblioteca/POR TARIA%20N%C2%BA%202.567.pdf. Accessed on May19, 2020.

Brasil. (2020a). Sistema Único de Saúde (SUS): estrutura, princípios e como funciona Availablefrom: https://www.saude.gov.br/sistema-unico-de-saude. Accessed on August 4, 2020.

Brasil. (2020b). Painel Coronavírus. Secretaria de Vigilância em saúde. Ministério da Saúde. 21 de mai. 2020. Availablefrom: https://covid.saude.gov.br/. Accessed on August 4, 2020.

Brasil. (2020c). CNES-Datasus. Ministério da Saúde. Available from: https://cnes2.datasus.gov.br/.

Brasil. (2020d). Saúde: Programa Diagnosticar para Cuidar – estratégia de ampliação da testagem da Covid-19 no Brasil. Governo do Brasil, 2020. Available from: https://www.gov.br/pt-br/not icias/financas-impostos-e-gestao-publica/500-dias/noticias-500-dias/saude-programa-diagnosti car-para-cuidar-2013-estrategia-de-ampliacao-da-testagem-da-covid-19-no-brasil. Accessed on June 09, 2020.

Cimerman, S., Chebabo, A., da Cunha, C. A., & Rodríguez-Morales, A. J. (2020). Deep impact of COVID-19 in the healthcare of Latin America: The case of Brazil. *Braz J InfectDis, 24*(2), 93–95. https://doi.org/10.1016/j.bjid.2020.04.005.

Cimini, Fernanda et al. (2020). Nota Técnica: Análise das primeiras respostas políticas do Governo Brasileiro para o enfrentamento da COVID-19 disponíveis no Repositório Global Polimap. Centro de Desenvolvimento e Planejamento Regional da UFMG – CEDEPLAR. Universidade Federal de Minas Gerais. 2020. Available from: https://www.cedeplar.ufmg.br/noticias/1242-nota-tecnica-analise-das-primeiras-respostas-politicas-do-governo-brasileiro-para-o-enfrentam ento-da-covid-19-disponiveis-no-repositorio-global-polimap. Acesso em: 20 mai. 2020.

Costa, J. S. M., & Reis, M. C. (2020). Uma análise da MP N° 936/2020 sobre os rendimentos dos trabalhadores e a renda domiciliar per capita. Nota Técnica N° 71. Diretoria de Estudos e Políticas Sociais. Instituto de Política Econômica Aplicada. Available from: https://www.ipea.gov.br/por tal/images/stories/PDFs/nota_tecnica/200603_nt_71_disoc.pdf.

Fiocruz. (2020a). Monitora COVID-19 – Fiocruz: Nota Técnica 04 Interiorização do Covid-19 e as redes de atendimento em saúde. Ministério da Saúde. Availablefrom: https://bigdata-covid19. icict.fiocruz.br/nota_tecnica_4.pdf. Accessed on June 15, 2020.

Fiocruz. (2020b). Monitora COVID-19. Ministério da Saúde. Available from: https://bigdata-cov id19.icict.fiocruz.br/.

Fiocruz. (2020c). Agência Fiocruz de notícias.InfoGripe: casos de SRAG estão acima do consid-erado muito alto.Available from: https://agencia.fiocruz.br/infogripe-casos-de-srag-estao-acima-do-considerado-muito-alto.

Fiocruz. (2020d). Agência Fiocruz de notícias.Observatório Covid-19: Taxas seguem altas em alguns estados. Available from: https://agencia.fiocruz.br/observatorio-covid-19-taxas-seguem-altas-em-alguns-estados.

Fiocruz. (2020e). Especialistas analisam a disponibilidade de leitos no país e discutem possibil-idades. Available from: https://portal.fiocruz.br/noticia/especialistas-analisam-disponibilidade-de-leitos-no-pais-e-discutem-possibilidades. Accessed on July 05, 2020.

Fiocruz. (2020f). Violência doméstica e familiar na pandemia COVID-19. Série Saúde mental e atenção psicossocial na pandemia COVID-19. Ministério da Saúde. Available from: https://www. arca.fiocruz.br/bitstream/icict/41121/2/Sa%C3%BAde-Mental-e-Aten%C3%A7%C3%A3o-Psi cossocial-na-Pandemia-Covid-19-viol%C3%AAncia-dom%C3%A9stica-e-familiar-na-Covid-19.pdf.

Fiocruz. (2020g). Suicídio na pandemia COVID-19. Série Saúde mental e atenção psicossocial na pandemia COVID-19. Ministério da Saúde. Availablefrom: https://www.arca.fiocruz.br/bitstr eam/icict/41420/2/Cartilha_PrevencaoSuicidioPandemia.pdf.

Frente pela vida. (2020). Plano nacional de enfrentamento a pandemia da COVID-19. Available from: https://www.abrasco.org.br/site/wp-content/uploads/2020/07/PEP-COVID-19_v2.pdf.

Fórum Brasileiro de Segurança Pública. (2020). Nota Técnica: Violência doméstica durante a pandemia de Covid-19 – ed. 2. Available from: https://forumseguranca.org.br/wp-content/upl oads/2020/06/violencia-domestica-covid-19-ed02-v5.pdf.

Fundação Getúlio Vargas – FGV. (2020). Impacto econômico do Covid-19: propostas para o turismo brasileiro. Centro de Estudos em Competitividade da FGV/EBAPE. Availablefrom: https://fgvprojetos.fgv.br/sites/fgvprojetos.fgv.br/files/01.covid19_impactoeconomico_v09_compressed_1.pdf.

Giovanella, L., Bousquat, A., Almeida, P. F. D., Melo, E. A., Medina, M. G., Aquino, R., & Mendonça, M. H. M. D. (2019). Médicos pelo Brasil: Caminho para a privatização da atenção primária à saúde no Sistema Único de Saúde? *Cadernos De Saúde Pública, 35,* e00178619. https://doi.org/10.1590/0102-311X00178619.

Instituto Brasileiro de Geografia e Estatística – IBGE. (2019b). Respiradores 2019. Available from: https://www.ibge.gov.br/estatisticas/investigacoesexperimentais/estatisticasexperimentais/27946-divulgacaosemanalpnadcovid1?t=downloads&utm_source=covid19&utm_medium=hotsite&utm_campaign=covid_19. Accessed on June 20, 2020.

Instituto Brasileiro de Geografia e Estatística – IBGE (2019a). Síntese de indicadores sociais : uma análise das condições de vida da população brasileira: 2019. Coordenação de População e Indicadores Sociais. Rio de Janeiro: IBGE, 130 p.

Instituto Brasileiro de Geografia e Estatística – IBGE. (2020a). Aglomerados Subnormais 2019: Classificação Preliminar e informações de saúde para o enfrentamento à COVID-19. Nota técnica. Available from: https://biblioteca.ibge.gov.br/index.php/biblioteca-catalogo?view=detalhes&id=2101717. Accessed on May18, 2020.

Instituto Brasileiro de Geografia e Estatística – IBGE. (2020b). Regiões de Influência das Cidades - REGIC. Available from: https://www.ibge.gov.br/geociencias/organizacao-do-territorio/redes-e-fluxos-geograficos/15798-regioes-de-influencia-das-cidades.html?edicao=27334&t=sobre. Acessed on June 20, 2020.

Instituto de Pesquisa Ambiental da Amazônia-IPAM, Coordenação das Organizações Indígenas da Amazônia Brasileira –COIAB. (2020). Não são números, são vidas! A ameaça da covid-19 aos povos indígenas da Amazônia brasileira. Available from: https://ipam.org.br/bibliotecas/nao-sao-numeros-sao-vidas-a-ameaca-da-covid-19-aos-povos-indigenas-da-amazonia-brasileira/. Accessed on July 28, 2020.

Institute of Global Health & Swiss Data Science Center. (2020). COVID-19 Daily Epidemic Forecasting. Available from: https://renkulab.shinyapps.io/COVID-19-Epidemic-Forecasting/_w_ff859a7c/?tab=ecdc_pred&country=Brazil.

Junior, R. R. F., & Santa Rita, L. P. (2020). Impactos da Covid-19 na Economia: limites, desafios e políticas. Cadernos de Prospecção, 13(2 COVID-19), 459. Doi: https://doi.org/10.9771/cp.v13i2.COVID-19.36183.

Malta, M., Rimoin, A. W., & Strathdee, S. A. (2020). The coronavirus 2019-nCoV epidemic: Is hindsight 20/20? EClinicalMedicine, 20. Doi: https://doi.org/10.1016/j.eclinm.2020.100289.

Observatório COVID-19. (2020). Análises comentadas. Available from: https://covid19br.github.io/analises.html?aba=aba5#. Accessed on June 15, 2020.

Ornell, F., Schuch, J. B., Sordi, A. O., & Kessler, F. H. P. (2020). "Pandemic fear" and COVID-19: Mental health burden and strategies. *Brazilian Journal of Psychiatry, 42*(3), 232–235. https://doi.org/10.1590/1516-4446-2020-0008.

Ortega, F., Behague, D. P. (2020). O que a medicina social latino-americana pode contribuir para os debates globais sobre as políticas da Covid-19: lições do Brasil. *Physis: Revista de Saúde Coletiva, 30*(2), e300205. Epub June 26, 2020. https://doi.org/10.1590/s0103-73312020300205.

Our World in Data. (2020). Statistics and Research. Coronavirus (COVID-19), Jun 15, 2020. Available from: https://ourworldindata.org/covid-deaths. Accessed on June 15th 2020.

PNUD: IPEA: FJP. (2016). Desenvolvimento humano nas macrorregiões brasileiras : 2016. – Brasília :, 2016. 55 p. il., gráfs., mapas color. ISBN: 978-85-88201-31-6. Available from: https://atlasbrasil.org.br/2013/data/rawData/19FEV_IDHM_WEB.pdf. Accessed on June 20, 2020.

Portal G1. (2020). Colapso no sistema de saúde em Manaus faz acompanhantes dormirem em chão de hospital. Jornal Nacional. Available from: https://g1.globo.com/jornal-nacional/noticia/2020/05/01/colapso-no-sistema-de-saude-em-manaus-faz-acompanhantes-dormirem-em-chao-de-hospital.ghtml. Accessed on July 05, 2020.

Prado, M. F. D., Antunes, B. B. D. P., Bastos, L. D. S. L., Peres, I. T., Silva, A. D. A. B. D., Dantas, L. F., ... & Bozza, F. A. (2020). Análise da subnotificação de COVID-19 no Brasil. *Revista Brasileira de Terapia Intensiva, 32*(2), 224–228.

Secretaria de estado de saúde do Amazonas – SUSAM. (2020). Manaus registra menor taxa de ocupação de leitos desde o início da pandemia, informa Susam. Available from: https://www.saude.am.gov.br/visualizar-noticia.php?id=4711. Accessed on July 05, 2020.

Sirkeci, I., & Yucesahin, M. M. (2020). Coronavirus and Migration: Analysis of Human Mobility and the Spread of COVID-19. *MigrationLetters, 17*(2), 379–398.

Supremo tribunal federal. (2020). Arguição de descumprimento de preceito fundamental 672 Distrito Federal. Available from: https://www.stf.jus.br/arquivo/cms/noticiaNoticiaStf/anexo/ADPF672liminar.pdf.

Tenente, L. (2020). Números mostram que Brasil ainda faz 'brutalmente' menos testes para coronavírus do que deveria; 'estamos no escuro', diz especialista. Portal de notícias G1, 12 de junho de 2020. Available from: https://g1.globo.com/bemestar/coronavirus/noticia/2020/06/12/numeros-mostram-que-brasil-ainda-faz-brutalmente-menos-testes-para-coronavirus-do-que-deveria-estamos-no-escuro-diz-especialista.ghtml. Accessed on June 13th 2020.

Vieira, B. (2020). Taxa média de ocupação das UTIs de hospitais municipais reservados para Covid-19 chega a 89% pela 1ª vez em São Paulo. Portal de notícias G1. Available from: https://g1.globo.com/sp/sao-paulo/noticia/2020/05/14/taxa-media-de-ocupacao-das-utis-de-hospitais-municipais-reservados-para-covid-19-chega-a-89percent-pela-1a-vez-em-sao-paulo.ghtml. Accessed on July 05, 2020.

Wang, C., Pan, R., Wan, X., Tan, Y., Xu, L., Ho, C. S., & Ho, R. C. (2020). Immediate psychological responses and associated factors during the initial stage of the 2019 coronavirus disease (COVID-19) epidemic among the general population in China. *International Journal of Environmental Research and Public Health, 17*(5), 1729. https://doi.org/10.1089/jpm.2020.0198.

World Health Organization (a). Novel coronavirus (2019-ncov): situation reports. Available from: https://www.who.int/emergencies/diseases/novel-coronavirus-2019/situation-reports/.

World Health Organization (b). WHO Coronavirus Disease (COVID-19) Dashboard. Available from: https://covid19.who.int/.

Wu, J. T., Leung, K., & Leung, G. M. (2020). Nowcasting and forecasting the potential domestic and international spread of the 2019-nCoV outbreak originating in Wuhan, China: a modelling study. Lancet. (Published online ahead of print.). Doi: https://doi.org/10.1016/S0140-6736(20)30260-9.

Zhang, C., Yang, L., Liu, S., Ma, S., Wang, Y., Cai, Z., ... Zhang, B. (2020). Survey of insomnia and related social psychological factors among medical staffs involved with the 2019 novel coronavirus disease outbreak. *Frontiers in Psychiatry, 11*, 1–9. Doi: https://doi.org/10.3389/fpsyt.2020.00306.

Impact of Coronavirus Disease (COVID-19) on Public Health, Environment and Economy: Analysis and Evidence from Peru

Pool Aguilar León, Kenji Moreno Huaccha, Judith Del Carpio Venegas, and Fiorela Solano Zapata

Abstract The pandemic due to the novel coronavirus, whose disease is called COVID-19, began and spread from Wuhan, China, with harmful results for public health and the economy. The measures adopted by the Peruvian government sought to mitigate the massive contagion and impact on vulnerable populations. However, despite the containment measures adopted, the overflow and collapse of health systems in the national territory has been observed. The chapter presents a brief reflection on the current state of SARS-COV-2's pandemic in Perú and analyzes the consequences from a healthcare, socioeconomic and environmental perspective. The pre-pandemic situation is detailed, as well as future gaps and opportunities.

Keywords COVID-19 · Coronavirus · Pandemics · Health · Environment · Socioeconomic

Introduction

The COVID-19 pandemic has been generating serious effects on global health and profound implications for economic growth and social development worldwide. In Latin America, Brazil was the first country to report a case in February 2020

P. Aguilar León (✉)
Cayetano Heredia National Hospital, Lima, Peru
e-mail: pool.aguilar@upch.pe

Faculty of Medicine, Cayetano Heredia Peruvian University, Lima, Peru

K. Moreno Huaccha
Inter-American Development Bank, Washington, USA
e-mail: moreno.kenji@gmail.com

J. Del Carpio Venegas
Wakaya Consultants and Ecoperuvian Consulting, Lima, Peru
e-mail: jcdelcarpiovenegas@gmail.com

F. Solano Zapata
Faculty of Medicine, Antenor Orrego Private University, Piura, Perú
e-mail: elicenesolanozapata@outlook.com

© The Author(s), under exclusive license to Springer Nature Switzerland AG 2021
R. Akhtar (ed.), *Coronavirus (COVID-19) Outbreaks, Environment and Human Behaviour*, https://doi.org/10.1007/978-3-030-68120-3_23

(Rodriguez-Morales et al. 2020) with a subsequent gradual continental expansion; heterogeneous in temporality and number of cases, and homogeneous in terms of socioeconomic outcome. At the end of September, South America registered more than 5 million cases and close to 200 thousand deaths (Dong et al. 2020).

The pandemic initially affected higher income population groups returning from abroad and their close contacts but spread rapidly to lower income populations that have less access to health services. Moreover, relatively low levels of investment in already stretched public health systems and lack of sufficient trained human health resources, mainly concentrated in capital cities, coupled with rapid population aging and rising prevalence of chronic diseases, have led to weak public primary health-care and hospital systems with suboptimal capacity of intensive care units. This represented a difficult context for confronting the situation.

Numerous environmental factors influence the outbreak and spread of epidemic or even pandemic events which, in turn, may cause feedbacks on the environment. The link between environment and the appearance of the COVID-19 pandemic occurs in two ways. On the one hand, the human–environment relationship, in the context of unsustainable lifestyles, with unsustainable patterns of consumption and production, increases the risk of zoonotic disease emergence (UN 2020). While, on the other hand, the emergence of the pandemic has caused environmental impacts and changes in environmental policy and management due to the natural effects of an outbreak and the mechanisms implemented to control it. Thus, in Peru, changes have been generated in the aspects of solid waste, generation of atmospheric pollutants, consumption patterns, deforestation activities and impact on the dialogue processes of socio-environmental conflicts.

Materials and Methods

The methodology developed aims to analyze the data provided by the Ministry of Health and biomedical databases about COVID-19 disease and its impact on Peru's public health through tables and graphs for assessing disease progression and related data. The keywords for searching were: COVID19, pandemic, socioeconomic impact and public health impact. The information presented in the study ranges from the first case registered in the country (March 2020) until September 30, 2020. Regarding environment perspective, the methodology followed included analyzing the available (by agost 2020) data provided by the Ministry of Environmental and the Ministry of Agriculture and Irrigation, as well as other national institutions that published environmental data as the Office of Public Defender, National Institute of Statistics and Informatics, National Agrarian Health Service and National Forest and Wildlife Service. The searched topics were: solid waste, deforestation, climate change, environmental politics, wildlife traffic, animal sanity and socio-environmental conflicts.

Healthcare Perspective

Disease Progression

Peru began its pandemic response preparations early, aware of the weaknesses of its under budgeted health system. On February 2, the Ministry of Health approved the National Plan for Preparation and Response against COVID-19 and a guideline for the management of suspected cases. However, there were delays and difficulties in purchasing PPE supplies, medical equipment, oxygen and supplies for laboratory molecular testing.

On March 6, the first confirmed case of COVID-19 was reported (Aquino and Garrison 2020). Subsequently, and responding to the growing number of cases, the government decreed the closure of schools and universities, the suspension of national and international flights and the declaration of health emergency with the concomitant closure of the country's borders. On March 15, the president declared a state of national emergency with a lockdown that placed strict controls on citizens' movements, except to purchase food or pick up medicines (Explorer 2020). Remote work was introduced, and only workers from critical sectors were allowed to commute.

Unfortunately, restrictions were not followed by all citizens, By March 30, the number of people detained at police stations for breaking the curfew reached 33,000, much higher than the reported number of infected people at the time (Explorer 2020). Additional measures were implemented in the following weeks, such as to wear face masks when outside the home, to restrict movement to specific days by gender just to purchase groceries and medications. However, those measures were difficult to maintain, partly supported by the precarious job security and informality (Taj and Kurmanaev 2020). These results are the consequences of a poor investment in health and education that goes back decades. Moreover, there was never a nationwide strategic communication plan to guide communication actions or health education or a mass public awareness campaign to motivate people to protect their health or change their behaviour and comply with preventive measures (García et al. 2020). This situation has worsened following the migration of thousands of Venezuelans to different cities. The emergence of migrant communities with inadequate living conditions, jobs and access to health services has further stretched the capacity of the nation's health system, which suffer from low levels of investment in public health and high levels of corruption.

Table 1 summarizes key socioeconomic characteristics. We emphasize that the shortages in the health sector are notable not only in terms of human resources but also in terms of logistics. Before pandemic, we had only 700 registered intensive care physicians and 820 intensive care units (ICU) beds with 900 mechanical ventilators. Hence, our country has had the lowest number of ICU beds per 100,000 inhabitants, below the average of 10 per 100,000.

One of the sustained weaknesses at the national level is the segmentation of the health system. Unlike other countries, Peru has several sectors: social insurance, public insurance and private insurance, in addition to the insurance of the armed

Table 1 Demographic, socioeconomic, health-related, and COVID-19-related characteristics of Perú

Population	33,089,123
GDP per capita (USD)	6941
GDP per capita (PPP, USD)	13,094
Population living in poverty (%)	16.8
Population living in extreme poverty (%)	3.7
Gini index	0.433
Physicians (per 10,000 people)	12.7
Nurses and midwives (per 10,000 people)	14
Hospital beds (per 10,000 people)	16
Current health expenditure (% of GDP)	5.1
Confirmed cases	818,297
Total COVID tests per million people	118,546
Reported deaths	32,535
Total confirmed number of deaths per million people	985

Sources World Bank, Economic Commission for Latin America and the Caribbean, United Nations Development Program and John Hopkins University-COVID-19 Dashboard
Note Last access from all sources, September 30, 2020

forces; each with its own autonomy, which determines an obstacle in the allocation of resources in addition to a greater workload for each health personnel who usually work for more than one sector.

During pandemic, ICU units increased to about 400%, mechanical ventilators by 2000% and ICU beds by 448%, in addition to an increased level of recruitment of physicians from all over the country.

At the end of September, Peru reached 818 thousand cases through two waves (Figs. 1 and 2), ranking as the third nation in South America, after Brazil and Colombia, with the major number of infected people (Worldometer 2020). Of the confirmed cases, 58.4% were adults with an incidence rate 10 times higher than children and adolescents. Hence, more than 12,000 Peruvians were hospitalized and about 1400 underwent mechanical ventilation (Ministry of Health 2020). Regarding mortality (Fig. 3), just over 32 thousand deaths have been notified (Ministry of Health 2020), with a lethality of 4.0%; placing the country as the second nation with the highest number of deaths per million inhabitants in the world (Worldometer 2020). The deaths due to COVID-19 might be higher than the official number reported due to two reasons. First, some deaths for which the real cause was COVID-19 may be underreported because there was not a positive test that confirmed the diagnosis, because the deaths reported by the government as COVID-19 deaths are only those with a positive test. Second, the pandemic can have an additional impact on deaths, that is, an indirect effect, increasing mortality for other diagnostics, caused by lower access to health. The elderly population is the most affected. Congregate settings

Fig. 1 Cumulative cases of COVID-19, between March and September. *Source* Ministry of Health

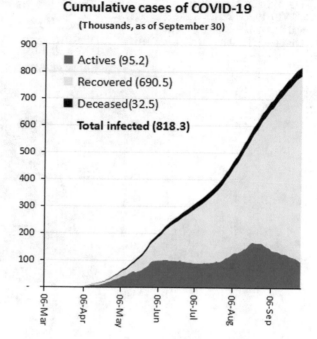

Cumulative cases of COVID-19

(Thousands, as of September 30)

■ Actives (95.2)

Recovered (690.5)

■ Deceased (32.5)

Total infected (818.3)

Fig. 2 Daily cases of COVID-19, between March and September. *Source* Ministry of Health

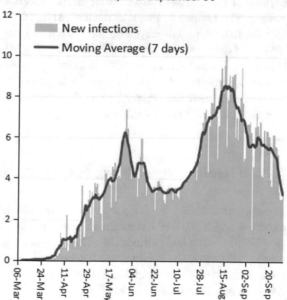

Daily cases of COVID-19

Thousands, as of September 30

New infections

Moving Average (7 days)

Fig. 3 Daily deaths from COVID-19, between March and September. *Source* Ministry of Health

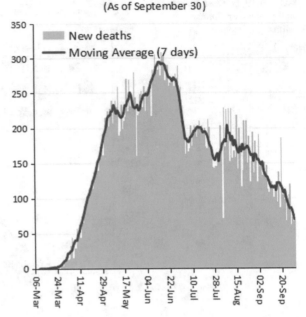

Daily deaths from COVID-19
(As of September 30)

favourable for increased transmission of the virus are present in the capital Lima because of rapid urbanization, which has produced a crowded city and widespread poverty. The regions that register the highest number of deaths, following Lima, are La Libertad, Piura, Lambayeque, Callao, Ica and Ancash, belonging to the macro-regions of the north and center of the country (Fig. 4). Despite the fact that the noti-fication of deaths was concentrated in hospitals, 11% (3172 deaths) were registered in homes, shelters and on public roads (Ministry of Health 2020). The magnitude of the pandemic has been compared with that observed in European hotspots, with the number of infected and deceased increasing rapidly in some cities.

Mental Health Issues During Pandemic

The current situation results in intense mental stress. People who have or have not experienced COVID-19 have mental disorders such as stress, anxiety and depression (Pfefferbaum and North 2020). There are also psychological disorders of work origin, such as those found in health personnel who work in contact with infected patients (Magnavita et al. 2020). Likewise, the population confined to their homes develops new emotional disorders (irritability, insomnia, fear, confusion, anger, frustration, boredom), which previously did not suffer from them, but which now persist even after the quarantine is lifted (Torales et al. 2020). In addition, recent studies

Fig. 4 COVID 19's
mortality rate by regions.
Source Ministry of Health

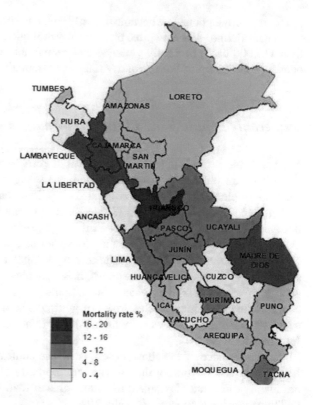

hypothesize that the virus not only affects the respiratory system, it could also be affecting the neurological system, which could cause the appearance of new mental illnesses (Yao et al. 2020).

In Peru, seven out of ten people experience a mental health problem, with anxiety symptoms being the predominant ones (Garay 2020). The most frequent symptoms are difficulties sleeping (55.7%), problems with appetite (42.8%), tiredness or lack of energy (44%), lack of concentration (35.5%) and suicidal thoughts or ideations (13.1%). Likewise, the fear of becoming infected and the consequences of transmitting it to the most vulnerable loved ones, such as parents and grandparents, are the most frequent consultations reported in psychological care through video calls and telephone. Mental health specialists specify three other triggers, such as the deep pain for the victims of COVID-19 within their families or closest circle, the lack of employment and the reduction in financial income (Redacción 2020).

The pandemic is causing an unprecedented mental health crisis in the Americas that is leading to an increase in domestic violence (Pan American Health Organization 2020). The quarantine encloses the victim and aggressor within four walls, increasing the tensions and associated risks in contexts of inequality, traditional gender roles and machismo (Hernandez 2020). Line 100 of the Ministry of Women and Vulnerable Populations became the emblematic service during the pandemic. Given the closure of the Women's Emergency Centers and the difficulty of going to the police stations to report, this telephone line became the only channel for victims of violence, mostly women in adulthood who reported physical-psychological aggressions. In total, more than 113,000 cases of family violence and sexual abuse were reported during the pandemic (Ministry of Women and Vulnerable Population 2020).

Vulnerable Populations: Elderly and Indigenous Communities

The pandemic is imposing a large death toll in Peru. Senior individuals, especially those who are older than 70 years of age, are being disproportionately affected, particularly elderly men (Munayco et al. 2020). Its lethality (15.72%) is 7.9 times higher than the adult population (Table 2). It is concluded that two out of every three deaths from COVID-19 occur in this age group (Ministry of Health 2020). One of the aggravating factors of the elderly's risk situation is the suffering of a chronic disease. In 2019, 39.4% of the country's population reported suffering from some chronic illness. In absolute figures, it is equivalent to 12 million 800 thousand people, of which 31.5% are older adults. Likewise, poverty is a multidimensional condition that also increases the risks of getting sick and dying. Last year, out of every 100 older adults, 15 were living in poverty (National Institute Of Statistics and Informatics 2019).

The importance of the above described in the context of COVID-19 lies in the fact that the risk factors for the severity and mortality of the disease in the country are mainly three and correspond to an age greater than 60 years, diabetes mellitus and hypertension (Ministry of Health 2020).

A further risk relates to the potential spread of SARS-CoV-2 infections to indigenous populations, especially the isolated groups in the Amazon. Lack of basic hygiene products, health professionals and hospital equipment and beds, poor transport infrastructure, language barriers or the difficulty to enforce self-isolation are just the tip of the iceberg of the numerous setbacks they face. Malnutrition, chronic and infectious diseases characterize its epidemiological profile, coupled with high fertility levels. Although efforts exist by the Ministry of Culture to ban access to protected areas and

Table 2 Lethality according to life stages, COVID-19, Peru 2020

Life stage	Infected	Deceased	Percentage	Lethality (%)
Children (0–11 years)	28,049	103	0.32	0.37
Adolescents (12–17 years)	21,451	44	0.14	0.21
Young (18–29 years)	152,497	287	0.88	0.19
Adults (30–59 years)	471,674	9,361	28.77	1.98
Elderly (60 and over)	144,626	22,740	69.89	15.72
Total	818,297	32,535	100	4.0

Source Ministry of Health

to deliver information on COVID-19 in 21 indigenous languages, policies show a lack of specificity to account for certain necessary actions, including adapted access to government aid, information on how to protect the elderly and how to adapt their burial rituals to minimize infection risks. Meanwhile, more than 15 thousand cases have been reported, 80% of which are Amazonian indigenous and the rest, residents of the Andes (Ministry of Health 2020).

Environmental Perspectives in Time of Covid-19

Risk of Zoonotic Outbreaks in Peru

Taking into account that the relationship between humans and the environment has been identified as one of the main causes of the current pandemic outbreak, it is important to analyze this causal relationship to reduce the probability of future zoonotic outbreaks emergence. Over 30 new human pathogens have been detected in the last three decades alone, 75% of which have originated in animals. What is more, 60% of all known infectious diseases in humans are zoonotic. Thus, the UN identifies as pathways for future pathogen transmission and potential future zoonoses the illegal animal trade, wet markets and ecosystems encroachment. So, to control these risks is important to work on stopping illegal trade, illegal wet markets, natural ecosystems encroachment and deterioration, as well as on the restoration of degraded ecosystems; and giving sustainable alternatives and protecting communities whose livelihoods might be threatened as a result (UN 2020).

To reduce the risk of future emerging diseases, it is important to regulate wildlife and their habitats, since they are the source of new pathogens, as well as livestock and domestic fauna, and they can act as amplifiers of emerging wild pathogens and as sources in themselves (UNEP 2020). Taking this into account could potentially be the place of origin for a new zoonotic outbreak. On the one hand, Peru is the second country with the largest extension of Amazonian forests in the world, after Brazil, with 60% of its extension covered by these (Ministry of Environment 2016); therefore, it is home to a large amount of fauna, which constitutes a potential source of pathogens not known to humans. The risk is even greater because between 2001 and 2018, 2.3 million hectares of Amazon forest have been lost (Ministry of Environment 2019), the main causes being the change of land use for pastures, agriculture, mining and human settlements (National Program for the Conservation of Forests for the Mitigation of Climate Change 2020). In addition to this, wildlife trafficking is also identified as a driver of emerging diseases in new regions and is a problem that affects the country. In Peru, as in Wuhan (China), traffic animals are commercialized in markets in major cities. This has been evidenced at national level (41 markets identified), mainly in the Amazon region, for different uses such as selling pets, use of parts or by-products in rituals, and folk medicine, as well as ornament and meat for consumption. (National Forest and Wildlife Service 2017).

On the other hand, the risk of the appearance of a new pathogen and zoonotic outbreak is greater due to the way in which livestock management occurs and the increase in the consumption of domestic animals in the country. In the country, there is a high rate of informality and a low rate of real compliance with regulations, including the existence of wet markets where the slaughter and commercialization of domestic and wild fauna occur without adequate biosecurity measures. This is the reason why the National Agrarian Health Service, the peruvian animal health authority, identifies as a problematic reality the introduction and spread of diseases in the livestock sector due to non-compliance with minimum sanitary standards, limited coverage in the care of zoonotic diseases and lack of control by the authorities (National Agrarian Health Service 2019). Furthermore, livestock production has intensified due to the increased demand for animal products. Between 2000 and 2018, even though the population increased by 19.6% (National Institute Of Statistics and Informatics 2019), the growth in production of animal products was up to 11 times that rate[1] (Ministry of Agriculture and Irrigation 2019).

Environmental Impact Due to Pandemic

Having already observed the importance of ecosystems protection to reduce the possibility of future zoonotic outbreaks, it is even more important not to neglect environmental issues within public policies and civil society interventions. This is even more important due to the fact that the health crisis has shown a decrease in environmental enforcement activities. Furthermore, due to the consequent economic crisis, it is possible that flexibilizations in environmental regulations are proposed to favour economic reactivation, as has happened in other countries.[2]

Solid Waste Management

A negative environmental impact has been observed regarding a greater generation of biocontaminated waste, a greater consumption of single-use plastic and the impact on the waste recovery chain, at the beginning of the social immobilization, due to the measures stricter social confinement. This happened in a context where there were already solid waste management problems, both on the side of disposal and the

[1]Production growth of poultry meat increased by 231%, pork meat by 77%, beef by 46%; while chicken egg production has increased by 153% and cow's milk by 129%.

[2]From example: (i) in USA, reporting rules for polluters have been relaxed during the pandemic, fuel-efficiency standards for new cars have been eased, implementation of estrict soot standards have been paralyzed, among others, (ii) in Alberta–Canada, environmental monitoring requirements have been suspended (iii) in Ontario–Canada, pandemic-related proposal were exempt from public consultation, this have raised concerns due to the transparency impairment that could enable government to allow projects proposals that are unrelated to the pandemic without the public knowing about them.

management of waste recovery. As an example of this, there is a coverage gap of 85% in adequate final disposal infrastructures, so that only half of municipal waste reaches a sanitary landfill (Office of Public Defender 2019). Thus, there are more than 1500 landfills and almost 2000 ha has been identified as degraded by solid waste, including riverbanks and beaches. (Office of Public Defender 2019). This problem not only needs to be solved by implementing more final disposal infrastructures, but also by reducing the generation of waste and inserting them into the recovery chain. The latter is particularly relevant since 63% of the waste that ends up in a sanitary landfill could have been recovered (Office of Public Defender 2019). In order to improve this situation, the government: (i) established regulations for single-use plastic[3]; and (ii) has an incentive mechanism, which encourages local governments to implement a recycling program to promote and collect recoverable waste at homes.[4]

Unfortunately, in the current context, several of the measures to prevent the spread of COVID-19, taken by the population, involve an increasing consumption of single-use containers and bags, as well as disposable PPE. In spite of this, the government has failed to implement a promotional strategy to reduce the generation of waste or to promote the reuse of materials, after their disinfection. In addition to this, during the first months of compulsory social immobilization, local recycling programs stopped, because waste recovery was not considered as an essential activity. The latter produced a disruption of the waste recovery chain, a socioeconomic vulnerability increase of formalized recyclers, who participated in the local recycling programs, and a negative effect on social awareness about the importance of waste segregation.

Hazardous waste, especially biohazardous, management deserves a special mention since waste classify in those types has been generated in much greater quantities due to the treatment of COVID patients, as well as the use of masks and other disposable PPE. This happens in a context where 19 of the 25 regions of the country do not have hazardous waste disposal infrastructures, and only 5% of biohazardous waste generated in health establishments is treated by autoclaving (Office of Public Defender 2020a).

---

[3]Regulations include the establishment of deadlines for the prohibition of: (i) the purchase, sale and use of single-use plastics in special areas such as protected natural areas; (ii) delivery of advertising and newspapers in plastic wrap; (iii) the manufacture of straws and plastic bags, that are not biodegradable, of a certain size and thickness; (iv) shops handing of non-reusable plastic bags; and (v) the manufacture, import and trade of bags and containers that are not biodegradable or reusable.

[4]The Peruvian Ministry of Economy and Finance has an Incentive Program for the Improvement of Municipal Management, whose objective is to improve the effectiveness and efficiency of public spending in the municipalities. Said incentive program, since 2011, includes a specific goal related to solid waste management. To achieve this goal, municipalities must implement recycling programs for segregation at the source and selective collection of solid waste in homes, where formalized recyclers' organizations have an important role. Recycling programs involve sensitizing and registering voluntary population to be part of the program through the segregation and periodic delivery of their usable waste to recyclers' organizations, which allows said waste to enter the recovery chain.

Climate Change's Agenda in the Current Context

Peru only generates 0.15% of greenhouse gases (GHG) produced worldwide (Ritchie 2017 the main source of emissions of these gases being land use change (45%), followed by the energy sector (30%) and the agricultural sector (16%) (Ministry of Environment 2019). Regarding mitigation of GHG generation, a positive environmental impact has been observed during the mandatory social isolation, since it was found that the accumulated deforested area in this time reached 7119 ha of forests, which is 28.7% less than what was registered in the same period during 2019 (Bosques Program 2020).

Moreover, in the pandemic context, the Congress approved the Annual Work Plan of the Climate Change Commission, which has as one of its objectives to contribute to the inclusion of climate and environment protection content in the post-COVID-19 economic reactivation and poverty reduction processes. Thus, it is intended that the economic reactivation policies in the face of the COVID-19 pandemic are aligned with the Nationally Determined Contributions (NDC); and that the characteristics of indigenous peoples, peasant producing communities and, in general, the population most vulnerable to climate change are considered and that special attention be paid to these population (Congress of the Republic 2020). Likewise, in July, the Ministry of the Environment ordered the creation of a permanent Multisectoral High Level Commission on Climate Change, which articulates the national government with subnational governments, and aims to propose adaptation and mitigation measures to climate change and NDCs, as well as issuing the technical report on NDCs every five years to the United Nations Framework Convention on Climate Change (UNFCCC).

Socio-environmental Conflicts

Finally, another environmental topic that was affected was the attention to socio-environmental conflicts, in a reality where, since 2006, half of the registered social conflicts are related to control, use and/or access to the environment and its resources. In the context of the crisis caused by the COVID-19 outbreak, in July 2020, the highest number of accumulated socio-environmental conflicts (129 cases) was identified in the last year. It is worth mentioning that the greatest recurrence of socio-environmental conflicts corresponds to conflicts related to extractive activities: mining (64%) and hydrocarbons (18%). Unfortunately, during quarantine, the dialogue spaces were suspended and the alternative of establishing remote dialogue spaces was delayed and has its limitations, which may generate more social tensions. (Office of Public Defender 2020b). Socio-environmental conflicts are a frequent and permanent problem in Peru which has often resulted in violence, injuries and even death of law enforcement officers and citizens. Lack of environmental management measures and/or the lack of adequate communication of these, as well as the genuine interest in citizen participation and local well-being by project managers, in a context where the government does not have firm protocols that require the population to be involved during project planning and allow for real citizen participation, results socio-environmental conflicts emergence and perpetuation. This situation becomes even more complicated and fosters an environment in which dialogue is difficult due to: (i) social preconceived notions regarding large extractive companies, that furnish the action of agitators; and (ii) the tendency to classify protesters as "terrorists," that facilitates the delegitimization of their requests (Bedoya 2015).

Socioeconomic Performance During Pandemic

After two decades of prudent fiscal and monetary management, Peru is one of the countries in the region with the strongest macro-economic position to face the crisis. Prior to the crisis, Peru had high levels of international reserves (29.7% of GDP) and financial assets (13.8% of GDP), a low public debt (26.8% of GDP), firmly anchored inflation expectations, as well as broad access to financial markets.[5] Likewise, to expand the capacity to finance crisis care, Peru suspended its macro-fiscal rules for 2020 and 2021. These aspects have allowed the implementation of a plan to attenuate the impacts of the crisis on economic activity and the well-being of the population. That plan has strategies that are equivalent to 20% of GDP, similar to the reaction of some developed economies (24% of GDP), and those focus on attending to the health emergency, the support of institutions and vulnerable populations, as well as the support and reactivation of the economy. These measures include sovereign guarantee programs to lower the cost of working capital credits to companies and the promotion of employment by the execution of public infrastructure works.

The initial impacts that Peru faced due to the COVID-19 crisis occurred in the external sector. The outbreak and spread of the virus caused the disruption of the logistics chains of productive inputs, especially with the main trading partners (China, the USA and Europe). At the same time, uncertainty about the world economy increased, which led to a drop in the prices of the main export commodities. The principal consequences of this shock are the accelerated growth of sovereign risk[6] and the deterioration of the trade balance.

On the other hand, according to the Oxford COVID-19 Government Response Tracker, the reaction of the Peruvian government to the crisis was one of the fastest and most energetic in the world. The first contagion was registered on March 6 and just ten days later, a rapid implementation of severe social confinement and distancing measures such as quarantine, curfew, closure of borders and educational centers, among others, began. However, although these measures served to mitigate

---

[5]Peru is one of the first emerging economies to issue debt during the pandemic (US$ 3 billion in bonds at historically low rates), which ensures resources to finance an important part of its actions to mitigate the crisis impacts.

[6]Peru's EMBIG rate increased from 116 points at the end of 2019 to 143 points in December 2020, with a peak of 372 points in March.

the spread of the virus,they also made economic activity work at only 45% of its capacity in the first months of lockdown.

The crisis has also been causing significant tensions in the social sphere, which could lead to the loss of the social advances obtained in the last decade. On the one hand, the labour market has been deeply weakened, since, in June, 23% of jobs were lost in the Metropolitan Lima area, especially in smaller firms. Likewise, those workers who kept their jobs have seen, on average, their income fall by 9.4% (INEI). This shock on employment could cause an increase in informality, which is already high (72% of total employment in 2019), and a greater precariousness of working conditions. On the other hand, there has been a high breach of social confinement measures, mainly due to overcrowding and the lack of certain equipment at home, which is more incisive in economically vulnerable groups.[7] Then, low financial inclusion[8] and incomplete household registration systems have hampered the effectiveness of the delivery of financial aids by the government. On the other hand, it should be noted that the Venezuelan migrant population has been excluded from all kinds of government aid, despite the fact that three-quarters of them do not have sufficient resources to acquire essential goods (Equilibrium—Center for Economic Development 2020). All these elements would cause changes in the country's socioeconomic structure due to a transfer of the middle class population toward a vulnerability situation (Castilleja 2020) and the probable increase in poverty to similar levels ten years ago (Lavado and Liendo 2020).

Fiscal accounts have been significantly affected. In 2020, fiscal revenues were reduced by 32% YoY due to the fall in imports, the reduction in the prices of export commodities, the weaker domestic demand and the tax reliefs provided by the government. Also, there is a risk of a persistent shock to tax collection if the informality and/or the tax evasion increase, as well as if new tax expenditures are implemented. In addition, the current crisis represents a substantial and immediate increase in public spending. Thus, the largest fiscal deficit in the last three decades would be recorded in 2020 (Central Reserve Bank of Peru 2020). As mentioned above, to allow a fiscal result of such magnitude, it was necessary to suspend fiscal rules, which could remain inactive for the next four to six years while the damages caused by the crisis are attended. The latter could imply risks on the credibility of the Peruvian macro-fiscal framework.

Finally, there are also impacts on the business fabric. Between the second and third quarter of 2020 at least 160 thousand companies were closed, mainly the smaller ones. This may be due to: (1) the disruption of supply chains, (2) the impossibility of operating normally due to social distancing measures and/or adapting activities to non-presential modalities and/or establishing virtual marketing channels, (3) changes in consumption patterns reflected in a greater preference for goods and services that

---

[7] In 2018, of the total of poor households, 9.4% lived in overcrowding and only 33% had a refrigerator, a clearly unfavourable situation compared to the case of high-income households (1.3% and 81%, respectively). The less ownership of refrigerators implies that the poorest households are forced to go to markets more frequently, which means a greater risk of contagion.

[8] As of June 2019, only 38% of the adult population has some type of bank account, and the financial system does not have a physical presence in 304 districts (16.2% of the total).

can be acquired in a remote manner (e.g.,, e-commerce grew during quarantine) or those suppose a lower risk of contagion, or (4) the lower purchasing power of the population, implying a negative demand shock. It should be noted that the closure of companies means a long-term shock on employment, since it implies the permanent destruction of jobs. The latter is particularly critical in a country with a very low degree of economic complexity such as Peru (The Atlas of Economic Complexity of Harvard University), since it reduces the possibility of workers to move quickly between different economic activities.

The described shocks caused, a cumulative contraction of the economy of -12.4% between January and November 2020, more deeply than in comparable countries. These impacts have been concentrated in the sectors that present the greatest difficulty in implementing sanitary protocols without considerably affecting the rhythm of production of goods and services such as tourism, construction, transportation and manufacturing.

Unfavourable conditions are likely to continue in the short and medium term. Therefore, it will be necessary to maintain economic support for vulnerable groups, which requires that transfer channels and targeting mechanisms be improved. Likewise, a substantial boost to job creation programs is urgently needed to make economic reactivation actions more effective. These and other measures will require an adequate fiscal space, which highlights the importance of proposing a reform on macro-fiscal rules.

Final Remarks

The Peruvian government faces the major challenge of balancing the introduction of compulsory and voluntary restrictions on population movement and other efforts to slow and prevent disease transmission while trying to contain the adverse impact of the pandemic on its economy and population, especially the financial and social fallout for the large proportions who are self-employed or work in the informal sector with few social protections. A major risk that needs to be addressed relates to the inconsistent compliance with social distancing, usage of face masks, isolation, quarantine, curfews and other voluntary restrictions on movement that could lead to uncontrolled re-emergence of infections.

We cannot ignore other impacts on public health such as possible increases in unintended pregnancies, mental health challenges, and domestic violence or worsening of chronic conditions due to overburdened health systems.

While the COVID-19 pandemic has had an unprecedented effect on society and the economy, to the contrary, it has helped repair some environmental damage (Shakil et al. 2020). Greenhouse gas emissions (GHG), nitrogen dioxide (NO_2), water pollution, noise pollution and pollution in beaches have reduced significantly due to full or partial lockdowns. Such restrictions have helped countries reduce their environmental pollution and improve air quality and quality of life. However, such containments also have negative consequences to the environment due to the increasing amount of domestic and medical waste that can be harmful and potentially transmit diseases to others unless appropriately treated.

In the economic sphere, it was evident that a country with accelerated economic growth and low foreign debt, faced with strict and early measures against the pandemic, had to reduce economic performance by less than half, with a detection of growth and above all with possibilities of decline at the global level and at the level of microenterprises with the advantage of informality that was already high with little assistance to foreign residents and low profits of certain microenterprises that resulted in bankruptcy. Moreover, the COVID-19 crisis has caused serious damage to the country's economy and social structure. In particular, the available projections predict a fall in GDP of more than 12%, as well as a substantial increase in poverty, unemployment and informality. Given this, Peru has been incorporating measures that, to date, are equivalent to 20% of GDP. In the medium and long term, it will be necessary to strengthen the role of public spending to maintain support for the most vulnerable groups through measures that boost private spending and promote employment, possibly implying the need for fiscal reform.

References

Aquino, M., & Garrison, C. (2020). Peru records first confirmed case of coronavirus, President Vizcarra says. *Reuters*. March 6. Available at: https://www.reuters.com/article/us-health-corona virus-peru/peru-records-first-confirmed-case-of-coronavirus-president-vizcarra-says-idUSKB N20T1S9.

Bedoya, C. (2015). Los conflictos socioambientales en el Perú y sus múltiples formas de entenderlos… y actuar en consecuencia, in La Trama, nr 46.

Bosques Program. (2020). Deforestación en Amazonía peruana se redujo en más de 28 % durante aislamiento social obligatorio. June 30. Available at: https://bosques.gob.pe/notasdeprensa/defore stacion-en-amazonia-peruana-se-redujo-en-mas-de-28--durante-aislamiento-social-obligatorio.

Castilleja, L. (2020) *The Andean middle class in the face of the Covid-19 shock* (Spanish). Inter-American Development Bank. Document for Discussion N° IDB-DP-00774. Doi: https://doi.org/ 10.18235/0002377.

Central Reserve Bank of Peru. (2020). *Inflation Report*. Lima, Peru. June 2020.

Congress of the Republic. (2020). Special Multiparty Commission in charge of the Follow-up and Formulation of Proposals for the Mitigation and Adaptation of Climate Change's Work Plan, Lima – Perú, p. 17.

Díaz-Cassou, J. (2020) *Country paper of COVID-19 in Peru* (Spanish). Inter-American Development Bank. Document for Discussion N° IDB-DP-00779.

Díaz-Cassou, J., Carrillo-Maldonado, P., & Moreno, K. (2020). *COVID-19: The impact of the external shock on the economies of the Andean region* (Spanish). Inter-American Development Bank. Document for Discussion N° IDB-DP-00779.Doi:https://doi.org/10.18235/0002457.

Dong, E., Du, H., & Gardner, L. (2020). An interactive web-based dashboard to track COVID-19 in real time. *the Lancet. Infectious Diseases, 20*(5), 533–534.

Equilibrium—Center for Economic Development. (2020). *Second National Opinion Survey "COVID-19 Quarantine in Venezuelan Migrant Population in Peru"— April, 2020* (Spanish). Retrieved from https://equilibriumcende.com/resultados-de-la-segunda-encuesta-nacional-de-opinion-cuarentena-covid-19-en-poblacion-venezolana-migrante-en-peru-abril-2020/.

Explorer A. (2020). Coronavirus in Peru- the latest updates. Available at: https://amazonas-explorer. com/is-there-coronavirus-in-peru/#Timeline_of_coronavirus_cases_in_Peru.

Garay, K. (2020). Covid-19: Siete de cada diez peruanos ven afectados su salud mental. *Diario El Peruano.* Available at: https://elperuano.pe/noticia-covid19-siete-cada-diez-peruanos-ven-afecta dos-su-salud-mental-100931.aspx

Garcia, P. J., Alarcón, A., Bayer, A., Buss, P., Guerra, G., Ribeiro, H., et al. (2020). COVID-19 Response in Latin America. *The American Journal of Tropical Medicine and Hygiene, 103*(5), 1765–1772.

Hernandez, W. (2020). La otra pandemia: Una mirada desde la evidencia. Radio Programas del Perú. Available at: https://rpp.pe/columnistas/wilsonhernandezb/la-otra-pandemia-una-mirada-desde-la-evidencia-noticia-1291081.

International Monetary Fund. (2020). *World economic outlook: A crisis like no other, an uncertain recovery.* Washington, DC. June 2020.

Lavado, P., & Liendo, C. (2020). *Covid-19, monetary poverty, and inequality* (Spanish). Foco Económico. Retrieved from https://focoeconomico.org/2020/05/29/covid-19-pobreza-moneta ria-y-desigualdad/.

Magnavita, N., Tripepi, G., & Di Prinzio, R. R. (2020). Symptoms in Health Care Workers during the COVID-19 Epidemic. A Cross-Sectional Survey. *International Journal of Environmental Research and Public Health, 17*(14), 5218.

Ministry of Agriculture and Irrigation. (2019). Anuario Estadístico Producción Pecuaria e Industria Avícola, Perú.

Ministry of Environment. (2016). *La conservación de bosques en el Perú (2011–2016)* (p. 180). Lima: Perú.

Ministry of Environment. (2019). Inventario Nacional de Gases de Efecto Invernadero del año 2014 y actualización de las estimaciones de los años 2000, 2005, 2010 y 2012, Lima – Perú, p. 317.

Ministry of Women and Vulnerable Population. (2020). Estadísticas - Consultas Telefónicas - Línea 100. Available at: https://www.mimp.gob.pe/contigo/contenidos/pncontigo-articulos.php? codigo=31.

Ministry of Health. (2020). Sala Situacional COVID-19 Perú. Available at: https://covid19.minsa. gob.pe/sala_situacional.asp.

Munayco, C., Chowell, G., Tariq, A., Undurraga, E. A., & Mizumoto, K. (2020). Risk of death by age and gender from CoVID-19 in Peru. *Aging, 12*(14), 13869–13881.

National Agrarian Health Service. (2019), Plan Operativo Institucional 2020.

National Forest and Wildlife Service. (2017). Estrategia Nacional para reducir el Tráfico Ilegal de Fauna Silvestre en el Perú 2017–2027, Lima - Perú, Ed 1, p. 72.

National Institute Of Statistics and Informatics. (2019). National Household Survey on Living Conditions And Poverty 2018. Available at: https://webinei.inei.gob.pe/anda_inei/index.php/cat alog/672.

National Program for the Conservation of Forests for the Mitigation of Climate Change – Ministry of Environment. GEOBOSQUES. (2020). Published online at https://geobosques.minam.gob.pe/ geobosque/view/cambio-uso.php.

Office of Public Defender. (2019). Informe Defensorial N° 181 – ¿Dónde va nuestra basura?: Recomendaciones para mejorar la gestión de los residuos sólidos municipales, Lima – Perú, Ed. 1, p. 265.

Office of Public Defender. (2020a). Serie Informes Especiales N° 24–2020-DP, Gestión de los residuos sólidos en el Perú en tiempos de pandemia por COVID – 19: Recomendaciones para proteger los derechos a la salud y al ambiente, Lima – Perú, p. 56.

Office of Public Defender (2020b). Reporte de Conflictos Sociales N° 197, Lima – Perú, p. 118.

Pan American Health Organization. (2020). Countries must expand services to cope with mental health effects of COVID-19 pandemic, PAHO Director says. Available at: https://www.paho.org/en/news/18-8-2020-countries-must-expand-services-cope-mental-hea lth-effects-covid-19-pandemic-paho.

Peruvian Institute of Economics. (2020). *Impact of the coronavirus on the Peruvian economy* (Spanish). Lima, Peru. Retrieved from https://www.ipe.org.pe/portal/informe-ipe-impacto-del-coronavirus-en-la-economia-peruana/.

Pfefferbaum, B., & North, C. S. (2020). Mental Health and the Covid-19 Pandemic. *The New England Journal of Medicine, 383*(6), 510–512.

Redacción EC. Essalud: temor de contagiar a seres queridos con COVID-19 es frecuente en consultas psicológicas. *Diario El Comercio.* Available at: https://elcomercio.pe/lima/sucesos/cor onavirus-peru-temor-de-contagiar-a-seres-queridos-es-frecuente-en-consultas-psicologicas-inf orma-essalud-nndc-noticia/?ref=ecr.

Ritchie, H., & Roser, M. (2017), CO_2 and Greenhouse Gas Emissions. Published online at OurWorldInData.org.

Rodriguez-Morales, A. J., Gallego, V., Escalera-Antezana, J. P., Méndez, C. A., Zambrano, L. I., Franco-Paredes, C., Suárez, J. A., Rodriguez-Enciso, H. D., Balbin-Ramon, G. J., Savio-Larriera, E., Risquez, A., & Cimerman, S. (2020). COVID-19 in Latin America: The implications of the first confirmed case in Brazil. *Travel medicine and infectious disease, 35,* 101613.

Shakil, M. H., Munim, Z. H., Tasnia, M., & Sarowar, S. (2020). COVID-19 and the environment: A critical review and research agenda. *The Science of the Total Environment, 745,* 141022.

Taj, M., & Kurmanaev, A. (2020). El virus exhibe las debilidades de la historia de éxito de Perú. NYtimes. Available at: https://www.nytimes.com/es/2020/06/12/espanol/america-latina/peru-coronavirus-corrupcion-muertes.html.

Torales, J., O'Higgins, M., Castaldelli-Maia, J. M., & Ventriglio, A. (2020). The outbreak of COVID-19 coronavirus and its impact on global mental health. *The International journal of social psychiatry, 66*(4), 317–320.

United Nations. (2020). A UN framework for the immediate socio-economic response to COVID-19, p. 49.

United Nations Environment Programme and International Livestock Research Institute. (2020). Preventing the next pandemic: Zoonotic diseases and how to break the chain of transmisión, Nairobi - Kenya, p. 82.

World Bank. (2020). *Global Economic Prospects, June 2020.* Washington, DC: World Bank. © World Bank. https://openknowledge.worldbank.org/handle/10986/33748. License: CC BY 3.0 IGO.

Worldometer. (2020). COVID-19 Coronavirus Pandemic. Available at: https://www.worldometers. info/coronavirus/.

Yao, H., Chen, J. H., & Xu, Y. F. (2020). Patients with mental health disorders in the COVID-19 epidemic. *The Lancet. Psychiatry, 7*(4), e21.

The Spread of COVID-19 Throughout Canada and the Possible Effects on Air Pollution

Fox E. Underwood

Abstract COVID-19 was declared a pandemic in March 2020, which is the same month we saw the spread of the virus throughout nearly all of Canada's provinces and territories. Although travelers to Canada from Asia and the Middle East tested positive in January and February, far more positive travelers arrived from the USA and Europe, suggesting a greater impact of COVID-19 in Canada. Following the surge of COVID-19 in March, Canada's most populated areas predominantly saw the highest absolute numbers of cases as well as cases per population. Smaller communities, and indigenous communities in particular, saw fewer cases. In comparing 2020 air pollution levels to 2019, there was a marginal decrease from 2019 to 2020 in O_3, PM_{10}, and $PM_{2.5}$ but not in CO and SO_2 following the pandemic shutdown in March. However, low numbers of monitoring stations and other factors hamper our ability to draw conclusions with great certainty.

Keywords Canada · COVID-19 · Air pollution · Particulate matter · Ozone · Carbon monoxide · Sulfur dioxide

Introduction

The spread of COVID-19 (Coronavirus Disease 2019) was declared a pandemic by the World Health Organization on March 11, 2020 (WHO Director-General 2020). In 2019, the presence of COVID-19 was detected in Europe (Chavarria-Miró et al. 2020, p. 2020.06.13.20129627) and South America (Fongaro et a. 2020, p. 2020.06.26.20140731), with patients positively identified in Asia in December (Wu et al. 2020, p. 265) and the first documented patient in Canada identified as a traveler returning to Canada in January from Wuhan, China (Zhao et al. 2020, p. 506). This chapter has three aims: to identify the origins of COVID-19 in Canada, to illustrate the spread of COVID-19 throughout Canada, and to compare air pollution levels of 2019 (pre-pandemic) to 2020 (pandemic).

F. E. Underwood (✉)
Departments of Medicine and Community Health Sciences, University of Calgary, Calgary, AB, Canada
e-mail: feunderw@ucalgary.ca

© The Author(s), under exclusive license to Springer Nature Switzerland AG 2021
R. Akhtar (ed.), *Coronavirus (COVID-19) Outbreaks, Environment and Human Behaviour*, https://doi.org/10.1007/978-3-030-68120-3_24

Methods

Data Sources

Canadian provincial and territorial governments report COVID-19 data. A group of academics (Berry et al. 2020, p. E420) have been compiling reported data into a freely available dataset: https://github.com/ishaberry/Covid19Canada.

Government-operated air pollution monitoring stations throughout Canada report data that are available through AirNow (AirNow 2020, https://www.airnow.gov/), which is a partnership of several US environmental and health agencies: https://files.airnowtech.org/?prefix=airnow/. The reported air pollutants available are carbon monoxide (CO) (8-h peak average, ppm), ozone (O_3) (1-h peak average, ppb), ozone (O_3) (8-h peak average, ppb), sulfur dioxide (SO_2) (24-h average, ppb), particulate matter 2.5 ($PM_{2.5}$) (24-h average, μg/m3), and particulate matter 10 (PM_{10}) (24-h average, μg/m^3).

A file that links 2016 Canada Census population from dissemination blocks to health regions has been created by Statistics Canada (Statistics Canada 2018): https://www150.statcan.gc.ca/n1/pub/82-402-x/2018001/corr-eng.htm. Province, territory, and health region boundary files were also provided by Statistics Canada (Statistics Canada 2019).

Data Preparation

Origin of COVID-19 in Canada

To identify the origins of COVID-19 in Canada, the proportion of cases diagnosed in people traveling to Canada and of cases diagnosed in people who contracted the virus by contact with infected people in Canada were explored. The countries and regions where people were traveling to Canada from were also described.

Distribution and Progression of COVID-19 Throughout Canada

To illustrate the spread of COVID-19 throughout Canada, a heat map of weekly total cases alongside cases per 100,000 persons across all health regions from January to July was created. Additionally, a choropleth map of health regions of cases per 100,000 persons was created, as well as a simplified topological map (Williams 2017, pp. 175–176; Underwood 2021, pp. 3–5), of Canada's provinces and territories that displays a bar code–style heat map inside, with each bar stripe representing a week's cases per 100,000 persons.

Comparison of 2019 and 2020 Air Pollution

To compare air pollution levels of 2019 (pre-pandemic) to 2020 (pandemic), line graphs of 2019 and 2020 values for each pollutant were examined, along with a table summarizing the differences. Pollution values were averaged (mean average) to the level of the country. Only monitoring stations present in both 2019 and 2020 were used for this comparison.

Results

Origin of COVID-19 in Canada

In January, all four COVID-19 cases in Canada were travelers returning from China. In February, 12 of the total 16 cases were travelers: four from China, seven from Iran, and one from Egypt. In January and February, destination provinces were British Columbia, a western province on the Pacific coast (13% of Canada's population), and the two large central provinces of Ontario and Québec, which together represent 60% of Canada's population. In March and April, virus-positive travelers from all regions of the world had returned to every province and territory in Canada except to the territory of Nunavut. The percentage of cases in returning travelers in March was 5.97%, with the following months reporting returning travelers as less than half a percent of cases (Table 1). Only 1,041 cases were confirmed as having contracted COVID-19 through contact with other Canadians, while 116,502 cases had no reported status. We are unfortunately only assuming that the vast majority of cases with no reported travel status contracted COVID-19 through local, community, contact. A total of 638 cases were confirmed to be people who were traveling to Canada. Table 1 additionally lists the number of non-travelers and number of cases with no reported travel status.

Although the earliest reported cases were of people traveling to Canada from Asia and the Middle East (Tables 1 and 2), in March, considerable numbers of travelers arrived from North America (115) and Europe (90). In fact, more virus-positive travelers returned from cruise ships (52) than from Latin America (43), the Middle East (46), or Asia (16).

Table 1 Cases in returning travelers: percentage of total cases

	January	February	March	April	May	June	July
Non-travelers, no reported status (#)	0	4	8059 (283, 7,776)	45,554 (320, 45,234)	38,182 (168, 38,014)	13,554 (157, 13,397)	12,190 (109, 12,081)
Travelers (#)	4	12	512	69	6	12	23
Travelers (%)	100	75	5.97	0.15	0.01	0.08	0.18

Table 2 Cases in returning travelers: returning from

	January	February	March	April	May	June	July
Asia	4	4	16	1			1
Europe			90	3		1	
Middle East		8	46				
North America (USA only)			115	8			3
Latin America			43	3			
Oceania			1	1			
Cruise ship			52				
Multiple regions visited			16				
Traveling from elsewhere in Canada			11	6	1	2	4
Region not reported			122	47	5	10	14

Distribution and Progression of COVID-19 Throughout Canada

In the choropleth map and topological map of Fig. 1 along with the heat maps of Fig. 2, we can see that the largest concentrations of absolute cases (i.e., over 500 cases in a week) were reported in and around some of the largest population centers in southern Ontario (Toronto, Peel) and Québec (Lanaudière, Laval, Montréal, Montérégie). In Ontario, this peak began in April and ended in early June. In Québec, this peak began a little earlier in late March and ended in early June. In Alberta (Calgary), three weeks in April saw more than 500 weekly cases reported.

After accounting for population, the largest concentration of cases per 100,000 persons (i.e., over 100 cases in a week) in Ontario disappeared entirely from Toronto and Peel. Instead, the less-populated region of Haldimand–Norfolk reported more than 100 cases per 100,000 persons in late May. In contrast, the large hot spot in Québec reduced but did not disappear: Lanaudière, Laval, Montréal, and Montérégie all reported more than 100 cases per 100,000 persons; however, this peak only lasted from April to early May. In Alberta, the South Health region reported two weeks in April of more than 100 cases per 100,000 persons. Around the same time, in the neighboring province of Saskatchewan, the Far North Health region reported more than 100 cases per 100,000 persons in the last week of April and the first week of May.

Over the entire period from January 1 to July 31, of the four regions in Canada (Territories, Western Canada, Central Canada, Atlantic Canada), the territories with 0.32% of the population had 0.016% of cases and 16.72 cases per 100,000 persons, the four Atlantic provinces with 6.63% of the population had 1.16% of cases and 58.07 cases per 100,000 persons, the four Western provinces with 31.55% of the population had 13.5% of cases and 142.70 cases per 100,000 persons, and the two Central provinces with 61.48% of the population had 85.3% of cases and 463.05 cases per 100,000 persons.

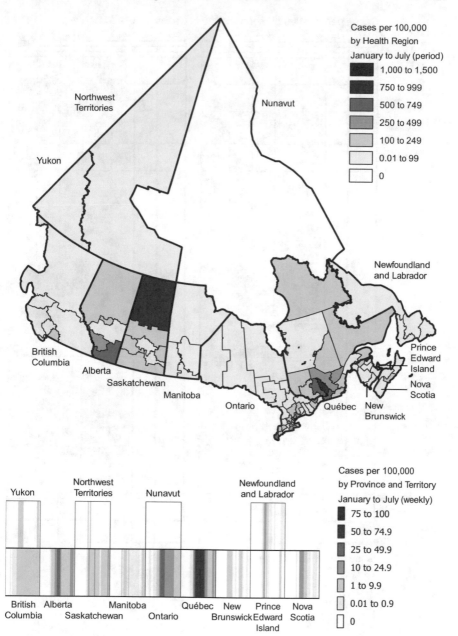

Fig. 1 COVID-19 cases throughout Canada: choropleth map (top) and topological map (bottom)

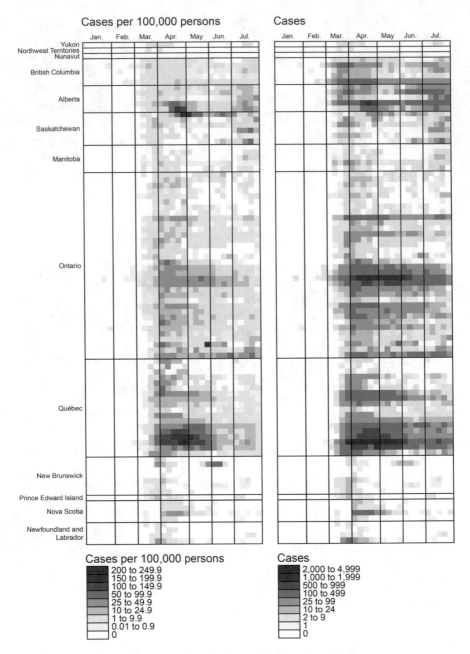

Fig. 2 COVID-19 cases throughout Canada: heat maps

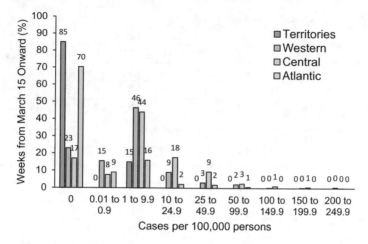

Fig. 3 COVID-19 cases throughout Canada: percentage of weeks

Nunavut is the only region in Canada with no reported cases between January 1 and July 31. Before mid-March, only three provinces had reported any cases (British Columbia, Ontario, and Québec). In mid-March, every province and territory except Nunavut reported cases. When considering only weeks from March 15 onward, the three territories reported 0 cases per 100,000 persons for 85% of the weeks and between 1 and 9.9 per 100,000 persons for 15% of the weeks (Fig. 3). Similarly, the three Atlantic provinces reported 0 cases per 100,000 persons for 70% of the weeks and between 0.01 and 9.9 cases per 100,000 persons for 25% of the weeks. Western Canada and Central Canada reported a more even spread with between 1 and 9.9 cases per 100,000 persons reported for 50% of the weeks.

Comparison of 2019 and 2020 Air Pollution

The monitoring stations in Canada that have reported data in 2019 and 2020 are present in every province and territory except for Yukon and Nunavut (Table 3). Additionally, two monitoring stations were present in 2020 that were not present in 2019, while one monitoring station was present in 2019 that was not present in 2020. Only monitors that were present in both 2019 and 2020 were compared. A single monitoring station can report more than one pollutant. As the trend of the 1-h peak average of O_3 was nearly identical to the 8-h peak average of O_3, and for reasons of space, only the 8-h peak average O_3 has been reported here.

As the week of March 15 was the first week that all provinces and territories had confirmed COVID-19 cases, only the sum of differences before and after that week was reported (Table 4). O_3 and $PM_{2.5}$ both decreased compared to 2019 across all periods considered. PM_{10} had been increasing before March 15 but then decreased

Table 3 Pollutants reported by monitoring stations

	2019					2020				
	CO	O_3	PM_{10}	$PM_{2.5}$	SO_2	CO	O_3	PM_{10}	$PM_{2.5}$	SO_2
Canada	10	140	6	128	16	9	142	5	128	15
Yukon	0	0	0	0	0	0	0	0	0	0
Northwest territories	0	2	0	0	0	0	2	0	0	0
Nunavut	0	0	0	0	0	0	0	0	0	0
British Columbia	1	26	3	40	4	1	26	2	40	4
Alberta	5	25	0	22	6	4	25	0	22	6
Saskatchewan	0	6	1*	4	0	0	6	1*	4	
Manitoba	0	2	3*	2	0	0	2	3*	2	
Ontario	0	28	0	23	0	0	29	0	24	
Québec	4	30	0	15	4	4	31	0	14	3
New Brunswick	0	6	0	8	0	0	6	0	8	
Prince Edward Island	0	3	0	3	1	0	3	0	3	1
Nova Scotia	0	7	0	7	1	0	7	0	7	1
Newfoundland and Labrador	0	5	0	4	0	0	5	0	4	0

*One monitoring station was located on the border of Saskatchewan and Manitoba. This monitoring station has been used for both provinces. This monitoring station was not counted twice when PM_{10} was averaged for all of Canada

Table 4 Sum of differences of 2020 minus 2019

	January 1–July 31		January 1–March 14		March 15–July 31	
Pollutant	Absolute	Percentages	Absolute	Percentages	Absolute	Percentages
CO (ppm)	0.08	−30.43	−0.15	−53.38	0.23	22.95
O_3 (ppb)	−28.12	−85.54	−10.62	−28.75	−17.49	−56.79
PM_{10} ($\mu g/m^3$)	−20.30	−1135.08	30.24	71.17	−50.54	−1206.26
$PM_{2.5}$ ($\mu g/m^3$)	−29.16	−607.45	−9.14	−173.89	−20.02	−433.55
SO_2 (ppb)	2.911	597.58	0.76	90.08	2.14	507.50

following March 15. By contrast, CO had been decreasing in January 1 to March 14 but then increased following March 15. As the sole exception, SO_2 increased in 2020 across all periods considered. This is corroborated in Fig. 4 where we can also see differences in shape (Fig. 4a shows 2020 minus 2019 values; Fig. 4b through Fig. 4f show 2020 and 2019 trends for each pollutant). Of note, CO peaked sharply at the end of May in 2019, whereas throughout June in 2020, we see a lower peak with a higher sustained value. In 2019, the same end-of-May spike can be seen in $PM_{2.5}$ in 2019, although without an increase in 2020 in June.

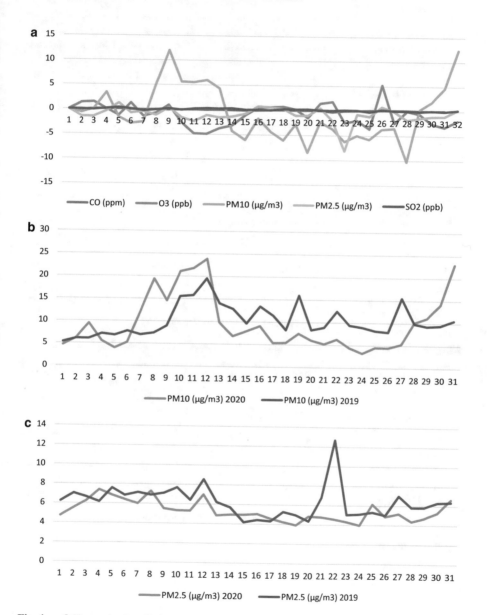

Fig. 4 a–f Change in air pollution from 2019 to 2020

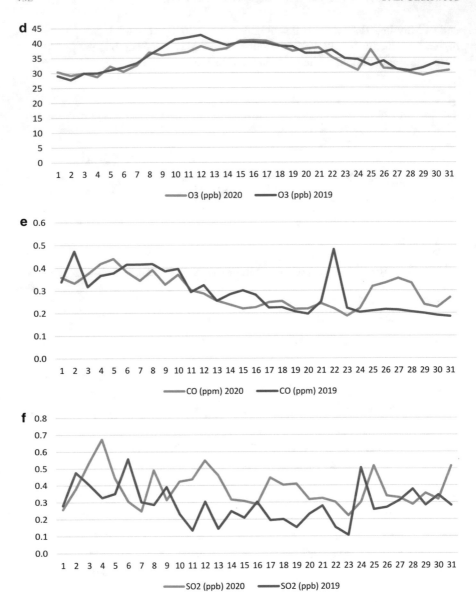

Fig. 4 (continued)

Discussion

Origin of COVID-19 in Canada

Although COVID-19 positive patients had been diagnosed in China in December 2019 (Wu et al. 2020, p. 265) and a small number of COVID-19 positive travelers had entered Canada in January and February from China (8), Iran (7), and Egypt (1), COVID-19 cases were only present in three of ten provinces in Canada until mid-March. Indeed, no community contact cases were present in January, while only four such cases were in February. In March, far more virus-positive travelers entered Canada from the USA (115) and Europe (90) than from Asia (16) and the Middle East (46), pointing to a greater impact of travelers from the USA and Europe. Even if all travelers of unknown origins (122) were added to cases from China and the Middle East (62), the total number of travelers is still fewer than the combined USA and Europe (205), with a potential range of 62 to 184 from Asia and the Middle East compared to a known total of 205 from the USA and Europe. For this reason, and with this current information, the impact of virus-positive travelers from the USA and Europe in March was likely the largest driver in the initial spread of COVID-19 throughout Canada. However, it is also possible that unreported cases were caused by contact with the travelers from China, Iran, and Egypt in January and February, with governments throughout Canada not testing large numbers of people until March. Additionally, the vast majority of cases did not have a marker for traveler or non-traveler, which means that we currently hold only an imperfect picture of how COVID-19 spread from other countries to Canada.

Distribution and Progression of COVID-19 Throughout Canada

British Columbia, Ontario, and Québec were the first provinces with known cases in January and February. Despite this earlier contact, only Ontario and Québec went on to report the highest number of cases per 100,000 persons in the period of January to July; in contrast, British Columbia never reported above 100 weekly cases per 100,000 persons. Indeed, even with later contact in March, Alberta and Saskatchewan surpassed British Columbia with weekly cases and total cases.

In considering the provinces and territories as four regions, the most populated regions had higher absolute numbers of total cases, which is expected. However, what was unexpected is that the most populated regions also had higher cases per population, albeit to a lesser degree than absolute cases. This observation usually holds true at the province and territory level, where provinces and territories with fewer than a million people had fewer cases per 100,000 persons. The exception is Nova Scotia (a population of 900,000), which had higher relative cases than British Columbia (a population of 4,600,000).

Despite Nova Scotia reporting a higher number of cases relative to its population than British Columbia, Nova Scotia had a higher number of weeks with no new cases at all in contrast to British Columbia which had reported cases every week since mid-February to late July. The remaining Atlantic provinces also reported several weeks with no new cases. The three territories remained almost untouched, reporting cases in only 7 of the 31 weeks in the study period. This potentially suggests a greater ability of regions with lower populations to suppress the spread of COVID-19, even if spikes in cases are occasionally greater relative to population than in regions with higher populations.

In this chapter, we looked at COVID-19 by province, territory, and health region. Given data of finer geographic regions, the pandemic should be examined further; for example, counties, communities of different sizes (i.e., cities, towns, villages), neighborhoods of different incomes, and indigenous communities. Banning (2020, p. E993) reported that as of August 6, 2020, the percentage of First Nations people living on reserves that tested positive "was one-quarter that of the general Canadian population" and that the fatality rate was only "one-fifth that of the general Canadian population." Perhaps smaller communities and communities further away from the largest population centers have fared better during the pandemic if only in terms of fewer positive COVID-19 cases.

Comparison of 2019 and 2020 Air Pollution

Since March 15, when COVID-19 cases began to surge and governments began responding with shutdowns, particulate matter in 2020 decreased substantially compared to 2019, with PM_{10} decreasing 1200% and $PM_{2.5}$ decreasing 433%. While PM_{10} was consistently lower each week from week 13 (the end of March) to week 28 (the start of June), $PM_{2.5}$ was alternatively lower and higher with only a brief period from week 21 (mid-May) to week 23 (early June) where the levels were consistently lower in 2020. The lower PM_{10} may indicate less vehicle exhaust from fewer vehicles on the roads following pandemic shutdowns. The lower $PM_{2.5}$ may have more to do with a forest fire or similar event causing a spike in $PM_{2.5}$ in 2019. This spike was also seen in 2019 for CO, possibly pointing to an event such as a forest fire. SO_2 in 2020 peaked in week 12 (mid-March) before sharply declining until week 14 and continuing to decline until week 16 (early April). However, a similar decrease in SO_2 around week 12 is present in 2019, suggesting no connection to any inactivity caused by pandemic shutdowns. Finally, no strong change in trend is seen immediately following mid-March (weeks 12–14) for either CO or O_3.

The most substantial limitation in this comparison is the low number of monitoring stations throughout Canada. For example, although PM_{10} may appear to have the strongest change in direction following mid-March, only five PM_{10} monitors were present in this dataset across all of Canada and only in three provinces. Only O_3 and $PM_{2.5}$ were represented by more than 100 monitors. CO, PM_{10}, and SO_2 were represented by fewer than 20 monitors. Future research could also compensate for forest fires, perhaps by weighting or by comparing only weeks with no forest fires in either time period. An additional suggestion is to compare pollution in 2020 to not only 2019 but also to, for example, 2-year, 3-year, 4-year, and 5-year averages of the previous years.

Adams (2020, pp. 4–5) examined levels of $PM_{2.5}$, nitrogen dioxide (NO_2), nitrogen oxides (NO_x), and O_3 in Ontario (the most populous province in Canada) and compared concentrations of pollutants in March and April 2020 to the past 5 years. NO_2 and NO_x saw a decline during the pandemic, likely a consequence of fewer vehicles on the roads and vehicle emissions being a larger source of NO_2 and NO_x than other sources are. $PM_{2.5}$ had no substantial change, with Adams speculating that any decline in $PM_{2.5}$ from traffic might have been offset by greater activity at home as larger numbers of people spend more time there. Finally, the reason for change in O_3 was uncertain: As the presence of O_3 will change from winter to spring, and as O_3 is produced in the presence of other air pollutants, complicating any determination. It is unclear at this point what effect the pandemic may have had on O_3. Adams notes that O_3 is something we will have to return to.

Stieb et al. (2020, p. 5) examined the association between $PM_{2.5}$ and COVID-19 incidence in Canada up to May 13, 2020. They found that higher $PM_{2.5}$ was associated with higher numbers of cases, particularly in regions where the population density was greater than the national median population density, although they say that their results should be interpreted with caution due to ecological fallacy and bias in their study design.

Conclusion

Despite the first virus-positive travelers arriving into Canada from Asia and the Middle East in January and February, the largest weekly number of cases across all provinces appeared throughout Canada in April and May, beginning a few weeks after travelers arrived in Canada from predominantly the USA and Europe. This implies a lesser impact from Asia and the Middle East as major sources of the spread of COVID-19 in Canada, although it is possible that more testing was done in March than in January and February, obscuring a true number of cases caused by Canadians in contact with travelers from Asia and the Middle East in January and February. However, as most cases were not listed as either traveler or non-traveler, future work is necessary to determine case travel status in the early months of January, February, and March if we are to better understand the spread of COVID-19 throughout Canada. As well, patient records of patients in January and February with complaints of pneumonia or similar symptoms should be examined to determine if there were more COVID-19 cases than we presently know of.

In considering the provinces and territories as four regions, the numbers of total cases as well as cases per 100,000 persons followed the population size of each region. Further, provinces and territories with fewer than a million people experienced a higher number of weeks with no reported cases, suggesting perhaps a greater ability of lower populated areas to mitigate the pandemic. However, a key limitation of this exploration was the differing percentage of population tested for COVID-19 in each province and territory. Future researchers should stratify or weight cases per population by the percentage of population tested. Future researchers should also examine differences in the spread of COVID-19 by using smaller geographic units, such as counties, cities, towns, and villages, and First Nations reserves.

Following the increase in COVID-19 cases in mid-March, PM_{10} was the only pollutant that had a visual change in trajectory in 2020 that was not also mirrored by 2019. Because of the low number of monitoring stations in this study, future researchers should obtain datasets of larger networks of PM_{10} monitors to confirm or deny this cursory observation. A comprehensive analysis would need to be done, taking into account industrial activity, forest fires, and regular seasonal changes in air pollution, to determine whether and how air pollution might have changed in response to the pandemic. Although, for reasons of space, the potential association between air pollution and the risk of COVID-19 was not examined, the severity of COVID-19 outcomes, or recovery rates, should all be studied further. Early work, with admitted study design issues, by Stieb et al. (2020, p. 5) has indicated a possible association between $PM_{2.5}$ and COVID-19 cases in Canada.

Addendum

This book chapter was written in the summer of 2020. Since then, both a second wave much larger than the first, as well as a serious outbreak in the territory of Nunavut, has occurred. See Underwood (2021), for a more recent and detailed breakdown by province and territory of the pandemic's spread throughout Canada.

For research examining the possible association between air pollution and COVID-19 cases, Goldberg et al. (2021, p. 110610) have written a critique of the work presented in Stieb et al. (2021, p. 5). Villeneuve et al. (2020, p. 095001) and Nicole (2020, p. 114005) have questioned the merits and methodology of several similar studies.

Keep your hearts warm to others. We have all lost something or someone during these trying times.

References

Adams, D. M. (2020). Air pollution in Ontario, Canada during the COVID-19 State of Emergency. *Science of the Total Environment, 742*, 1–5. Doi: https://doi.org/10.1016/j.scitotenv.2020.140516

AirNow. (2020). AirNow. Retrieved from https://www.airnow.gov/.

Banning, J. (2020). Why are Indigenous communities seeing so few cases of COVID-19? *Canadian Medical Association Journal, 192*(34), E993-994. https://doi.org/10.1503/cmaj.1095891.

Berry, I., Soucy, J.-P.R., Tuite, A., & Fisman, D. (2020). Open access epidemiologic data and an inter-active dashboard to monitor the COVID-19 outbreak in Canada. *Canadian Medical Association Journal, 192*(15), E420–E420. Doi: https://doi.org/10.1503/cmaj.75262.

Chavarria-Miró, G., Anfruns-Estrada, E., Guix, S., Paraira, M., Galofré, B., Sáanchez, G., … Bosch, A. (2020). Sentinel surveillance of SARS-CoV-2 in wastewater anticipates the occurrence of COVID-19 cases. *medRxiv*, 2020.2006.2013.20129627. Doi: https://doi.org/10.1101/2020.06.13.20129627.

Fongaro, G., Hermes Stoco, P., Sobral Marques Souza, D., Grisard, E. C., Magri, M. E., Rogovski, P., … Rodriguez-Lazaro, D. (2020). SARS-CoV-2 in human sewage in Santa Catalina, Brazil, November 2019. *medRxiv*, 2020.2006.2026.20140731. Doi: https://doi.org/10.1101/2020.06.26.20140731.

Goldberg, M. S., Villeneuve, P. J. (2021). An ecological analysis of long-term exposure to PM2.5 and incidence of COVID-19 in Canadian health regions. *Environmental Research, 194*, 110610. Doi: https://doi-org.ezproxy.lib.ucalgary.ca/10.1016/j.envres.2020.110610.

Nicole, W. (2020). Air of uncertainty: Can we study pollution and COVID-19 in the midst of a pandemic? *Environmental Health Perspectives, 128*, 114005. Doi: https://doi.org/10.1289/EHP8282.

Statistics Canada. (2018). Health Regions: Boundaries and Correspondence with Census Geog-raphy. Retrieved from https://www150.statcan.gc.ca/n1/pub/82-402-x/2018001/corr-eng.htm.

Statistics Canada. (2019). 2016 Census—Boundary files. Retrieved from https://www12.statcan.gc.ca/census-recensement/2011/geo/bound-limit/bound-limit-2016-eng.cfm.

Stieb, D. M., Evans, G. J., To, T. M., Brook, J. R., & Burnett, R. T. (2020). An ecological anal-ysis of long-term exposure to PM2.5 and incidence of COVID-19 in Canadian health regions. *Environmental Research, 191*, 1–7.

Underwood, F. E. (2021). Using topological maps to explore the COVID-19 pandemic in Canada. *Cartographica: The International Journal for Geographic Information and Geovisualization, 56*(1). Doi: https://doi.org/10.3138/cart-2020-0024.

Villeneuve, P. J., Goldberg, M. S. (2020). Methodological considerations for epidemiological studies of air pollution and the SARS and COVID-19 coronavirus outbreaks. *Environmental Health Perspectives, 128*, 095001. Doi: https://doi.org/10.1289/EHP7411.

WHO Director-General. (2020, March 11). WHO Director-General's opening remarks at the media briefing on COVID-19—11 March 2020. Retrieved from https://www.who.int/dg/speeches/detail/who-director-general-s-opening-remarks-at-the-media-briefing-on-covid-19---11-march-2020.

Williams, C. (2015). *Doing International research: Global and local methods.* London: SAGE.

Wu, F., Zhao, S., Yu, B., Chen, Y. M., Wang, W., Song, Z. G., … Zhang, Y. Z. (2020). A new coronavirus associated with human respiratory disease in China. *Nature, 579*(7798), 265–269. Doi: https://doi.org/10.1038/s41586-020-2008-3.

Zhao, N., Liu, Y., Smargiassi, A., & Bernatsky, S. (2020). Tracking the origin of early COVID-19 cases in Canada. *International Journal of Infectious Diseases, 96*, 506–508. https://doi.org/10.1016/j.ijid.2020.05.046.

Public Health Perspective of Racial and Ethnic Disparities During SARS-CoV-2 Pandemic

Michele Kekeh and Muge Akpinar-Elci

Abstract Health professionals have warned about the threats of flu (influenza)-like pandemics, for decades. As of February 2021, over 109 million people have been infected with COVID-19 worldwide. The USA counts for one-fifth of all infection cases worldwide. In the USA, racial and ethnic minority groups, such as African Americans and Hispanics, have been disproportionately impacted compared to their white counterparts due to their work and living conditions, lack of healthcare access and lower health insurance, long-standing structural forms of discrimination, pre-existing economic and social inequalities, and high rates of underlying chronic conditions. The inability to socially distance due to crowded living and working conditions, inability to take time off from work or work from home, and the lack of adequate testing and tracing protocols have put low-income communities, including service workers, farmworkers, and frontline workers, at high risks of infection. Worldwide, the inability to socially distance due to crowded living and working conditions, inability to take time off from work or work from home, and the lack of adequate testing and tracing protocols have put low-income communities, including service workers, farmworkers, and frontline workers, at high risks of infection. Some countries, however, have been less impacted than others due to good public health practice and early implementation of social measures, such as the ban on international travels, rigorous testing, implementation of contact tracing protocols, and social distancing. For many vulnerable communities, these measures must be rethought of and adapt to the specific local community.

Keywords COVID-19 · Infectious diseases · Racial disparities · Social distancing · Public health measures · Vulnerable populations

M. Kekeh (✉) · M. Akpinar-Elci
Center for Global Health, College of Health Sciences, Old Dominion University, Norfolk, VA, USA
e-mail: mkekeh@odu.edu

© The Author(s), under exclusive license to Springer Nature Switzerland AG 2021
R. Akhtar (ed.), *Coronavirus (COVID-19) Outbreaks, Environment and Human Behaviour*, https://doi.org/10.1007/978-3-030-68120-3_25

Introduction

Coronavirus disease 2019, more commonly known as COVID-19, is a novel coronavirus first identified in Wuhan, Hubei province, in the People's Republic of China. COVID-19 is caused by the severe acute respiratory syndrome coronavirus 2 (SARS-CoV-2) (Word Health Organization [WHO] 2020a). Among human populations, the virus is easily transmitted from person to person via inhalation. Small droplets expelled from the nose or mouth of infected individuals through sneezing, speaking, or coughing can quickly infect others in proximity. Disease transmission also occurs when individuals touch surfaces or objects on which viruses are present and then touch their nose, eyes, or mouth, allowing the virus to travel into the body. Clinical manifestations appear between 2 and 14 days after exposure. The most frequently reported symptoms of the disease include fever, cough, headache, sore throat, loss of taste, nasal congestion, gastrointestinal distress, pneumonia, and shortness of breath (Guan et al. 2020).

According to the Centers for Disease Control and Prevention (CDC), anyone is at risk of contracting the virus without safety measures. As of February 2021, over 109 million people have been infected by the virus worldwide (COVID Statistics 2021). The USA accounts for one-fifth of all cases worldwide(Johns Hopkins University and Medicine [JHU] 2020). About 80% of infected people recover from the disease without needing hospital care or treatment; however, older adults (individuals over 65 years old) and people with underlying health conditions, such as those who are immunocompromised or have diabetes, are at an increased risk for severe outcomes attributable to the disease (The Centers for Disease Control and Prevention [CDC], 2020a).

Health professionals have warned about the risk of an influenza-like pandemic, such as the COVID-19 pandemic, for decades. Despite these warnings, health professionals and governments worldwide were unprepared for the severity and seriousness of the COVID-19 outbreak. Unlike previous epidemics, this pandemic has been hallmarked by the dangers of asymptomatic disease and disease transmission. The high death rate associated with COVID-19 and peoples' ability to be infected and remain asymptomatic make this disease unlike any recorded in modern history. For reasons that are presently not well understood, the virus was able to infect individuals without causing clinical symptoms. It resulted in the rapid spreading of the virus among those infected without them knowing it. About 40% of infected people are asymptomatic. By the end of January 2020, only a month after the first case was identified, the disease spread from China to 20 other countries. The World Health Organization (WHO) Emergency Committee declared COVID-19 a global health emergency at the end of January 2020 (Velavan and Meyer 2020). Thus, the international community mobilized and began critical research to assess knowledge about the virus, developing global approaches and preparedness plans, creating platforms to enable collaborations, and accelerating interventions. The present chapter examines racial and ethnic disparities, related causes, public health and social measures, and reasons for lack of adherence to social distancing among vulnerable populations throughout the COVID-19 pandemic.

Racial and Ethnic Disparities

Although COVID-19 was initially thought to pose the same serious threat to all populations, present data has shown that disparities in afflicted populations exist. According to the National Center for Disaster Preparedness (NCDP), racial and ethnic minority groups, such as African Americans and Hispanics, are being impacted by COVID-19 disproportionally compared to other racial and ethnic groups. Although African Americans account for 13.4% of the US population, they represent 27.5% of all COVID-19 positive cases (NCDP 2020). His panics represented 18.5% of the US population in 2019; however, they account for 29.4% of the COVID-19 cases (CDC 2020b).

Health and social inequalities have been identified as important factors in explaining the disproportionate impacts of COVID-19 cases, hospitalizations, and deaths among racial and ethnic minorities (Karaye and Horney 2020). In New York City, a major epicenter for COVID-19 in the USA, for example, most cases of the disease were observed in working-class neighborhoods, comprising of primarily racial and ethnic minority groups, undocumented workers, and service workers, such as people working in grocery stores and public transportation, as well as people working in manufacturing and construction (van Dorn et al. 2020). In Utah, where a notable workplace COVID-19 outbreak occurred, about 58% of people impacted by the disease were service workers employed in sectors such as wholesale, construction, and manufacturing (Bui et al. 2020). Service workers have a high chance of contracting the virus due to increased exposure and close contact with the public, constant physical contact with coworkers, and their inability to self-isolate through work-from-home initiatives (NCDP 2020). Individuals employed in this sector tend to live in low-income communities and in older housing, struggle with food insecurity as a result of little to no access to supermarkets, lack active lifestyle opportunities, and are more likely to deal with pollution.

According to the Agency for Healthcare Research and Quality (AHRQ), minorities such as African Americans and Hispanics are more likely to work and reside closer to factories or refineries, which make them susceptible to the health effects of pollution, such as asthma and other lung diseases (AHRQ 2015). Poor living conditions and increased exposure to pollutants put minority groups at higher risk of chronic diseases, such as diabetes, hypertension, obesity, asthma, and cardiovascular diseases, compared to Whites. Research indicates that people with underlying chronic diseases are more likely to have an adverse outcome after contracting the virus, including death (van Dorn et al. 2020). Specifically, individuals who smoke and who have conditions such as diabetes, hypertension, asthma, and cardiovascular diseases experience the worst outcomes of COVID-19.

In addition to adverse environmental conditions and increased risk of chronic diseases plaguing racial and ethnic minorities in the USA, these groups have faced more challenges in accessing healthcare services historically. A lack of access to healthcare services, a lack of adequate health insurance, and lower health insurance enrollment affect racial and ethnic minorities' ability to seek out medical care. Racial and ethnic minorities face increased challenges in accessing healthcare services, providers, and preventive care, primarily due to the cost of care, when compared to Whites (CDC 2020c). Racial discrimination over healthcare availability has been studied for decades. The literature has explained extensively the health disparities and their underlying causes. The most frequently reported barriers to healthcare access include lack of transportation and child care, inability to take time off from work, conflicting work schedule, lack of time due to caring for parents, cultural differences, language barriers, low health literacy, and lack of trust in the healthcare system (CDC 2020c).

Although health disparities and inequities between racial and ethnic groups have been well studied, and significant problems have been identified across these groups, significant disparities also exist within groups. Immigrants, or those individuals born outside of and now presently residing in the USA, are an increasingly vulnerable population. The term "immigrant" includes foreign-born undocumented personnel, naturalized citizens, green card holders, refugees, asylum seekers, and those with temporary residency status residing in the USA (Castañeda et al. 2015; Cruz et al. 2009). Adding the first and second generation together, presently, there are 43.7 million immigrants in the USA, comprising 13.5% of the total population. Upon their migration to typically unfamiliar host countries, migrants face various difficulties that adversely impact their satisfaction, including language and social hindrances, lodging and work issues, low financial status, and absence of clinical protection (Castañeda et al. 2015; Cruz et al. 2009).

Issues such as lack of access to healthcare coverage and type of employment increase their vulnerability to suffer from chronic conditions and make them particularly vulnerable to diseases like the SARS-CoV-2 virus. Language barriers often prevent immigrants from acquiring health insurance and inhibit their ability to access healthcare services. Health insurance can be complicated, often riddled with complexities in deductibles, copays, and coverage, creating further challenges and barriers to access for this population. Language barriers also limit immigrants' abilities to obtain accurate health information. While there may be many sources of information on the prevention and mitigation of infectious diseases, in the case of COVID-19 quarantine guidelines, a lack of adequately translated materials limits individuals' abilities to protect themselves from adverse health outcomes. In such cases, individuals often rely on social media for information, which can be misleading, erroneous, and possibly dangerous during serious threats like the COVID-19 pandemic (Shadmi et al. 2020; Stokes et al. 2020).

Improving the present understanding of the disease's distribution by race, ethnicity, and working groups is an essential step in designing effective public health interventions, preventive strategies, and general actions to improve health outcomes among vulnerable groups. Without a comprehensive understanding of those groups disproportionately afflicted by COVID-19, formulated health interventions will be able to yield only limited success, and these populations will continue to suffer from preventable health outcomes. Early on during the COVID-19 pandemic, many communities did not collect demographic data, which is essential in providing assessments of how disease afflicts populations and determining groups at increased vulnerability for disease. Although there were likely many reasons such data was not collected, including the rapid and uncontainable spread of disease, public health professionals in the future, including health providers and other health entities, must collaborate to determine critical data that needs to be collected during public health crises.

COVID-19 and Stigma

Stigma is a negative attitude or belief toward a person or a group of persons due to a circumstance or event. Stigma is often intended to devalue members of a specific group in various ways, including in-person social interactions. According to the WHO, stigma is both a major cause of discrimination and a significant barrier. It affects people's self-esteem, disrupts their family, and limits their ability to socialize (WHO 2014). Public health professionals are very concerned about stigma and its related consequences, especially during public health crises. Stigma and discrimination have been serious issues throughout the COVID-19 pandemic and have been fueled by a lack of adequate knowledge about the virus and how it spreads, as well as a fear-fueled need to identify an entity to blame. In some of the worst cases of stigma, in the USA and Europe, those identifying as Asian or thought to be of Asian descent were discriminated against and even blamed for the occurrence and spread of COVID-19 (WHO 2020b).

The US president referred to COVID-19 as the "Chinese virus," perpetuating an already damaging stigma and inciting further discrimination against Asian descendants. Because the earliest cases of disease and community spread were identified in China, early rationalizations and beliefs were that only those traveling to China and neighboring countries could be affected. These beliefs, and ultimately the fear of disease spread, resulted in international travel bans worldwide that were ultimately unsuccessful. The first cases of the virus in the USA arrived via air travel and led to substantial cases of morbidity and mortality across the country (Laurencin and McClinton 2020).

Healthcare workers, recovered individuals and their families, and people who have been released from quarantine have also faced discrimination and stigmatization throughout the pandemic (Villa et al. 2020). Individuals in stigmatized groups may feel socially excluded, isolated, and depressed. Stigma can also affect their health and well-being and deprive them of jobs and resources to take care of their families. Stigma makes it more difficult to control the spread of the disease by discouraging individuals with symptoms from seeking testing and treatment and adopting healthy behaviours such as isolation. Providing facts and educating people about COVID-19 are crucial to reducing stigma. Health professionals must always protect the privacy and confidentiality of all cases, especially as they conduct contact tracing investigations. Community leaders must use all appropriate avenues, including news media and social media, to appreciate healthcare workers, speak out against negative language, and support individuals impacted by the damaging stereotypes (WHO 2014).

Public Health and Social Measures

Many countries worldwide have implemented public health and social measures to curb the contagion. These measures consist of movement measures, physical and social distancing measures, personal or individual measures, and special protection for vulnerable communities (WHO 2020c). These measures aim to control the pandemic through the slowing of disease transmission and reduction in mortality within and across countries. Movement measures, widely adopted across countries throughout the pandemic, include guidelines to limit local, national, and international travel. As an example of implementation, early in the epidemic, many countries in the European Union, except for Sweden, implemented a complete lockdown to stem the spread of the disease (Primc and Slabe-Erker 2020). China implemented a complete lockdown of the epicenter of the outbreak, the city of Wuhan, and took emergency response measures in other provinces across the country (Zhang et al. 2020). New Zealand, Italy, and Germany enforced mandatory lockdowns until COVID-19 cases were reduced and under control. In New Zealand, the complete lockdown allowed the government to manage its borders, provide testing, and allow for disease surveillance. Many lockdowns mandated that only essential businesses, such as grocery stores and medical facilities, remain open, temporarily banning recreational shopping, and other activities in order to minimize close contact between persons (Brooks 2020).

Physical and social distancing measures helped prevent the transmission between individuals, including the closure of schools and universities, the closure of non-essential businesses, implementing work-from-home initiatives, cancelation and banning of large social gatherings, including things like concerts and parties, complete closures of bars and restaurants or operating these services at limited capacity, and implementation of a minimum distance of at least 6 feet (2 m) between people in public or shared spaces (WHO 2020c). At the individual level, measures aimed to limit person-to-person transmission were highly encouraged and later mandated and enforced as countries and states saw fit, including the wearing of facial masks covering the nose and mouth, encouraging hand hygiene, environmental cleaning, disinfection at home, and social distancing (CDC 2020a).

Other measures to curve the spread of disease included rigorous disease testing, self-isolation, quarantine, and contact tracing. These measures helped identify potentially infected contacts to quarantine them early to manage further contamination and exposures. Countries such as South Korea, New Zealand, China, and Canada have brought their cases down drastically due to the implementation of testing and contact tracing. Special protection measures were also put in place to address concerns about the spread of disease to and among populations at increased risk of both contracting and suffering severe health consequences attributable to COVID-19. Special protection measures included strategies and means to minimize frontline workers' risk of exposure to the disease, shelter in place orders and mandates, limited visitations for older adults, especially those in long-term care facilities like retirement homes, and the identification of and planning for individuals at higher risks (WHO 2020c).

The decision to adopt and implement specific public health and social measures should depend on local epidemiology and the ability to identify and manage cases and contacts. The WHO has developed guidelines to assist countries and their decision-makers in deciding which measures they should adopt, how to apply the measures, and how to adjust them based on their countries' contexts (WHO 2020c). Each decision should consider (1) whether or not the epidemic is controlled, (2) whether or not the local health system can cope with a resurgence of COVID-19 cases after the implementation of a measure, and (3) whether or not there are surveillance measures in place to trace and manage new cases. Although the measures discussed are crucial in the mitigation of disease, implementation of such strategies should always consider the public health aspects and other factors, such as the economy, food security, human rights, and public sentiment.

Social Distancing and Complexities Among Vulnerable Populations

Globally, the COVID-19 pandemic has affected all countries. However, the burden of disease has afflicted countries differently. Countries such as South Korea, Canada, and China reported lower rates of infection and cases of death, primarily due to proactive public health measures implemented early in the pandemic. Social distancing has been one of the most significant measures in place, reducing the likelihood of person-to-person contact and virus transmission. In many low-income communities and across developing countries, the concept of social distancing is both far-fetched and seemingly impossible. In the USA, New York State, home to the most populous city in the nation, harbored the highest number of cases of the disease. Other cities, such as Chicago, Detroit, New Orleans, and California City, recorded many cases. These cities are a melting pot of cultures and groups; African Americans, Hispanics, Chinese, and other minorities are well represented here. Because these communities tend to live in close groups, they are more vulnerable to contagious diseases. Social or physical distancing is challenging in these areas (Laurencin and McClinton 2020).

For example, the overcrowding in New York and high rental often force people into unconventional living situations that put them at further risk for disease. The clusters of high COVID-19 rates in New York early were in Brooklyn, Staten Island Flushing, East Bronx, University Heights, and Queens. The relative risks of positive tests ranged from 1.15 to 1.38 (Cordes and Castro 2020). The median household income in these neighborhoods is about $48,000 (Buchanan et al. 2020). In the University Heights neighborhood, the median household income is approximately $22,000 (Goldbaum and Cook 2020).

Similarly, the global prison population is more likely to be affected by COVID-19 due to the enclosed and crowded facilities and limited medical resources to test, trace, isolate, or quarantine them (Cloud et al. 2020; Shadmi et al. 2020). The homeless population also faces a high risk of infection due to their living environments and lack of regular access to necessary hygiene supplies, making it easy for the virus to be transmitted from one person to the other (Tsai and Wilson 2020).

Farmworkers, especially those who work in meatpacking facilities, are at high risk of infection because they work in proximity with other people, share cramped housing, and travel to the fields in crowded vans. The CDC developed a set of measures to prevent SARS-CoV-2 transmission among farmworkers, including site risk assessment, guidance to ensure social distancing measures, prevention practices, training and education about the disease, and enforcement protocols (CDC 2020d). In low-income countries, the burden of COVID-19 among the most vulnerable populations cannot be overstated. It is common for extended family members to live in one household, making it difficult for social distancing. In addition, unsanitary living conditions with a lack of water and soap for regular hand washing put people at high risk of infection (van Dorn et al. 2020; Yancy 2020; Poole et al. 2020). Furthermore, the weak healthcare system with low resource capacity, which added to the dire testing and contact tracing shortfalls, and the lack of transparency in the data on the pandemic's actual impacts constitute significant challenges in the fight against the virus (International Rescue Committee 2020). The lack of adequate personal protective equipment puts frontline workers at higher risk of infection, especially healthcare workers (Behzadifar et al. 2020).

In many communities in developing countries, it is essential to understand the populations' economic challenges to determine potentially effective policies and target specific contexts and challenges. At the beginning of the pandemic, many developed countries provided economic assistance, such as large stimulus packages, reduced interest rates, and social safety nets to citizens (Bodewig et al. 2020). In many countries in the developing world, no assistance was provided to populations that were already living in poverty, lacked resources to fulfill basic needs, and were unable to afford bulk groceries or necessities to survive over a period of uncertainty (Maringira 2020). In sub-Saharan Africa, 80% of the population lives in dire social and economic conditions and is involved in low-wage employment. Asking these households to adopt public health measures such as staying home from work leads to an economic shock, putting their livelihoods at risk.

The COVID-19 pandemic has revealed how inadequately prepared many developed countries were, and the developing world is addressing the most vulnerable needs. Many of the preventive approaches must be carefully assessed before applying them to vulnerable populations due to their specific contexts and challenges (Maringira 2020). These challenges should be considered, and context-driven public health measures that include social protection programs should be developed to address the developing world's needs instead of applying a blanket set of standards for all countries.

Summary

Although many health professionals knew of the high probability of the occurrence of a flu-like pandemic, many countries were not adequately prepared against the virus. In the recent Ebola outbreak, countries collaborated and galvanized their efforts to support communities most impacted in the African region. With COVID-19, the lack of preparedness left each county to fight against the virus on its own. In the USA, without federal leadership, states were left to implement preventive measures according to their local context. Despite having one of the world's most advanced healthcare systems, the USA has had the highest mortality and morbidity ratios worldwide. Minority groups and low-income communities have experienced the worst outcomes of COVID-19. For an adequate fight against the virus, it is essential to have rigorous testing and contact tracing protocols to identify cases and isolate them. In addition, comprehensive policies should be implemented to help vulnerable communities recover inclusively and equitably.

"The manuscript provided has been read by all the authors. The requirements for authorship have been met, and each author believes the manuscript attached represents honest work."

References

Agency for Healthcare Research and Quality. (2015). *2014 National Healthcare Quality and Disparities Report*. US Department of Health and Human Services. Rockville, MD. Retrieved from http://www.ahrq.gov/sites/default/files/wysiwyg/research/findings/nhqrdr/nhqdr14/2014nhqdr.pdf.

Behzadifar, M., Ghanbari, M. K., Bakhtiari, A., Behzadifar, M., & Bragazzi, N. L. (2020). Ensuring adequate health financing to prevent and control the COVID-19 in Iran. *International Journal for Equity in Health*, Vol. 19. https://doi.org/10.1186/s12939-020-01181-9.

Bodewig, C., Gentilini, U., Usman, Z., & Williams, P. (2020). *COVID-19 in Africa: How can social safety nets help mitigate the social and economic impacts? World Bank Blogs*. Retrieved from https://blogs.worldbank.org/africacan/covid-19-africa-how-can-social-safety-nets-help-mitigate-social-and-economic-impacts.

Buchanan, L., Patel, J. K., Rosenthal, B. M., & Singhvi, A. (2020). A month of coronavirus in New York City: See the hardest-hit areas. Retrieved from https://www.nytimes.com/interactive/2020/04/01/nyregion/nyc-coronavirus-cases-map.html.

Bui, D. P., McCaffrey, K., Friedrichs, M., LaCross, N., Lewis, N. M., Sage, K., Barbeau, B., Vilven, D., Rose, C., Braby, S., Willardson, S., Carter, A., Smoot, C., Winquist, A., & Dunn, A. (2020). Racial and ethnic disparities among COVID-19 cases in workplace outbreaks by industry sector—Utah, March 6–June 5, 2020. *MMWR. Morbidity and mortality weekly report, 69*(33), 1133–1138. https://doi.org/10.15585/mmwr.mm6933e3.

Castañeda, H., Holmes, S. M., Madrigal, D. S., Young, M.-E. D., Beyeler, N., & Quesada, J. (2015). Immigration as a social determinant of health. *Annual Review of Public Health, 36*(1), 375–392. https://doi.org/10.1146/annurev-publhealth-032013-182419.

Cloud, D. H., Ahalt, C., Augustine, D., Sears, D., & Williams, B. (2020). Medical isolation and solitary confinement: Balancing health and humanity in US jails and prisons during COVID-19. *Journal of General Internal Medicine, 1.* https://doi.org/10.1007/s11606-020-05968-y.

Cordes, J., & Castro, M. C. (2020). Spatial analysis of COVID-19 clusters and contextual factors in New York. *Spatial and Spatio-temporal Epidemiology, 34,* 100355). https://doi.org/10.1016/j.sste.2020.100355.

Covid Statistics. (2021). *Coronavirus/COVID-19 worldwide cases live data & statistics.* Retrieved from https://covidstatistics.org/.

Cruz, G. D., Chen, Y., Salazar, C. R., & Le Geros, R. Z. (2009). The association of immigration and acculturation attributes with oral health among immigrants in New York City. *American Journal of Public Health, 99*(Suppl. 2), S474. https://doi.org/10.2105/AJPH.2008.149799.

Dong, E., Du, H., & Gardner, L. (2020). An interactive web-based dashboard to track COVID-19 in real time. *The Lancet Infectious Diseases, 20,* 533–534. https://doi.org/10.1016/S1473-309 9(20)30120-1.

Goldbaum, C., & Cook, L. R. (2020). *They can't afford to quarantine. So, they brave the subway.* Retrieved from https://www.nytimes.com/2020/03/30/nyregion/coronavirus-mta-subway-riders.html.

Guan, W., Ni, Z., Hu, Y., Liang, W., Ou, C., He, J., … Zhong, N. (2020). Clinical characteristics of 2019 novel coronavirus infection in China. *New England Journal of Medicine.* https://doi.org/10.1101/2020.02.06.20020974.

International Rescue Committee. (2020). *As confirmed COVID cases more than double in July across African countries, a lack of testing in crisis-affected contexts keeps responders in the dark about the real spread of the disease, warns IRC.* Retrieved from https://www.rescue.org/press-release/confirmed-covid-cases-more-double-july-across-african-countries-lack-testing-crisis.

Johns Hopkins University & Medicine. (2020). *COVID-19 dashboard by the center for systems science and engineering (CSSE) at Johns Hopkins University (JHU).* Retrieved from https://coronavirus.jhu.edu/map.html.

Karaye, I. M., & Horney, J. A. (2020). *The impact of social vulnerability on COVID-19 in the U.S.: An analysis of spatially varying relationships. American Journal of Preventive Medicine.* https://doi.org/10.1016/j.amepre.2020.06.006.

Laurencin, C. T., & McClinton, A. (2020). The COVID-19 pandemic: A call to action to identify and address racial and ethnic disparities. *Journal of Racial and Ethnic Health Disparities, 7*(3), 398–402. https://doi.org/10.1007/s40615-020-00756-0.

Maringira, G. (2020). Covid-19: Social distancing and lockdown in black townships in South Africa. *Social Science Research Council.* Retrieved from https://kujenga-amani.ssrc.org/2020/05/07/covid-19-social-distancing-and-lockdown-in-black-townships-in-south-africa/.

National Center for Disaster Preparedness. (2020). *Racial disparities and COVID-19.* Retrieved from https://ncdp.columbia.edu/ncdp-perspectives/racial-disparities-and-covid-19/.

Poole, D. N., Escudero, D. J., Gostin, L. O., Leblang, D., & Talbot, E. A. (2020). Responding to the COVID-19 pandemic in complex humanitarian crises. *International Journal for Equity in Health, 19,* 41. https://doi.org/10.1186/s12939-020-01162-y.

Primc, K., & Slabe-Erker, R. (2020). The success of public health measures in Europe during the COVID-19 Pandemic. *Sustainability, 12*(10), 4321. https://doi.org/10.3390/su12104321.

Shadmi, E., Chen, Y., Dourado, I., Faran-Perach, I., Furler, J., Hangoma, P., ... Willems, S. (2020). Health equity and COVID-19: Global perspectives. *International Journal for Equity in Health, 19*(1). https://doi.org/10.1186/s12939-020-01218-z.

Shu, Y., & McCauley, J. (2017). GISAID: Global initiative on sharing all influenza data—From vision to reality. *Eurosurveillance, 22,* 1. https://doi.org/10.2807/1560-7917.ES.2017.22. 13.30494.

Stokes, E. K., Zambrano, L. D., Anderson, K. N., Marder, E. P., Raz, K. M., El Burai Felix, S., ... Fullerton, K. E. (2020). Coronavirus disease 2019 case surveillance—United States, January 22–May 30, 2020. *MMWR. Morbidity and Mortality Weekly Report, 69*(24), 759–765. https://doi.org/10.15585/mmwr.mm6924e2.

The Centers of Disease Control and Prevention. (2020a). *Coronavirus disease 2019 (COVID-19).* Retrieved from https://www.cdc.gov/coronavirus/2019-ncov/hcp/non-covid-19-client-interaction.html.

The Centers for Disease Control and Prevention. (2020b). *Demographic trends of COVID-19cases and death in the U.S. reported to CDC.* Retrived from https://www.cdc.gov/covid-data-tracker/index.html#demographics.

The Centers for Disease Control and Prevention. (2020c). *Health equity considerations and racial and ethnic minority groups.* Retrieved from https://www.cdc.gov/coronavirus/2019-ncov/community/health-equity/race-ethnicity.html.

The Centers for Disease Control and Prevention. (2020d). *Agriculture and employers.* Retrieved from https://www.cdc.gov/coronavirus/2019-ncov/community/guidance-agricultural-workers.html.

Tsai, J., & Wilson, M. (2020). COVID-19: A potential public health problem for homeless populations. *The Lancet Public Health, 5,* e186–e187. https://doi.org/10.1016/S2468-2667(20)30053-0.

van Dorn, A., Cooney, R. E., & Sabin, M. L. (2020). COVID-19 exacerbating inequalities in the US. *The Lancet, 395*(10232), 1243–1244. https://doi.org/10.1016/s0140-6736(20)30893-x.

Velavan, T. P., & Meyer, C. G. (2020). The COVID-19 epidemic. *Tropical Medicine & International Health, 25,* 278–280. https://doi.org/10.1111/tmi.13383.

Villa, A., Jaramillo, E., Mangioni, D., Babdera, A., Gori, A., & Raviglione, M. C. (2020). *Stigma at the time of the COVID-19 pandemic.* Retrieved from https://www.clinicalmicrobiologyandinfection.com/action/showPdf?pii=S1198-743X%2820%2930477-8.

World Health Organization. (2014). *Stigma and discrimination.* Retrieved from https://www.euro.who.int/en/health-topics/noncommunicable-diseases/mental-health/priority-areas/stigma-and-discrimination.

World Health Organization. (2020a). *Naming the coronavirus disease (COVID-19) and the virus that causes it.* Retrieved July 8, 2020, from https://www.who.int/emergencies/diseases/novel-coronavirus-2019/technical-guidance/naming-the-coronavirus-disease-(covid-2019)-and-the-virus-that-causes-it.

World Health Organization. (2020b). *Addressing human rights as key to the COVID-19 response.* Retrieved July 30, 2020, from https://www.who.int/publications/i/item/addressing-human-rights-as-key-to-the-covid-19-response.

World Health Organization. (2020c). *Public health criteria to adjust public health and social measures in the context of COVID-19.* Retrieved from https://www.who.int/publications/i/item/public-health-criteria-to-adjust-public-health-and-social-measures-in-the-context-of-covid-19.

Yancy, C. W. (2020). COVID-19 and African Americans. *JAMA—Journal of the American Medical Association, 323,* 1891–1892. https://doi.org/10.1001/jama.2020.6548.

Zhang, S., Wang, Z., Chang, R., Wang, H., Xu, C., Yu, X., et al. (2020). COVID-19 containment: China provides important lessons for global response. *Frontiers of Medicine, 14*(2), 215–219. https://doi.org/10.1007/s11684-020-0766-9.

Geographic Patterns of the Pandemic in the United States: Covid-19 Response Within a Disunified Federal System

Glen MacDonald

Abstract COVID-19 arrived in the United States of America (US) early from multiple sources. Although federal border restrictions were implemented, the magnitude of international travel into the country rendered it impossible to stop the arrival of the virus at multiple locations. Community spread began quickly, and the high personal mobility of US via air travel and personal vehicles allowed for rapid dispersal across the 48 conterminous States. The subsequent per capita infections and mortality has varied markedly for individual states The former President Trump of the United States abrogated much of the pandemic public health responsibilities to the states and provided mixed-messaging on the pandemic. This contributed to a disunified and politicized response, with individual states having differing policies. Public opinion and behaviour also became divided. A unified, federally led, science-based response may have reduced COVID-19 cases and deaths. However, it can be argued that the federal system and its division of powers, disunified in this case, allowed a number of states to pursue public health measures that attenuated the severity of the pandemic in their jurisdictions without the executive branch of the federal government being able to limit measures.

Keywords United States of America · COVID-19 · Public health · Federal system · States · Politics · Geography

Introduction

The United States of America (US) is one of the most economically and technologically advanced nations in the world. According to the World Bank (2020d), the US was ranked number one in gross domestic product (GDP) in 2019 with a total GDP of over $21 trillion (all estimates in $US at roughly current rates). The per capita GDP of the US places it amongst the top nations (World Bank 2020c). Annual per capita healthcare spending is estimated at $10,246 (World Bank 2020b). In view of these

G. MacDonald (✉)
Department of Geography, UCLA, Los Angeles, CA 90095, USA
e-mail: glen@geog.ucla.edu

© The Author(s), under exclusive license to Springer Nature Switzerland AG 2021
R. Akhtar (ed.), *Coronavirus (COVID-19) Outbreaks, Environment and Human Behaviour*, https://doi.org/10.1007/978-3-030-68120-3_26

451

immense economic resources, and expenditures on healthcare, it is seemingly incongruous that the US leads the world with the number of COVID-19 (SARS-CoV-2) infections with 30,095776 cases and 545,188 deaths as of 1st April, 2021 (World Health Organization 2021). This represents almost 20% of all known COVID-19 cases and 19% of all known COVID-19 deaths from a country which has only about 4% of the world's population. The numbers of infections and deaths in the US speak of a tragic national failure. This chapter considers the performance of the US in light of the geographic diversity imparted at the state level by its federal structure and related politics. Geographic differences in the evolution of the pandemic in the US reflect many factors, but the fact that the US has a politically polarized federal system with important divisions of power between the central federal government and the individual states is an overarching feature of the behaviour of the US in the pandemic. The focus in this chapter will primarily be upon geographic diversity in COVID-19 responses and impacts that are detectable at the state level, and how this diversity may reflect variations in federal and state-level responses and associated political factors. It is recognized that differences exist in COVID-19 testing over time and space (Perniciaro and Weinberger 2020), but it will be assumed that the Centers for Disease Control data are broadly representative of the general trends and patterns.

Some Geographic Facets of the US Federal System Relevant to the Pandemic

The US federal system designates a high degree of intrastate authority to the 50 states. The federal government retains significant authority, and in areas of external relations, such as borders and immigration, it is unilaterally empowered. In areas of health, medicine, and internal emergency measures related to pandemic responses, the delegation of authority is complex (National Conference of State Legislatures 2014; Holt et al. 2019; Rasmussen and Goodman 2018). The US Constitution empowers the federal government to provide for the general welfare of United States. The US Public Health Service, the Centers for Disease Control and Prevention, and the Food and Drug Administration (FDA) are federal agencies. The federal government also has broad authority to act when a national emergency is declared by the President. These range from external border restrictions, allocation of emergency funds, the reassigning of public health workers who are paid by the federal funds, quarantining, providing of medical supplies etc. States also have very strong independent powers to protect the public health of their citizens. These include substantial internal "policing powers" that allow the states to regulate the activities of private citizens and commercial interests in order to protect public health. The federal system of the US provides for a high degree of geographic diversity. Individual states may have very different regulations from their neighbors. The polarization of American politicians and the electorate between the Democrat and Republican parties often enhances such differences (de Bruin et al. 2020).

In addition to the constitutionally prescribed federal regulatory structure, there is also the impact of the so-called bully pulpit to influence policy decisions by policy makers and the adherence to those policies by the public (Forester and McKibbon 2020). Government leaders at the federal, state, and local level can by their pronouncements and actions convey persuasive messages of appropriate or inappropriate responses to public health crises. There are also positive feedbacks that operate between politicians and their constituents that can reinforce the direction of policy and the behaviour of both leaders and followers (Campbell 2012; Rudden and Brandt 2018).

The United States displays a huge degree of geographic diversity in demographic, socioeconomic, and political conditions that could influence the epidemiology of an infection such as COVID-19 (Abedi et al. 2020; Zhang and Schwartz 2020). The population density for the 48 conterminous states varies from over 1000 people per sq mile to less than 10 per sq mile (US Census 2020a). States such as California and Massachusetts have 90% of their population classified as urban, while other states such as Mississippi and North Dakota have populations that are 60% or less urban (US Census 2020b). Racial composition is similarly diverse. Per capita GDP ranges from ~$75,000 in New York to ~$35,000 in Mississippi (Duffin 2019). Geographically variably environmental factors such as air temperature and humidity may influence airborne pathogens such as COVID-19 and this has received study (e.g., Qi et al. 2020; Wu et al. 2020). This chapter will focus on political factors with the understanding that this is only one facet of the complex diversity of the US, and the variables listed above often have interrelationships with political partisanship and dynamics. At the time of writing, 24 states had Democrat governors and 26 had Republican governors. Urbanized states and counties often favour Democrats, whereas rural areas favour Republicans (Scala and Johnson 2017). States with Democrat controlled governments typically have a higher per capita GDP. That gap has been growing an in 2018 Democrat controlled states had a per capita GDP that was $12,010 higher than Republican controlled states (Green 2020).

Geographic Andtemporal Patterns of Initial Spread, Increasing Infection Rates, and Mortality

SARS-CoV-2, now commonly known as COVID-19, was identified as the source of an outbreak of a virulent pneumonia in Wuhan, China, at the end of 2019. In late January 2020, a case of COVID-19 was confirmed in the state of Washington (Harcourt et al. 2020). In the period of January 21–February 23, 2020, 14 US cases were identified in a widely distributed set of states ranging from California and Arizona in the southwest, to Wisconsin and Illinois in the Great Lakes region, and Massachusetts on the east coast (Fig. 1). From that point onward, the infection spread

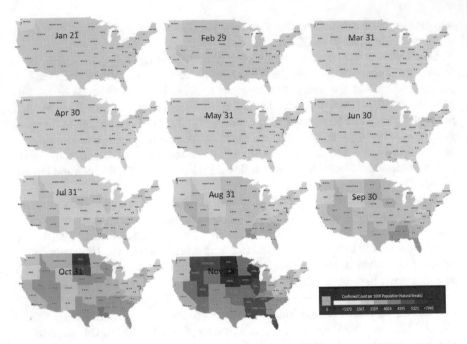

Fig. 1 Confirmed COVID-19 case counts per capita (measured as cases per 100,000 population) for the 48 states of the conterminous United States from the documented first case in late January, 2020 up to November 18, 2020 (extracted from US COVID ATLAS daily maps Nov 22, 2020 https://geodacenter.github.io/covid/map.html)

rapidly throughout the country. By the end of March, confirmed cases of COVID-19 were present throughout the conterminous US (Fig. 1). Cases related to internal community spread were identified in late February in California and Washington (Jorden et al. 2020).

The initial geographic dispersion of COVID-19 across the US was followed by increasing rates of infection and associated mortality (Fig. 2). An initial national peak in daily reported cases was experienced in April, 2020, when over 30,000 new cases were typically reported daily (Fig. 2). New cases declined slightly during the later spring, but surged again during the summer. Another slight decline occurred in the late summer and was followed by a precipitous increase to almost 200,000 new cases being reported daily by mid-November (Fig. 2). COVID-19 deaths may lag confirmation of infection by two to eight weeks (Baud et al. 2020). The initial April peak in COVID-19 daily infections corresponds with the highest reported daily mortality figures (CDC 2020). Over 2000 deaths daily were reported on a number of days during this period. On November 19, 2020, the daily death toll for the US again exceeded 2000 for the first time since the spring.

The national trends in daily reported cases mask the large degree of geographic variability that exists in the evolution of the pandemic in the US. Unlike the initial spread of the disease, which occurred quickly and impacted all states within a matter

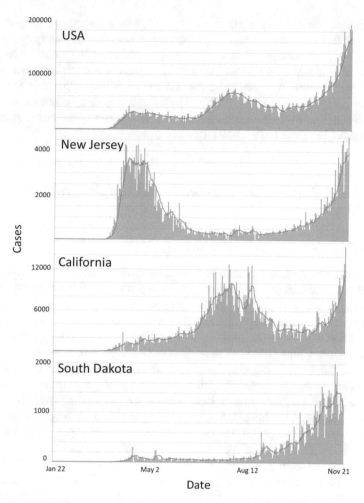

Fig. 2 Daily COVID-19 case counts for the entire United States, and the states of New Jersey, California and South Dakota. Red line is the 7 day moving average (time series from CDC Data Tracker Nov 22, 2020 https://covid.cdc.gov/covid-data-tracker/#trends_dailytrendscases)

of weeks, individual states have had very different temporal trends in community spread and resulting daily caseloads (Fig. 2). Some states, such as New Jersey, experienced their first peaks in COVID-19 cases in April. Others such as California had their first peak cases in August. A number of other states such as South Dakota are experiencing their peaks in daily reported cases and mortality at the present time after having had low number for the earlier portion of the pandemic (Fig. 2; CDC 2020). At the time of writing, many states are experiencing daily rates of infection that eclipse earlier peaks and this aggregate total is reflected in the national statistics (Fig. 2). The likely causes of the differences in the temporal evolution of the pandemic in different states are complex and beyond the scope of this chapter. The magnitudes of cumulative per capita infection rates have also varied geographically (Fig. 1). At this time, the highest per capita infection rates (typically measured as cases per 100,000 population) have come to be located in the central interior and south of the US. North and South Dakota being notable geographic centers in this regard (Fig. 1). These two states also currently (November, 2020) have the highest seven day mortality rates (CDC 2020). The overall trajectory of cases and deaths in the US suggests in the near term the country will retain it dubious distinction of being a global leader in COVID-19 cases and deaths (CDC 2020).

Federal and State Policy Responses and the Geographies of Covid-19

Introduction, Initial Spread and Responses

The early arrival and rapid geographically dispersed occurrences of COVID-19 across the US (Fig. 1) are concordant with the high degree of international travel to and from the country. There were over 240 million passengers flying between the US and the rest of the world between September 2018 and September 2019 (US Department of Transportation 2020). The earliest cases of COVID-19 in the US appear associated with travel from China (Harcourt et al. 2020). However, this appears to have been followed by subsequent importations of genetically distinctive strains from Europe (Jorden et al. 2020). On February 2, 2020, the US placed restrictions on travel from China. On March 13, 2020, a travel ban on non-citizens traveling from Europe went into effect. However, by this point, community spread was already well established in the US (Jorden et al. 2020). International border controls and travel restrictions are the purview of the federal government. It might be argued that earlier border restrictions by the federal government may have slowed introduction of COVID-19. However, one modeling study estimated that even a 90% travel restriction would have had only modest impacts on disease spread if employed alone (Chinazzi et al. 2020).

Following its arrival in the conterminous US, COVID-19 spread rapidly (Fig. 1). It is recognized that high degrees of internal mobility promote the rapid spread of epidemics (Merler and Ajelli 2010). The US has the greatest number of passengers carried by air travel of any country in the world (Word Bank 2020a). Rapid dispersion of COVID-19 by internal air travel is an anticipatable result (Fauver et al. 2020; Watts et al. 2020). Although the federal government has the ability to protect and regulate interstate travel and commerce, US courts have upheld arguments that states can restrict the entry of persons, animals or goods that are likely to be injurious to citizens (Studdert et al. 2020). Starting in mid-March of 2020, a number of states began to take such measures. Imposing self-quarantines on those arriving from out of state was common (Studdert et al. 2020). Despite these measures, case loads and mortalities rose rapidly through March and into April. Particular hotspots in terms of COVID-19 cases and mortality were the New York City region and New Jersey

(Figs. 1 and 2). In the end, travel restriction orders were difficult to enforcein the US where about 92% of households report owning personal motor vehicles (Bureau of Transportation Statistics 2020). Beginning in May, a number of states eased or removed interstate travel restrictions.

Subsequent Evolution of the Pandemic and Responses

In the face of mounting COVID-19 cases and deaths, President Donald J. Trump declared a National Emergency on March 13, 2020. Federal social distancing guidelines were also issued. It was clear that the crisis now hinged on slowing community spread. However, it is during this time that mixed-messaging from the federal government occurs. In the early period of community spread, the Surgeon General had recommended against the widespread public use of masks (Goldberg et al. 2020). However, on April 3, 2020, the Center for Disease Control made a public recommendation to wear masks. It is now clear that the federal government agencies knew or strongly suspected the aerosol transmission of COVID-19 no later than early February. On February 7, 2020, in a private interview, President Trump himself told the journalist Bob Woodward that the virus was transmitted through the air and was deadly. Based on a subsequent interview of the Surgeon General, it seems that the apparent misguidance by the federal government on the use of masks was in part done to help keep supplies of masks available for healthcare professionals. Medical professionals called for the federal government to invoke its power to increase the supply or personal protective equipment (Cohen and van der Meulen Rodgers 2020).

After initially promising strong federal action, declaring an emergency and issuing social distancing guidelines, the former President subsequently advised state governments to make their own arrangements for medical personal protective equipment, ventilators, and public health strategies. The President indicated that the federal government would recede to a back-up role in battling the community spread and treatment of pandemic victims (Yeh and Woodward 2020). There was a failure to develop a national COVID-19 testing program, along with the failure to develop a national program to provide medical workers with personal protective equipment (Todd 2020). It has been argued that the lack of a uniform national testing program confounded efforts to track, understand, and mitigate early community spread, potentially contributing to the high cases and mortality in the US (Todd 2020). However, significant federal attention has been directed at evaluation of drug therapies and development/distribution of vaccines. The federal government also maintained international travel restrictions.

Political leaders have considerable power through the ability to use the bully pulpit to sway public opinion and behaviour (Forester and McKibbon 2020; Hahn 2021). Polling from March of 2020 shows that some 90% of Republicans trusted the President to provide reliable information on the pandemic (Hahn 2021). In early February, 2020, the President began to tell the American people that the COVID-19 pandemic would simply go away. That message was repeated at least 38 times from early February onward (Wolfe and Dale 2020). There was no evidence to support these optimistic statements, and indeed as the data show (Figs. 1 and 2), the pandemic has not miraculously disappeared. The White House pronouncements helped fuel a rapid politicization of the crisis. By late March, divisive politicization was apparent in the messaging from congressional members (Halpern 2020). Thus, very early in the pandemic, the likelihood or non-likelihood of following CDC or other health professional guidelines for stemming the COVID-19 pandemic showed significant splits along party lines, with Republicans being less likely to follow such guidance (Clements 2020).

As the pandemic progressed, federal messaging emanating from the White House continued to be disruptive in terms of adherence to public health practices. On April 3, 2020, the President publicly stated that he was not going to use a mask personally (Hahn 2021).The use of masks to avoid transmitting or contracting illness is common in a number of Asian countries, but almost previously unknown in the US, and viewed with suspicion and disdain by some (Ma and Zhan 2020). Mask wearing became a political statement. Halpern (2020) argues that this pitted effective public safety behaviours against perceptions that they were an infringement on personal liberty. Such perceptions stoked street protests against masks and other public health measures. The White House also sought to undercut public health regulations and messaging from federal agencies such as the CDC. Halpern (2020) cites the example of CDC issued guidelines for reopening schools safely, and President Trump and Education Secretary Betsy DeVos dismissing CDC guidelines for reopening. There is not enough space in this chapter to recount all of the public messaging from the White House that was at odds with many public health officials and accepted public health policies. A review of such statements and their impacts is provided by Hatcher (2020). Public acrimony also erupted between the President and his administration, and some governors. On November 15, 2020, one of the President's COVID-19 senior advisors, Dr. Scott Atlas, exhorted Michigan residents to "rise up" against public health measures in that state. This statement is particularly striking in view of the fact that Michigan Governor Gretchen Whitmer had recently been the target of a politically motivated kidnapping plot. Earlier that month President Trump threatened to withhold any COVID-19 vaccine from New York State, reflecting ongoing disputes between the President and the Democrat Governor of New York, Andrew Cuomo (Solender 2020).

In terms of federal response to COVID-19, it can be argued that the line was crossed from disruptive messaging to physically placing people in harm's way as the Presidential election season began in the summer of 2020. Eschewing public health advice and practices, President Trump chose to host over 60 public rallies from June to November 2020 in a number of different states. At these rallies, masking and social distancing was minimal. Bernheim et al. (2020) estimate that 18 of the early rallies may have led to 30,000 cases of COVID-19 and 700 deaths. A smaller gathering for incoming Justice Amy Coney Barrett was held at the White House and Rose Garden on September 26, with little use of masks or social distancing. The reception resulted in a number of COVID-19 infections amongst attendees. It has been identified as a likely superspreader event, wherein conditions allow a small number of infected individuals to produce an enhanced number of secondary infections amongst others (Majra et al. 2020).

Although agencies such as the CDC, Surgeon General and National Institute of Health attempted to provide scientifically informed and rational guidance, the lack of coherent federal leadership, and ultimate descent into knowing endangerment of human health, by the President meant that states would ultimately exert much of the regulatory power in response to the COVID-19 pandemic. This was largely done in an independent and uncoordinated fashion. At the state level, emergency health orders of this nature generally emanate from the office of the governor. On March 19, 2020, California became the first state to issue a stay-at-home order (Fig. 3). There was variability in both the timing of individual states taking action and the nature of their orders (Moreland et al. 2020; Fig. 3). Some states, such as California, instituted early and relatively restrictive stay at home orders, while others, such as

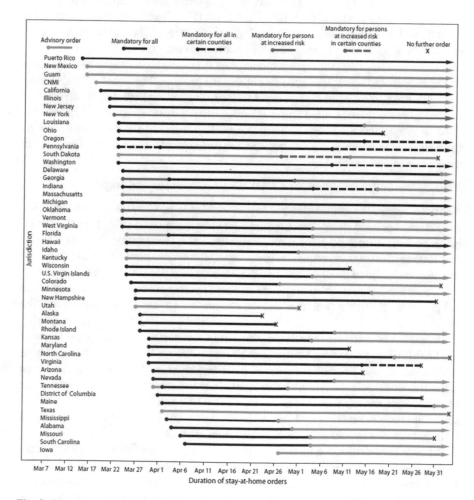

Fig. 3 Duration and type of COVID-19 stay at home orders by US states and other jurisdictions (graph from CDC Nov 22, 2020 https://www.cdc.gov/mmwr/volumes/69/wr/mm6935a2.htm)

Iowa, enacted orders late and many took the form of advisories only. Finally, there was wide variability in the date of termination of orders and advisories, with some states ending such measures in April, and many ending or decreasing the scope of their orders in May (Fig. 3). However, from July onward, per capita case rates continued to increase in many areas (Fig. 1).

Mask ordinances have also been uncoordinated and uneven. Between April 8, 2020 and May 15, 2020, the governors in 15 states had signed mandatory mask directives (Lyu and Wehby 2020). Many other states eventually followed suit, but 15 states still have no mandatory mask regulations or had only limited restrictions (Ballotpedia 2020). One study concluded that the implementation of the mask mandates in early adapter states may have averted more than 200,000 COVID-19 cases by May 22, 2020 (Lyu and Wehby 2020). There is a general geographic correspondence between states with high per capita rates of COVID-19 infection (Fig. 1) and states which currently have limited to no masking directives (Ballotpedia 2020). The distribution of such states runs roughly from the northern interior to the south (Fig. 1).

The net results of the community spread of COVID-19 and the successes or failures of the combined federal and state responses are presented in Table 1. The data in the Table are current as of November 20, 2020. The average cumulative number of confirmed cases for the United States as measured as the average rates for the states is 3795. However, there is a large range in these values from state to state. North Dakota has a current cumulative case rate per 100,000 residents of 9212 versus Vermont with a case rate of 528 (Table 1).

Politics, the Federal System, and COVID-19 Responses and Impacts

As noted above, the cumulative per capita incidence of COVID-19 displays wide geographic variance when measured on a state by state basis (Fig. 1 and Table 1). States such as North and South Dakota have cumulative case rates of 9212 per 100,000 and 7905 per 100,000, respectively. In contrast, many states have less than one half to less than one third of those cumulative confirmed cases per 100,000 residents. What role might the federal system of governance and the sharply bipartisan politics of the US played in these infection rates and the geographic variations observed within the country?

A number of analyses have highlighted the deep partisan divide in the US over the COVID-19 pandemic (Altman 2020; Green 2020; Green et al. 2020). These differences include perceptions regarding the seriousness of the pandemic, or even if the pandemic is real (Calvillo et al. 2020). The differences also include the more likely acceptance of, and adherence to, COVID-19 mitigation measures such as social distancing, masking and stay at home orders by Democrats as compared to Republicans (Gollwitzer et al. 2020; Hsiehchen et al. 2020).

As discussed earlier, the messaging and the actions of the (former) President had increasingly been in conflict with public health advice and practices (Hatcher 2020). Members of congress, governors and other political leaders beside the President can also have a role in persuading citizens to adhere to or eschew public health guidance (Green et al. 2020). They are influenced both by the policies of their party leader and by the demands of their constituents. Beyond the bully pulpit, governors also have the

Table 1 November 20, 2020, COVID-19 cases per capita (measured as cases per 100,000 population) by US states and the political affiliation of the governors. New York City has been separated from the rest of New York State by CDC. States with a Democrat governor have a mean of value of 3289.080 cases (including New York City) and states with a Republican governor have a mean of 4283.192 cases. Difference is significant at $p \leq 0.05$ (data from CDC Nov 20, 2020 https://covid.cdc.gov/covid-data-tracker/#cases_casesper100k)

State	Cases per 100,000 Since Jan 2020	Political party of Governor
North Dakota	9212	R
South Dakota	7905	R
Iowa	6327	R
Wisconsin	6160	D
Nebraska	5664	R
Utah	5251	R
Idaho	5015	R
Montana	4878	D
Illinois	4877	D
Tennessee	4846	R
Mississippi	4702	R
Arkansas	4640	R
Alabama	4622	R
Louisiana	4549	D
Wyoming	4530	R
Minnesota	4454	D
Oklahoma	4451	R
Rhode Island	4441	D
Kansas	4417	D
Florida	4231	R
Nevada	4214	D
Missouri	4208	R
Georgia	4156	R
Indiana	4117	R
Arizona	4005	R
South Carolina	3957	R
Texas	3696	R
New Mexico	3537	D
New York City*	3485	D
Alaska	3378	R
Kentucky	3321	D
New Jersey	3297	D
Colorado	3210	D

(continued)

Table 1 (continued)

State	Cases per 100,000 Since Jan 2020	Political party of Governor
North Carolina	3132	D
Delaware	3118	D
Michigan	3114	D
Maryland	2931	R
Massachusetts	2926	R
Ohio	2794	R
Connecticut	2782	D
California	2678	D
New York*	2607	D
Virginia	2505	D
Pennsylvania	2256	D
West Virginia	2071	R
Washington	1824	D
Oregon	1453	D
New Hampshire	1200	R
Hawaii	1174	D
Maine	744	D
Vermont	528	R

regulatory powers in their states to either take action or not. Gusmano et al. (2020) suggest that the party affiliation of the governor was one of the strongest influences on state responses as the pandemic unfolded. One extreme example of this is Republican Governor Kristi Noem of South Dakota who was a strong supporter of President Trump and his policies and has advocated against mask directives or banning of large public gatherings (Pilkington 2020). Governor Noem supported the holding of a 500,000 person motorcycle rally in Sturgis, South Dakota in August 2020 despite the obvious dangers of spreading COVID-19 (Pilkington 2020). One study estimates that due to this event, the counties that contributed the highest numbers of rally attendees experienced increases in COVID-19 cases on the order of 7.0–12.5% (Dave et al. 2020). All 15 states which currently have no mask mandates have Republican governors (Ballotpedia 2020). In fairness, it should be noted that South Dakota may be considered an extreme case and not all Republican state executives have dismissed scientific public health guidance and policies. For example, 10 Republican states do have mask mandates (Ballotpedia 2020). Nor can it be said that the only large public gatherings held in contravention been at the behest of Republicans. The summer of 2020 saw a number of large Black Lives Matter protests promoted by progressives and Democrats in general (see a discussion of this relative to COVID-19 by Kampmark 2020).

At this time, there is a significantly higher number of cumulative per capita COVID-19 cases in states with Republican governors (4283) versus states with Democrat governors (3289) (Table 1). The political affiliation of the governors is not a perfect predictor of case rates, and some states with Republican leadership have laudably low case rates (Table 1). Although the political affiliation of the governor has been pointed to as an important determinant of state actions (Gusmano et al. 2020), to fully understand the role of political affiliation relative to other factors that might drive state by state differences in COVID-19 policies and influence per capita case load requires detailed and careful multifactorial study beyond the scope of this chapter. Differences in population density, economic status, education, radial composition, access to medical care and COVID-19 testing, etc. are all important variables to be examined. One recent state-level analysis of some of these factors by Hsiehchen et al. (2020) concluded that variables such as degree of urbanization, percent of essential workers, and population size did not have the predictive value of party affiliation in terms of changes in mobility behaviours during the pandemic. However, more analyses are required at finer-grained spatial and demographic scales to fully resolve the questions of differential responses to the pandemic and resulting impacts of COVID-19.

What then might we say in reference to COVID-19 about the federal system in the US and the division of public health powers between the central government and the states? Certainly, this creates the potential for wide geographic diversity in the responses to a crisis such as the pandemic. On November 17, 2020, Anthony Fauci, Director of the National Institute of Allergy and Infectious Diseases, stated "We need some fundamental public health measures that everyone should be adhering to, not a disjointed, 'One state says one thing, the other state says another thing,'." In this, he may be only half right. If there was a consistent and effective, science-based federal program for COVID-19 response, and consistent messaging in support of that program, a more unified national response would be desirable. This might decrease the impact of the pandemic and perhaps lessen some of the geographic differences in its severity. However, under the current conditions, when the actions and the messaging by the federal executive branch have been contradictory, disruptive and dangerous from a public health perspective, having the freedom for some states to take stronger and more aggressive measures despite the apparent wishes of the President and the White House has likely attenuated the effects of the pandemic in those states. State actions to confront the pandemic based upon sound public health guidance have not all been uncoordinated, nor restricted solely to Democrat run states. In August, the Democrat and Republican governors of Louisiana, Maryland, Massachusetts, Michigan, North Carolina, Ohio, and Virginia joined together in an effort to secure their own supply of COVID-19 tests (Todd 2020). Without the division of public health powers in the US federal system, states might have been more constrained in their responses by a federal executive branch.

Finally, it must be noted that the COVID-19 pandemic is still evolving in the US, and the relative performance of the nation and the states in battling the disease may change going forward. President Trump lost the 2020 election and replaced by Democrat candidate Joe Biden in January 2021. President Biden has been much more adamant about the use of masks, social distancing, restricting rallies, and other measures related to COVID-19. It is clear there will be a readjustment of federal policy and presidential messaging. In addition, several promising vaccines have now been released. Given the fluidity of the situation and its continuing evolution at the time of writing, this chapter should be seen as a very broad-scale snapshot of a few selected elements of complex and changing situation.

References

Abedi, V., Olulana, O., Avula, V., Chaudhary, D., Khan, A., Shahjouei, S., et al. (2020). Racial, economic, and health inequality and COVID-19 Infection in the United States. *Journal of Racial and Ethnic Health Disparities, Sept, 1,* 1–11. https://doi.org/10.1007/s40615-020-00833-4.

Altman, D. (2020). Understanding the US failure on coronavirus—an essay by Drew Altman. *BMJ, 370.* https://doi.org/10.1136/bmj.m3417.

Ballotpedia. (2020). Documenting America's path to recovery: November 17, 2020. https://news.ballotpedia.org/tag/coronavirus/. Accessed November 22, 2020.

Baud, D., Qi, X., Nielsen-Saines, K., Musso, D., Pomar, L., & Favre, G. (2020). Real estimates of mortality following COVID-19 infection. *The Lancet Infectious Diseases, 20*(7), 773. https://doi.org/10.1016/S1473-3099(20)30195-X.

Bernheim, B. D., Buchmann, N., Freitas-Groff, Z., & Otero, S. (2020). The effects of large group meetings on the spread of COVID-19: The case of Trump Rallies. *Stanford Institute for Economic Policy Research,* Working Paper, No. 20-043. https://siepr.stanford.edu/sites/default/files/publications/20-043.pdf.

Bureau of Transportation Statistics. (2020). *Household, individual, and vehicle characteristics.* https://www.bts.gov/archive/publications/highlights_of_the_2001_national_household_travel_survey/section_01. Accessed November 22, 2020.

Campbell, A. L. (2012). Policy makes mass politics. *Annual Review of Political Science, 15,* 333–351.

Calvillo, D. P., Ross, B. J., Garcia, R. J., Smelter, T. J., & Rutchick, A. M. (2020). Political ideology predicts perceptions of the threat of COVID-19 (and susceptibility to fake news about it). *Social Psychological and Personality Science, 11,* 1119–1128.

CDC. (2020). *CDC COVID data tracker.* https://covid.cdc.gov/covid-data-tracker/. Accessed November 22, 2020.

Chinazzi, M., Davis, J. T., Ajelli, M., Gioannini, C., Litvinova, M., Merler, S., y Piontti, A. P., Mu, K., Rossi, L., Sun, K., & Viboud, C. (2020). The effect of travel restrictions on the spread of the 2019 novel coronavirus (COVID-19) outbreak. *Science, 368,* 395–400.

Clements, J. M. (2020). Knowledge and behaviours toward COVID-19 among US residents during the early days of the pandemic: Cross-sectional online questionnaire. *JMIR Public Health and Surveillance, 6*(2), e19161. https://doi.org/10.2196/19161.

Cohen, J., & van der Meulen Rodgers, Y. (2020). Contributing factors to personal protective equipment shortages during the COVID-19 pandemic. *Preventive Medicine, 141,* 1–7.

Dave, D. M., Friedson, A. I., McNichols, D., & Sabia, J. J. (2020). The contagion externality of a superspreading event: The Sturgis motorcycle rally and COVID-19. *National Bureau of Economic Research.* Working Paper, 27813. https://doi.org/10.3386/w27813. https://www.nber.org/papers/w27813.

de Bruin, W. B., Saw, H. W., & Goldman, D. P. (2020). Political polarization in US residents' COVID-19 risk perceptions, policy preferences, and protective behaviours. *Journal of Risk and Uncertainty*. https://doi.org/10.1007/s11166-020-09336-3.

Duffin, E. (2019) Per capita U.S. real gross domestic product (GDP) by state. *Statista*. https://www.statista.com/statistics/248063/percapita-us-real-gross-domestic-product-gdp-by-state/. Accessed November 22, 2020.

Fauver, J. R., Petrone, M. E., Hodcroft, E. B., Shioda, K., Ehrlich, H. Y., Watts, A. G., et al. (2020). Coast-to-coast spread of SARS-CoV-2 during the early epidemic in the United States. *Cell*. https://doi.org/10.1016/j.cell.2020.04.021.

Forester, J., & McKibbon, G. (2020). Beyond blame: Leadership, collaboration and compassion in the time of COVID-19. *Socio-Ecological Practice Research, 2,* 205–216.

Goldberg, M., Gustafson, A., Maibach, E., Ballew, M., Bergquist, P., Kotcher, J., Marlon, J., Rosenthal, S., & Leiserowitz, A. (2020). Mask-wearing increased after a government recommendation: A natural experiment in the U.S. during the COVID-19 pandemic. *Frontiers in Communication, 5,* 44. https://doi.org/10.3389/fcomm.2020.00044.

Gollwitzer, A., Martel, C., Brady, W. J., Pärnamets, P., Freedman, I. G., Knowles, E. D., et al. (2020). Partisan differences in physical distancing are linked to health outcomes during the COVID-19 pandemic. *Nature Human Behaviour, 4,* 1186–1197.

Green, A. (2020). *The growing divide: red states vs. blue states Georgetowm public policy review.* http://gppreview.com/2020/02/21/growing-divide-red-states-vs-blue-states/. Accessed November 20, 2020.

Green, J., Edgerton, J., Naftel, D., Shoub, K., & Cranmer, S. J. (2020) Elusive consensus: Polarization in elite communication on the COVID-19 pandemic. *Science Advances, 6*(28), eabc2717. https://doi.org/10.1126/sciadv.abc2717.

Gusmano, M. K., Miller, E. A., Nadash, P., & Simpson, E. J. (2020). Partisanship in initial state responses to the COVID-19 pandemic. *World Medical & Health Policy*. https://doi.org/10.1002/wmh3.372.

Hahn, R. A. (2021). Estimating the COVID-related deaths attributable to President Trump's early pronouncements about masks. *International Journal of Health Services, 51,* 14–17.

Halpern, L. W. (2020). The politicization of COVID-19. *AJN The American Journal of Nursing, 120,* 19–20.

Harcourt, J., Tamin, A., Lu, X., Kamili, S., Sakthivel, S. K., Murray, J., et al. (2020). Severe acute respiratory syndrome coronavirus 2 from patient with coronavirus disease, United States. *Emerging Infectious Diseases, 26,* 1266–1273.

Hatcher, W. (2020). A failure of political communication not a failure of bureaucracy: The danger of presidential misinformation during the COVID-19 pandemic. *The American Review of Public Administration, 50,* 614–620.

Holt, J. D., Ghosh, S. N., & Black, J. R. (2019). Legal considerations. In S. A. Rasmussen & R. A. Goodman (Eds.), *The CDCField epidemiology manual* (pp. 263–280). New York: Oxford University Press.

Hsiehchen, D., Espinoza, M., & Slovic, P. (2020). Political partisanship and mobility restriction during the COVID-19 pandemic. *Public Health, 187,* 111–114.

Jorden, M. A., Rudman, S. L., Villarino, E., Hoferka, S., Patel, M. T., Bemis, K., Simmons, C. R., Jespersen, M., Johnson, J. I., & Mytty, E. (2020). Evidence for limited early spread of COVID-19 within the United States, January–February 2020. *Morbidity and Mortality Weekly Report,69,* 680–684. https://www.cdc.gov/mmwr/volumes/69/wr/mm6922e1.htm.

Kampmark, B. (2020). Protesting in pandemic times: COVID-19, public health, and black lives matter. *Contention, 8,* 1–20.

Lyu, W., & Wehby, G. L. (2020). Community use of face masks and COVID-19: Evidence from a natural experiment of state mandates in the US. *Health Affairs, 39,* 1419–1425.

Ma, Y., & Zhan, N. (2020). To mask or not to mask amid the COVID-19 pandemic: How Chinese students in America experience and cope with stigma. *Chinese Sociological Review*. https://doi.org/10.1080/21620555.2020.1833712.

Majra, D., Benson, J., Pitts, J., & Stebbing, J. (2020). SARS-CoV-2 (COVID-19) Superspreader events. *Journal of Infection*. https://doi.org/10.1016/j.jinf.2020.11.021.

Merler, S., & Ajelli, M. (2010). The role of population heterogeneity and human mobility in the spread of pandemic influenza. *Proceedings of the Royal Society B: Biological Sciences, 277*, 557–565.

Moreland, A., Herlihy, C., Tynan, M. A., Sunshine, G., McCord, R. F., Hilton, C., Poovey, J., Werner, A. K., Jones, C. D., Fulmer, E. B., & Gundlapalli, A. V. (2020). Timing of state and territorial COVID-19 stay-at-home orders and changes in population movement— United States, March 1–May 31, 2020. *Morbidity and Mortality Weekly Report, 69*, 1198–1203. https://www.cdc.gov/mmwr/volumes/69/wr/mm6935a2.htm.

National Conference of State Legislatures. (2014). Responsibilities in a public health emergency. https://www.ncsl.org/research/health/public-health-chart.aspx (Accessed February 11, 2021).

Perniciaro, S. R., & Weinberger, D. M. (2020). Variations in state-level SARS-COV-2 testing recommendations in the United States, March–July 2020. *medRxiv*. https://doi.org/10.1101/2020.09.04.20188326.

Pilkington, E. (2020). Kristi Noem rigidly follows Trump strategy of denial as Covid ravages South Dakota. *The Guardian*. https://www.theguardian.com/us-news/2020/nov/19/kristi-noem-trump-strategy-denial-covid-ravages-south-dakota.

Qi, H., Xiao, S., Shi, R., Ward, M. P., Chen, Y., Tu, W., et al. (2020). COVID-19 transmission in Mainland China is associated with temperature and humidity: A time-series analysis. *Science of the Total Environment*. https://doi.org/10.1016/j.scitotenv.2020.138778.

Rasmussen, S. A., & Goodman, R. A. (Eds.). (2018). *The CDC field epidemiology manual*. New York: Oxford University Press.

Rudden, M., & Brandt, S. (2018). Donald Trump as leader: Psychoanalytic perspectives. *International Journal of Applied Psychoanalytic Studies, 15*, 43–51.

Scala, D. J., & Johnson, K. M. (2017). Political polarization along the rural-urban continuum? The geography of the presidential vote, 2000–2016. *The ANNALS of the American Academy of Political and Social Science, 672*, 162–184.

Solender, A. (2020). Here's why Trump and Cuomo are feuding over a Coronavirus vaccine. *Forbes*. https://www.forbes.com/sites/andrewsolender/2020/11/14/heres-why-trump-and-cuomo-are-feuding-over-a-coronavirus-vaccine/?sh=4a8618df3167.

Studdert, D. M., Hall, M. A., & Mello, M. M. (2020). Partitioning the curve—Interstate travel restrictions during the COVID-19 pandemic. *New England Journal of Medicine, 383*. https://doi.org/10.1056/NEJMp2024274.

Todd, B. (2020). The US COVID-19 testing failure. *AJN The American Journal of Nursing, 120*, 19–20.

US Census. (2020a). *2010 census: Population density data*. https://www.census.gov/data/tables/2010/dec/density-data-text.html. Accessed November 12, 2020.

US Census. (2020b). *2010 census urban and rural classification and urban area criteria*. https://www.census.gov/programs-surveys/geography/guidance/geo-areas/urban-rural/2010-urban-rural.html. Accessed November 16, 2020.

U.S. Department of Transportation. (2020). *U.S. international air passenger and freight statistics*. https://www.transportation.gov/office-policy/aviation-policy/us-international-air-passenger-and-freight-statistics-report-third. Accessed November 16, 2020.

Watts, A., Au, N. H., Thomas-Bachli, A., Forsyth, J., Mayah, O., Popescu, S., et al. (2020). Potential for inter-state spread of Covid-19 from Arizona, USA: Analysis of mobile device location and commercial flight data. *Journal of Travel Medicine*. https://doi.org/10.1093/jtm/taaa136.

World Bank. (2020a). *Air transport passengers carried*. https://data.worldbank.org/indicator/IS.AIR.PSGR?most_recent_value_desc=true. Accessed November 15, 2020.

World Bank. (2020b). *Current health expenditure per capita (current $US)*. https://data.worldbank.org/indicator/SH.XPD.CHEX.PC.CD?most_recent_value_desc=false. Accessed November 12, 2020.

World Bank. (2020c). *GDP per capita (current US$)*. https://data.worldbank.org/indicator/NY.GDP. PCAP.CD?most_recent_value_desc=true. Accessed November 12, 2020.

World Bank. (2020d). Gross domestic product 2019. https://databank.worldbank.org/data/dow nload/GDP.pdf. Accessed November 12, 2020.

World Health Organization. (2021). Corona virus (COVID-19) dashboard. https://covid19.who.int. Accessed April 1, 2021.

Wolfe, D., & Dale, D. (2020). 'It's going to disappear': A timeline of Trump's claims that Covid-19 will vanish. *CNN*. https://www.cnn.com/interactive/2020/10/politics/covid-disappearing-trump-comment-tracker/. Accessed November 16, 2020.

Wu, Y., Jing, W., Liu, J., Ma, Q., Yuan, J., Wang, Y., et al. (2020). Effects of temperature and humidity on the daily new cases and new deaths of COVID-19 in 166 countries. *Science of the Total Environment*. https://doi.org/10.1016/j.scitotenv.2020.139051.

Yen, H., & Woodward, C. (2020). AP fact check: Trump, 'wartime' pandemic leader or 'backup'? AP 2020. https://apnews.com/a64cf7fd5095d4d3b002dc4830e32119. Accessed November 16, 2020.

Zhang, C. H., & Schwartz, G. G. (2020). Spatial disparities in coronavirus incidence and mortality in the United States: An ecological analysis as of May 2020. *The Journal of Rural Health, 36,* 433–445.

Conclusion and Suggestions

Rais Akhtar

Abstract Epidemics and pandemics of diseases with high mortality had always been a major menace of humanity. Both developed and developing countries suffered immensely with cholera, plague, typhus, smallpox, and tuberculosis. Coronavirus or COVID-19 caused a similar health disaster in 2020 that devastated the social, economic, and health structure of the world. Globally, as of 1st April, 2021, there have been 128 million confirmed cases of COVID-19, including 2.6 million deaths, reported to WHO. Some countries have not reported a single case of Covid-19 or are declared free from the infection. With such devastating experience, we will be able to make better sense of COVID-19 in terms of how we can learn from this pandemic experience-identify our weaknesses in handling global health crisis. We can utilize the strategies adopted successfully in some countries and to ensure all communities are adequately cared for, and reducing disparities, and have access to the resources including COVID-19 vaccine that will contribute to the overall well-being of our society.

Keywords Cholera · Plague · WHO · Mental health · Environmental change · Human behaviour · Vaccine

Introduction

COVID-19 health disaster has resulted in a greater economic contraction, with loss of employment, the corresponding increase in poverty and inequalities, that give rise to new challenges for sustainable development and social peace.

In addition, the effects of the pandemic on people's mental health—which can be affected for different reasons: stress, social isolation, family losses, economic losses or fear of being infected by them and/or their families—make this situation a real challenge for the health sector and for society as a whole. Food insecurity concerns are lurking in many countries. The COVID-19 pandemic is affecting food

R. Akhtar (✉)
International Institute of Health Management and Research, New Delhi, India
e-mail: raisakhtar@gmail.com

systems directly through impacts on food supply and demand and indirectly through decreases in purchasing power and in the capacity to produce and distribute food WFP 2020). CBS News project that more than 50 million Americans facing hunger in 2020, projections show (CBS News 2020).

Some countries have not reported a single case of COVID-19 or are declared free from the infection that are Kiribati, Marshall Islands, Micronesia, Nauru, North Korea, Palau, Samoa, Solomon Islands, Tonga, Turkmenistan, Tuvalu, and Vanuatu. With the exception of North Korea and Turkmenistan in east and central Asia, all are Pacific Ocean countries. This is one area of further investigations as to why the two Asian countries are free from COVID-19, and the infection does not spread in Pacific Island countries. Besides this, there are wide spatial variation in the intensity of COVID-19, and some countries, for instance New Zealand, are able to control the infection, while European countries, including United Kingdom and the USA, are in the grip of another wave of COVID-19 infection with greater intensity leading to extensive lockdown in these countries. Thus, geographical explanation with focus on: time and space is a contemporary interdisciplinary perspective on spatial and temporal processes and events such as social interaction, social and environmental change, and changing human behaviour.

Varied Geographic Regions

Some examples of regional perspective of COVID-19 cases, taking into account geographical variations, are relevant in this context. As stated by Eberhard focusing on Pacific Islands, a unique geographic entity, economically, the pandemic and the measures to contain the crisis pose major challenges to PICTs. While negative consequences are already felt and will continue to impact the economies, the major goal amidst the crisis is to limit the negative effects. Obviously, the extent of the economic damage depends on the duration of restrictions on international travel and trade, and the level of containment of domestic economic fallout resulting from lockdowns. Pacific governments will need to carefully consider and evaluate the health risks associated with any lifting of restrictions against any economic benefits. A further spread of the virus within the region might even have more severe impacts on the economy than living with the restrictions of public life.

According to Mei-Huil, Taiwan has adopted proactive strategies for epidemic surveillance, border controls, community-based prevention and control, stockpiling of medical supplies, and health education and outreach to prevent the spread of the disease and maintain the health and safety of the Taiwanese population. Using lessons learned from the SARS epidemic, both the Taiwanese government and the public have adopted proactive strategies and precautionary measures to address COVID-19. Tight border controls, rapid mobilization by the public and private sectors, prompt decision-making and implementation, well-coordinated distribution of medical supplies, and transparent information and advanced digital technology are helping Taiwanese society to combat COVID-19 effectively.

As Svetlana and colleagues assert that geographical factors such as the remoteness from major international airports of the region, the transport network of the region, population density, and the proportion of the urban population in the region played a leading role in the spread of the infection across Russia. Further study of the role of various factors including those related to the environment, climate, cultural and social characteristics of the population, as well as profile of the health status of local populations, will help better understand the possibilities of controlling infectious diseases. In addition, it could help identify the most vulnerable regions and population groups for targeted health improvement measures. A separate task for further research is to analyse the effectiveness of restrictive measures, in turn, depending on many factors: the level of education of the population, ethnic and religious specifics, and the degree of well-being.

In a mountainous country like Peru, Pool Konrad Aguilar León and colleagues focus on the environmental aspect. It was pointed out how the link between humans and the environment creates conditions for the outbreak of zoonotic diseases and how Peru's characteristics, such as great biodiversity, deterioration of ecosystems, fauna trafficking, increase in livestock production and inadequate animal health, generate the risk of a zoonotic outbreak in its territory. Likewise, the authors described the various effects that the COVID-19 outbreak in the country generated in different environmental aspects such as waste generation, deforestation activities, environmental policies, and the approach to socio-environmental conflicts.

At the same time, understanding the spread of COVID-19 in a dense metropolis such as Jakarta, Kuala Lumpur, Bangkok or Mumbai and their geo-ecology and pattern of human behaviour of surrounding area requires a thorough examination of how points of crowding become spaces where super-spreader events occur.

What COVID-19 holds for a vast country like Australia in the future is obviously impossible to tell with certainty. Total eradication of the virus though is unlikely, stabilization at low case and death figures probably the best that can be achieved in such a geographically large and diverse country. The author suggests that sensible social distancing, rigorous personal hygiene and other anti-COVID-19 behaviours will continue to be essential and improved. It will also clearly be vital to maintain careful screening for COVID-19 infection in those entering Australia. Dr.Anthony Fauci, virus expert from USA, appreciated the mitigation strategies to combat COVID-19 spread in Melbourne (Zhou 2020).

We must identify some areas of research in the future. Since the coronavirus originates in Wuhan, China, a centre of wildlife trade, thus in order to prevent the next epidemic and pandemic related to high-risk human–wildlife interactions and interfaces that led to the emergence of SARS-CoV and of 2019-nCoV must be investigated. WHO team of experts visited Wuhan in early February, 2021 and paid a visit to Wildlife market and the Wuhan Institute of Virology. The team of experts concluded, "The coronavirus is unlikely to have leaked from Chinese lab. And is more likely to have jumped to humans from an animal" (H.T. 2021). To focus on research and investment in three areas: (1) surveillance among wildlife to identify the high-risk pathogens they carry; (2) surveillance among people who have contact with wildlife to identify early spillover events; and (3) improvement of market biosecurity regarding the wildlife trade. As the emergence of a novel virus anywhere can impact the furthest reaches of our connected world, international collaboration among scientists is essential to address these risks and prevent the next pandemic (Daszak et al. 2020).

Air pollution has also been associated with COVID-19. Air pollution increases the risk of respiratory diseases, and the COVID-19 is mainly a disease of the respiratory system; however, in the chapter on Canada, Underwood saw only very weak evidence of an association between air pollution, in the form or O_3, and COVID-19 cases. A single ecological correlation analysis cannot be used to make a conclusion on the association between air pollution and an individual's susceptibility to an infectious disease. That would be ecological fallacy. This correlation analysis should instead be viewed as a hypothesis-generating tool: all factors that can lower a person's immune system should be investigated in further detail, as lowered immune systems increase an individual's risk of infection with COVID-19.

Government policies to address the economic and social challenges focus on upgrading health care system to cope with emergence of new virus and vector-borne diseases that are likely to appear. Pharmaceutical companies have invested heavily in vaccine development in the USA and in the United Kingdom. Pfizer, Moderna and Johnson & Johnson (Janssen) vaccines have been authorized for emergency-use. In fact, United Kingdom has already authorized Pfizer for its emergency use. Oxford vaccine AstraZeneca, Russian vaccine Sputnik V, Chinese Sinovac vaccine, Indian vaccine Covishield are being used in different countries. However, there remains uncertainty with respect to the large-scale roll-out of COVID-19 vaccines and access to all segments of population both in developed and developing countries. Dwindling supply of Covid-19 vaccines is being experienced in several European countries (Adler 2021). Faced with the shortage of vaccines EU decided not to permit vaccine export until the requirements of European Union countries are fulfilled. Italy was the first country to block the vaccine shipment to Australia. United Kingdom and the World Health Organization have criticised the decision of European Commission. Hungary becomes the first EU country to use Chinese vaccine against COVID-19.

Region-Based Research

It is evident from the diffusion of COVID-19 that a cooler climate is congenial for its spread. The opposite is true of malaria plasmodium which prefers a warmer climate and enhanced warming causing occurrence of altitudinal rise of malaria in highlands and sub-mountainous areas. Based on the rainfall worldwide, Brazilian researchers confirm COVID-19 cases increase with greater precipitation. For each average inch per day of rain, there was an increase of 56 COVID-19 cases per day. However, no association was obtained between rainfall and COVID-19 deaths (Guite 2020). Links between COVID-19 cases and temperature are less certain. Some studies reported link between temperature and COVID-19, while others have not. However, higher temperatures are associated with a lower number of cases in Turkey, Mexico, Brazil, and the U.S. (Guite 2020). There is need to conduct regional studies in geographically diverse regions in both developed and developing countries to look into the association between temperature and other environmental determinants with the origin and diffusion of COVID-19.

Research must also focus as to why northern hemisphere is impacted by coronavirus than the southern hemisphere. Regional studies can elaborate the significance of socio-geographical and human behavioural determinants in the outbreak and intensity of COVID-19 in particular regions.. and its spread.. Attempt should also be made to study the impact of this pandemic on mental stress of population in different socio-economic and cultural scenarios.

With such devastating experience, we will be able to make better sense of COVID-19 in terms of how we can learn from this pandemic experience-identify our weaknesses in handling global health crisis. We can utilize the strategies adopted successfully in some countries and to ensure all communities are adequately cared for, by reducing disparities, and have access to the health care resources including COVID-19 vaccine that will contribute to the overall well-being of our society.

References

Adler, K. (2021, January 29). Covid: why is EU's vaccine rollout so slow? BBC

CBS NEWS. (2020, November 24). *More than 50 million Americans facing hunger in 2020,* projections show.

Daszak, P, Olival, K. J., & Li, H. (2020). A strategy to prevent future epidemics similar to the 2019-nCoV outbreak. *Biosaf Health, 2*(1), 6–8.

Guite, H. (2020, August 16). How does weather affect COVID-19? *Medical News Today.*

World Food Programme (WFP). (2020). *Food security and vulnerability assessment in Armenia.* Yerevan, Armenia.

Zhou, N. (2020, October 29). Dr. Fauci praises Australia's coronavirus response and Melbourne's face mask rules. *The Guardian*

Index

Printed in the United States
by Baker & Taylor Publisher Services